GEOLOGY IN ENVIRONMENTAL PLANNING

Arthur D. Howard

Professor Emeritus of Geology
Stanford University

Irwin Remson

Professor of Applied Earth Sciences
and Geology
Stanford University

With contributions by

W. R. Dickinson
T. H. Hughes
R. H. Jahns
E. Just
J. O. Leckie
G. G. Mader
B. M. Page
G. A. Parks
H. J. Ramey, Jr.

geology in

McGraw-Hill Book Company · New York · St. Louis
Madrid · Mexico · Montreal · New Delhi · Panama

environmental planning

San Francisco • Auckland • Bogotá • Düsseldorf • Johannesburg • London
Paris • São Paulo • Singapore • Sydney • Tokyo • Toronto

This book was set in Times Roman by Black Dot, Inc.
The editors were Donald C. Jackson and James W. Bradley;
the designer was Anne Canevari Green;
the production supervisor was Dominick Petrellese.
The drawings were done by J & R Services, Inc.
R. R. Donnelley & Sons Company was printer and binder.

GEOLOGY IN ENVIRONMENTAL PLANNING

Library of Congress Cataloging in Publication Data

Howard, Arthur David, date
    Geology in environmental planning.

    Includes bibliographies and index.
    1. Geology. 2. Environmental protection.
I. Remson, Irwin, joint author. II. Title.
QE33.H68   550   77-9320
ISBN 0-07-030510-2

Frontispiece

**The earth, a minor planet with limited air, fresh
water, and material resources, and already overtaxed
by a burgeoning population. A tiny closed system,
our island home for better or worse. (View from the
moon. *NASA photograph.*)**

To
Julia and Edna,
our gracious
and most-patient
wives

Yesternight the squire had me to dine. A memorable meal. Much good talk over a bottle of port so rare that I felt unworthy of it. We came to the old topic, our boyhood days. How sparkling clear was our little river then, how sweet the air. And today, said the squire, the village is become a town and devil take it, a town that stinks. The tannery. The brickworks. The market. The everburning hill of public refuse. The shambles. They choke us with their noxious stenches, while our once highly healthy stream receives effluent which no man of sensibility would stoop to describe. Rouse up, he ended, and preach that cleanliness cometh next to Godliness. I answered mildly thus: I have passed these offenders and they all sing the same tune. Says the tanner, "If folk want my leather, they must abide the slight odor of my tannery. But go chide the brickworks." Says my churchwarden, "If folk want my bricks, they must abide the slight smell which goes with my baking of them. But what of Perkins, the tanner?" Says our worthy council clerk, "If the tannery and the brickworks can abate their nuisances, maybe they can tell me how the market shambles and the garbage fire can be made to smell like attar of roses." I doubt, said I, that the sermon you suggest would be of practical effect, but I will ponder it. Walking through the snow to my parsonage I suddenly bethought that the great dung heaps on the squire's own manor farms drained into our erstwhile sparkling stream. Said I to myself mirthfully: If ever I preach that sermon, the text shall be from St. Matthew, Chapter 7, Verse 3: "Why beholdest thou the mote that is in thy brother's eye but considereth not the beam that is in thine own?"

Reverend Samuel Puldom's Commonplace Book, *"Notes for Sermons near Christmas of 1818."*

contents

preface

Ecology, the interrelationship of organisms and the environment, involves the earth's lands, waters, and air and the myriad organisms, including man, which inhabit these realms. Study of these realms is of such broad scope and so complex as to defy satisfactory treatment in a single text of acceptable size. This book, therefore, is restricted primarily to one part of the science of ecology, the interrelationship of man and the geologic environment both on a local and a global scale. We consider this to be the scope of environmental geology.

The geologic environment includes the topography, the mantle of soil and other loose materials, the bedrock below, the natural processes that modify the landscape, and contributory factors, such as vegetation or permanently frozen ground, that influence the processes in operation. We find it useful to discuss environments under diverse headings, emphasizing process, climate, or distinctive setting. We do this recognizing that most processes are active in all environments, differing only in relative importance.

We have included chapters on environmental law (Chapter 15), land-use planning and geology (Chapter 16), private development (Chapter 17), and a chapter (18) describing a student group project on land-use planning in which many factors are considered. The student project imparts a sense of reality to what otherwise might be a largely academic experience. It acquaints the student with the diverse types of environmental data (geological and nongeological) that are considered in land-use planning and the ways in which such data may be presented and used effectively. Comparable study projects are possible elsewhere.

This text is intended for the general college undergraduate and the informed lay reader and as a primer for the urban planner. It provides sufficient geologic background information to accommodate the nonscientific reader. Geologic terms are defined where they first appear, and the pertinent pages are in boldface type in the index. In recognition of the trend toward the metric system, all notations are in both metric and English units.

The book is divided into four parts as follows:

Part One An introduction to environmental geology, including brief advance summaries of the natural processes that cause change in the landscape, the role of man in initiating or accelerating change, and the problems of environmental planning.

Part Two Detailed discussions of geologic processes of widespread environmental impact and of a number of distinctive environments. Each chapter includes consideration of natural and artificially created hazards, the constraints on human use, the types of preparatory and remedial procedures, and the status of long-range planning.

Part Three Discussions of several special environmental problems and of special investigative techniques. The special problems relate to geochemistry and environmental impact, the management of solid mineral resources and solid wastes, and the energy industries. The special techniques relate to the investigation and management of hydrologic units and some of the techniques of remote sensing used in environmental investigations. These include air photography and imagery from infrared and radar scanning devices. A number of stereopairs (stereograms) are included. These overlapping air photos provide three-dimensional views when examined under a pocket stereoscope. Although the individual photos of each pair suffice to illustrate the subject, the three-dimensional view enhances the portrayal.

Part Four Discussions of environmental law, land-use planning and geology, private development, and presentation of a student group project on environmental land-use planning.

The text is adaptable to different course lengths. Short courses could omit the chapters on geochemistry and the environment (Chapter 10), hydrologic management (Chapter 13), remote sensing (Chapter 14), and environmental law (Chapter 15). If a student project is not contemplated, Chapter 18 may be deleted.

The text emphasizes that it is often desirable to reject a site for human occupation rather than combat nature. Many current environmental programs are designed primarily to prepare against or remedy damage at sites previously selected because of economic or topographic suitability. Rejection of such sites may be justified if preliminary investigation suggests a high degree of risk, aesthetic degradation, or the possibility of costly preventive or remedial measures. It must also be borne in mind, however, that no place on earth is hazard-free and that Nature itself is constantly modifying the environment.

This book is intended to acquaint the reader with the character of the geologic environment, the hazards it presents, and the kinds of measures that may be taken to accommodate to it. It is a survey of environmental geology and its use in the planning and implementation of land-use developments. Students intending to specialize in the subject will find it necessary to follow with more specialized courses. Planners, developers, architects, engineers, and those with simple educational curiosity, however, should find the present text adequate for the background they seek.

We are indebted to the following for constructive suggestions: Kenneth Crandall, Robert Kovach, Ron Lyon, Otto Bittencourt Netto, Troy Péwé, Ernest Rich, George Thompson, L. Jan Turk, John Moss, Harold Pelton, and Charles Waag. Special thanks are due Kenneth H. Crandall and Richard H. Jahns for valuable suggestions in the preparation of the chapter on the energy industries (Chapter 12).

Arthur D. Howard
Irwin Remson

GEOLOGY IN ENVIRONMENTAL PLANNING

PART ONE

introduction

CHAPTER ONE

the natural environment and man

INTRODUCTION

The Environmental Problem

Until populations stabilize, growth pressures will force man into relatively undeveloped environments. Such invasion involves tampering with the landscape and disturbing the habitats of the flora and fauna. These are matters of concern.

Opposing environmental views range from total exclusion of man from as yet undeveloped terrain to unrestricted development. It is generally recognized that population pressures and the many needs of society require urban expansion, the founding of new communities, and the occupation of isolated sites for the development of needed resources. It is also recognized, however, that irresponsible development has resulted in severe environmental degradation and a steadily declining quality of life for many people.

Many environmentalists and land developers recognize that compromise is necessary. Environmentalists provide a valuable service by exerting pressure to ensure that the environment is not needlessly damaged and that, wherever feasible, it is restored to as near its original condition as possible.

An example of environmental restoration is provided by the new open-pit mine of the American Smelting and Refining Company in Manchester Township, New Jersey. The ore mineral is ilmenite, from which titanium is derived. The ilmenite occurs as one of the heavy minerals in sands close to the surface. Production began in 1973.

The mining was begun by excavating below the ground-water level to create a lake. A dredge (see Figure 11-4) excavates the area in front, extracts the 4 percent or so of heavy minerals, and spreads the clean waste sand behind. The lake thus moves ahead with the dredge. Because the entire process is mechanical, the sand that is restored to the pit is clean, and the waters seem to be unchanged biologically and chemically. The topsoil is removed ahead of the dredge and stockpiled for return to the reclaimed area. Cover grasses and other vegetation are planted as the operation progresses. These reclamation procedures leave the area practically as it was prior to extraction of the valuable resource. Although the situation here is uniquely adapted to rehabilitation, it does illustrate the new environmental approach.

As an illustration of environmental concerns at the community level, we will consider the case of Walnut Creek, California. It is one of the many American communities undergoing rapid urbanization. The residents are caught up in a changing life-style, and conflicts between opposing interests have developed. The local government is attempting to meet state and federal regulations, respect property rights, and preserve the environmental amenities that make Walnut Creek a pleasant place in which to live.

All too often, real estate developments are approved on the basis of inadequate studies of geology. We shall follow up the Walnut Creek example, therefore, with a hypothetical case, typical in whole or part of numerous actual examples, to illustrate how a belated understanding of the geology of an area may force radical revision of earlier plans.

Example of a Community in Transition

Walnut Creek, California is 24 km (15 mi) east of San Francisco Bay. The 1970 population of 23,000 was located largely on an orchard-covered plain. The plain is flanked by ridges that are covered by oak-grass assemblages and used primarily for grazing. The area has a predominantly rural aspect and great visual beauty.

The main transportation link between Walnut Creek and Oakland is a multilane freeway. A recently completed rapid transit system links Walnut Creek to Berkeley and San Francisco by high-speed rail service. Consequently, Walnut Creek is undergoing rapid development as a commuting residential community for San Francisco, Oakland, and Berkeley. Real estate developers are subdividing available acreage on the plain and are extending their operations into the hills.

The development of residential subdivisions is resisted by the residents and others interested in preserving the rural environment. Unquestionably, the beauty of Walnut Creek as a rural community is jeopardized. Unless restricted, developers usually prepare large, flat building sites, often by leveling small hills, terracing slopes, and filling small valleys, with results that are considered aesthetically unpleasant by many residents. Flora and fauna are destroyed, and beautiful rural land is turned into monotonous subdivisions.

The growing population is introducing problems of water supply, waste disposal, water and air pollution, traffic congestion, and noise pollution. Additional problems stem from expansion of development into the surrounding hills.

The developers point out that they are responding to increasing population pressures and social needs. They complain that land is often purchased and subdivisions planned under one set of regulations, and the regulations are then changed so that it is no longer economically feasible to develop the properties.

Because of the detailed environmental investigations in Walnut Creek, a significant number of geological problems have been identified. These include hazards from ground displacements, earthquake shaking, landslides, and soils which expand on wetting. Other geologic problems relate to foundation materials, erosion, flooding, sedimentation, and ground water. It should not be concluded from the above that Walnut Creek is a geologically undesirable area in which to build. Rather, the reverse is true because the hazards have been identified and planning and management decisions take cognizance of this information.

Under present regulations, the tract builder in Walnut Creek provides a draft of an environmental impact report. The city authorities have a verifying study made by an impartial contractor whom they employ. The city planning department uses this information to develop an environmental impact report as required by the state of California. The purchaser now has assurance that his or her home is reasonably safe from geologic and other hazards. However, the population continues to grow, subdivisions keep expanding into the hills, wildlife is being eradicated, pollution is increasing, and the rural flavor of the community is being lost.

A Hypothetical Example of the Geological Impact on Planning

Land-use planning involves a variety of factors, each of which plays an important role in determining the best use of the environment. These factors can be divided into four categories:

1. Physical (and chemical), both natural and artificial
2. Economic, such as the demand for stores and offices
3. Social, such as the needs of all social groups
4. Matters of public interest, such as sanitation and other services

Many planning programs neglect the details of the natural physical environment. The major elements of the natural environment, such as mountains, ridges, hills, valleys, plains, natural harbors, and forests, are easily recognized. However, the more subtle characteristics and dynamics of the natural environment may be overlooked. The following hypothetical example illustrates how the detailed characteristics of the natural environment may affect the planning of a city.

Setting and Early Development. The planning area in Figure 1-1a includes a mountain range, hills, and highly productive agricultural land on the valley floor. The demand for agricultural products, which the planning area exports, has resulted in modest development of housing for farm workers and merchants, stores and offices to supply goods and services to the community, and establishment of a church and a school.

The growth of an adjoining metropolitan area has made the planning area attractive for suburban residential housing, and a few small residential developments have already been built. To date, pressures in the public interest as represented by the local government have been minimal, and land-use planning and regulation have been aimed primarily at such matters as water supply and road construction. Now, with the pending influx of population, the community must prepare a general plan (see Chapter 16) to guide future growth.

Initial Plan. Figure 1-1b shows the initial plan prepared by the community. The local officials and residents believe that the proposed land use best satisfies the public interest as well as the economic, social, and physical factors. The expanded commercial area is along level land where there is good access to the main road. Institutional facilities, such as churches, a city hall, schools, and a library, are provided near the commercial area on level land with good access. The need for multifamily housing is to be met in medium density residential areas convenient to circulation and on readily developable level land. The hills and mountains, which are more difficult to develop, are reserved as low-density residential areas. A portion of the mountainous area is set aside as a large, open-space preserve for the community. With this plan, the community believes it has a sound guide for future growth.

Physical Environmental Data. After experiencing some minor landslides following adoption of the plan, the community has decided that a more thorough investigation of the geology is warranted. A study is conducted which reveals the complex geologic setting shown in Figure 1-1c.

A fault (a fracture along which the ground has moved, as in Figures 8-7 and 8-14) has long been known by geologists to cross the area, branching at its north end. The presence of the

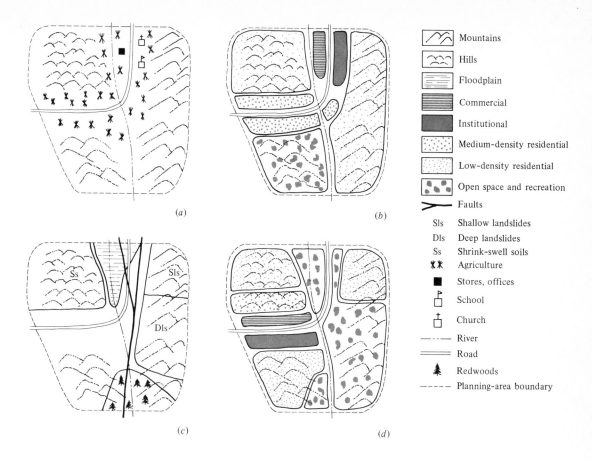

FIGURE 1-1. **Hypothetical example of the geological impact on planning.** (*a*) **Setting and early development;** (*b*) **initial plan;** (*c*) **physical setting with geological and other data; and** (*d*) **revised plan.**

Legend:

- Mountains
- Hills
- Floodplain
- Commercial
- Institutional
- Medium-density residential
- Low-density residential
- Open space and recreation
- Faults
- Sls — Shallow landslides
- Dls — Deep landslides
- Ss — Shrink-swell soils
- ✗✗ — Agriculture
- ■ — Stores, offices
- School
- Church
- River
- Road
- Redwoods
- Planning-area boundary

fault had never come to the attention of the town officials. Geologic evidence suggests that the fault has been active in the recent past. It was also learned that a substantial portion of the valley floor has been subject to flooding. Also, upon closer examination, the mountain area was found to have many landslides. Some landslides are so deep as to make stabilization of the ground infeasible, while others are shallow enough to make stabilization technically and economically possible. The low-lying hills were found to include soils which shrink when dry and swell when wet (shrink-swell soils). These soils require removal prior to construction of buildings, or else special foundations

must be prepared. Finally, an outstanding grove of giant redwood trees in the southern part of the area was judged to be an environmental asset.

Revised Plan. The new information indicated that the general plan was in conflict with the natural environment. The town therefore undertook a major revision of the plan.

The revised plan (Figure 1-1*d*) keeps development clear of the active fault, the floodplain, and the active deep landslides. The areas of shallow, manageable landslides and shrink-swell soils are set aside for residential development, with the medium-density housing located closest to the commercial and institutional areas. Developers are to be required to undertake detailed geologic and engineering studies and propose solutions for stabilization of the landslides if development is to take place. Septic tanks will not be permitted in landslide areas because they would aggravate the sliding problem by introducing water into the ground. Proposals are made for ensuring that development programs solve the shrink-swell-soils problem. Because land along the fault and in the flood-plain is set aside as open space, the commercial, institutional, and medium-density residential areas are relocated to level or gently sloping land free of these constraints. Finally, the town open-space preserve is relocated to include the stand of redwood trees. Regulation and acquisition programs are recommended as the next steps in implementing the plan.

Summation. While this example is hypothetical, comparable geologic situations are widespread. Serious mistakes in planning have resulted in large losses of property and even of life. By recognizing environmental hazards and resources and by planning accordingly, such mistakes can usually be avoided. To do this, the proper expertise in earth sciences and other natural science and engineering fields must be introduced into the comprehensive city planning process.

The revised plan responds to the hazards and amenities with appropriate revisions to land-use proposals. Importantly, the text of the plan also emphasizes that to ensure adherence to the recommendations, the town must adopt ordinances and procedures to guarantee that development is carried out as prescribed by the plan. This requires successively more detailed proposals by developers as well as means for evaluation and either approval or rejection of these proposals (see Chapter 17).

THE GEOLOGIC ENVIRONMENT

Now that we have illustrated the importance of geology in the planning process, a broad overview of the geologic environment is in order.

Global Aspects

The Earth's Envelopes. Most human activities are restricted to the outermost rind of the earth, the *lithosphere*. Only in a few deep mines have we penetrated to depths of as much as 2000 or 3000 m (6500 to 10,000 ft). Indirectly, by means of oil wells, we have probed as deep as 9.2 km (5.7 mi).

The envelope of air, the *atmosphere*, supplies the oxygen and other gases needed for the survival of organisms, for natural processes that contribute to the development of soil and the creation of some vital mineral ores, and to serve as a shield against harmful solar radiation. The oceans and the surface and subsurface waters of the lands make up the *hydrosphere*, equally vital to human existence.

From the global point of view, pollution of the earth's air and waters is a threat to all life. At present, the possibility of radioactive pollution is creating the most concern. There is also the threat of pollution on a global scale from wind-dispersed insecticides, oil spills on the sea, increasing amounts of carbon dioxide in the atmosphere, and possible damage to the ozone shield that protects us from most of the sun's ultraviolet radiation. Our chief concern, however, is with more localized geological hazards.

Nature of the Lithosphere. The lithosphere, on whose surface we live and from whose depths we derive many of the materials of civilization, extends to depths of a few kilometers under the ocean floors and to as much as 60 km (35 mi) under the continents. This rigid layer which rests on a weaker, partially molten zone is commonly referred to as the *crust* of the earth. The lithosphere under the oceans consists of a dark, originally molten rock known as *basalt*. The continents consist largely of the light-colored rock *granite*, also once molten. We do not see granite everywhere because it is hidden under other rocks. Where this cover has been breached, however, as in the cores of mountain ranges or in deep canyons, granite is commonly revealed.

Basalt and granite are the most abundant representatives of the class of rocks known as *igneous*—rocks that were once molten.

All rocks exposed to the weather or to erosive action are broken down into loose sediment or partially dissolved. The loose sediment is spread out in layers by streams, winds, glaciers, waves, and currents and the dissolved materials precipitate on the sea floor. The layers eventually solidify to become members of the second great class of rocks, the *sedimentary rocks*. Examples are *conglomerate*, consisting of solidified gravel; *sandstone*, or solidified sand; *shale*, a solidified mud; and *limestone*, commonly a solidified lime mud.

Throughout earth history, large areas of the crust have been subjected to high pressures and temperatures during repeated episodes of mountain making. Under these severe conditions, often accompanied by chemical reactions, the rocks have been changed into new rock types, the *metamorphic rocks*. The characteristics of these rocks differ considerably, depending on the relative importance of pressure, temperature, and fluids in the metamorphism, the intensity of the metamorphism, and the nature of the original rocks. For example, a shale subject to only moderate pressures becomes the metamorphic rock *slate*. Wherever pressures persisted beyond this stage, visible flakes of mica appeared in the slate and grew larger until the rock consisted of a mass of oriented mica flakes. Such a rock is known as *schist*. Other rocks, with more variable compositions than shale, may become the metamorphic rock *gneiss*, in which layers or strings of light and dark minerals alternate. Sandstones are generally metamorphosed to the very resistant light-colored rock *quartzite*, and limestones are converted to *marble*.

The three major classes of rocks, igneous, sedimentary, and metamorphic, as well as the soils that are derived from them, make up the lithosphere. The landscape is the face of the lithosphere.

The landscape is constantly changing in response to a variety of earth processes. These processes and the changes they bring about are introduced here briefly as a groundwork for the discussions that follow.

Concept of Change

The landscape changes in response to geologic and climatic stimuli. Its appearance at any one time represents a fleeting stage in a continuing conflict between internal processes which tend to elevate the lands and external processes which tend to wear them down. Although the results of most such changes are generally imperceptible and become visible in the landscape only after centuries or millennia, individual local events such as landslides, earthquakes, and volcanic eruptions may take place very rapidly and constitute serious environmental hazards.

Present-day geologic processes, of course, can affect soils and rocks of any geologic age. Thus, the environmental geologist is necessarily concerned with the characteristics of ancient as well as modern earth materials.

The age of earth materials is determined by a variety of methods. Through the years, a geologic time scale (Figure 1-2) has evolved which we shall refer to on numerous occasions.

Humans often induce change or accelerate the process of change. Their problem is not to bring environmental change to a halt, a generally impossible task, but to adapt to the environment and to occupy it with the least physical and aesthetic damage. To do so, they must be familiar with earth processes so that they may avoid or minimize damage to the terrain as well as to life and property.

The following brief summary of the processes of change will serve as an introduction to the more detailed discussions in the chapters of Part Two. We shall review first the processes of tectonics and volcanism that construct new landscapes and then those destructive processes that wear them down.

Tectonics. *Tectonics* includes the processes involved in the displacement and crumpling of segments of the earth's crust, generally resulting in elevation of the land. These enormous forces have succeeded in breaking up the crust of the earth into six or seven major segments or plates (see Figure 8-15) which slowly separate, grind past, or pass under each other. In so doing, they account for most crustal deformation. The movement of these plates and the resulting deformation is known as *plate tectonics*. Plate tectonics is discussed in Chapters 8 and 9.

Tectonic forces may locally raise domelike mountain masses, some very large like the Black Hills of South Dakota, 130 km (80 mi) long, and others quite small, like the oil-producing domes along the Gulf Coast that may be less than 1.5 km (1 mi) across. Figure 1-3

GEOLOGIC TIME SCALE			Time (millions of years before present)
Era	Period	Epoch	
Cenozoic	Quaternary	Holocene	
		Pleistocene	
			2
	Tertiary	Pliocene	5
		Miocene	24
		Oligocene	39
		Eocene	53
		Paleocene	65
Mesozoic	Cretaceous		136
	Jurassic		
			190
	Triassic		
			225
Paleozoic	Permian		280
	Carbon-iferous — Pennsylvanian		
	Carbon-iferous — Mississippian		345
	Devonian		395
	Silurian		430
	Ordovician		500
	Cambrian		570
	Precambrian		3600 +

FIGURE 1-2. **Geologic time scale. The absolute dates are determined from the extent of disintegration of certain radioactive atoms. The 5-million-year date for the Miocene-Pliocene boundary represents a recent radical revision. To get some idea of the immensity of geologic time, assume a building 300 m (1000 ft) high to represent the age of the earth. A sheet of paper on the roof would represent all recorded history, approximately 25,000 years.**

shows an eroded dome several kilometers across. Elsewhere, the layered rocks of the earth have been crumpled into folds by lateral pressures (Figure 1-4). In some places, the crumpling has created important mountain ranges such as the Appalachian Mountains of eastern United States and the Jura Mountains of Europe. The activity takes place extremely slowly, over millions of years. Elsewhere, as in the southwestern United States and southwestern China, the internal forces have repeatedly broken (faulted) the earth's crust locally, culminating in lofty block mountains, The Sierra Nevada of California (see Figure 8-10) and the Wasatch Range of Utah (see Figure 3-10) are examples. In parts of the Rocky Mountains and the Alps, great slabs of the earth's crust have

FIGURE 1-3. **Eroded dome. Mauritania, West Africa.** *(Photo by U.S. Air Force.)*

been shoved bodily upward and outward over lower terrain in front (Figure 1-5). In still other places, segments of the earth's crust move past each other along long faults known as *rifts*. The San Andreas Rift in California is a currently active rift (see Figure 8-8).

Of the tectonic movements, only those involving faulting take place rapidly enough to create environmental hazards. In the 1906 San Francisco earthquake, for example, the ground on opposite sides of the San Andreas Rift suddenly shifted horizontally as much as 7 m (21 ft). The hazards result not only from the actual rupture but also from ground-shaking and earthquake-induced landslides and ocean waves.

Volcanism. Volcanic activity includes violent explosive eruptions as well as the quiet upwelling of lava. Explosive eruptions have shattered some volcanoes and hurled cubic kilometers of debris into the air. Dense incandescent clouds of dust-laden gas sometimes rush down the slopes of volcanoes and obliterate all life in their paths. Huge waves generated during eruptions of coastal volcanoes have inundated low-lying coastal regions, with great loss of life and extensive erosion. The heavy rains that accompany many violent eruptions often convert the loose ash on volcanic slopes into muddy slurries which rush downslope as destructive mudflows. Volcanic ash from major eruptions has temporarily altered the world's climate. The eruption of lava, while a relatively quiet phenomenon, has submerged many valleys and buried entire villages. These and other aspects of volcanic activity are treated in Chapter 9, on volcanic environments.

Weathering. In opposition to the forces of tectonism and volcanism which tend to build up the landscape are other processes which tend to wear it down. The more subtle of these "destructive" processes are included under the term "weathering." Some of the weathering processes break solid rock into smaller particles

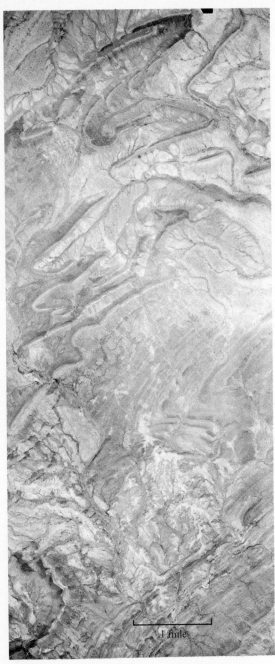

FIGURE 1-4. **Rock layers crumpled by internal forces and eroded to contorted ridges and valleys. Marathon Basin, Texas.** *(Photo by Army Map Service.)*

(*physical disintegration*); others convert it chemically into new products (*chemical decomposition*). The most important product of weathering is vegetation-supporting soil. *We shall use the term "soil," however, in the engineering sense for all loose material overlying bedrock, whether capable of supporting vegetation or not.* In many exposures, soil grades imperceptibly downward into solid unweathered rock. Weathering is discussed in Chapter 2.

Mass Movements. Downslope movements of soil and rock under the action of gravity alone are termed *mass movements.* The movements may be very rapid, as when rock falls from a cliff or when a mass of rock or earth slides catastrophically down into the lowlands. Or, movement may take place so slowly as to be imperceptible except by long-term observations. These slow movements are referred to as *creep.* In some areas, widespread creep is a greater total hazard to structures than are the more rapidly moving but localized landslides. Mass movements and their environmental impacts are also discussed in Chapter 2.

Erosion. Whereas landslides and other mass movements are propelled downward by gravity alone, the *processes of erosion* require moving media such as running water, waves and currents, wind, or glacial ice.[1] These *agents of erosion*, together with mass movements and deposition, are the great levelers of the earth's landscape. They move soil and rock debris from highlands to lowlands. If unhindered, they would level the highlands, fill up the lowlands, and reduce the earth to a topographically smooth sphere. Tectonism and volcanism, however, continually create new landscapes. The agents of erosion, the processes they are involved in, and the environmental problems they create are discussed in Chapter 3 to 7.

Deposition. The products of erosion ultimately end up in the sea, the earth's settling basin. The march to the sea, however, may be interrupted for intervals of a few hours or millions of years.

---

[1]Some geologists include mass movements in erosion.

FIGURE 1-5. **Part of the Lewis thrust fault, Montana. The black line marks the trace of the low-angle fault. The upper block, with Chief Mountain as the easternmost erosion remnant, consists of Precambrian rocks more than half a billion years old. This enormous sheet was thrust eastward to its present position overlying Mesozoic rock layers formed about 100 million years ago. (U.S. Geological Survey.)**

Brief interruptions are illustrated in many rivers where sand may settle to form bars, only to be picked up and moved by the next flood. Sediment brought to the sea by rivers, or eroded from the coast by waves, may be trapped temporarily in deltas, beaches, spits, and bars, landforms that are especially susceptible to erosion by the sea. In contrast, sediment dropped in lowlands beyond a mountain front may accumulate as extensive plains and remain for geologic epochs. Such is the origin of the American Great Plains.

Soil and rock debris eroded by former glaciers still mantles large areas of the earth. Much of this debris forms distinctive landscapes and gives rise to a variety of environmental problems based on topography and material constituents.

Wind deposits cover extensive areas of the earth. In arid regions, where wind is active and much loose sediment is available, environmental problems are caused by drifting sand and dust storms. Modern and ancient sand fields require special adjustments due to the uneven topography, the usually low ground-water level, and the generally infertile soil. Wind-transported dust, which merely mantles preexisting topography, generally becomes fertile soil but is prone to rapid erosion.

The nature of the sediments deposited by the agents of erosion, the distinctive landforms they assume, and the environmental problems they present are discussed throughout Part Two.

Changes in Sea Level. The level of the sea with respect to the lands is changing constantly, albeit at a very slow rate. Worldwide changes are due to fluctuations in the level of the sea itself. The most recent fluctuations are related to the waxing and waning of great ice sheets during Pleistocene time (Figure 1-2). During times of maximum glaciation, when much of the ocean's water was stored on the land as ice, sea level stood as much as 130 m (425 ft) lower than now. During interglacial stages, when glacial melting had progressed further than at present, sea level may have stood 60 to 90 m (200 to 300 ft) higher than now.

To complicate matters, while sea level was fluctuating worldwide, tectonic forces were locally elevating some coastal areas and depressing others. Thus, along many coasts, old marine shorelines have been lifted high above the levels reached by the Ice Age seas. In other places, high terraces change elevation along the coast, locally descending beneath the sea. Coastal warping is indicated. In higher latitudes, the weight of the vast Ice Age glaciers of northern North America and Europe depressed the earth beneath them. As these glaciers melted, the ground rose again, an adjustment that is still going on. This accounts for some tilted shorelines and for raised beaches, deltas, and other shoreline features.

Geochronology: Dating the Past

It is often difficult or impossible to predict potential dangers from some environmental hazards because of the absence of historical records or the obliteration with time of field evidence. Information may be lacking for great floods, for exceptional ocean storms, for volcanic eruptions, and for certain other events. It thus becomes necessary to infer potential hazards from the record of ancient events preserved in the rocks. For example, Chapter 9 tells how the potential hazards from future eruptions of presently dormant Cascade volcanoes have been evaluated by determining the dates and areal extent of ancient ash deposits and lava and mudflows.

A widely used method for dating ancient events involves *radioactive decay*, the process whereby some elements, without outside stimuli, disintegrate into other elements, releasing energy in the process. An important property of radioactive elements is that they have fixed half-lives. The *half-life* is the time required for half of the atoms to change to new atoms. This same amount of time is required for half of the remaining radioactive atoms to disintegrate, and for half of the remainder, and so on. By measuring the proportion of new to old atoms, and knowing the half-life of the radioactive element, one can determine when disintegration began, or the time of formation of the element. If the element is in an igneous rock, the derived

date indicates the time of formation in the original melt, often exceeding 1 billion years.

Carbon 14 is widely used for dating deposits less than 50,000 years old. It is an *isotope* of carbon. Isotopes are atoms of the same element differing in weight. Carbon 14, unlike the far more abundant carbon 12, is radioactive, with a half-life of about 5700 years. It is formed when cosmic rays, on entering the atmosphere, collide with and alter the atomic structure of nitrogen atoms. The proportion of carbon 14 to the common carbon 12 in the atmosphere is about 1 in a trillion. All living organisms, animal and vegetable, absorb and store carbon in their body tissues. Plants breathe it in as carbon dioxide, and animals eat the plants. As long as the organisms remain alive and the intake continues, the ratio of carbon 14 to carbon 12 remains constant. When the organism dies, however, the intake stops, but the disintegration of the carbon 14 continues, thereby slowly changing the ratio. By determining the diminished carbon-14 to carbon-12 ratio, the scientist can determine how long ago the organism died. Thus, the date of death of trees buried in ancient flood deposits, in volcanic ash, in mudflows, and in other deposits can be determined. If the death was due to the catastrophic event which produced the deposit, the event itself is thereby dated.

There are other methods for dating geologically recent deposits, for example, tree-ring dating, or *dendrochronology*. Each pair of rings constitutes a single year's growth. One ring records the portion of the year in which growth is rapid, generally the summer or a wet season, whereas the companion ring records the season of retarded growth. Thus, each pair of rings is a record of 1 year's weather. The succession of pairs in a 2000-year-old sequoia provides a 2000-year climatic record. If a 50-year-old tree had been buried in a landslide 100 years ago, the date could be determined by matching its succession of rings with the equivalent portion of the tree ring succession of an older living tree in the area.

Although most of the dates that appear in this book are based on the radioactive technique, some have been established by other means, as we shall see.

Engineering Properties of Earth Materials

Earth materials include all consolidated and unconsolidated rock matter. We are especially interested in the earth materials that provide the natural foundations for structures and affect surface topography. By *bedrock* we mean the firm rock below the mantle of soil. By *soil* we mean all unconsolidated material, material that can be removed without blasting.

Engineering Properties of Bedrock. It is vital to know whether the bedrock on which a structure is to be placed has the strength to withstand the forces that will be exerted on it. Can it, for example, withstand the weight of a heavy structure without compacting? This is a question of *compressive strength*, a measure of the resistance to a force tending to decrease its volume. Or, in the case of a structure built on a rock slope, how well can the bedrock withstand *shearing*, that is, the slipping of one part of the rock mass past another? Resistance to such shearing motion is *shear strength*. Finally, how competent is the rock to resist forces tending to pull it apart and create cracks and fissures. This type of strength is *tensile strength.*

To measure the compressive strength, the pressures on a small, unconfined laboratory sample are gradually increased until the rock breaks. Igneous rocks and some quartzites and sandstones have high compressive strengths.

Shear strength is also measured by application of a vertical force, but here the cylindrical sample, in a rubber jacket, is placed in a cylinder containing a fluid which is also under pressure and provides lateral support to the sample. The applied stress at the time of failure under these confined conditions is the shear strength.

Tensile strength is measured by placing a stone slab across two supports and increasing the weight of a load placed on the center until the slab fails. Tensile strengths of rock and soil are very low, roughly 10 percent of the compressive strength; hence, structures that are expected to experience tensile stresses are not made of rock but of steel or reinforced concrete.

Under sufficient pressure, many rocks deform by slow *plastic flow*. The release of pressure may result in flow in reverse. This is rarely a concern of the environmental geologist.

Engineering Properties of Soils. The engineering properties of soils and their behavior in response to engineering structures are reasonably well known. Yet foundation failures are common. The reason is that the properties of soil are usually not uniform in distribution and may change with time. Soils may consist of layers, each with different properties. The layers may change properties laterally and vertically so that measurements at one site may not be valid a short distance away. The situation is complicated if the soil layers are inclined, folded, or fractured, characteristics which affect the strength of the mass as a whole. The nature and configuration of the underlying bedrock are other important factors. The engineer can measure the soil properties, but the distribution and configuration of the soil and bedrock units are problems for the geologist.

The engineering properties of soils vary with texture (size of particles), shape of particles, mineral composition, moisture content, degree of consolidation, and degree of uniformity.

Textural analysis involves passing the soil through screens of different mesh sizes to determine the relative amounts of clay, silt, sand, and gravel. The proportions of different particle sizes affect the potential degree of compaction and the *permeability* (capacity to transmit fluids).

The shapes of grains are important in that rounded grains or pebbles provide a less stable surface than one underlain by angular grains. The shapes of grains also affect soil *porosity* (percentage of pore space).

Composition is important in several ways. For example, one type of clay swells when wet, whereas another does not. Swelling clays are responsible for the cracking of many foundations and for many landslides. Also, clay containing much organic matter is less stable than clay derived solely by weathering of rock minerals.

Moisture, if present in large amounts, may cause landslides and other mass movements. Part of the water can be easily drained; other parts are firmly held in the soil. The moisture content can be determined by drying the sample in an oven and measuring the loss in weight.

Ordinarily, texture and moisture content are determined first. These tests may be followed by tests for plasticity, shrink-swell potential, load-bearing capacity (soil strength), soil density, and soil permeability.

Shrink-swell or *expansive soils* are common in the western United States but rare in the East. The common adobe clay is generally susceptible to shrink-swell in the upper 1 m (3 ft) or so.

Load-bearing capacity, or soil strength, may be measured in the field by driving a cone or circular piston into the soil under controlled conditions. In the laboratory, a cylindrical sample is loaded until it fails.

Soil density is basically the weight per unit volume of a given sample. Soil density is important if excavated soil is to be dumped on slopes elsewhere.

Soil permeability is important for drainage purposes, as where a septic tank is to be emplaced. Permeability may be measured by forcing water through a soil sample.

MAN THE INTRUDER, ACCELERATOR, AND CATALYST

The above brief introduction to the natural processes that affect the geologic environment emphasizes the concept of natural, inexorable change. The environment will continue to change, often for the worse from the human point of view, and such changes will occur whether people occupy the environment or not. Human intrusion, however, often accelerates the rate of change and, in many instances, initiates new changes.

Consider, for example, the human role in slope failure. In many regions, slopes are steep and in a precarious state of rest. Heavy rains, undermining by streams, and earthquakes constantly cause masses of soil and rock to move downslope. The movements are rapid, as in landslides, or imperceptibly slow, as in creep. These sensitive slopes are being occupied to an increasing degree. Improper treatment often disturbs the delicate equilibrium and initiates landslides, reactivates old slides, or accelerates creep. The added weight of a structure, the weight of the displaced soil and rock from grading operations, the cutting of steep slopes, and interference with drainage may destroy the uneasy equilibrium and result in failure sooner than would have occurred under natural conditions. The structural damage from slope failures in the San Francisco Bay region during the winter of 1968–1969 is assessed at $25,400,000.

Coastal environments similarly illustrate the adverse effects that may follow human activities. A community may erect a groin (see Figure 4-19) for protection against beach erosion or a jetty (see Figure 4-21) to provide a safe harbor for shipping. The obstruction, however, by trapping the sand moving along the shore, may deprive areas farther along of the sand which formerly compensated for erosional losses. Without this source of replenishment, beach erosion may be accelerated downdrift from the area protected by the groin or jetty, and serious damage may result.

A situation common on river floodplains also illustrates the far-reaching effects of local modification of the natural environment. A given community builds high levees to prevent local overflow during floods. By preventing the natural overflow at this site, however, more water than normal is passed downstream. Thus, the flood situation downvalley is aggravated.

People not only accelerate many natural processes but may introduce new hazards to an area. Undercutting slopes in constructing roads or in preparing flat ground for buildings often initiates landsliding. Excessive withdrawal of ground water or intensive underground mining may cause subsidence or collapse of the surface. Exposing bare ground through overgrazing, deforestation, or dissemination of smelter fumes may initiate gullying and badland development. Farming in semiarid environments exposes the ground to the wind, which may remove much of the fertile topsoil and create problems of drifting sand and dust. Damming of rivers causes sediment deposition in reservoirs and may initiate erosion downvalley by the now underladen waters. Preventing stream-borne sediment from reaching the sea may cause coastal erosion. Interfering with Arctic tundra may initiate environmental changes due to thaw of the frozen ground. And mining operations may initiate other environmental hazards (see Chapters 11 and 12). Thus, careful planning is vital if new environmental hazards are to be avoided. The expected consequences of occupation of a site may justify abandonment of the project or radical reduction in its scope.

ADDITIONAL READINGS

Cargo, D. N., and B. F. Mallory: "Man and His Geologic Environment," Addison-Wesley Publishing Company, Inc., Reading, Mass., 1974.

Flawn, P. T.: "Environmental Geology," Harper & Row, Publishers, Incorporated, New York, 1970.

Legget, R. F.: "Geology and Engineering," McGraw-Hill Book Company, New York, 1962.

———: "Cities and Geology," McGraw-Hill Book Company, New York, 1973.

Longwell, C. R., R. F. Flint, and J. E. Sanders: "Physical Geology," John Wiley & Sons, Inc., New York, 1969.

McKenzie, G. D., and R. O. Utgard, eds.: "Man and His Physical Environment," Burgess Publishing Company, Minneapolis, 1972.

Nichols, D. R., and C. C. Campbell, eds.: "Environmental Planning and Geology," U.S. Geological Survey and U.S. Department of Housing and Urban Development, Washington, D.C., 1971.

Tank, R. W., ed.: "Focus on Environmental Geology," Oxford University Press, New York, 1973.

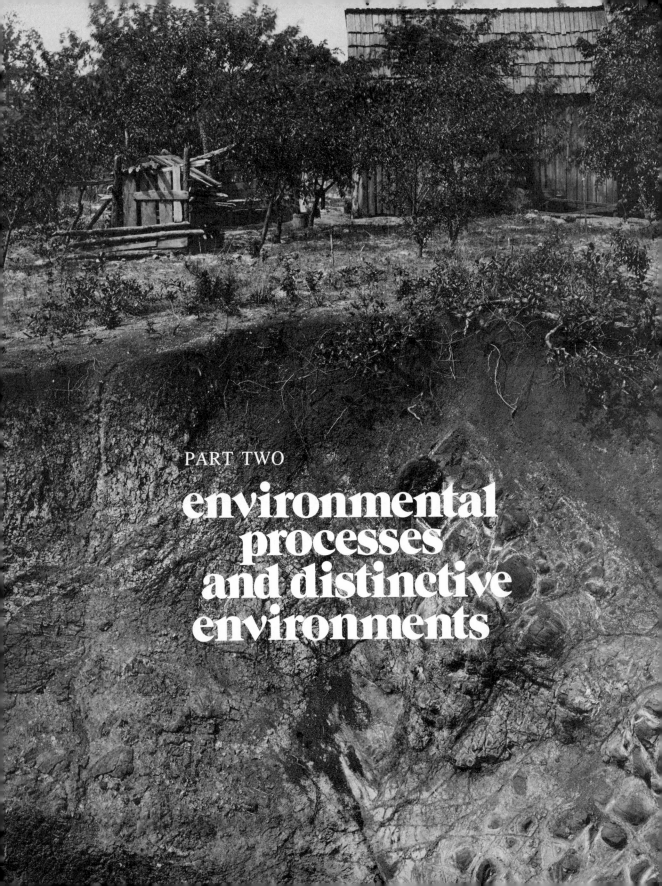

PART TWO

environmental processes and distinctive environments

The geologic environment includes the topography, the mantle of soil, the bedrock below the soil, the natural processes that modify the landscape, and contributory factors, such as vegetation or permanently frozen ground, that influence the effectiveness of the processes in operation. We find it useful to discuss the environments under diverse headings that emphasize process, climate, or distinctive setting. We do this recognizing that most processes are active in all environments, differing only in relative importance. For example, in Chapter 2 we discuss situations wherein the environmental impacts of mass wasting are especially important. Similarly, in Chapter 3 we discuss situations wherein the environmental impacts of hydrologic processes are especially important.

Unless discussions specifically relate to special climatic environments, temperate humid environments are to be assumed.

CHAPTER TWO

mass wasting and the environment

INTRODUCTION

The Wasting Lands

On June 28, 1974, on the Villa Vicencio highway 150 km (95 mi) east of Bogota, Colombia, an enormous mass of ground detached itself from the mountainside and rushed downward with a fearsome roar. More than 730 m (800 yd) of the twisting highway were buried by the debris, which entombed 20 vehicles. Two hundred lives were lost, and the river below was dammed to form a lake that now threatens catastrophic flooding. The threat is not an idle one. Two months before, on April 25, a landslide in the Mantaro Valley of central Peru impounded a lake 30 km (19 mi) long. After a few weeks, the waters broke through the barrier and rushed downvalley in great waves. Twenty villages, abandoned in advance of the flood, were obliterated. In northern Italy, on October 9, 1963, 2100 people lost their lives when a huge landslide dropped into the deep but narrow reservoir behind the Vaiont Dam. Enormous water surges overtopped the dam and laid waste the valley below.

Similar disasters have occurred throughout history. Yet these events are only part of the broader problems resulting from gravitational influences on earth materials.

Although catastrophic landslides and rockfalls capture the

imagination, much of the land finds its way to lower levels piecemeal. Thus a home in Utah is destroyed when a river bluff collapses, a Pennsylvania barn drops 20 ft in a few hours as the ground collapses into a cavern or sinkhole, and freeway traffic in Los Angeles is interrupted as a large slide obstructs the right of way. Some movements are extremely slow. A home owner may helplessly watch the destruction of his or her home as the ground moves under it by inches per day or feet per year. The movements are often perceptible only by careful measurements, by the progressive development of strains and cracks in structures, or by malfunctions in installations such as pipelines and canals.

Movements of material by gravity alone are *mass movements*. They involve falling, sliding, flowing, and subsidence, and their expressions include falls, avalanches, landslides, flows, creeping soils, and collapsing or "settling" ground. *Mass wasting* refers to the lowering or wasting of the land surface by mass movements.

Damage by Mass Movements

Mass movements cause loss of life, property damage, and disruption of activities. During the present century, the world death toll has been decreasing because of improvements in understanding natural processes, in predicting where and when undesirable events are likely to occur, and in responding to warnings. But property damage is increasing because of population clustering, increases in number and size of structures, and ill-advised decisions on location, design, and protection of such works.

For proper perspective, let us consider the costs of major natural disasters in the United States. Approximately 18,600 deaths can be attributed to floods, hurricanes, tornadoes, earthquakes, and mass movements during the half century 1925–1975. Nearly half of this total was due to tornadoes. The death toll from earthquakes was only 590 for this period, and the loss of life directly attributable to mass movements was fewer than 100.

The record for property damage is startlingly different. During the period 1925–1975, floods, hurricanes, tornadoes, and earthquakes

in the United States accounted for $20 billion in damage, based on estimates unadjusted for changes in dollar value. But landslides and subsidence alone accounted directly for $75 billion in damage. In California, damage from landslides and subsidence is projected at $10 billion for the period 1970–2000. This is a substantial fraction of the $55 billion estimated for damage from all geologic hazards in the State during the same period.

In landslide terrain, the ground may rupture, fissure, tilt, fragment, subside, move laterally, or be flooded. Structures may crack, shear, collapse, or be submerged in debris or water. Remedies range from repair to abandonment. Many mass movements generate legal actions, reduce property values, increase insurance rates, interfere with normal living, interfere with agriculture and mining, interrupt or impede transportation and communications, and eliminate some options for future land use. Costs from indirect physical effects can be heavy, as when uncontrolled fires follow landsliding that ruptures water mains, when flooding results from avalanching into a reservoir, or when silting of a reservoir is accelerated by mass movements on adjacent slopes.

Most mass movements involve loose materials rather than bedrock. Hence it is appropriate to consider briefly the origin of loose materials, the causes of their instability, the mechanics of movement, the topographic forms diagnostic of movement, and the methods of investigation and treatment of moving ground.

ROCK WEATHERING AND SOIL FORMATION

Rocks are converted to loose debris by *weathering*, which involves either physical disintegration, in which the chemical composition is not altered, or chemical decomposition, in which chemical changes result. Most rock weathering involves both physical and chemical processes.

Physical Disintegration

A potent mechanism for disintegrating rocks in mountains and other seasonally cold environ-

FIGURE 2-1. **View up Lake Canyon, west of Mono Lake, California. Note long talus slopes on both sides of the canyon and numerous avalanche chutes, as in left foreground.** *(Eliot Blackwelder.)*

ments is *frost wedging*, or *frost shattering*. The disruption is due to the expansive force that accompanies freezing of water in fractures and other pore spaces.

On level surfaces, the broken rock fragments may accumulate as widespread blankets known as *block fields*. On steep slopes, blocks are easily detached to expose new rock faces to further frost action. The accumulating debris forms *talus slopes* at the foot of cliffs (Figure 2-1).

Under many conditions, the molecules of dissolved substances in water come out of solution as minute crystals which slowly increase in size. This is the process of *crystallization*. Crystallization can generate pressures sufficient to disrupt rocks. In addition, some substances may *hydrate*, that is, absorb water, resulting in expansive pressures.

Cleopatra's Needle, an Egyptian obelisk in New York City (Figure 2-2), is an example of rapid weathering of stone in an inhospitable environment. The granite obelisk was erected on the Nile River opposite Cairo about 1500 B.C. It was toppled by Persian invaders about 500 B.C. For five centuries it lay partially buried in moist sediments of the Nile, suffering some damage from water absorption and crystallization before being moved to Alexandria. In 1880, it was moved to New York City and immediately began to scale. The accelerated weathering resulted from penetration of moisture, with rupture by frost wedging and by hydration of the substances that had been introduced in Egypt. After 10 years the obelisk was treated with sealants to halt the decay, but the inscriptions were already largely obliterated.

Disintegration can also result from *expansion* and *contraction* of rocks exposed to extreme variations in daily or seasonal temperature. Temperatures in some desert regions reach 80°C (176°F) in the sun, and winter temperatures drop well below freezing. These temperature variations, repeated over months or years, cause rocks to split along surfaces of inherent weakness. When sun-heated rocks are deluged in a desert cloudburst, the rapid chilling assists in disruption. Even more effective are the sudden changes when rocks are heated by forest fires and cooled rapidly in rainstorms.

Wedging by growing plants also ruptures rocks, especially where fractures are already present. Steep slopes are especially susceptible. Ground-dwelling animals such as ants, rodents, and worms assist in physical disintegration by dislodging particles of partially weathered bedrock.

Some rocks that formed at great depths expand when the pressures confining them are relieved by erosion of the overlying rocks. The expanding rock shrugs off great curved plates

FIGURE 2-2. **Cleopatra's Needle, in Alexandria, Egypt (*a*), and in Central Park, New York City (*b*). (*Courtesy the Metropolitan Museum of Art.*)**

(*a*)   (*b*)

or exfoliation shells (Figure 2-3). This type of *exfoliation* is most common in igneous rocks such as granite.

Chemical Decomposition

Just as iron rusts and statues become defaced with time, so rocks react with chemicals in air and water and slowly decompose. Chemical decomposition yields products that differ markedly from the original rock.

Air and water contribute to decomposition either directly or indirectly as carriers of corrosive substances. The atmospheric substances most commonly involved in chemical weathering are oxygen, water, and carbon dioxide, but other gases, such as chlorine and oxides of sulfur and nitrogen, also contribute. The combination of oxygen with iron-bearing minerals is *oxidation*, a reaction that imparts rusty colors to rocks and soils. The formation of weathering products through absorption of water by rock minerals is *hydration*, which commonly produces clay. *Solution* is the dissolving of minerals by fluids. Some rocks, such as limestone, are particularly susceptible to solution and may be honeycombed with caverns and display a pitted topography known as *karst* (Figure 2-4). Organisms, including bacteria as well as higher animals and plants, engage in *biologic processes* that cause rock decay. Some of these generate potent acids. The decomposition of rocks usually results in an increase in volume. If these products form below the rock surface, where moisture is retained longest, the development and expansion of clays and other minerals may pry loose small exfoliation shells.

Chemical weathering dominates in warm, moist climates because most chemical reactions are accelerated by heat and water. Physical weathering is relatively more important in cold and dry climates. Topography is a factor in that rock fragments do not remain on steep slopes long enough for significant chemical weathering. Physical weathering thus dominates on steep slopes. On gentle slopes, where the products of weathering remain in place longer, chemical decomposition can proceed to greater depths. The depth of chemical weathering in some moist, humid regions with relatively flat ground is locally as much as 50 m (165 ft) (Figure 2-5).

FIGURE 2-3. Exfoliation of granite resulting from relief of pressure. (*a*) Exfoliation shells south of Tenaya Lake, Yosemite National Park. *(Arthur D. Howard.)* (*b*) Flexed shell suggesting expansion may still be in progress. *(François Matthes, U.S. Geological Survey.)*

Soil

All loose materials are classed as *soil* by the civil engineer whether they are capable of supporting vegetation or not. Unless otherwise stipulated, "soils" will be so used in this text. Many soils are zoned, or separated into layers or horizons (Figure 2-6). The A horizon, the uppermost layer, has had most of the soluble substances leached (dissolved) from it and moved downward, along with much of the clay. Owing to the loss of material, the A horizon

(topsoil) is soft, porous, and friable. Materials removed from the A horizon are deposited 0.3 to 0.6 m (1 to 2 ft) below in the B horizon. Because clays and other materials have been added, the B horizon is usually dense, tough, and relatively impermeable. Below the B horizon is the C horizon, a zone of partly decomposed rock that grades downward into fresh bedrock.

Time Factor in Rock Weathering and Soil Formation

The rate of weathering depends upon the physical and chemical environment and the characteristics of the rock that is under attack. In high mountains, numerous outcrops of rock that were polished by ice thousands of years ago are

FIGURE 2-4. **Limestone plain with numerous sinkholes, some containing ponds. Near Park City, Kentucky.** *(W. Ray Scott, National Park Concessions, Inc. Courtesy Kentucky Geological Survey.)*

FIGURE 2-5. **Decomposed rock. Hard cores are left in the decomposed mass. Wadesboro, North Carolina.** *(Bailey Willis.)*

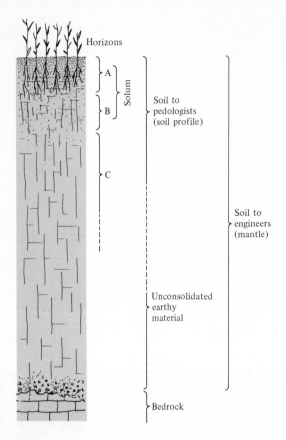

FIGURE 2-6. **Generalized soil profile common in humid temperate climates.** *(After A. C. Orvedal, 1952, "Frost Action in Soils," Washington, D.C., National Academy of Sciences, National Research Council, Highway Research Board, Special Report 2.)*

relatively unweathered (Figure 2-7), whereas much decay can be seen in other environments on rock surfaces that are much younger. Thus much of the statuary of Venice, Italy, is already weathered beyond repair.

One way to study rates of weathering and soil formation is to examine tombstones for which the dates of erection are known. Limestone tombstones in Great Britain, for example, have weathered 2.5 cm (1 in.) in 240 years. Studies reveal comparable rates of weathering for limestone tombstones in Massachusetts and

New York, but slower rates for comparable monuments in New Mexico and Arizona where the climate is drier.

Weathering in warm, humid climates can be astonishingly rapid, especially in loose materials. Within 45 years of the eruption of Krakatoa in 1883, 35 cm (13.8 in.) of soil had formed on the volcanic dust that covered a neighboring island in the East Indies. Much slower rates of soil formation characterize cold or arid environments.

CAUSES OF MASS MOVEMENTS

Ground or *slope failure* occurs when the gravitational force exceeds the resisting forces due to strength and cohesion of the material, friction between it and adjacent materials, and buttressing or other external support.

The causes of failure are *inherent* or *superimposed*. Inherent factors relate to the material and its physical setting and are *passive* in that they remain constant over long periods of time. The principal inherent factors are:

1. *Properties and distribution of minerals and other constituents.* For example, some minerals change dimension or become slippery when wet.

2. *Structural features.* These include layering, the parallel arrangement of platy minerals (such as mica flakes), and the presence of fractures.

3. *Moisture content.* Water adds weight, contributes to buoyancy, and decreases the cohesion of clay minerals.

4. *Topography.* Many slopes are already in delicate equilibrium so that slight disturbances cause failure.

5. *Vegetation.* Plant growth adds to stability by retarding downward infiltration of water and by stabilizing the ground through its root systems. It does add weight, however.

Superimposed factors are *active* in that they increase the gravity-related force or decrease the resisting forces. They trigger mass movements where inherent causes are already present. The superimposed factors include:

FIGURE 2-7. **Glacially polished and grooved granite near Johnson Lake in the Sierra Nevada, California. Although the outcrop has probably been exposed for thousands of years, weathering has been slight.** *(H. W. Turner, U.S. Geological Survey.)*

1. *Deterioration of materials.* Weakening of materials can be caused by weathering, alteration, and solution along fractures or other paths of fluid circulation, rupturing by frost action, the prying of plants, and earthquake shaking.

2. *Increases in moisture content.* These changes affect the weight, volume, internal water pressure, and internal cohesion of the materials. They may result from increased precipitation or the melting of snow. Filling or draining of reservoirs, irrigation, watering of residential areas, and leakage from septic tanks and artificial ponds may also be important.

3. *Overloading.* Overloading can occur through addition of water or through superimposed dumps, fill, or artificially created structures.

4. *Removal of underlying or lateral support.* Natural buttresses or other supports can be weakened or removed by stream and wave erosion, by surface excavation, and by underground solution or mining.

5. *Earthquakes.* Earthquake shaking can reduce friction and dislodge masses of soil and rock.

6. *Other factors.* Additional factors include volcanic eruptions; collapse of underground caverns; the weight of grazing cattle, trains, or deep snow; and shaking by mine or quarry blasts.

PROCESSES AND EXPRESSIONS OF MASS MOVEMENTS

It is convenient to distinguish mass movements in terms of their dominant styles of behavior. Thus *falls* are free drops from steep or overhanging cliffs, *slides* are movements along one or more slip surfaces of shear failure, and *flows* involve slow to rapid movements of soil in which the materials are deformed.

Combinations of movement result when one kind of behavior gives way to another during a failure. Rockfalls often convert to avalanches on lower slopes, and many slide masses convert to flows. The term "landslide" is commonly used for combinations of falling, sliding, and flowing.

Mass movements are considered *active* if there is evidence of present or recent movement and *inactive*, *dormant*, or even *dead* if movement has ceased. An active landslide can be stabilized, either naturally or through human activities, just as a dormant landslide can be reactivated.

Soilfalls and Rockfalls

Characteristics. Soilfalls and rockfalls are free drops of materials from precipitous slopes. They range from small earth clods or rock fragments to enormous masses. Soilfalls are most abundant along fresh banks and cliffs formed by vigorous stream cutting, wave erosion, or artificial excavation. Soilfalls generally involve small amounts of debris because soil, being unconsolidated, does not ordinarily stand in high cliffs. An exceptionally large soilfall occurred from the sea cliff at Pacific Palisades, California (Figure 2-8).

FIGURE 2-8. **Soilfall from sea cliff at Pacific Palisades, California. The 30-m- (100-ft-) high cliff of unconsolidated but cohesive sediments was formed by wave action prior to construction of the highway across the beach. November, 1966.** *(Courtesy Teledyne Geotronics.)*

FIGURE 2-9. (*a*) **Huge rockfall from limestone cliffs of Hsi Shan (West Mountain), west side Kunming Lake, South of Kunming, Yunnan Province, China.** (*b*) **The scar of an older, smaller rockfall appears in left center. The large fracture back of its crest indicates instability and a threat to the community below. 1945.** (*Arthur D. Howard.*)

Rockfalls, unlike soilfalls, may be very large because of the great height of many rock cliffs (Figure 2-9). Separation of rock masses occurs along steeply dipping fractures or along the contacts between rock layers. Rock falls are commonly triggered by natural or artificial undermining or by earthquake shaking. In California, enormous rockfalls took place in Yosemite Valley during the earthquake of 1872, which centered in Owens Valley more than 100 km (60 mi) to the southeast.

Investigation and Treatment. If cliffs are of unconsolidated materials, investigation should focus upon the strength of the materials, both dry and wet, the inclination and distribution of fractures, and the occurrences of springs, seeps, and other expressions of ground-water circulation. If cliffs are of rock, surfaces of potential separation such as fractures and weak layers should be investigated. And wisdom dictates a search for evidence of previous falls. Evidence is obvious in some places (Figures 2-8 and 2-9) where fresh debris occurs below scars in the cliff above.

Where material falls repeatedly from cliffs as individual fragments or small masses, devices are employed to minimize the impact or to provide warning when falls occur. In some situations, wire meshes are draped over exposures. Electrified slide-detection fences are used along some highways and railroads. In other situations, walls divert the debris from important structures (Figure 2-10). *Prevention* of rockfalls can require even greater ingenuity. Protective devices on the faces of the steep cliffs of Rio de Janeiro include concrete and masonry walls, concrete supporting pillars, and metal bolts that pin threatening rock masses to the slopes (Figure 2-11).

Slides

Characteristics. Slides involve slippage on a surface of rupture created when the downslope force due to gravity (shear stress) exceeds the resisting forces (shear resistance). Resistance is controlled by the slope of the slide surface, its form, its roughness, the presence or absence of weak materials or weak structural features, and the amount and distribution of ground water.

Planar surfaces of sliding may be layers of weak material, or the planes separating soil or rock layers, or fractures. Figure 2-12 shows rock failure along planar surfaces in Montana.

FIGURE 2-10. Concrete guard walls protect the Calderwood powerhouse of the Aluminum Company of America from rockfalls. Little Tennessee River, Tennessee. *(Courtesy Aluminum Company of America.)*

Slabs of rock have slid down the inclined bedding planes toward the highway and railroad cuts.

Curved surfaces of rupture and slippage, commonly scoop-shaped, develop where masses of soil or weak rock slump away from the enclosing material. The upper part of the slide mass in Figure 2-13 is an example of failure by slumping. Many slump blocks show backward rotation on curved surfaces of shear (see Figure 2-16).

Many slopes on unconsolidated materials show small, steplike slumps or terracettes that result from undercutting at the base, or from the repeated passage of grazing animals (see Figure 2-29).

FIGURE 2-11. **Rockfall prevention. P—supporting pillars; B—rock bolts; R—bolted concrete ribbon; D—wall to divert drainage and stop rolling rocks. Rio de Janeiro, Brazil.** *(Arthur D. Howard.)*

FIGURE 2-12. **Planar rock slides along highway and railroad near Glacier Park, Montana. The rock layers dip toward the observer.** *(Montana Highway Department.)*

FIGURE 2-13. **A typical slide-flow complex in glacial deposits, western Wyoming.** *(Eliot Blackwelder.)*

Some slides take place on horizontal surfaces. These *translatory*, or *spreading*, *slides* (Figure 2-14) are due to the outflow of weak subsurface materials, causing a settling and spreading of the overlying sediments. Such ground failure occurred at Anchorage, Alaska, in 1964, when seismic shaking reduced the strength of a buried layer of clay.

Slide movements range in velocity from barely perceptible to hundreds of kilometers per hour. Most slumping occurs at slow to moderate rates. The landslide of 1881 at Elm, Switzerland, on the other hand, attained a velocity of 320 km (200 mi) per hour.

Topographic Expression. The downward movement at the head of a slump or translatory slide forms a scarp (steep slope or cliff) (see Figure 2-34). This is the *main scarp*, or *headscarp*, and its rim is the *crown*. The crown would match up with the *head* of the slide mass if the mass were restored to its original position. If the movement is outward as well as downward, a fissure may open up. As it widens, a segment of ground may subside into it to form a *graben* (see Figure 2-14).

The snout of a slide commonly has a well-defined *toe* (see Figure 2-34) or a low pressure ridge (see Figure 2-14) where the mass moves upward at its terminus.

Examples of Damaging Slides. In 1958 a spectacular slide occurred on the sea cliff at Pacific Palisades in southern California (Figure 2-15). This failure was the latest in a series extending back through prehistoric time. The large, arcuate "bite," extending back into the terrace surface to the right of the most recent slide, resulted from even larger scale slumping that more than once caused relocation of the rim road and led to abandonment of the winding road. The 1958 failure, which followed lengthy fall and winter precipitation, involved a mass of soil and rock 450 m (1500 ft) wide and 275 m (900 ft) from front to back.

Many small landslides are due to inadequate preparation of sites before construction. A small destructive landslide, typical of many in California, is shown in Figure 2-16. The homes had been built on a bedrock ridge which

FIGURE 2-14. **Translatory slide. Anchorage, Alaska.** *(After W. R. Hansen, U.S. Geological Survey.)*

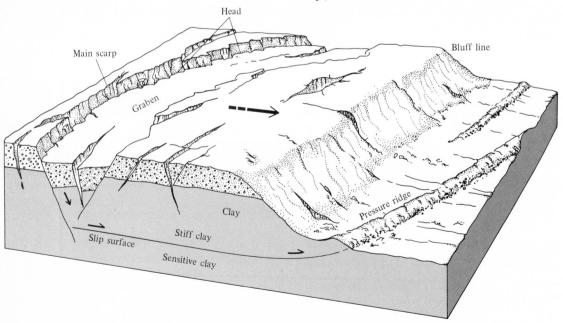

FIGURE 2-15. **Ancient and recent slide masses along a 50-m (160-to-170-ft-) high sea cliff at Pacific Palisades, California. Alluvial deposits (loose sediment deposited by running water) are exposed in the uppermost parts of the cliff at the left. These form a thick blanket over older shale and sandstone visible near the center of the view.** *(Courtesy Teledyne Geotronics.)*

had been flattened by grading and widened through the lateral placement of carefully compacted fill. Unfortunately, the slope materials on which the fill had been emplaced had been migrating downslope for centuries.

The Portuguese Bend landslide, in the Palos Verdes Hills, is the most damaging landslide in California. The present sliding, which destroyed more than 100 homes and a recreational facility, began in 1956 and is still continuing.

The failure involves a broad, platterlike mass of deformed, soft, clay-bearing strata and ancient slide debris 1100 m (3600 ft) in width and 1300 m (4200 ft) from head to toe. The main scarp is 150 m (500 ft) above sea level. The slowly moving mass (Figures 2-17 and 2-18), with a maximum thickness of 80 m (260 ft), is a composite of earlier landslide materials and newly detached bedrock that is breaking up as it moves. The easterly one-third of the mass is sliding along weak, clay-rich layers in shaly bedrock and the remainder across ancient slide deposits that rest upon bedrock. The easterly part has moved farther and more rapidly than the rest, from which it is separated by a near-vertical zone of shearing in the vicinity of Portuguese Canyon (Figure 2-18).

The present movement is only the youngest in a long sequence of slides, as indicated by numerous older slip surfaces and by ancient topographic features that include subdued scarps, swales, and hummocky ground (Figure 2-17).

The sliding was triggered by loading of the headward part of the complex with debris from road construction, and the slope must have been near failure at the time. Continued movement may be due in part to increases in ground water, which suggests that dewatering and im-

FIGURE 2-16. Results of slope failure, including both gliding and slumping. Palos Verdes Hills, California. The view shows a part of the nearly vertical main scarp beneath and on the near side of the house in the right foreground. The slide mass has moved toward the observer and toward the left, its upper and middle parts downward, and its lowermost parts outward in broad, fissured bulges. July 1960. *(R. H. Jahns.)*

FIGURE 2-17. Northward view of Portuguese Bend area, southern California, showing large slide complex prior to reactivation in 1956, when bedding-plane failure began to propagate from the vicinity of the new roadcuts in the right distance. Nearly all buildings in the foreground and middle distance have since been removed. July 1955 *(Courtesy Teledyne Geotronics.)*

provements in surface drainage might be a useful step toward stabilization.

The maximum horizontal displacement of the slide mass from 1956 to 1970 has been 40 m (140 ft), and maximum vertical displacement, near the northeastern corner, has been 12 m (40 ft). Horizontal velocities decreased from initial rates of 3.6 to 7.3 m (12 to 25 ft) per year in 1956 and 1957 to 1 to 4 m (3 to 14 ft) per year in 1961. The latter rates remained constant during the period 1961–1968, after which there was a dramatic rise. The increase may reflect accumulation of ground water within the complex. It correlates well with trends in rainfall. Acceleration of movement occurs within hours of heavy precipitation, and a velocity peak of 1.5 to nearly 4 times the current rate is generally achieved within 48 hours. The movement at Portuguese Bend is accelerated by some earthquakes, but the effects are measured in millimeters and are short-lived.

FIGURE 2-18. **Sketch of the Portuguese Bend slide late in 1957. Arrows indicate principal directions of slide movement, hachured lines the main scarp and secondary scarps, solid lines the major groups of ground cracks and fissures, and short arrows the toes of the main slide mass and two large slide units within it. (R.H. Jahns)**

Flows

Characteristics. Flowage involves a myriad of small-scale internal movements rather than the sliding of large masses. Rapid flow causes a churning known as *turbulent flow*, whereas in very slow flow the upper portions move a trifle faster than the lower parts, but without turbulence. It is as though the moving materials consisted of thin layers shearing slowly over one another. This is *laminar flow*.

Many mass movements of the flowage type start as falls or slides at higher levels. In the common type of landslide shown in Figure 2-13, the slump blocks are broken up as they rise over the foot of the slide and further movement is by flowage.

Flowing masses contain particles of nearly all sizes and vary in content of water and air. Steepness of slope is a major factor in rate of flow, but on a given slope, flows with the highest content of water move fastest.

Flows that leave valleys spread out as fans, lobes, or tongues on reaching level ground. Their forward parts are marked by wrinklelike ridges roughly parallel to their arcuate snouts,

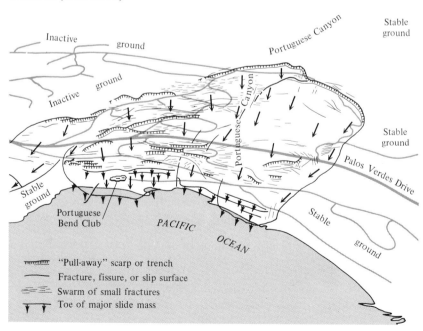

and lateral ridges that parallel the margins of the flow masses are common.

Rapid Flowage. *Debris Avalanches.* Many falls and slides of rock and soil which start as coherent units on very steep slopes break up into streams of debris following narrow tracks down the mountainside. These exceptionally rapid flows are *debris avalanches.* Because succeeding avalanches follow the same tracks, the scraping and pounding may create troughlike chutes (Figure 2-19).

Some debris avalanches are dry and owe their mobility not only to the steep slope but to included air and small particle sizes. Dense

FIGURE 2-19. Avalanche chute, Sequoia National Park. *(François E. Matthes, U.S. Geological Survey.)*

clouds of dust generally accompany dry avalanches, and the outrushing air may be intense enough to blow down trees and destroy buildings. Most avalanches are wet, and their mobility is promoted by included water or watery slurries of snow, ice, or rock particles.

Many debris avalanches are triggered by mild earthquakes, the vibration of railway traffic, a sonic boom, or even the weight of animals. Three examples of large earthquake-triggered avalanches in Alaska will illustrate modes of behavior as well as the wisdom of avoiding sites in or below the tracks of such mass movements.

During the Alaska earthquake of March 27, 1964, an avalanche of rock, soil, snow, and vegetation moved rapidly down the south slope of Puget Peak, in the southeastern part of the state. It began when a large rockfall near the summit dropped onto snow-covered talus, with an impact velocity exceeding 100 km per hour (60 mi per hour). The enormous mass of debris rushed down a timbered slope at speeds that approached 80 km (50 mi) per hour, tilting, stripping, and snapping trees. The resulting track, 100 to 300 m (about 330 to 1000 ft) wide and more than 2100 m (6900 ft) long, was formed in less than 2 minutes. Its vertical range was 1200 m (3900 ft). The avalanche stopped when its snout entered the sea.

Another example associated with the Alaska earthquake of 1964 is the avalanche that spread over the surface of Sherman Glacier 35 km (22 mi) northeast of Cordova on the Gulf of Alaska (Figure 2-20). The debris traveled as much as 5 km (3 mi) from its origin and spread across the broad glacier in a sheet 8.5 sq km (3.3 sq mi) in area, about 1.3 m (4.3 ft) thick, and 10 million cu m (13 million cu yd) in volume. The debris came to rest upon undisturbed snow that had fallen on the glacier during the preceding winter, and this and other considerations led investigators to conclude that the avalanche traveled on a cushion of air.

A third example in southeastern Alaska occurred at Lituya Bay in 1958. A mass of rock with a volume of 30 million cu m (39 million cu yd) was dislodged from a steep slope by a mild earthquake and fell into the bay from a maximum height of 1000 m (3300 ft). The abrupt dislodgement of water generated an enormous wave that peeled away the forest, soil, and

FIGURE 2-20. Debris avalanche across part of Sherman Glacier, southeastern Alaska. The debris was derived from a large rockfall (extreme upper right) triggered by the Alaska earthquake of March 27, 1964. *(A. S. Post, U.S. Geological Survey.)*

loose rock from the slopes of islands and walls of the bay to a height of 520 m (1700 ft) above sea level (see Figure 4-13). Older scours indicate that similar devastating waves were generated in Lituya Bay at earlier times.

Dry Flows. As noted, some debris avalanches are dry flows. Many volcanic avalanches are also dry in that steam rather than water is present during movement. These are high-velocity incandescent clouds of rock fragments, dust, and congealing pellets of lava swirling about in a matrix of steam and other gases. These *glowing avalanches* are discussed in Chapter 9.

Slow or rapid dry flowage of fine-grained materials can occur on a large scale. Most susceptible to such mass movements are the wind-deposited silts which cover large areas of the earth. In Kansu Province, China, in 1920, earthquakes converted such silts into dry fluids consisting of solid particles and air. Huge silt flows rushed down into the valleys, burying entire villages. Normally these silts are sufficiently coherent to stand in vertical faces (see Figure 5-18).

Debris Flows. Slower moving than avalanches, *debris flows,* are mixtures of solid fragments, water, and air. A watery debris flow in which at least half the solid material is smaller than sand size is referred to as a *mudflow.* There are transitions from extremely coarse, bouldery flows (Figure 2-21) to mudflows that contain few boulders or none at all (Figure 2-22).

Debris flows behave as heavy fluids and have astonishing capabilities for transporting large rock fragments and other bulky objects. Rates of movement range between 2 and 40 km (about 1 to 25 mi) per hour.

FIGURE 2-21. **Typical chaotic assemblage of blocks, boulders, and smaller fragments in a debris-flow deposit in an alluvial fan near Pala, California.** *(R. H. Jahns.)*

FIGURE 2-22. **Lower part of a mudflow that issued in July 1917 from Surprise Canyon in the left-center distance. Death Valley area, California. The debris, the latest major addition to the large alluvial fan, rests upon the darker surface of older mudflow deposits.** *(A. M. Johnson.)*

Like avalanches, debris flows commonly start as falls and slides on higher ground, and some are avalanches that have slowed down in their lower courses. They can also be generated by mobilization of loose materials during torrential rainstorms.

Though infrequent at any one locality, debris flows are widespread in arid and semiarid regions. Surprise Canyon (Figure 2-22) repeatedly experiences devastating debris flows. In 1874, the mountain mining camp of Panamint City was wiped out by a cloudburst-induced debris flow. The flow swept away the center of the town, roared down into Panamint Valley thousands of feet below, and dumped the remains of the settlement half way across the valley.

In 1943, a large desert debris flow northeast of Parker, Arizona, was preceded by a cloudburst in a nearby mountain range. At the canyon mouth, the mass of dark reddish brown, water-lubricated debris, presented a front 11 m (35 ft) high. Blocks of rock more than 3 m (10 ft) across cascaded down the front and were quickly overridden. On reaching the flat valley floor, the irresistible mass moved like wet concrete. Patches of soil, mats of brush, branches and trunks of trees, and large boulders bobbed up and down. The flow spread over the valley floor and stopped 1.5 km (1 mi) from the canyon mouth, with a steep front 5 m (15 ft) high.

Such events cause great damage in settled areas. Farmers and urban dwellers along the base of the Wasatch Range in Utah have suffered repeatedly from invasion of their homes, fields, and orchards by debris flows. In southern California during the New Year's storm of 1934, many lowland residents had their properties covered by debris flows that originated in the nearby San Gabriel Mountains on steep slopes that had been denuded by fire a short time before. In 1965, part of a large waste pile at a coal mine near Aberfan, Wales, was mobilized to form a debris flow that caused more than 100 deaths.

Of special importance in settled areas are small debris flows that develop during heavy rainstorms from local slumping of natural and artificial slopes. Many start as earth slides which transform into moist earthflows (Figure 2-23). Although these flows are small in volume

FIGURE 2-23. **Earthflow at foot of slump area. Coast of California, south of San Francisco. (Arthur D. Howard.)**

and travel short distances, they have been responsible for millions of dollars in damage during single storms.

Slopes that are covered with volcanic ash provide especially favorable sites for mudflows. Heavy rains readily mobilize the fine debris into channeled runoff. Figure 2-24 shows Lassen Peak after its eruptions in 1914 and 1915, with large mudflows attributed to water spewed out of the crater during the eruptions. Forests were destroyed along the paths of the flows, and huge boulders were carried down into the lowlands.

Liquefaction. Certain water-saturated fine-grained materials are *sensitive* in that they quickly convert to fluid masses when they are physically disturbed. Most sensitive are the *quick clays*, which are capable of spontaneous liquefaction and movement. The rapid conversion of an apparently solid clay to a thick fluid is promoted by saturation with water, by the leaching out of certain cementing substances, or by vibratory disturbance from traffic, explo-

sions, pile driving, or earthquakes. A simple test for clay sensitivity is to mold a sample in the fingers and subject it to vibrations to see if it becomes fluid.

A common site for mass movements induced by liquefaction is a terrace of sedimentary materials underlain by sensitive clays. Movement of the clays can lead to sliding of the overlying materials on a large scale (Figure 2-14). Large blocks of the clay, along with roads, houses, and other structures may move off with the flow in raftlike fashion (Figure 2-25).

Liquefaction flows commonly dam rivers. The St. Thuribe flow along a tributary of the St. Lawrence River 65 km (40 mi) west of Quebec, Canada, dammed Rivière Blanche to a height of 8 m (25 ft) for a distance of 3 km (2 mi). The transporting power of clay flows is comparable to that of debris flows. In an earthquake-triggered occurrence in Japan, boulders the size of a large room were moved.

FIGURE 2-24. Mudflows of volcanic debris from
the eruptions of 1914 and 1915, Lassen Peak,
California. *(B. F. Loomis.)*

FIGURE 2-25. Destruction at Nicolet, Quebec. This
failure by lateral spreading in sensitive marine clays
took place without warning on November 6, 1955.
Property loss was $2,000,000 and three people died.
The house was once at the level of the unbroken
ground above it. *(National Photography Collection,
Public Archives of Canada/ The Gazette, Montreal.)*

Although sensitive clays are common in formerly glaciated regions, they may occur anywhere. An area of clayey sand was involved in a terrace flow near Greensboro, Florida, in April 1948 (Figure 2-26). The flow occurred after a year of heavy rainfall, including 50 cm (16 in.) 30 days preceding the movement.

Many cities in glaciated terrain, including New York, Boston, Chicago, Detroit, Seattle, Anchorage, and Seward, are faced with liquefaction problems. Liquefaction is also of concern in unglaciated areas subject to strong earthquake shaking. This applies to the San Francisco Bay region, where many structures are located on natural and artificial fill.

Investigation and Treatment. Because rapid flowage is commonly involved in slide-flow complexes, and liquefaction is common in settlement and subsidence, investigation and treatment of these phenomena are deferred to later sections.

Slow Flowage or Creep. *Nature and Causes.* The slow downslope flowage of soils is *creep.* The moving materials, including weathered bedrock, may be a few centimeters to tens of meters thick. Average downslope rates range from a fraction of a centimeter per year to 10 cm (4 in.) or more per week, depending upon slope, climate, and characteristics of the materials.

In creep, gravity works in a variety of subtle ways to achieve the downslope movement. Some of the processes involved are illustrated in Figure 2-27. Others include slippage between soil grains when disturbed, human activities, and the growth and thaw of soil ice.

FIGURE 2-26. **Earthflow near Greensboro, Florida, in flat-lying partly indurated clayey sand. Length of slide from scarp to trees in foreground is 275 m (900 ft). Vertical interval from top to base of scarp, 14 m (45 ft); and from top of scarp to toe, 18 m (60 ft).** *(Photo by R. H. Jordan. From D. J. Varnes, Landslide Types and Processes, in E. B. Eckel, ed.: "Landslides and Engineering Practice.")*

Types of Creep. Soil creep is the slow flow of unconsolidated materials and weathered bedrock. The materials are streamed out in laminar flow (Figure 2-28a) that commonly is accentuated by trains of hard, resistant fragments (Figure 2-28b). *Solifluction*, a common creep process in cold regions, is a viscous, porridgelike flow of soil that becomes water-saturated in the thaw season (see Figure 7-21).

One type of *rock creep* is illustrated by the bending of weathered rock strata and the incorporation of fragments of the layers in the moving soil (see Figure 2-31). Another type involves large blocks of rock separated from a cliff face and resting on a sloping surface. Such blocks move imperceptibly downslope as a result of thermal expansion and contraction, freezing and thawing of water, and other processes.

Talus creep is most rapid in cold climates, where it is due primarily to freezing and thawing of contained water. In warm, dry climates, the process is slower and due primarily to daily or seasonal temperature changes. In humid regions, additional factors include plant activities, downslope migration of weathering products, removal of fine particles by water filtering through the mass, and downslope settling. Spe-

cial expressions of talus creep known as *stone stripes* (Figure 2-29) are due to the downslope streaming of debris from rock exposures.

Rock-glacier creep is the slow movement of rock fragments in a form resembling a glacier (Figure 2-30). The fragments are derived from heavy accumulations of talus at the head of a valley. Freeze and thaw within the accumulating mass cause the downvalley movement.

Recognition, Investigation, and Treatment. Creep movements are so slow that they rarely can be noted except by repeated surveys or automatic recording devices. Their cumulative effects, however, may be easily recognized (Figure 2-31). Those most readily observed at the surface are progressively opening cracks in curbs, pavement, and walls; tilted or bent trees; tilted or displaced poles, posts, tombstones, retaining walls, and foundations; and shifts in alignment of reference features such as fences, roads, railways, and canals. Many masses of creep materials assume distinctive forms such as stream- and glacierlike concentrations of rock fragments, and some are expressed by

FIGURE 2-27. **Common causes of creep. The arrows in 1, 2, 3, 5, and 6 indicate relative displacements in uphill and downhill directions. The dotted arrows in 7 indicate direction of filling of cavities. The pattern in 8 represents the zigzag movement of soil due to swelling and contraction. Item 9 illustrates the tendency for burrowing animals to heap excavated earth on the downslope sides of burrows. (Arthur D. Howard.)**

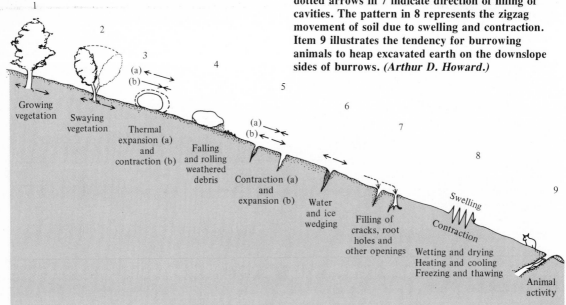

(a)

(b)

FIGURE 2-28. Evidences of creep. (*a*) Granite boulders in soil softened by chemical decay and deformed into pancake shapes by creep. In the lower part of the exposure, the boulders retain their original form. Nearer the surface, the section of each deformed boulder appears as a long, thin, white lens. *(S. R. Capps, U.S. Geological Survey.)* (*b*) The layer of angular rock fragments below the soil is creeping downslope from an exposure upslope. California Coast Ranges, west of Stanford University. *(Eliot Blackwelder.)*

wavelike turf rolls on slopes and at the undercut banks of roads and fields. In many three-dimensional exposures of creep, a downslope flexing of weathered rock layers is revealed (Figure 2-31), as well as deformation and ribbonlike stretching of decayed boulders (Figure 2-28*a*) and the streaming of rock fragments (Figure 2-28*b*).

Engineers find creeping soils difficult to handle, especially where the rates of creep can be significantly changed by human activities.

For example, the soils on many natural slopes have long been creeping at average rates of less than 1 millimeter per year, but these rates can be increased as much as one hundred-fold if the slopes are improperly loaded with buildings or masses of fill.

The unstable soils on many slopes are spotty in their distribution, commonly forming narrow bands and patches along the floors of swales. Where feasible, such localities should be avoided in favor of more stable sites nearby. It is sometimes practical, however, to design structures so that they bridge across ground that is subject to creep, or to use pilings that

FIGURE 2-29. Stone-stripes and terracettes on a 33° slope along the Yakima River, 16 km (10 mi) north of Yakima, Washington. The stripes start at basalt exposures. *(C. F. Sharpe, "Landslides and Related Phenomena.")*

extend down to firm materials, thereby allowing the unstable ground to creep between these supports (Figure 2-32). Another alternative is to strip off all unstable soil prior to construction. Care must be taken, however, to ensure that the excavation does not remove vital support from the slope above and increase creep rates on the upper slope.

When a creep soil problem appears after construction, the creeping materials can sometimes be slowed or stabilized by subsurface drainage or by the injection of cementing materials.

FIGURE 2-30. A rock glacier in the Copper River region of Alaska. The glacier is supplied with rock from the talus-covered slopes above. *(F. H. Moffit, U.S. Geological Survey.)*

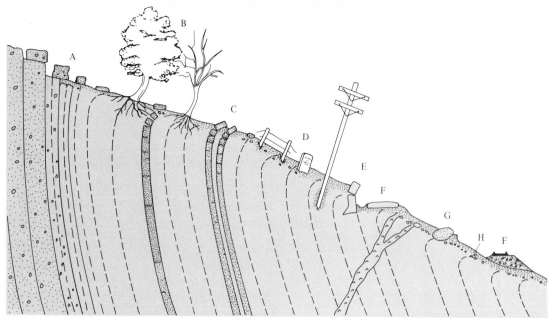

FIGURE 2-31. **Evidences of creep. (A) Moved blocks; (B) trees with curved trunks concave upslope; (C) downslope bending and drag of rock layers, fragments present beneath soil elsewhere on the slope; (D) displaced posts, poles, and monuments; (E) broken or displaced retaining walls and foundations; (F) roads and railroads moved out of alignment; (G) turf rolls downslope from creeping boulders; (H) stone-line at base of creeping soil. (A) and (C) represent** *rock creep*; **all other features are due to** *soil creep*. *(After C. F. Sharpe, "Landslides and Related Phemonema.")*

FIGURE 2-32. **Residence bridging a swale mantled by creeping soil, Los Altos Hills, California. The creeping soil is up to 3 m (10 ft) thick beneath the structure, which is supported by steel columns.** *(Arthur D. Howard.)*

Rock glaciers can also threaten structures, but most of these moving masses are in uninhabited parts of glaciated mountains. Some rock streams resemble rock glaciers but consist of coarse alluvium washed downvalley during floods. The highway between Dehra Dun and Mussoorie, in northern India, skirts and crosses such a river of rock that is nourished by landslide debris at the head of the valley. To forestall destruction of the highway by this rock stream, which becomes active during the monsoon floods, masonry walls and spur dikes have been constructed (Figure 2-33). The growing rock stream also threatens a bridge a short distance below the view shown.

FIGURE 2-33. **The Kalagarh landslide and rockstream, northern India. View north. The rocks in the lower part of the landslide scar are sheared and fractured slates and shales. The upper rocks are quartzites. Dip is into the slope. The slides are due to undermining of the quartzite by more rapid erosion of the fractured rocks below. The Dehra Dun Mussoorie highway with its protective works is in the lower right.** *(Arthur D. Howard.)*

Slide-Flow Complexes

Characteristics and Examples. Many mass movements involve combinations of falling, sliding, and flowing. Slide-flow complexes are the most common examples of what are ordinarily referred to as *landslides*. One of the simplest types involves slumping at its head and flowage in its lower course (Figure 2-13). Figure 2-34 illustrates the terminology applied to this type of landslide. The main slippage takes place on a concave surface of rupture, and the displaced material rides out over the end (foot) of this surface onto the slope below.

When many small complex landslides of the kind shown in Figure 2-13 occur in urban areas during a severe rainstorm or an especially rainy season, the total damage may be great. Larger slide-flow complexes have correspondingly greater capacity for damage, and maximum hazard arises when some regional triggering action sets off many such landslides.

In the Madison Canyon landslide of August 1959, in Montana west of Yellowstone National Park, the source materials were fractured metamorphic rocks that were dislodged by earthquake shaking from the steep south wall of the canyon. The enormous mass of rock broke up as it rushed downslope at a velocity of 160 km (100 mi) per hour. Momentum swept parts of the debris 130 m (430 ft) up the opposite wall of the canyon, and great masses of debris cluttered the canyon floor, creating Earthquake Lake (Figure 2-35).

Another example of a large slide-flow complex is the Vaiont slide in northern Italy. This slide was abetted by the inclination of sedimentary strata containing clay layers and by numerous fractures, faults, and solution cavities, all inclined toward Vaiont Reservoir. Erosion of the deep canyon in which the reservoir is located had long ago removed lateral support for the inclined and broken rocks. The first warning of ground instability came in 1960, when a mass of 700,000 cu m (about 900,000 cu yd) slid down the south side of the reservoir near the dam (Figure 2-36a). This failure was followed for nearly 3 years by rock creep over an extensive area of adjacent ground. Creep rates averaged 1 cm (less than 1/2 in.) per week.

Heavy rains in August and September of

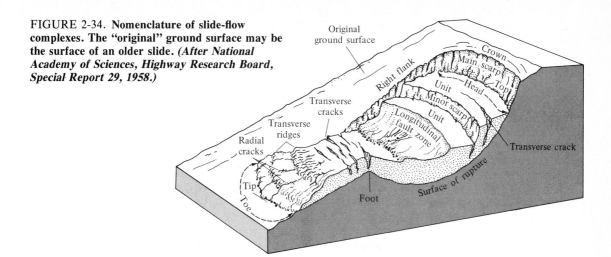

FIGURE 2-34. **Nomenclature of slide-flow complexes. The "original" ground surface may be the surface of an older slide.** *(After National Academy of Sciences, Highway Research Board, Special Report 29, 1958.)*

1963 added to the weight of the creeping mass and caused swelling of clay minerals along bedding and fracture surfaces and in some of the rock layers. Furthermore, filling of the reservoir had raised the ground-water level along the canyon, creating a buoyancy effect in the inundated rocks along the valley side. These factors markedly increased the rates of creep, which by mid-September were as much as 1 cm per day, and attempts were made to lower the level of the reservoir. These were only partially successful, and the rock mass encroached upon the reservoir at rates that increased to 80 cm (about 32 in.) per day just before the ultimate failure.

In less than 60 sec on October 9, 1963, 240 million cu m (315 million cu yd) of rock roared down into the deep reservoir, which contained 120 million cu m (160 million cu yd) of water. The impact created tremors that were recorded as far away as Brussels, Belgium. The canyon was blocked for 1.8 km (1.1 mi) with debris 400 m (1300 ft) thick and reaching a height of 150 m (490 ft) above the reservoir level (Figure 2-36b). The momentum of such rapid sliding forced an enormous mass of rock, water, and air up the opposite canyon wall to heights of 260 m (850 ft) above the reservoir level. The rapidly displaced contents of the reservoir were thrown into huge waves that overtopped the dam by 100 m (330 ft) and rushed down the canyon. The main surge of water was 70 m (230 ft) high at 1.6 km (about a mile) below the dam. It swept down

the valley of the Piave River, destroying everything in its path and killing 2100 people in less than 7 minutes.

Recognition, Investigation, and Treatment. *Recognition.* The first approach to the problem of mass movements is the recognition of former

FIGURE 2-35. **Landslide in Madison Canyon from the mountainside in the left background after an earthquake in August 1959. Madison River was dammed to form the lake in the foreground.** *(J. R. Stacy, U.S. Geological Survey.)*

slope failures. Many landslides, even though dormant, are defined by fresh-appearing scarps, toes, hummocky surfaces, and other features of the kinds shown in Figures 2-13 and 2-34. Most readily identified are recent rotational slides with backward-canted blocks of ground, tilted trees, displaced roads, and disturbed surface drainage.

In time, the topographic expression of inactive landslides becomes obscure. The scarps weather and erode, fissures disappear, depressions on the inner sides of rotated blocks fill with sediment and vegetation, and the hummocky topography is softened and smoothed. Thus, many landslides which may be only temporarily inactive, can be easily overlooked. Figure 2-37 illustrates such a situation. A proposed road was to loop around each of the projecting ridges on its ascent to the upland.

FIGURE 2-36. **Scene of Vaiont disaster.** *(Courtesy G. A. Kiersch.)* **(a) View of Vaiont Dam, world's second highest at 265.5 m (875 ft), in 1962. Just beyond the dam, to the right, the dip of the strata toward the reservoir is revealed. Beyond it is the relatively small landslide mass of 1960. (b) View shortly after the slide of October 9, 1963. The mass of debris has all but obliterated the reservoir.**

(a)

(b)

FIGURE 2-37. **Influence of landslides and creeping soil on routing of new highway below Pacheco Pass, Coast Ranges of California west of Tracy. C—creep; L—landslide.** *(California Division of Highways.)*

The route would have taken the road across the landslide mass at L, which was not recognized in preliminary surveys because the landslide topography is very faint. It is clear in the photograph because the sun was at an angle to provide shadows of the faint surface undulations on the landslide and on the adjacent slopes undergoing slow downhill creep. Recognition of the landslide led to abandonment of the projected loops in favor of a series of large cuts and fills to provide a more direct and stable route to the upland, as sketched on the photograph.

Once a landslide is recognized, adjacent areas should be scrutinized for evidences of additional unstable ground. If the landslide proves to be a unique feature, an alternate construction site should be available nearby. For example, the original pier site in Figure 2-38 was discarded when a large landslide mass was recognized from the arcuate reentrant in the cliff, tension cracks, and anomalous topographic features. A safer site was selected a short distance away.

Recognition of landslide *potential* at a site can be difficult. A rapid survey of all identifiable slides may indicate a slope angle below which they do not occur and above which they are common. Steepness of slope, however, is by itself reliable only where ground materials, moisture content, vegetative cover, natural or artificial undercutting, and other pertinent factors are equal or nearly so. Deep cracks, growing fissures, and indications of creep can provide preliminary warnings of potential slope failure.

Investigation. Investigation of a slide complex begins with careful mapping to determine the

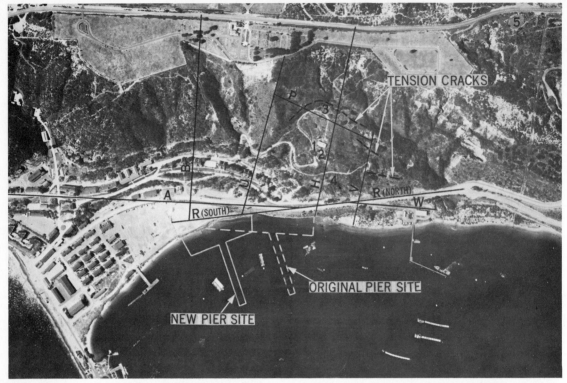

FIGURE 2-38. **Recognition of the landslide to the left of the words "tension cracks" prompted changing the pier site. Note the structures already on the landslide.** *(U.S. Navy photograph.)*

extent of the unstable ground. The resulting map should show the configuration of the ground surface by contour lines (lines of equal elevation). In addition, longitudinal and cross profiles may be required for portrayal of the topography and geology.

Basic objectives of topographic and geologic mapping include characterization of the slide complex in three dimensions, establishing its relationship to the surrounding and underlying ground, determining the depth to the principal surface of rupture, determining the factors responsible for its occurrence, and monitoring its movements if it is active.

If a landslide shows continuing slow movement, its progress can be traced by means of survey control stations. The simplest procedure is to drive stakes or rods into the ground for checking by periodic surveys from instrument stations on stable ground outside the slide. Strain gauges also are employed to measure displacements. A stake firmly set in a landslide mass, for example, is connected by a wire to a recording device on stable ground.

Geologic investigation of large landslides ordinarily involves mapping and study of both soil and bedrock. The soil characteristics to be recorded are structure (e.g., dense, loose, prismatic), permeability, and composition, with special attention to slide-promoting clay minerals. The bedrock is examined for lithology and structure, with special attention to weak layers and fractures. Hydrologic factors that require attention include the distribution of stream channels, springs, seeps, and exposed permeable layers that could conduct surface water below ground.

Possible triggering influences are evaluated through examination of (1) weather and climate records to determine any correlation between sliding and periods of rainfall or freezing and thawing, (2) ground-water data before and after sliding, (3) earthquake records, for correlation between sliding and earthquakes, and (4) records of excavation and other activities that may have overloaded or undermined the slope.

Subsurface exploration is necessary to determine three-dimensional relationships and to provide samples for laboratory tests of shear strength, sensitivity, and other properties. Relatively undisturbed soil samples can be obtained by driving tubes into the ground and withdrawing them with the enclosed soil. Clam-shell devices also are used to recover soil samples, as are drill holes, pits, and trenches.

Ground-water observation wells provide data on water levels and pressures, and directions of flow can be traced by means of injected dyes. Geophysical techniques can be used to obtain data on the number and thicknesses of soil layers and depths to firm bedrock.

Treatment. Three approaches in dealing with actual or potential landslide sites are avoidance, prevention, or correction. Avoidance may involve either relocation of a project or adoption of a design to circumvent the hazard. Thus a broad landslide complex can be avoided through relocation (Figures 2-37 and 2-38), whereas a narrow creep or landslide tongue can be bridged (Figure 2-32).

As shown in Table 2-1, methods for prevention and correction are aimed at reducing the driving force (shear stress), increasing the resisting forces (shear strength), or both. Reduction of shear stress may require excavation to decrease the weight of slide materials, to reduce slope declivity (slope angle), or to remove unstable ground. Flattening or benching of slopes reduces weight and decreases the overall slope angle. Removal of materials from the headward parts of slides reduces weight and shear stress, and in rotational slides the shear resistance can be increased by transferring these materials to the toe of the slide.

Stabilization of a slide complex often involves control of external and internal drainage. Water falling or flowing on the surface must be led off and standing water eliminated. Paved ditchs, conduits, and drainage benches are used for this purpose. Some slopes require smoothing to eliminate traps where water may be ponded. Cracks and fissures are sealed with clay, asphalt, or other impermeable material to prevent penetration of water. At some localities it has been expedient to coat exposures with asphalt or other impermeable materials, or to cover with plastic sheeting.

Subsurface water reduces shear resistance and should be removed by horizontal tunnels or perforated pipes, vertical wells, or trenches that are backfilled with coarse, permeable materials.

Restraining structures to increase shear resistance are most effective when emplaced before, rather than after ground failure has begun. The most common types are masses of rock or compacted earth placed as loads or buttresses. *Loads* are positioned on the outer parts of slide masses to counteract upward bulging. *Buttresses* are lateral supports placed against the lower margins of slide masses, fill embankments, or excavated slopes to prevent downslope movement.

Walls, slope pavements of concrete or masonry, and cribs or bins of timber, steel, or concrete filled with rocks or earth (Figure 2-39) are most useful where the volume of unstable ground is small.

On soil slopes, the driving of piles is often successful in preventing mass movements, but such pinning rarely controls active slides. This is because the moving debris bypasses the piles, tilts or breaks them, or moves them off.

The shear resistance of an unstable mass can be increased by internal hardening through injection of cement, asphalt, or special chemicals. Freezing has been employed to forestall movement of loose, porous, water-saturated materials until permanent control could be established.

There remains the problem of treatment of landslide-dammed lakes. Slide materials are generally loose and so susceptible to rapid erosion that catastrophic flooding can result if the dam is breached. Alternatively, rapid seepage through the landslide debris can enlarge passages by solution or the removal of fine sediment, thereby increasing the possibility of a breakout below the crest.

Seepage through dams of landslide debris can be controlled by injecting stabilizing materials or by facing the upstream sides with cement, clay, or other impermeable materials. Where such measures are not feasible, it may

Table 2-1. Methods for prevention and correction of landslides. (From A. W. Root, Prevention of Landslides, chapter 7 in E. B. Eckel, ed., "Landslides and Engineering Practice." See Additional Readings)

Effect on Stability of Landslide	Method of Treatment	General Use		Frequency of Successful Use[1]			Position of Treatment on Landslide[2]	Best Applications and Limitations
		Prevention	Correction	Fall	Slide	Flow		
Not affected	I. Avoidance methods:							
	A. Relocation	×	×	2	2	2	Outside slide limits	Most positive method if alternate location economical
	B. Bridging	×	×	3	3	3	Outside slide limits	Primary highway applications for steep, hillside locations affecting short sections
Reduces shearing stresses	II. Excavation:[3]							
	A. Removal of head	×	×	N	1	N	Top and head	Deep masses of cohesive material
	B. Flattening of slopes	×	×	1	1	1	Above road or structure	Bedrock; also extensive masses of cohesive material where little material is removed at toe
	C. Benching of slopes	×	×	1	1	1	Above road or structure	Relatively small shallow masses of moving material
	D. Removal of all unstable material	×	×	2	2	2	Entire slide	
Reduces shearing stresses and increases shear resistance	III. Drainage:							
	A. Surface:							
	1. Surface ditches	×	×	1	1	1	Above crown	Essential for all types
	2. Slope treatment	×	×	3	3	3	Surface of moving mass	Rock facing or pervious blanket to control seepage
	3. Regrading surface	×	×	1	1	1	Surface of moving mass	Beneficial for all types
	4. Sealing cracks	×	×	2	2	2	Entire, crown to toe	Beneficial for all types
	5. Sealing joint planes and fissures	×	×	3	3	N	Entire, crown to toe	Applicable to rock formations
	B. Subdrainage:							
	1. Horizontal drains	×	×	N	2	2	Located to intercept and remove subsurface water	Deep extensive soil mass where ground water exists
	2. Drainage trenches	×	×	N	1	3		Relatively shallow soil mass with ground water present
	3. Tunnels	×	×	N	3	N		Deep extensive soil mass with some permeability
	4. Vertical drain wells	×		N	3	3		Deep slide mass, ground water in various strata or lenses
	5. Continuous siphon	×		N	2	3		Used principally as outlet for trenches or drain wells

Category	Method						Portion of slide treated	Remarks
Increases shearing resistance	**IV. Restraining structures:**							
	A. Buttresses at foot:							
	1. Rock fill	×	×	N	N	1	Toe and foot	Bedrock or firm soil at reasonable depth
	2. Earth fill	×	×	N	N	1	Toe and foot	Counterweight at toe provides additional resistance
	B. Cribs or retaining walls	×		3	3	3	Foot	Relatively small moving mass or where removal of support is negligible
	C. Piling:							
	1. Fixed at slip surface	×	×	N	N	3	Foot	Shearing resistance at slip surface increased by force required to shear or bend piles
	2. Not fixed at slip surface	×	×	N	N	3	Foot	
	D. Dowels in rock	×	×	3	3	3	Above road or structure	Rock layers fixed together with dowels
	E. Tie-rodding slopes	×	×	3	3	3	Above road or structure	Weak slope retained by barrier, which in turn is anchored to solid formation
Primarily increases shearing resistance	**V. Miscellaneous methods:**							
	A. Hardening of slide mass:							
	1. Cementation or chemical treatment							
	(a) At foot	×	×	3	3	3	Toe and foot	Noncohesive soils
	(b) Entire slide mass	×	×	N	3	N	Entire slide mass	Noncohesive soils
	2. Freezing	×		N	3	3	Entire	To prevent movement temporarily in relatively large moving mass
	3. Electroosmosis	×		N	N	3	Entire	Effects hardening of soil by reducing moisture content
	B. Blasting		×	N	3	N	Lower half of landslide	Relatively shallow cohesive mass underlain by bedrock
								Slip surface disrupted; blasting may also permit water to drain out of slide mass
	C. Partial removal of slide at toe			N	N	N	Foot and toe	Temporary expedient only; usually decreases stability of slide

[1] 1 = frequently; 2 = occasionally; 3 = rarely; N = not considered applicable.

[2] Relative to moving or potentially moving mass.

[3] Exclusive of drainage methods.

53

FIGURE 2-39. **Concrete crib to prevent land movement, Arcata, California.** *(T. W. Smith, California Division of Highways.)*

be advisable to lower the lake to a safe level or to drain it by controlled deepening of the outlet channel or by excavating a bypass tunnel.

Settlement and Subsidence

The ground surface has been lowering in many large cities, such as Tokyo, London, Venice, Houston, and Mexico City, as well as in many agricultural areas. Surface depression may be a single drastic event, it may be episodic, or it may be progressive. Greatest damage has been caused by the cumulative effects of slow, progressive subsidence.

Lowering of the ground surface can be accomplished by *settlement* or *subsidence*. *Settlement* results from compaction of the ground by loads naturally or artificially imposed upon them; *subsidence* is collapse of the ground due to development of subsurface voids or reduction in volume of subsurface materials either naturally or artificially.

Settlement. *Nature, Causes, and Examples.* Compressible soils of low strength are most susceptible to settlement during and after the placing of loads upon them. Such soils include peat and other organic materials, clays and some coarser grained sediments, and artificial fill materials that were insufficiently compacted when emplaced. Thus settlement has been most common in depositional environments that include deltas, floodplains, tidal flats, old lake bottoms, and other areas of loose sediment.

Settlement results from compaction of water-saturated sediments as additional sediments accumulate above them, and it may continue long after such deposition ceases. The included water is forced elsewhere, and the sediment becomes more tightly packed. Such changes can be initiated or hastened if fill or heavy structures are placed on the surface.

The most famous example of heavy loading on weak foundation materials is the Leaning Tower of Pisa, in northwestern Italy. The tower was based on a circular slab only 19 m (63 ft) in diameter and began to tilt during early stages of construction. The differential settlement, due to compression of a thick layer of clay 8.5 m (28 ft) beneath the foundation, continued long after construction was completed in 1350. Repeated attempts were later made to stabilize the tower, and only after much costly work was its survival assured.

Settlement has also affected the Washington Monument in Washington, D.C., built during the period 1848–1880. The settling totals 15 cm (6 in.), or nearly 0.1 percent of its height.

Older buildings in Mexico City have long been settling into the weak lake beds on which they are built, and many are leaning or being wracked by differential movements. Old cathedrals in Europe, most of them with foundations of rock rubble or old timber, have required costly repairs. The worst damage occurs where parts of a structure are underlain by hard, stable materials, and other parts by weak materials that compress with time. Extreme examples of differential settling are a large grain elevator in Manitoba and a cracking tower in a Texas oil refinery, both of which tipped over when they were first filled. The aggregate cost of such damage to major structures, however, is small compared with that required for repair of highways, railroads, canals, sewers, utility lines, houses, and business buildings that have differentially settled.

Investigation and Treatment. Appraisal of the potential for ground settlement requires exploration and sampling of the subsurface materials and determination of their engineering properties. Where the thickness of materials susceptible to compaction is 15 m (50 ft) or less, heavy structures can be supported on piles or caissons that rest on underlying stable ground. Where the compressible materials are thicker, alternative designs can provide for distribution of the load over considerable areas of ground to ensure that settlement is uniform and will not tilt or wrack the structures.

Rates and amounts of settlement can often be forecast with considerable accuracy, especially for regularly layered soils and homogeneous masses of artificial fill. Where tidal flats are to be reclaimed, for example, fill is commonly pumped into diked enclosures as slurries of fine-grained solids in water. These slurries and the underlying natural materials settle at predictable rates that decrease with time. Thus it is practical to leave the fill exposed for months or even years, after which the surface can be developed, taking into account the remaining small amount of future settling. Conversely, where compressible materials are irregularly distributed, the site should not be used for structures that cannot stand differential settlement. Owners of homes built on old sanitary landfills or accumulations of loose waste from mines and quarries will attest to this.

Subsidence. Ground subsidence results from volume changes without benefit of a superposed load. Its surface expression may be similar to that of settlement. Subsidence can occur over periods of seconds or centuries and can range from millimeters to tens of meters.

Withdrawal of Ground Water. The most widespread cause of extensive subsidence is the pumping of ground water. Such fluid withdrawal from clay-bearing sediments, soft silty limestones, and other loose, fine-grained materials leads to volume reductions. Subsidence also results when water is extracted from an aquifer (a water-bearing formation) that is confined by impermeable formations. The confined water is under pressure which is reduced when water is withdrawn. The lowered water pressure results in compression of the aquifer and the confining clay layer.

Industrialization of the Koto district of Tokyo, with its heavy demands on ground water, has caused a drop in the ground-water level of 20 m (65 ft) and a ground subsidence of 2 m (6.5 ft) since 1923. The rate of sinking has subsequently been reduced through control of pumping. In 1966, an embankment was constructed to prevent flooding, but additions to this barrier will be required if subsidence continues.

Venice, Italy, is built on an island at the northern periphery of the Po delta. It has been subsiding at an average rate of 20 cm (8 in.) per century, with a cumulative total of 3 m (10 ft) since it was founded more than 1500 years ago. The present rate of 2 mm (0.08 in.) per year is surpassed at points farther south along the Adriatic coast (Figure 2-40). The city now is frequently inundated during storms, and the famous Piazza San Marco is awash several times a year.

The combination of settlement and subsidence in Venice reflects natural compaction of the loose sediments on which it is built, loading by heavy structures, and depletion of ground water by 7000 wells. The rate of subsidence correlates with the rate of pumping. Plans are under way to shield the city from the sea by dikes and locks at the inlets to the lagoon of Venice so that inundation can be prevented even though sinking of the ground continues. To combat the sinking, withdrawal of ground water is being controlled, and limitations are being imposed on the number and heights of buildings erected on the island and nearby mainland.

In Mexico City, subsidence due to water withdrawal has reached 7.5 m (25 ft) in places, and the fixed steel casings of many wells now protrude above the ground (Figure 2-41). Many older buildings have tilted and ruptured from differential compaction of the weak, water-bearing sediments on which the city is built. These sediments include clays, sands, volcanic ash, and organic materials, most of which were laid down in ancient Lake Texcoco. During recent years the annual subsidence has been reduced to several centimeters, mainly through control of ground-water pumping. Detailed subsurface investigations and good foundation design have permitted the erection of lofty buildings such as the 43-story Latino Americana Tower.

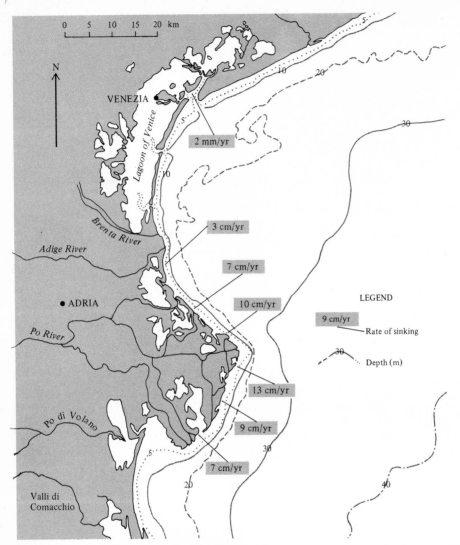

FIGURE 2-40. **Sinking along the Adriatic coast correlates with loading caused by deposition at the mouth of the Po River and locally with pumping of groundwater.** *(After H. W. Menard, "Geology, Resources, and Society: An Introduction to Earth Science," W. H. Freeman and Co. After B. Nelson, Soc. Economic Paleontologists and Mineralogists, Spec. Publ. No. 15, 1970, pp. 152–184.)*

An example of subsidence of agricultural land due to withdrawal of ground water is provided by California's Santa Clara Valley, south of San Francisco Bay. The subsidence averages 1 m (3 ft) over an area of 260 sq km (100 sq mi), with a maximum of 2.5 m (8 ft) (Figure 2-42). Damage to wells has exceeded $4 million, and the cost of levees and other remedial works to prevent flooding has been $10 million. The subsidence has resulted from compaction of near-surface aquifers as demonstrated by well casings, solidly emplaced at depth, that now protrude above the depressed ground surface. The rate of subsidence has varied with the rate of lowering of the artesian pressure.

Subsidence on a larger scale has occurred in the San Joaquin Valley of California, where irrigation accounts for one-fifth of all the ground water pumped in the United States. During the past 50 years, water levels in wells dropped 15 m (50 ft) or more over large areas, with local maximums of 120 m (400 ft). Surface subsidence has ranged from 1 to 9 m (3 to 30 ft). Damage to wells, canals, bridges, and other structures has been extensive.

Withdrawal of Oil and Gas. Subsidence due to production of hydrocarbon fluids is similar to that caused by withdrawal of ground water. The first reported occurrence was in the Goose Creek oil field of Texas, where production in the early 1920s was accompanied by cracking of the low-lying ground adjoining Galveston Bay. The ground sank as withdrawal continued, and an extensive shallow, saucerlike depression sagged below the level of the bay and was permanently inundated.

Many additional occurrences have since been described, especially in Japan, in the Lake Maracaibo area in Venezuela, and in California and Texas. In Japan, sections of Niigata have sunk below sea level since the beginning of large-scale gas production in the late 1940s. Once the cause of the subsidence was recognized, production rates were cut back, and the affected subsurface formations were repressured by introduction of water.

Significant subsidence has been reported from 22 oil and gas fields in California. The Wilmington field has become famous because of the magnitude and economic impact of the surface deformation. The subsidence centers about Long Beach Harbor and extends into adjacent parts of Los Angeles Harbor and the city of Long Beach. Damage exceeds $100 million.

Development of the Wilmington oil field began in 1938, and by 1941, 40 cm (1.3 ft) of subsidence had occurred at the east end of Terminal Island, an artificially created island. During a period of 2 decades the sinking spread over an area of 65 sq km (25 sq mi) and reached a maximum of 9.5 m (31 ft).

Most investigators attribute the subsidence to reduction of fluid pressures as oil was withdrawn from deep sedimentary rocks, thereby permitting compression of the porous layers under the weight of the rocks above.

The subsidence not only damaged many oil

FIGURE 2-41. **Settlement of Mexico City illustrated by protrusion of well casing originally flush with surface. The height of the casing was 4.2 m (14 ft) in 1958 and 5 m (16 ft) in 1975. Because *differential* settlement is minimal, damage is not apparent. (*Courtesy Ing. R. J. Marsal.*)**

wells, but reversed the flow of sewers and storm drains and submerged wharves, sea walls, levees, bridges, and much valuable land. The ground movements deformed and broke streets, curbs, railroad tracks, pipelines, and drains, and jammed bridges so that they could not be opened for marine traffic. Perhaps the only advantageous effect was a deepening of the harbor bottom. Some of the depressed ground was subsequently filled in, and new retaining walls were constructed to protect the U.S. Navy shipyard and other installations from flooding (Figure 2-43).

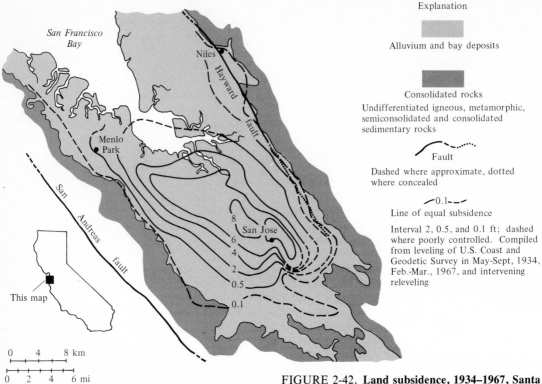

FIGURE 2-42. **Land subsidence, 1934–1967, Santa Clara Valley, California.** *(Modified from open-file map of the U.S. Geological Survey, from California Geology, California Division of Mines, August 1971.)*

As subsidence continued, estimates of its ultimate magnitude reached 22 m (72 ft), and the necessity for control became increasingly evident. Artificial repressuring of the oil zones by water injection began in 1958. The injected waters were pumped from shallow aquifers. Subsidence has been arrested over most of the affected area. The surface has rebounded slightly in some places but cannot be expected to return to its original position. The repressuring may eventually involve 300 injection wells, 1 million barrels of injected water per day, and costs exceeding $10 million. The costs, however, should be more than compensated for by increased recovery of oil, estimated at one-half billion barrels.

Dewatering of Mines and Reservoirs. Subsidence in many mining areas is caused by the dewatering of old mine workings and loss of fluid support. Collapse of dewatered mines is most common where the workings are in soils, in shales or other relatively soft rocks, or in

harder rocks that are greatly fractured. The shallow tunnels that have been excavated in loose, gravelly deposits in search of gold, tin, and other minerals are especially liable to failure.

Some remarkable occurrences of subsidence have been reported from the Far West Rand mining district, 65 km (40 mi) west of Johannesburg in the Union of South Africa. A major dewatering program was begun in 1960 to permit the extension of deep gold mining, and almost immediately large depressions began to develop at the surface (Figure 2-44). Within 5 years, eight holes more than 46 m (150 ft) wide and 27 m (90 ft) deep had been formed, involving heavy property damage and loss of life.

The draining of reservoirs for inspection, repairs, or new construction or the natural lowering of water level during long dry periods

FIGURE 2-43. View of the U.S. Naval Shipyard at Long Beach, California, showing the retaining wall necessary, as a result of ground subsidence, to keep out the sea. *(Official photograph, U.S. Navy.)*

can have effects akin to those of heavy pumping of ground water. If the ground-water level in adjacent ground is thereby lowered, subsidence may result.

Wetting and Drying of Soils. In arid regions, where the water table is low and overlying soils are prevailingly dry, the moisture-deficient materials may be actually compacted when water is introduced by irrigation, storm runoff, or leakage from wells and canals. Such *hydrocompaction*, or *hydroconsolidation*, occurs in loose, open-textured soils whose particles are weakly held together by films of clay. In the most susceptible materials, such as alluvial and wind-blown silts, the water loosens the bonds between the particles, so that the weight of the overburden compacts the soil layers at the

FIGURE 2-44. Johannesburg, South Africa. Collapse pits that developed suddenly in the West Reef area, after deep pumping had started for mine drainage. Scale is provided by trees in background. *(R. F. Legget, "Cities and Geology," McGraw-Hill Book Company, 1973.)*

Mass Wasting and the Environment   59

expense of former pore space. Hydrocompaction is restricted to moisture-deficient soils within 60 m (200 ft) of the ground surface.

Penetration of surface water in the dry San Joaquin Valley of California has resulted in subsidence of as much as 1.5 m (5 ft) over large areas, with a maximum of 6 m (20 ft). Because of such subsidence, irrigation systems may be rendered ineffective, and fields, highways, bridges, and utility lines may require releveling. Figure 2-45 shows subsidence effects of irrigation and obstruction of drainage at road embankments.

Subsidence also results from the *drying of soils* with high contents of shrinkable clay or organic materials. Many river deltas and other swampy areas are underlain by peat and fine-grained sediments rich in organic matter. When such lands are reclaimed for agriculture, dikes are constructed to exclude open water, and the basins are then pumped dry. Dewatering of the sediments results in shrinkage and subsidence of the partly decomposed plant materials as they dry out. The Fens area of England, the Rhine delta of the Netherlands, and the Sacramento–San Joaquin delta of California are

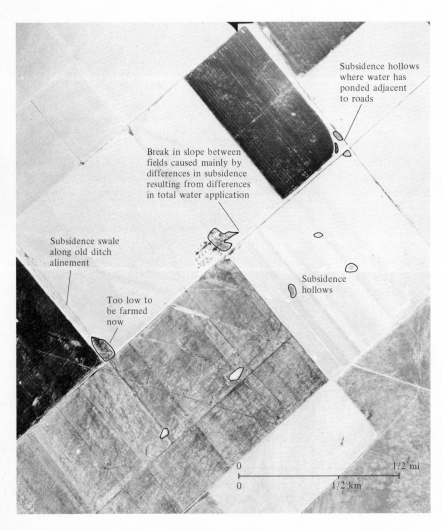

FIGURE 2-45.
Hydroconsolidation subsidence. San Joaquin Valley, California. (Courtesy Francis Riley and Howard Chapman, U.S. Geological Survey. Annotation by William B. Bull.)

FIGURE 2-46. **Home lost in collapse of limestone sink, Bartow, Florida, 1967.** *(U.S. Geological Survey.)*

areas where large-scale reclamation has been accompanied by subsidence locally greater than 5 m (15 ft).

Removal of Subsurface Solids. Surface subsidence can be caused by the removal of sediment at depth, either selectively or in wholesale fashion. Downward or laterally moving water, for example, can remove fine particles from loose subsoils. Such action develops narrow subsurface passages which may be enlarged to widths of meters. The roofs of such cavities sometimes collapse to yield steep-sided depressions at the surface, and remnants of the roofs may form natural bridges. In cold regions, subsidence results from the melting of ice masses in frozen ground.

The solution of rocks such as limestone results in subsurface tunnels and caverns. Occasionally the roof of such an opening collapses, carrying structures with it (Figure 2-46). On December 2, 1972, a giant sinkhole developed near Montevallo, Alabama. Its width reached 130 m (425 ft) and its depth, 46 m (150 ft).

Solution also progresses downward from the surface, forming shallow furrows, pits, or broader depressions. The surface expressions of solution are sinkholes, many of which have circular or elliptical outlines (see Figure 2-4).

Two restricted types of subsidence result from the natural burning of coal beds and the oxidation of sulfide ore bodies. Both involve loss of volume, with attendant subsidence.

Caves formed by the battering of waves along a rugged coastline can be beaten far back into the cliff faces and enlarged until their roofs collapse to form steep-sided openings in the ground surface (see Figure 4-28).

Earthquakes, by jostling sediments that vary laterally in porosity, may cause uneven ground subsidence. They may also impart fluid-like qualities to fine sediments whose outflow causes subsidence of the ground above.

The roofs of active mine workings may collapse during mining operations or after they have ceased. The problem is most serious where the workings are shallow and extensive and where the overlying ground is weak, such as in Paris, which is underlain by limestone used as building material from Roman times until the nineteenth century. The labyrinth of workings that resulted from this extraction lies 2 to 35 m (6 to 110 ft) below the surface and has resulted in subsidence in many places.

The mining of coal has caused subsidence in many parts of the world. The effects have ranged from sudden opening of large cavities to long-continued uneven sagging of the surface over extensive areas. The property damage from mining subsidence has exceeded $10 million in single housing tracts in the Pittsburgh and Scranton districts of Pennsylvania.

Where salt has been mined underground on a large scale, as in Kansas, New York, Michigan, and Ontario, collapse of old workings is a potential hazard. Part of the difficulty has been due to yielding of the salt pillars left as support in mined-out areas. Subsidence has been more extensive where salt has been extracted by solution techniques. This is because all the salt is removed. Solution by ground water after the mines are abandoned extends the hazards beyond the original operations. In England, where salt has been extracted from the subsurface for more than 2000 years, the most drastic surface effects have been due to collapse of old workings exposed to circulating waters from more recent solution mining.

Tectonic Subsidence. Tectonic subsidence, accompanying deformation of the earth's crust, occurs either as sudden, long-separated events or as slow, continuous movements. The sudden events are usually associated with earthquakes. During the Japanese earthquake of 1923, the ground was depressed over thousands of square kilometers, with a maximum subsidence of 1.5 m (5 ft). The Alaska earthquake of 1964 was accompanied by downwarping over an area of 125,000 sq km (48,000 sq mi) that included Kodiak Island, Kenai Peninsula, parts of Prince William Sound, and the Chugach Mountains.

Other areas subside under the weight of accumulating sediments, glacial ice, or water.

Investigation and Treatment. Subsidence caused by natural underground solution, the removal of fine sediment by ground water, the collapse of sea caves, or tectonic movements is probably beyond human control. Even so, suspected areas should be examined carefully to assist in decisions as to whether a site should be avoided or is safe enough for the planned land use. There are undoubtedly many unknown caves and many long-forgotten mine workings which are potential hazards. Geophysical exploration by electrical, gravimetric, and seismic methods are promising discovery techniques. Probing by radar has been successful in several places, and infrared imagery holds promise (see Chapter 14).

Any area underlain by limestone or other soluble rocks is potentially hazardous. The ground between sinks and caverns is not necessarily secure, because collapse may expand beyond existing openings. Gravity surveys are especially useful in detection of such caverns. If an opening is present, the pull of gravity is less than that in an area of solid rock, and the difference can be registered on a sensitive gravimeter.

Where ground materials are subject to volume changes through hydroconsolidation, care must be taken to avoid excess wetting by uncontrolled runoff or by leakage from wells and canals. Alternatively, the ground can be thoroughly wetted and precompacted before structures are placed upon it. This was done for a major aqueduct along the west side of the San Joaquin Valley in California.

Subsidence due to withdrawal of subsurface fluids can be reduced or arrested by reduction of pumping or by recharging the ground with imported water. In the case of petroleum, restoration of subsurface pressures by water injection is commonly the only viable solution. In the dewatering and reclamation of deltas, swamps, and other low-lying lands, potential subsidence should be appraised for proper design of canals, ditches, drains, roads, and structures for the prevention of flooding. Most subsidence is largely irreversible, and a return to original ground levels cannot be expected.

In parts of England, subsidence attending extraction of salt has become so serious that solution mining has been strictly regulated, and measures have been adopted for maintaining underground pressures. In many old limestone workings under Paris, underpinnings have been installed to support construction overhead. In a few densely populated areas with high property values, fill has been added to old mine workings through holes drilled from the surface, and the fill has been bonded with cement. In Johannesburg, South Africa, new structures have been designed with special foundations to accommodate subsidence that might attend future mining beneath the city. The alternative is to move all structures to a different locality. This was done years ago in Minnesota, when the entire town of Hibbing was moved to permit the mining of the iron ore beneath it.

ADDITIONAL READINGS

American Association of State Highway Officials: Standard Recommended Practice for the Classification of Soils and Soil-Aggregate Mixtures for Highway Construction Purposes, in "Standard Specifications for Highway Materials and Methods of Sampling and Testing," Washington, D.C., 1955, pp. 45–51.

Birkeland, P. W.: "Pedology, Weathering, and Geomorphological Research," Oxford University Press, New York, 1974.

Eckel, E. B., ed.: "Landslides and Engineering Practice," Highway Research Board Special Rept. 29, NAS-NRC Publ. 544, National Academy of Sciences, Washington, D.C., 1958.

Hunt, C. B.: "Geology of Soils," W. H. Freeman and Company, San Francisco, 1972.

Keller, W. D.: "The Principles of Chemical Weathering," Lucas Brothers, Columbia, Mo., 1957.

Krynine, D. P., and W. R. Judd: "Principles of Engineering Geology and Geotechnics," McGraw-Hill Book Company, New York, 1957.

Sharpe, C. F. S.: "Landslides and Related Phenomena," Columbia University Press, New York, 1938.

Terzaghi, K.: Mechanism of Landslides, in "Application of Geology to Engineering Practice," Berkey Volume, Geological Society of America, Inc., 1950, pp. 83–123.

U.S. Waterways Experiment Station: "Unified Soil Classification System," Tech. Memo. 3-357, 3 volumes, Vicksburg, Miss., 1953.

Winkler, E. M.: "Stone: Properties, Durability in Man's Environment," Springer-Verlag New York Inc., New York, 1973.

Zaruba, Q., and V. Mencl: "Landslides and Their Control," American Elsevier Publishing Company, Inc., New York, 1969.

CHAPTER THREE

hydrology
and the environment

INTRODUCTION

In 1972, two widely separated areas of the United States were ravaged by flood waters. The first occurred on June 9 and 10, when record rains, locally as much as 37 cm (15 in.) in 6 hours, led to the failure of a dam holding back Canyon Lake in the Black Hills of South Dakota. This sent a 1.8- to 2.4-m (6- to 8-ft) wall of water down Rapid Creek and through Rapid City. Property damage was $160 million and 237 people died. On June 19, Hurricane Agnes entered Florida and dropped 105 cu km (25.1 cu mi) of water, causing devastating floods over wide areas of the Atlantic seaboard, particularly in Pennsylvania, New York, Maryland, and Virginia. A storm of this magnitude occurs on the average only once in 500 to 2000 years. The damage totaled $3.1 billion, but the early-warning system held the death toll to 117. The Agnes floods created in many Americans a sense of futility that was magnified during the awesome Mississippi River flood in 1973, during which vast areas of the valley were inundated by waters rushing through more than 300 breaks in the levees.

In the three devastated areas the stunned populaces wondered whether such catastrophic events could ever be controlled. Many questions were raised: "Is occupation of floodplains advisable? Why is it impossible to predict these devastating events? Why cannot adequate protective and constraining structures be built along the rivers and streams? Why are warning systems often inadequate for such floods?"

While the paths of hurricanes can often be predicted a day or two in advance, they are susceptible to sudden changes in direction and intensity. Furthermore, the amount of rainfall from a given storm in any particular area is difficult to assess.

To protect all rivers and streams against the rare catastrophe would require an almost limitless number of costly dams, levees, and other flood-protective structures, even if there were no objections to their proliferation. Furthermore, most of these huge, costly structures would outlive their engineering lives without ever fulfilling their designed roles. The expectation of complete protection is unrealistic; there will always be extraordinary events caused by unusual combinations of events. The best that can be done is to guard against the expectable periodic assaults of nature and either avoid areas where major disasters may strike or provide warning systems to minimize the impacts. Unfortunately, in the Rapid City flood, many residents failed to respond to warnings.

Many hazardous sites are occupied by people aware of potential catastrophes but naively convinced that they and their property will be spared. People must learn to adjust to life among the rivers and streams, to recognize and understand their vagaries, and to avoid or minimize their possible destructive effects.

The destructive aspect of rivers is not the only hydrologic concern. Flood ravages can be repaired, but water is vital for agriculture, industry, the household, and life itself. Careful planning is necessary for wisest use. Fortunately, the coterminous United States has a surplus of water; we use only 20 percent of the annual amount available. The problem is distribution; some areas have a superabundance, others, a serious deficiency. In addition, seasonal and cyclic variations in precipitation create local problems. Finally, increasing volumes of wastes are rendering many waters unfit for most uses.

We are herein primarily concerned with water as an agent in modification of the natural environment and with its conservation and management as a natural resource. Pollution is discussed elsewhere.

THE HYDROLOGIC CYCLE

Hydrology is the study of water from the time it falls on the land until it is discharged into the sea or returned to the atmosphere. Hydrology involves both surface water and ground water. To appreciate the behavior of the waters of the land, an understanding of the hydrologic cycle is necessary (Figure 3-1).

Energy provided by the sun evaporates water from land and sea; the vapor is transported elsewhere by the winds and falls as *precipitation* (rain or snow), again either on land or sea. Rivers return part of the precipitation to the oceans. The return may be delayed, as when surface waters evaporate before reaching the sea or enter temporary ground-water storage. Regardless of the delays, the sea is the ultimate repository. The water that temporarily flows on or below the surface of the ground is generally available for water supply.

The bulk of the water involved in the hydrologic cycle is evaporated from the oceans. Some of this water falls back into the sea, but 381 cm (150 in.) is carried inland annually over the United States. About 305 cm (120 in.) of this land-borne moisture is transported back over the oceans as part of the atmospheric circulation. The remaining moisture, 76 cm (30 in.), falls on the land largely as rain and snow.

At this point it becomes necessary to define a number of basic terms. *Evaporation* is the conversion of water or ice to vapor. Much precipitation may evaporate in the air or on the surface of vegetation or buildings before reaching the ground. Part of the precipitation that reaches the ground may infiltrate to become *soil moisture* (moisture partially filling the voids in the upper part of the soil). Another part may percolate deeper and become *ground water* (water completely saturating the ground below the soil moisture zone). The upper surface of the ground water is the *water table*.

Water that does not enter the ground becomes *surface runoff* and eventually discharges into the oceans. In cold regions, snow may accumulate to form glaciers whose meltwaters contribute to runoff. In dry climates, or in dry seasons, a considerable part of the runoff may be lost by evaporation and by *transpiration* (exhalation of vapor by vegetation). Of the 76 cm (30 in.) of precipitation that falls annually

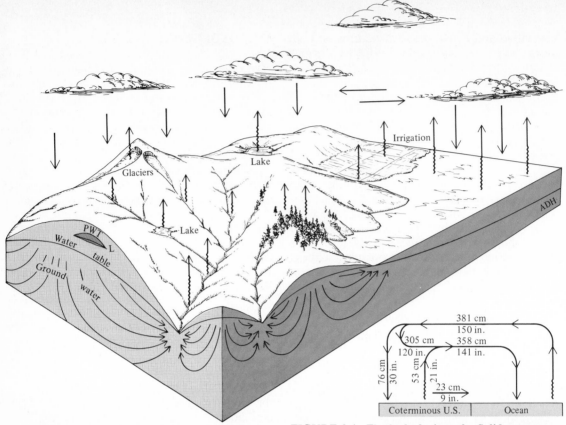

FIGURE 3-1. **The hydrologic cycle. Solid arrows—precipitation; wrinkled arrows—evaporation and transpiration. The small diagram shows approximate values for the hydrologic cycle for the contiguous 48 United States, using data from the U.S. Geological Survey.**

over the United States, 53 cm (21 in.) returns to the atmosphere by evaporation and transpiration, collectively referred to as *evapotranspiration*. Almost all the 23 cm (9 in.) which is not returned directly to the atmosphere returns to the sea via surface runoff, and a smaller part becomes ground water.

Water will freely enter a well that extends below the water table into the saturated zone but will not enter one limited to the unsaturated zone above. Figure 3-1 shows that the water table is a subdued replica of the topography. The figure also shows that small layers of impermeable material (L) may block water from descending to the water table and create small perched water tables (PWT). The long, curved arrows in the sides of the figure indicate general directions of ground-water movement, ultimately forming seeps or springs in streams, lakes, or the ocean.

The water table fluctuates, rising in periods of heavy precipitation and falling in times of drought. Expectably, the water table is at a

greater depth in arid regions than in humid regions. Where valleys are eroded deeply, they may intersect the water table and have access to a permanent supply of water from springs. Should the water table drop in times of drought, these valleys become dry. Streams that flow at some times and not at others are *ephemeral streams*. The basins of permanent lakes also intersect the water table. There are other lakes, however, which are above the water table and contain water only after rains, the waters of which either evaporate or disappear into the ground.

In many regions, considerable amounts of water are confined under pressure to particular permeable layers known as *artesian aquifers*, or *confined aquifers* (see Figure 13-8). Where these are cut by valleys, they give rise to springs on the valley sides or in the stream bed. Other confined aquifers may lie deep below the surface, and their water may be attainable only by deep drilling.

HYDROLOGIC PROBLEMS AND CONTROLS

Sound environmental planning must consider all water-related problems such as water supply, flooding, erosion, ground failure, undesirable sedimentation, and water quality. These problems are discussed below.

The Federal Water Resources Council has divided the United States into 21 regions on the basis of regional hydrologic characteristics. Eighteen of the regions are in the coterminous 48 states, and one each in Alaska, Hawaii, and Puerto Rico (Figure 3-2). The Great Basin is unique in that it has no permanent integrated drainage system. Others, such as the Missouri and Ohio regions, are dominated by a single great river system. The Colorado and Mississippi watersheds differ in their upper and lower reaches and are each separated into two regions. Others, such as the South Atlantic–Gulf

FIGURE 3-2. **Map of the United States showing Water Resources Council regions. Upper and lower numbers are, respectively, surface-water and ground-water withdrawals in billions of gallons per day in 1970.** (*After U.S. Geological Survey.*)

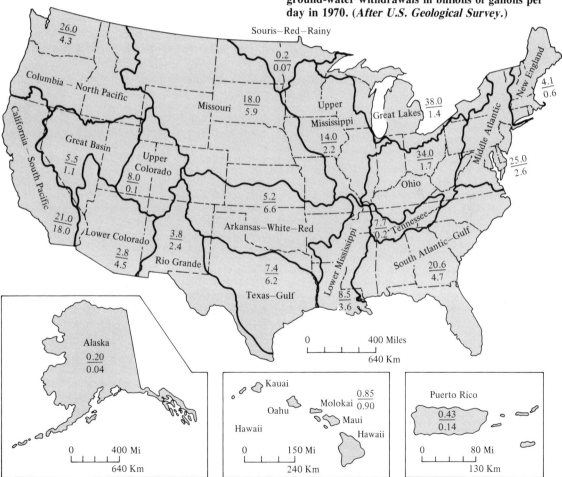

region, include multiple watersheds with similar characteristics. Regional water data are published by the U.S. Geological Survey.

Water Supply

As noted above, the United States has abundant water, but poor distribution and management create problems. Too often, land-development schemes pay little or no attention to water supply. Many metropolitan areas, too, have not made sufficient allowance for growth.

Although the variables involved in the water budget of an area are numerous and their interrelations complex, reasonable approximations of the budget are possible, provided data are available. The problem of determining the amount of water available on a continuing basis in a particular area involves measurement of input and output as well as the amount that may be diverted without infringing on the rights of others.

Surface Water. *Surface Runoff.* The surface runoff in a drainage basin depends on annual precipitation, its distribution throughout the year, duration of storms, intensity of precipitation (amount per unit time), evapotranspiration rates, depth to the water table, permeability of the ground, vegetative cover, steepness of slopes, characteristics of the stream channels, and human activities.

The volume of direct surface runoff is total precipitation minus evapotranspiration and infiltration. Precipitation may be measured by rain and snow gauges, and surface runoff may be measured by stream gauges. The difference between the two measurements approximates evapotranspiration and infiltration.

Rainfall distribution, intensity, and duration may be far more important than annual averages. If a rainfall of 100 cm (40 in.) is evenly distributed throughout the year, it will cause less flooding and erosion than if the 100 cm fell in a few brief storms of high intensity. The duration of storms is also important. The rain from a mild storm of a few minutes duration may largely be retained by the vegetative canopy or the soil so that little runoff results. If the rainfall persists, however, the soil becomes soaked, and any excess rain will run off. If the

distribution of rainfall is such that even storms of moderate intensity come in rapid succession, the soil may remain soaked and unable to accept more water. Increased runoff results. On the other hand, warm, dry weather leads to increased evapotranspiration between storms, drier soils, and less surface runoff.

The depth to the water table influences surface runoff in that the higher the water table, the thinner the unsaturated zone and the sooner it becomes wet enough to shed excess water. Highly permeable ground encourages infiltration of surface waters; ground with low permeability promotes surface runoff. Vegetation, until its foliage becomes soaked, may shield the ground and hence prevent runoff, and by forming a mat it may help retain water that does reach the ground. On the other hand, absence of vegetation may cause excessive runoff. Steep slopes, because of the high velocities imparted to surface runoff, decrease infiltration and result in heavy runoff. The characteristics of stream channels, whether straight, meandering, braided, smooth, rough, deep, or shallow, influence friction and control the rate of flow. Human activities, such as destruction of vegetation, improper agricultural practices, and the paving of large areas of ground, prevent infiltration and contribute to runoff.

Conservation of Runoff. The *planting of vegetation* provides a spongelike mat of soil and vegetation which is very effective in restraining runoff. *Dams* retain runoff for domestic and industrial supplies, for flood control, for power development, for irrigation, and for groundwater recharge. Some dams are multipurpose. Large dams, such as Hoover and Grand Coulee, affect much larger areas than the environmental dams herein considered. We are primarily concerned with dams located within the confines of communities, real estate developments, farms, and individual homesites.

Much of the protest against large dams stems from the fact that they drown out valuable flatlands, scenic canyons, and white-water rivers. Advantages and disadvantages must be weighed in the construction of any dam. In the case of public dams, an overriding consideration should be whether they provide the greatest good for the greatest number of people by contributing to flood control and by providing water for domestic use, for crops and animal husbandry, for industry, and for generation of electrical energy. However, account must also be taken of the effects on recreation, aesthetics, wildlife, and the enjoyment and use of the land by the public.

Many people objected to the construction of Glen Canyon Dam on the Colorado River because much of the scenic beauty of the area would be submerged. Others pointed out that few people had the hardihood to hike up the canyon into the colorful plateau lands it dissects. They point out that now, thousands of visitors to Lake Powell, the reservoir behind Glen Canyon Dam, enjoy superb views from scenic points or travel by boat up the intricate maze of canyons, whose unsubmerged upper levels still present magnificent vistas. The influx of visitors to the Glen Canyon region will, however, create serious environmental problems unless suitable regulatory measures are enacted and enforced. Furthermore, the existence of wild canyon lands is part of the natural heritage that should be handed down to future generations. The destruction of aesthetic resources must be fully justified.

Small dams present problems which, in total, may have greater ecological impact than the relatively few large dams. Small reservoirs, like all small lakes, have relatively short lives because of filling by sediment and vegetation. They may last only a few years and, prior to final filling, may become virtual cesspools.

Communities and developers exploit natural lakes and, where none exist, commonly create lakes. The term "real estate lakes" is coming into vogue for these lakes. They are desired for recreation and aesthetics. Too often, however, the assumption seems to be that once a lake is created it retains its original qualities indefinitely. Nothing is further from the truth. Lack of suitable drainage may foul the waters, and uncontrolled influx of sediment and pollutants may lead to rapid filling and algal growth (see Chapter 13). Thoughtlessly conceived dams on one property may harm property downstream.

Farm dams are built to provide water for cattle, irrigation, and recreation. Every small dam conserves surface runoff within its small drainage basin. They occur in such great numbers, particularly in drier parts of the country, that they also help reduce the magnitude of floods on the larger streams. Many opponents of large flood-control dams on major rivers suggest as an alternative the building of great numbers of small dams on tributary streams. This is not always a viable alternative because considerable portions of drainage basins drain directly into major streams and not into tributaries of significant size.

A farm dam should not seriously diminish the flow of water to individuals downstream. Many farm dams are in small valleys that are dry much of the time, and the impounded waters are largely the rainwaters that fall within the area between one dam and the next.

Proper *farming and forestry practices* prevent wasteful runoff and erosion. One harmful practice which cannot easily be avoided is the destruction of vegetation by cutting or burning in preparation for cultivation. *Overgrazing* is another common cause of excess runoff. If grass is cropped too close, it is less of a hindrance to runoff, and the topsoil and remaining grass may be removed by erosion. Serious gullying may take place on slopes or on flats if gullies extend inland from a steep embankment.

Indiscriminate *forest cutting* also leads to erosion and flooding. Lumbering practices have improved in recent years through selective cutting and replanting. Sometimes the removal of trees and ground litter for one purpose may create problems of another nature. Thus, in the winter of 1973–1974, many eucalyptus trees were killed by winter freeze in the Berkeley Hills of California. Because of the fire hazard, the trees were chopped down and all debris was cleared from the slopes. Heavy rains in the following spring caused severe erosion, mudflows, and landslides.

Finally, a serious cause of vegetative destruction and accelerated runoff in some areas is the *chemical fumes* released into the atmosphere by coke ovens, smelters, and other industrial plants (Figure 3-3). Fumes from smelters may lead to the creation of *badlands* (severely eroded and gullied areas) even in humid, normally forested localities. This cause is being eliminated by increasingly strict pollution controls. Concern about the long-range

FIGURE 3-3. **Soil acidity caused by fumes from beehive coke ovens killed the vegetation and caused the gullying on this Pennsylvania hillside.** (***U.S. Dept. Agriculture.***)

effects of automobile fumes on vegetation, and therefore on runoff, should diminish with strict emission control laws.

Legal measures to conserve runoff are for the most part coincidental in laws designed to protect landowners from erosion, flooding, and sedimentation. Many legal conflicts over surface waters stem from uncertainty over water rights. In many states, riparian landowners, that is, those on the shores of lakes or streams, enjoyed rights to unlimited use of the water. With increasing demands on water, however, and with recognition of the needs of others particularly in times of drought, the *riparian doctrine* proved unrealistic, and judicial rulings and statutes have modified it to ensure fairer distribution of water. Modern law also prohibits wasteful use of water.

Where surface water is scarce, the doctrine of *appropriative rights* has prevailed. The first person to appropriate the water has first right to it. The water need not be on the appropriator's land; it may be miles away. Successive appropriators have lesser rights until no water is left except the occasional surplus flood flow. This doctrine, too, has been modified by rulings and statutes aimed at more equitable water distribution. Environmental law is considered in Chapter 15.

Desalinization and Cloud Seeding. Because these techniques have greatest application in dry regions, we will defer details to Chapter 5. Suffice to say that *desalinization* is the conversion of seawater or other saline waters to fresh water. *Cloud seeding* involves the dispersal of fine particles in clouds to stimulate rain.

Cloud seeding may prove practical in humid regions in times of drought. It has also shown promise in reducing the destructive force of hurricanes. For example, in 1969, Hurricane Debbie was seeded with massive amounts of silver iodide on August 18 and 20. Wind velocities were reduced 31 and 15 percent, respectively. The seeding was done off the coast to avoid increasing flood hazards and to reduce hurricane storm waves and coastal damage. The decision to seed a hurricane involves great responsibility for public officials. Property owners might sue on the grounds that the seeding caused them damage, perhaps by diverting the path of the storm toward them.

Evapotranspiration and Leakage Control. Because evapotranspiration loss is most serious in arid environments, discussion of details is deferred to Chapter 5. Evapotranspiration may also contribute to water shortages in humid regions during long, dry summers or years of subnormal precipitation.

Leakage from reservoirs is distinguished from *seepage* on the basis of magnitude. A certain amount of seepage under and around a dam and around a reservoir is expected. Leakage, however, applies to an abnormally large escape of water from fissures or other openings in rocks or from permeable layers in the soil.

Leakage may be due to rock characteristics such as fracturing and solubility, to unusually high permeability, or to the presence of buried, sediment-filled channels which offer escape routes for the water. Solution is a very active process in limestone. If it is well advanced, there may be underground caverns and other large passageways. Even if these are filled with sediment, water may leak through the sediment when the reservoir is full. Remedies generally involve injection of cement through drill holes (*grouting*).

The Hale's Bar Dam on the Tennessee River was built on soluble limestone. Many solution cavities were found during excavation. About 5000 tons (almost 5 million kg) of cement were used to grout the open cavities. Despite these measures, leakage occurred under the dam after completion. Water was lost through sinkholes upstream and boiled out of the ground downstream. Unsuccessful attempts were made to fill the sinkholes and caverns with cement, rock, gravel, clay, and other materials. The foundation was finally sealed by drilling holes 27 m (90 ft) deep and grouting with 11,000 barrels of hot liquid asphalt.

In areas once overridden by glaciers, some valleys were filled to overflowing with sediment so that no surface evidence of their presence remains. Serious leakage may occur through such *buried valleys* to which reservoir waters have access. The remedy generally involves sealing off the permeable formations.

Where leakage is anticipated from canals and ditches, they are often lined with impervious clay, concrete, or asphalt. Plastic and rubber sheeting have not proven suitable for these purposes, although they are successful in sealing small ponds.

Besides involving loss of water, leakage around and under a small dam or reservoir may so weaken its foundation as to cause failure. Or, leakage from canals or ditches may cause subsidence due to hydroconsolidation.

Effects of Dams, Reservoirs, and Lakes. Most of the sediment carried into a reservoir by a stream is deposited as a delta at the head of the reservoir. Large quantities of fine material, however, may be spread throughout the reservoir, generally by density currents which hug the bottom. The greater density of these currents is usually due to the suspended sediments, although factors such as temperature differences may contribute.

The almost flat slope of the delta surface replaces the originally steeper stream gradient buried below. The stream flowing toward the reservoir and encountering this reduced gradient slows down and deposits part of its sedimentary load. Thus, the deposition that originated at the head of the reservoir progresses upvalley and buries formerly exposed portions of the valley floor. Below the dam, the water is now relatively free of sediment, and the energy that was formerly used to transport the sediment becomes available for erosion. Bottom lands may be trenched, and gullies may ramify outward from the main stream.

Reservoirs may be entirely filled with sediment (Figure 3-4) unless countermeasures are taken. One such measure is dredging, a costly operation. Another is sluicing, that is, periodically opening the outlet gates so that the escaping torrent of water will flume out some of the sediment. Some reservoirs are provided with an adjoining basin (desilting basin) into which the sediment is deposited before the water enters the main reservoir. This basin must be periodically cleaned out.

Large reservoirs may have useful lives of centuries. The useful life of Lake Mead, behind Hoover Dam near Las Vegas, Nevada, is expected to be 300 years and that of Elephant Butte Reservoir on the Rio Grande in New Mexico, 150 years. It is estimated that 21 percent of the nation's municipal reservoirs will have useful lives of less than 50 years because of sediment filling, 25 percent will last 50 to 100 years, and 54 percent will last more than 100 years. Thus almost half of the municipal reser-

FIGURE 3-4. **Mono Reservoir on Mono Creek in the mountains inland from Santa Barbara, California. The dam was constructed between 1935 and 1937. A range fire prior to construction had denuded much of the watershed, raising the specter of accelerated erosion and rapid sedimentation. Heavy rains in the next two winters before vegetation was reestablished led to rapid silting of Mono Reservoir. This example illustrates the ultimate fate of all reservoirs. The numerous channels on the fill surface suggest that the plain is subject to flooding. (*U.S. Dept. Agriculture.*)**

voirs will have outlived their usefulness within a century unless desilting measures are taken. For example, New Lake Austin on the Colorado River of Texas lost 95.6 percent of its capacity in 13 years.

A side effect of reservoir silting is the creation of flood hazards on the sedimentary fill. The sedimentary plain left after complete filling of the reservoir behind the Furnish Dam on the Umatella River in Oregon led to occasional flooding of the Union Pacific right-of-way after only 16 years. To eliminate the flood hazard, Union Pacific purchased the entire reservoir site and breached the central part of the dam. The river eroded down to the new level and then opened out a wide valley below the surface of the original fill.

Small natural and artificial lakes on golf courses, in parks, and in real estate developments may last only a few years because of filling by sediment (Figure 3-5), by accumulating vegetation, by excess fertilizers and other chemicals, by ground-water transport of sewage wastes and dissolved solids, and by growth of algae and other organisms. This aging process is known as *eutrophication*. Dissolved materials from septic tanks may promote the growth of algae and other organisms, even though pathogenic (disease-causing) bacteria have been eliminated in passage through the soil. Many golf course and real estate lakes rapidly become fouled with algal scum. Obviously, provision must be made for adequate drainage and for control of all wastes that might find their way into a lake.

The most serious problem with artificial lakes is filling by sediment eroded from adjacent slopes. The cost of removal of the sediment carried into the lake may be considerable. Careful planning is therefore essential. Erodibility maps should be prepared based on slope declivity and resistance of surface materials. Type and density of development should then be programmed so that each parcel of land is exposed to the least erosion.

Large-scale earth moving, as in grading, installation of sewer, gas, and other lines, road construction, and preparation of parking lots, paths, and building pads should be completed *before* a lake is constructed. Furthermore, this work should not be done during the rainy season, and slopes should not be left bare any longer than is necessary. Wherever possible, construction should be restricted to gentle slopes. Construction should proceed progressively, and each segment of the project should be graded and covered immediately with rapidly growing grass rather than left bare until completion of the whole project (Figure 3-6). Where the latter is unavoidable, the lake shoreline and stream banks should be protected against erosion by lining with concrete or masonry or, if the slopes are gentle, by immediate sodding.

If sedimentation cannot be avoided, it may be advisable to construct sedimentation basins along the feeder streams above the lake. These will require periodic cleaning. Disposal sites for the sediment must be provided. It may be advisable to provide diversion paths for flood waters into neighboring streams which do not flow into the lake and which do not create problems for others.

Effects of Water Diversion. The diversion of streams for irrigation, water supply, power generation, and mining may create serious problems A classic example on a large scale was the diversion in 1900 of Colorado River waters into Imperial Valley, California, for irrigation purposes. The floor of Imperial Valley is 84 m (273 ft) below sea level. A canal was constructed from Yuma, Arizona, to the bottom of the basin. In 1905, 1906, and 1907, floods overtopped the headworks, and the waters, following the steep gradient of the canal, cut a trench

FIGURE 3-5. **Sediment accumulation can turn lakes into mudflats. (*U.S. Dept. Agriculture.*)**

FIGURE 3-6. **Grading more land than needed for immediate construction caused severe erosion and sedimentation in this area. (*U.S. Dept. Agriculture.*)**

locally 24 m (80 ft) deep and swelled the previously small Salton Sea to many times its former size and depth. The river was finally brought under control at the headworks, but a railroad had to be relocated 15 m (50 ft) higher and as much as several kilometers from its former position.

Diversion of local waters may deprive inhabitants downstream of needed water. For many years the Paiute Indians in the Pyramid Lake area 48 km (30 mi) north of Reno, Nevada, have been fighting the continued withdrawal of water from the Truckee River which feeds the lake. The diverted water helps satisfy the growing needs of Reno and its environs and is increasingly used for irrigation. Shrinkage and increased salinity of Pyramid Lake, however, are endangering the fishing on which the Paiute Indians depend. In November 1972, a federal court order required the Department of Interior

to maintain Pyramid Lake at its present level. One consequence has been the reduction of water withdrawals for the Truckee-Carson irrigation district.

Application of water of good quality may help maintain fertile soils by leaching out excess salts. Diversion of this water may deprive the soils of this cleansing action. Diminished water downstream may also lead to lowering of the water table, which in turn may cause the drying up of wells and subsidence of the ground. Diversions for irrigation may even create problems in the recipient area. If the water table approaches too close to the surface in dry regions, it may create salinity problems detrimental to crops.

Diversions of water on a grand scale may have serious climatic effects. Russian scientists have proposed grandiose plans for southward diversion of some of the great rivers of Siberia. The dam for one project alone would create a lake with an area of 250,000 sq km (96,000 sq mi). There is concern over the global climatic effects of such large-scale hydrologic manipulations.

A South American Great Lakes System has been proposed by damming and interconnecting rivers of the interior. Aggregate area of the proposed lakes would be 670,000 sq km (260,000 sq mi). In Africa, a Congo Sea has been proposed, to be impounded by a mile-long dam on the Congo River at Kisangani (see Figure 6-12). The lake would have an area of close to 1 million sq km (380,000 sq mi), nearly equal to the combined areas of the Baltic, Black, and Caspian Seas.

When one reflects on the many harmful effects of impounding the Nile behind the High Aswan Dam in Egypt, it becomes obvious that careful planning must precede the initiation of such major projects. The High Aswan Dam not only flooded valuable agricultural land but, because of great fluctuations in lake level, prevented attainment of a stationary shoreline. Thus, agriculture close to the lake is hazardous, and no lakeshore communities or ports can safely be established.

The expanse of fresh water in the lake basin and its tributary streams, and the proliferation of irrigation canals and ditches as far as the Mediterranean Sea, has led to the spread of schistosomiasis, a debilitating disease due to infestation of the bladder and intestines by a flatworm whose spiny eggs cause hemorrhage and tissue damage. If the eggs find their way into fresh water, they hatch to larvae which may be ingested by a certain variety of freshwater snail. The larvae leave the snails as minute needlelike forms which easily penetrate the skin of humans entering the water and migrate to visceral organs, maturing in their passage. Because these snails favor the calm waters of sluggish streams, reservoirs, ditches, and canals, their spread represents an adverse impact on the regional ecosystem.

Other adverse effects from construction of the dam have appeared downvalley. The bottom lands, deprived of their annual increment of new sediment, are deteriorating in quality.

The Nile, now underloaded in regard to sediment, is eroding its bed and banks, and gullying is spreading. To check the erosion and trap the sediments, a series of downstream dams has been built at great expense. The loss of chemical nutrients formerly carried downstream in floods has affected the fishing industry even as far as the Mediterranean coast. Schools of fish which were attracted to the delta coast by plant growth fertilized by these nutrients have diminished considerably.

Ground Water. *Origin and Relative Importance.* Figure 3-7 shows the important groundwater areas in the contiguous 48 states.

Subsurface waters accumulate by infiltration of rainwater, meltwater, and the water of streams, ponds, lakes, and reservoirs (see Figure 3-1). At a certain depth, depending on climate and type of earth materials, the ground is saturated. The water in the saturated zone, as well as the water confined to deep aquifers, is ground water.

The volume of underground water in the contiguous 48 states is almost 7 times that of all surface waters, including the United States portions of the Great Lakes. It totals more than 125,000 cu km (30,000 cu mi). Billions of gallons of ground water are withdrawn each day. Eighty percent of rural needs, 25 percent of municipal needs, 7 percent of industrial requirements, and 25 percent of irrigation needs are supplied from ground water. In arid regions in particular, it provides the bulk of usable water. Penetration of deep aquifers has opened the southwestern deserts to agriculture and habitation. In the Sahara, new oases have sprung up around wells tapping deep aquifers. In dry southern California, in spite of aqueducts bringing huge volumes of water from the Sierra Nevada and northern California, ground water satisfies 40 percent of present needs.

Ground water is also important in humid regions where surface supplies are either inadequate or unusable. Thus, the inadequacy of surface waters in large parts of Florida forces reliance on subsurface reserves. Miami is almost totally dependent on deep aquifers.

Slow, steady seepage of ground water contributes to streamflow and maintains the flow of perennial streams during periods between

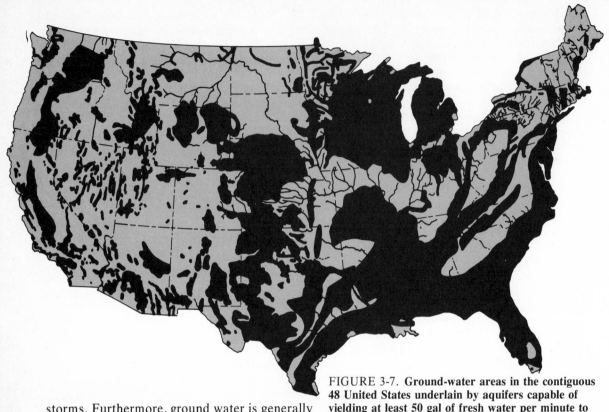

FIGURE 3-7. **Ground-water areas in the contiguous 48 United States underlain by aquifers capable of yielding at least 50 gal of fresh water per minute to individual wells. (*After U.S. Dept. Agriculture.*)**

storms. Furthermore, ground water is generally a more dependable source of irrigation water than surface water.

The comparative volumes of stored surface and subsurface water are not the dominant factors in determining relative importance. The degree to which the waters circulate freely and renew themselves is more important. Ground water, in proportion to its volume, has relatively insignificant circulation. Thus, withdrawals in excess of recharge cause a more or less permanent storage loss, or "mining," of water.

Some ground-water reservoirs have areas of less than 112 sq km (40 sq mi); others cover thousands of square kilometers. The depletion of underground reserves is more common in arid regions because needs are greater than in humid regions, and natural replenishment is small. Corrective measures include prevention of waste, curtailment of drilling, controlled pumping, reclamation of used waters, replenishment by the spreading or injection of surplus stream water, and importation of water as a substitute for ground water. Only when water

levels in wells remain constant has an approximate equilibrium between use and replenishment been attained.

Ground-water systems have three characteristics that make them valuable even in areas of adequate surface-water supplies. First, some confined aquifers extend great distances and underlie many properties. Farmers tapping the Dakota Formation in the Great Plains are withdrawing water that entered the formation hundreds of years ago at its outcrop in the mountains far to the west. Second, ground-water systems are natural storage reservoirs that are relatively free of evaporation losses. Finally, they improve water quality (1) by filtration of undesirable elements during percolation, (2) by maintenance of uniform temperature, (3) by retention of undesirable elements on soil particles by molecular attraction, and (4) by allowing time for short half-life radioactive substances to decay during slow percolation.

Conservation. Figure 3-8 shows overdrawn ground-water reservoirs in the western United States as of 1951. Only small changes are reported to have taken place since that time.

Inasmuch as subsurface water is derived by infiltration of surface water, the retardation of surface runoff is of first importance in ground-water conservation. An additional conservation measure is *control of withdrawal rates.* Environmental planning requires the collection of ground-water data and prediction of the amounts that may be withdrawn without exceeding rates of replenishment and without leading to undesirable side effects.

Conservation of subsurface water may involve *control of vegetation* which is wasteful of water. Because vegetation is of serious concern as a user of water in dry regions, the details are discussed in Chapter 5. Suffice to say that many plants are avid drinkers and have long roots that tap the water table even at depths of 9 m (30 ft) or more. Furthermore, in agriculturally marginal subhumid or semiarid areas, a choice of crops based on differences in water needs may mean

FIGURE 3-8. **Overdrawn ground-water reservoirs in western United States as of 1951. Although no later map is available, no exceptional changes in the situation are reported. (*After H. A. Thomas "The Conservation of Ground Water," McGraw-Hill Book Company, 1951.*)**

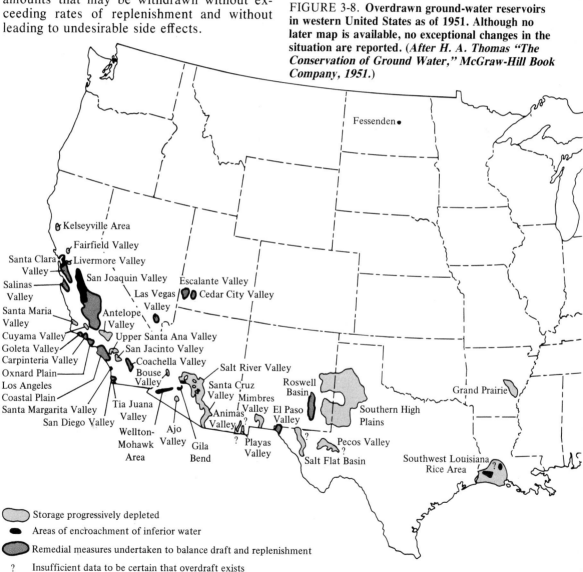

Fessenden

Kelseyville Area
Fairfield Valley
Santa Clara Valley
Livermore Valley
San Joaquin Valley
Salinas Valley
Escalante Valley
Las Vegas Valley
Cedar City Valley
Santa Maria Valley
Antelope Valley
Cuyama Valley
Goleta Valley
Upper Santa Ana Valley
Carpinteria Valley
San Jacinto Valley
Oxnard Plain
Coachella Valley
Bouse Valley
Los Angeles Coastal Plain
Salt River Valley
Santa Cruz Valley
Roswell Basin
Grand Prairie
Santa Margarita Valley
Mimbres Valley
El Paso Valley
Southern High Plains
Tia Juana Valley
San Diego Valley
Animas Valley
Wellton-Mohawk Area
Ajo Valley
Gila Bend
Playas Valley
Pecos Valley
Salt Flat Basin
Southwest Louisiana Rice Area

Storage progressively depleted

Areas of encroachment of inferior water

Remedial measures undertaken to balance draft and replenishment

? Insufficient data to be certain that overdraft exists

the difference between a profitable or unprofitable season.

Legal measures dealing with ground water are more difficult to conceive and apply than those dealing with surface waters. The difficulty is partly due to the fact that the details of ground-water movement are not always understood. Ground water generally permeates rock fractures and soil pores and only rarely moves as a well-defined stream; hence, the doctrine of riparian rights is inapplicable. Furthermore, because of the vast areas underlain by single aquifers, the number of individuals and communities affected by use of the ground water is large. Determination and apportionment of rights among large numbers of users is very difficult.

The tendency is to give local, state, and federal agencies power over utilization of ground-water resources. Laws vary from state to state. Many state engineers have the power to issue licenses for large-capacity wells.

Artificial Recharge. Where ground water has been extracted at too rapid a rate, creating shortages, *artificial recharge* should be considered. Water may be imported and spread over the land or fed into wells. In artesian areas, water may be forced under pressure into wells penetrating the aquifer, thereby restoring lost pressures.

Saltwater Encroachment. A major problem in coastal areas is *saltwater encroachment* (Figure 3-9). The encroachment may extend thousands of feet inland, causing abandonment of wells. The encroaching salts include common sodium chloride and salts of calcium, magnesium, potassium, and other elements. In general, 1000 ppm (parts per million) of dissolved solids classifies the water as saline. Sea water has a salinity of 35,000 ppm.

Saline water, because of the dissolved solids, is heavier than fresh water; hence it wedges under fresh water in both surface and subsurface waters. In tidal estuaries, the salt water occupies the lower levels in the channel, forming a wedge that advances and recedes with the tides. Fresh water entering the sea may retain its identity at the surface for considerable distances.

Along coasts, excessive ground-water withdrawal permits the heavier seawater to encroach under the fresh water, the boundary or interface descending inland. If an artesian aquifer is involved, the reduction of pressure due to heavy pumping permits the salt water to invade the aquifer.

The water supply of Miami, Florida, comes from the Biscayne Aquifer, a prolific water-bearing limestone that underlies much of southern Florida. A well field was developed in 1896 and 1907, but heavy withdrawals resulted in saltwater encroachment, forcing abandonment of the field in 1925. Wells were drilled in a new area in 1925, but again saltwater encroachment resulted. Controlled withdrawal, and the construction of barriers against saltwater encroachment in drainage canals, have halted further contamination.

Freshwater discharge into the large drainage canals in the Miami area normally prevents salt water from encroaching more than 3 to 5 km (2 to 3 mi) inland. When rainfall is subnormal, however, the lessening of the canal discharge permits greater encroachment in the canals. Thus, a salt-water wedge invaded the

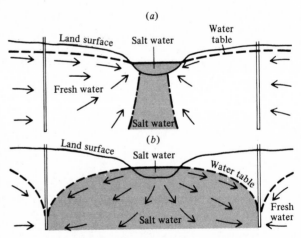

FIGURE 3-9. **Ground water near a tidal stream that carries salt water.** *(a)* **Movement under natural conditions before pumping.** *(b)* **Movement during ground-water pumping. A cone of depression surrounds each pumped well, the water table is depressed, and salt water encroaches into the aquifer.** (*After U.S. Dept. Agriculture.*)

Miami Canal for 16 km (10 mi) in 1939. The salt waters spread laterally and contaminated ground water in adjacent portions of the Miami supply field. Heavy rains subsequently recharged the aquifer, drove the salt water out, and forced the salt water in the canal back toward the sea. A control that has proved effective is the use of gates in tidal canals close to the sea. These are closed when there is danger of saltwater intrusion.

Ground-water extraction is the most common cause of salt-water encroachment. Lowering of the water table by pumping creates a cone of depression (see Figure 3-9), which may expand into an area of salt water and induce saltwater movement toward the pumping well. To prevent the formation of a cone of depression, skimming tunnels are being substituted for wells in Hawaii. They are excavated below the water table to skim off the upper water over a fairly large area.

Saltwater encroachment is not exclusively a coastal problem. Saline waters have many sources, including inland bodies of salt water such as the Great Salt Lake, dry lake floors with saline deposits, evaporation ponds, salt water entombed in sediments and released by deep wells, underground salt deposits, saline springs, irrigation waters, and industrial waste water.

Ground Subsidence. Ground subsidence is another result of hydrologic mismanagement. The subject was discussed in Chapter 2.

Earthquakes. It has been known for years that earthquakes affect water levels in wells. Only recently, however, has it come to be recognized that hydrologic factors may stimulate earthquakes.

The origin of the stresses that cause earthquakes is considered in Chapter 8. In some places, these stresses may be so close to the critical level that any additional hydrologic stress may stimulate an earthquake. Some believe that the weight of a dam and its impounded water may supply the additional stress, or seeping water under pressure may lubricate the fault surface and stimulate movement.

The injection of chemical waste fluids from the Rocky Mountain Arsenal north of Denver into formations at a depth of more than 3400 m (11,000 ft) triggered earthquakes in the Denver area. Although injection ceased in February 1966, a new equilibrium has not yet been attained, and earthquakes continue, but as trivial shocks.

Floods

Much of the property damage from natural disasters in the United States is caused by floods. This is due not only to the frequency of floods but to the concentration of a significant part of the population in areas susceptible to flooding. Most urban development is on floodplains. In New York State, 79 percent of the 330 communities having populations greater than 2500 have flood problems. The U.S. Water Resources Council has predicted that flood damage will increase in spite of the many flood-control programs under way. Yet, human development of floodplains continues at an accelerating pace.

Causes. Floods were common long before humans appeared. Even today, most floods are independent of our influence and result from heavy runoff due to intensity or persistence of rain, rapid melting of snow, or destruction of vegetative cover by lightning-induced fires.

But humans have contributed to the problem. Fires started by campers denude large areas of protective forest cover, and ill-conceived practices in lumbering, agriculture, and industry contribute to the devastation. Urbanization, by sealing off water-absorbing ground with buildings and pavement, or by oversteepening slopes, adds to flood hazards.

Flood Hazards. Physical hazards involved in floods include direct inundation, caving of river banks, landslides, erosion of new channels, debris and mudflows, and deposition of sediment.

Inundation may result in loss of life and property damage on floodplains. Caving of banks along floodplain channels undermines buildings, roads, and other structures. Landslides in steep mountain valleys result from accelerated stream erosion, particularly on the outside of curves. New channels are often eroded when floodwaters overtop channel

FIGURE 3-10. **In semiarid lands, the mudflow is a characteristic and destructive form of mass movement. This photograph shows the steep fault face of the Wasatch Mountains at Centerville, north of Salt Lake City, Utah. Grazing on the mountains has depleted the vegetal cover and accelerated runoff. Rains of July 1930 washed down so much soil and loose rock in Parrish Canyon, in the center, that the stream emerged at the canyon mouth as a mass of flowing mud. The thick mudflow carried boulders weighing up to 200 tons. The whitish fan-shaped patch below the canyon mouth is the area covered by mudflow debris.** (*U.S. Dept. Agriculture.*)

banks. These flood channels may cross built-up areas, destroying buildings and washing out roads, railroads, and utility lines.

Debris and mudflows result when floodwaters, laden with sediment, lose water by underground seepage and from evaporation on entering the lowlands beyond the mountains. They gradually convert to porridgelike flows of water-saturated sediment (Figures 3-10 and 3-11). The flows may fill channels and spread over neighboring ground, penetrating and crushing buildings. Twelve people lost their lives by burial in debris and mudflows in the floods of 1969 in southern California.

Sediments carried downvalley fill reservoirs, lakes, and ponds. Annual losses due to reduced reservoir capacities are $100 million in the United States. Flood sediment reaching the sea causes shallowing of harbors and reduces the area of open water. More than $125 million dollars is spent each year to dredge sediment from United States harbors and waterways. Where floodwaters introduce large volumes of sediment into coastal or lake waters, the volume may be too great to be dispersed by longshore currents, and beaches may expand seaward. This may strand piers and other shore structures inside the new shoreline.

Flash-Flood Warnings. The National Weather Service is urging greater use of a successful automatic flash-flood warning system. The stimulus for automatic systems was the loss of 153 lives in Virginia during Hurricane Camille in 1969. At that time, 70 cm (27 in.) of rain fell in 8 hours, causing catastrophic floods.

The first automatic alarm system was installed in Wheeling, West Virginia, in May 1972. Others are now located at Plainfield, New Jersey; Chester, Pennsylvania; Wooster, Ohio; Waynesboro, Virginia; Spring City, Tennessee; and Rosman, North Carolina. By the time this text becomes available, two others will be in operation in Maryland near Washington, D.C.

Each system has three electrically connected stations. An upstream station has a float device which activates an electric current when the water reaches a critical level. The signal is transmitted to an intermediate station downstream where telephone service is available, and from there it is transmitted to a continuously staffed alarm station such as a police station or firehouse.

Controls. Flood-control measures include dams, levees, floodways, overflow basins, channel modifications, and land-use management.

The flood-control dams and reservoirs store floodwaters for controlled release later. The levees and channel modifications are designed to provide sufficient channel capacity to carry peak flood flows. Channel modifications involve dredging, removal of debris, straightening, smoothing, revetments, and spur dikes. Land-use management includes improved forestry, agricultural, and urbanization practices. We will consider these in order.

Flood-Control Dams. Flood-control dams occur in all sizes, are located in diverse topographic environments, and may be constructed of earth or concrete. Many are multipurpose, designed not only for flood protection but also for furnishing water for irrigation, generation of electric power, and recreation.

FIGURE 3-11. **Mudflow of 1941 at Wrightwood, southern California. The mountain resort was partly inundated by stony mudflows for more than a week because of rapid melting of winter snow. Debris was transported 24 km (15 mi) on gradients less than 1° near the terminus. Velocities of the surging flow fronts ranged up to 16 km (10 mi) per hour.** (*Courtesy R. P. Sharp and L. H. Nobles.*)

Flood-control reservoirs require management procedures different from reservoirs used primarily for water supply, power generation, or recreation. These three purposes require that the reservoir be kept full, particularly to provide water for drought periods. For flood control, the reservoir should be kept at low levels to provide storage for floodwaters. The uncertainties of occurrence of droughts and floods make management difficult. To add to the flood-control problem, a typical river system may include a number of reservoirs of different functions.

Many dams, particularly in the arid Southwest, are located in lowlands beyond the mountain valleys. This type of dam forms a large arc across a shallow stream or dry wash to enclose a relatively large basin (see Figure 3-16). Where these dams impound streams, lakes are formed, but the levels are kept low by controls at the outlet. The lakes generally provide recreational facilities. Dams across dry washes enclose dry basins. The land is generally used for parks, golf courses, riding trails, and comparable recreational facilities. In both types, the basins are intended for flood-control storage; hence, permanent occupation, as for residences, is prohibited.

Differences of opinion exist as to the value of the large flood-control dams. There is increasing agreement that environmental as well as economic factors must be considered in planning the use of flood-prone areas. Federal agencies are now required to prepare two feasibility plans for each project. One plan considers the economic benefits; the other covers the environmental aspects. Then, in addition to the plan for a big dam, a second plan is submitted, presenting an alternate approach to the flood-control problem. The alternate plan might suggest leaving the floodplain relatively undeveloped, with considerable park and open space that would not be damaged much by floodwaters.

A distinction must be made between dams that protect established cities from floods and dams that keep water out of largely undeveloped floodplains so that they can be urbanized. There is increasing support for maintaining undeveloped floodplains as natural overflow areas.

Levees. The construction of artificial levees, or the improvement of natural levees, is designed to enlarge channel capacity and contain all but the exceptional flood (Figure 3-12).

Protection of bottom lands from floods cannot be undertaken as piecemeal projects by individual landowners or communities. Individual property owners or communities cannot finance as effective flood-control measures as regional agencies can. Even though a community has the means to construct and maintain adequate levees, floods may overtop less adequate structures upstream and come in behind the "adequate" levees. Furthermore, the prevention of overflow in adequately protected stretches means that increasing volumes of floodwater are passed on downvalley to flood less adequately protected areas.

Until passage of the first Flood Control Act in 1917 which committed the federal government to comprehensive flood-control programs, levee construction was a hit-and-miss proposition in the Mississippi Valley. The levee system was a frail line of defense even against minor floods, and the more the river was hemmed in locally, the more devastating the floods became downstream. As for the exceptional great flood, there seems to be little hope of control even today. During the great Mississippi River flood of 1973, there were more than 300 breaks in the levee line.

In many regions, levees thought to be of adequate height have been overtopped in exceptional floods. In November 1950, levees along the right bank of the American River near Sacramento, California, were overtopped, and 2000 hectares (ha) (5000 acres) of residential and agricultural land were flooded. The completion of Folsom Dam in 1956, 32 km (20 mi) upstream from Sacramento, has reduced the flood hazard. As part of the same project, the levees which were overtopped in 1950 were set back and enlarged, thereby increasing channel capacity.

Floodways. Floodways are emergency escape routes for floodwaters at densely populated centers (Figure 3-13). To protect Cairo, Illinois, and reduce flood heights, a diversion route was provided between the levee on the west side of the Mississippi River and a setback levee 8 km

FIGURE 3-12. Levees and concrete-lined channel for flood control, Branciforte Creek and San Lorenzo River at Santa Cruz, California. (*U.S. Army Corps of Engineers, San Francisco District.*)

FIGURE 3-13. Excess waters are diverted through spillways to floodways. Bonnet Carre spillway on the Mississippi River 32 km (20 mi) northwest of New Orleans. Flood waters are diverted to Lake Pontchartrain north of the city. (*Mississippi River Commission, U.S. Army Corps of Engineers.*)

(5 mi) farther west. Floodwaters follow this floodway from a point opposite Cairo to New Madrid 65 km (40 mi) downvalley, where a point of reentry is provided. This floodway saved Cairo from extensive damage in 1937.

In some areas, floodwaters are diverted down abandoned channels or into neighboring rivers on the floodplain. Controlled openings through the levee must be provided at the points of diversion and reentry.

Overflow Basins. Overflow basins are usually backswamp areas confined by levees of the master and tributary streams and the valley sides. Some natural basins may be continuously occupied by floodwaters; others are dewatered to provide sunken agricultural land. In some instances, artificial construction has enclosed a low area to form a basin. During floods, waters are diverted into these basins, diminishing the volume continuing downvalley.

Channel Modifications. Channel modifications for flood control include dredging, straightening, smoothing, protecting the banks from erosion, and clearing the channel of debris.

Dredging enlarges channel capacity and thereby reduces flood heights. Dredging is expensive, and construction of levees is preferred.

The disposal of dredging spoils may create problems. In San Francisco Bay, much previously dumped land waste is now mixed with the bottom sediments. The waste includes sewage, oil and grease, industrial refuse, and toxic heavy metals such as lead, zinc, mercury, and copper. These sediments when dredged from shipping channels, were deposited elsewhere in the bay. Because of the possibility of contamination of food fish, the Bay Area Regional Water Quality Control Board in 1972 prohibited further dumping of dredge spoil within the bay. Problems created by these orders led to a temporary halt in dredging by the Corps of Engineers, and this would have crippled the shipping industry if continued. The Water Board now usually grants dredging permits with permission to dump the spoil elsewhere in the bay if it can be shown that the dredging is essential, that land disposal sites are not available, and that the environmental impact is slight. Possible beneficial uses of the spoil are being investigated. These include restoring eroded shorelines, creating new islands and wildlife marshes, and building up subsiding agricultural lands of the bay delta.

The problem of disposal of dredge spoils may be even more acute farther inland. The spoils must either be disposed of on the floodplain or carried away at considerable expense.

The *straightening of channels* (Figure 3-14) by eliminating meanders (looplike bends in stream channels) contributes to flood control in two ways. First, it eliminates overbank floods on the outside of curves, against which the swiftest current is thrown and where the water surface rises highest. Secondly, the shortened course increases the gradient and velocity, and the floodwaters erode and deepen the channel, thereby increasing its flood capacity. The sediment from the induced erosion does not present noticeable problems, but the containment of flood waters at one place increases the flood hazard downstream.

A program of channel cutoffs was initiated along the Mississippi River in the early thirties. By 1941, it had lowered flood stages by more than 2 m (6 ft) at Vicksburg, Mississippi, and 4 m (12 ft) at Arkansas City, Arkansas. By 1950, the length of the river between Memphis, Tennessee, and Baton Rouge, Louisiana, 600 km (380 mi) downvalley, had been reduced 270 km (170 mi) as a result of 16 cutoffs.

The *smoothing of channels* reduces friction and increases flow. Irregularities that retard flow increase flood heights upstream. The smoothing involves removal of projections, filling of reentrants, and removal of vegetation.

Protection against bank erosion, especially on the outside of curves, is provided by *revetments* (Figure 3-15). These are protective armors to shield the banks and sides of levees. They may consist of mats of willow or other brush, lumber, dumped or fitted rock fragments, concrete blocks, or other materials. The most effective revetment is the articulated mattress of concrete slabs 0.3 by 1.3 m (1 by 4 ft) and 0.08 m (3 in.) thick held together by reinforced fabric and wire. In places, vertical walls of wood piling, rock, or concrete are employed to stop bank caving. In other places, *spur dikes*, commonly consisting of rows of piles 0.6 m (2

Ashbrook
cutoff,
1935

Tarpley
cutoff,
1935

Leland
cutoff,
1933

FIGURE 3-14. **Cutoffs shorten distances, speed flow, and cause channel deepening. Mississippi Valley. (*Mississippi River Commission, U.S. Army Corps of Engineers.*)**

ft.) apart and fastened together, are constructed perpendicular to or oblique to the banks to trap the sediment carried by the river. Their use is inadvisable, however, where even a slight hampering of flood flow or reduction in channel capacity might lead to overtopping of the banks.

Clearing of channels of loose debris, mostly trees and other vegetation, reduces the chances of their snagging and forming growing obstructions. Any obstacle to free flow will act like a partial dam and raise the flood level upstream.

Land-Use Management. In the preceding discussion, we have considered the direct control of flood runoff. Measures are also taken to reduce the flood hazard at its source by preventing excess runoff. One cause of excess runoff is destruction of forest cover by natural or humanly created fires. Lightning-induced fires cannot be prevented, but it is becoming

increasingly necessary to prohibit human entry into forested areas during dry seasons. Rapid reforestation after denudation is essential. Responsible agricultural practices, such as contour plowing and the prevention of overgrazing, also help to reduce surface runoff.

Where considerable ground is covered during *urbanization*, the resulting increased runoff must be controlled. This may involve enlarging natural drainageways, straightening courses, clearing brush, building levees, or constructing drainage tunnels.

In floodplains where urbanization has not advanced far, strict zoning regulations are in order. These would restrict urbanization of floodplains in favor of agricultural and recreational uses.

An outstanding example of floodplain plan-

FIGURE 3-15. **Stone paving, with articulated concrete mattress extending below water line, for bank and levee protection. Mississippi Valley.** (***Mississippi River Commission, U.S. Army Corps of Engineers.***)

ning is the Chattahoochee Corridor study by the Atlanta Regional Commission in Georgia which led to landmark state legislation in 1973. The commission was authorized to review all development within 600 m (2000 ft) of the river for an 80-km (50-mi) stretch of the river through metropolitan Atlanta. Land most vulnerable to flooding was to be reserved as open space. Two scenic and historic sites were to become, respectively, a state park and a scenic-historic park. In order that future development harmonize with the natural environment, the regional commission was to review all proposals for development. Problems of erosion, sedimentation, and pollution had to be satisfactorily resolved. Land use was to be assessed on the basis of vegetation, soils, hydrology, slopes, aspect (scenic), and geology.

In Houston, Texas, far-sighted land developers have led the way with careful planning to increase preservation of streamways as greenbelts and to provide for attractive development with much open space. Raleigh, North Carolina, has the Capital City Greenway Plan, with an open-space system penetrating throughout the city as "green fingers" and mostly following natural stream courses.

Flood Control in Urbanized Southern California. Human influence on the hydrologic environment is well illustrated in urbanization. In Los Angeles and neighboring counties, urbanization has expanded over alluvial fans, old river bottoms, and modern floodplains. Older developed areas, protected by a variety of flood-control facilities, suffered little damage in the exceptional floods of 1969 in comparison with more recently developed areas where flood-control facilities have failed to keep pace with development.

The hazards induced by urbanization depend on the characteristics of the stream channels in different parts of the environment. Beyond the mouths of most mountain canyons are *alluvial fans*, fanlike accumulations of clay, silt, sand, and gravel that extend down to the lowlands (see Figure 5-4). Before flood control, storm-swollen, debris-laden waters pouring out of the mountains ran in ill-defined shifting channels on the fans and spread widely over the plains beyond. The torrents inundated large areas and dumped great volumes of debris. To protect lives and property in urbanized areas, flood-control facilities of several types have been employed. These include (1) reservoirs to hold back the floodwaters (Figure 3-16), (2) basins to trap debris that would otherwise spread over the fans and valley floors, (3) diversion channels to lead sediment-laden floodwaters into areas where the excess water can be absorbed into the ground and the sediment deposited, (4) realignment, enlarging, and paving of permanent channels to lead off excess runoff, (5) artificial levees to increase channel capacity, and (6) in the mountains, measures to retard runoff and erosion, such as the prevention and control of forest fires, the reseeding of denuded areas, and the enactment of regulations to control land use.

As noted, the more recently urbanized areas in southern California were the most heavily damaged because flood-control measures had not kept pace with the urban expan-sion. An important lesson is that flood-control planning and construction of facilities should precede urbanization. Furthermore, the planning and implementation cannot be left to local communities without consideration of the problems of adjacent communities and the larger regional problems. Another lesson is that flood problems, including the inevitable landslides and mudflows, can never be completely solved; only accommodation can be achieved. In many areas, even attempts at accommodation should be avoided, and the areas should be left in open space or set aside for recreation, subject to closure when floods threaten.

Flood-Hazard Maps. A knowledge of areas likely to be flooded is needed to cope with

FIGURE 3-16. **Whittier Narrows Reservoir protects 24,000 ha (60,000 acres) of the Los Angeles metropolitan area from floodwaters draining more than 1300 sq km (500 sq mi) of mountainous area.** (*U.S. Army Corps of Engineers, San Francisco District.*)

flood-hazard problems. The flood-hazard map meets the need for this type of information and assists in planning and management of floodplains.

A flood-mapping program has been underway in the Chicago area since 1961. The program covers the six-county metropolitan area and involves the Northeastern Illinois Planning Commission, the state of Illinois, and the U.S. Geological Survey, Nearly the entire area is now covered by flood-hazard maps. Had such programs been in existence before the present intense urbanization of floodplains, many flood problems might have been avoided.

The flood-hazard base maps are standard U.S. Geological Survey topographic quadrangle maps $7^{1}/_{2}$ minutes by $7^{1}/_{2}$ minutes in latitude and longitude and covering, at the latitude of Chicago, 148 sq km (57 sq mi). The scale is 1:24,000. The areas covered by particular floods of record are outlined. Also shown are the locations of drainage divides, gauging stations, and flood-crest gauges. Figure 3-17 is a portion of one of the Chicago-area maps. The maps are accompanied by explanatory texts, tables, and graphs. One set of graphs shows the flood profiles along the major stream, from which the user can tell how high the water rose at any point. Other graphs show probable frequency of flooding at selected gauging stations. These frequencies can be expressed as probabilities so that the flood hazard at a particular property can be rated, for example, as having a 5 percent chance of being inundated each year. The maps and charts are also' useful in answering questions such as, "At what elevation may one build a home where the flood risk is once in 25 years?"

The maps have proved useful in floodplain management and have led to detailed floodplain zoning. The zoning is delegated to local governments, although county governments may exercise zoning powers in unincorporated areas.

The U.S. Geological Survey and the American Society of Civil Engineers have cooperated in an analysis of 26 floodplain cities. The study indicates for each city the area that would be inundated by a flood of the magnitude that occurs on an average of once every 100 years (the *"100-year flood"*). The percentages of the urbanized areas of the floodplains judged to be subject to a 100-year flood range from a high of 81 percent for Monroe, Louisiana, to a low of 2.4 percent for Spokane, Washington.

The U.S. Geological Survey has established the Land Information and Analysis Office to complete land-use mapping from satellite imagery of the entire nation in 5 years. The maps cover nine general categories: urban and built-up land, agriculture, rangeland, forestland, water, wetlands, barren land, tundra, and perennial snow and icefields. Although designed mainly to assist local, state, federal, and private planners and decision makers in providing for best use of the land and other resources, the maps have already proved useful in emergency situations such as flooding. Thus, current satellite imagery plus professional interpretational assistance was provided the governor of Louisiana to help determine the proportion of each land-use category that was flooded in the great Mississippi Valley flood of April 1973. This enabled the governor to decide whether to request federal disaster relief and provided farmers and other flood victims with a basis for disaster-relief applications.

Federal Role in Flood Control. Prior to 1917, federal flood-control responsibility was limited to measures that benefited navigation, and this was the only role assigned to the Mississippi River Commission when established by Congress in 1879. It took the great Mississippi River floods of 1912, 1913, and 1916 to arouse national interest, culminating in the first flood-control act in 1917. This assigned to the U.S. Corps of Engineers responsibility for flood-control work on the Sacramento and Mississippi Rivers.

The federal government was authorized by the Flood Control Act of 1917 to pay up to two-thirds of the cost of levees, with the remaining one-third to be contributed by local interests. The latter were also to contribute rights-of-way and to maintain levees after completion. Results proved unsatisfactory because the degree of protection afforded was in proportion to the ability of local interests to raise the necessary funds. Weak links in the levee system ruined the effectiveness of the system as a whole. This was illustrated in the calamitous flood of 1927. Out of this disaster came the new Flood Control Act of 1928, the first really

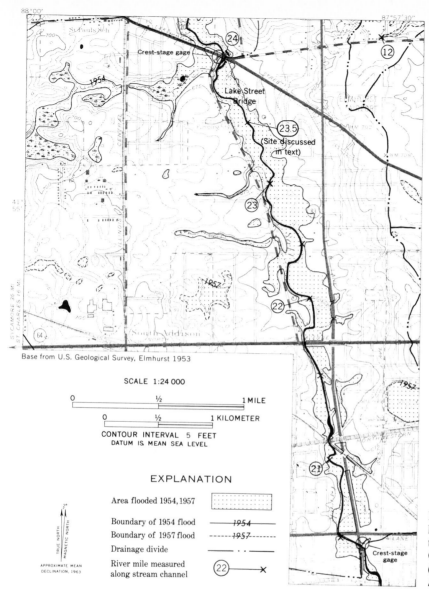

Base from U.S. Geological Survey, Elmhurst 1953

SCALE 1:24 000

CONTOUR INTERVAL 5 FEET
DATUM IS MEAN SEA LEVEL

EXPLANATION

Area flooded 1954, 1957	
Boundary of 1954 flood	—— *1954* ——
Boundary of 1957 flood	------ *1957* ------
Drainage divide	—— · · ——
River mile measured along stream channel	㉒ ——✕

FIGURE 3-17. **Flood-hazard map of part of the Elmhurst quadrangle, Chicago area, Illinois.** (*After U.S. Geological Survey.*)

comprehensive flood-control act. The implemented program has withstood the tests of two exceptional floods, in 1937 and 1945, but not the catastrophic flood of 1973. At the present time, the Mississippi Valley is reasonably well protected against the *average flood* by levees, bypasses, and diversion channels. Additional construction and improvements continue.

The Corps of Engineers has the responsibility for all aspects of flood control. Each flood-control project, except for certain small im-

provements and emergency work, must be authorized by Congress. Because of rapid urbanization of floodplains, the 1960 Flood Control Act authorized the Corps of Engineers to map areas susceptible to floods of various magnitudes and frequencies, to establish criteria for use of floodplains, to disseminate the information to all interested parties, and to provide engineering advice in local programs aimed at reducing flood hazards.

Prior to 1968, the only relief for victims of

flood damage had been special disaster loans. In 1968, Congress established the National Flood Insurance Program and broadened the program in 1969. This is a voluntary insurance program to provide limited financial relief to victims of flood disasters. Insurance rates were made reasonable through federal subsidy. To be eligible for the insurance, communities in flood-prone areas are expected to enact land-use and control measures for rational use of floodplains.

The 100-year flood has been adopted by the Federal Insurance Administration (FIA) as a standard for identification of flood-hazard areas and as a base flood elevation for local land-use control. Fourteen states and Puerto Rico now require the 100-year flood standard for floodplain management purposes, and 24 other states use the 100-year flood standard administratively.

The 100-year flood is not the greatest flood that an area may experience. In 1972, 50 percent of the 45 officially declared flood disasters in the United States were equal to or exceeded the 100-year standard. Communities, therefore, should be alerted to the maximum-potential event even though insurance rates based on such a flood would be prohibitive.

In 1973, a revised Flood Disaster Protection Act amended the earlier act as follows:

1. The limits of insurance coverages were doubled or tripled.
2. Identification of flood-prone areas and the dissemination of information regarding them is provided for.
3. State and local communities, as a condition of federal aid, are required to participate in the flood insurance program and to adopt adequate floodplain ordinances.
4. The purchase of flood insurance is required of all property owners who are being assisted by federal programs in the acquisition or improvement of land and facilities.

Flood insurance also covers damage from landslides directly related to flooding and damage from shoreline erosion due to exceptional flood-related storms.

To date, 15,000 communities have been notified by FIA that they are flood-prone and should apply for program coverage. Each has been provided with a map tentatively identifying hazardous areas. Acceptance in the program requires that the community, in its application, indicate an adoption of floodplain management regulations.

SOIL EROSION

General Considerations

The magnitude of the soil-erosion problem in the United States is dramatically indicated in Figure 3-18. Careless management has resulted in the washing away of agricultural soils which have taken decades or centuries to form, has scarred and laid waste scenic slopes, and has damaged roads, railroads, and other structures.

Most soil erosion results from uncontrolled runoff. In areas protected by vegetation, there is a partial balance between slow loss of soil by rain, rill wash, and creep and the creation of new soil by weathering of the rocks below. The loss is more rapid on steep slopes and in areas where the vegetation is short and sparse than on gentle slopes or in areas where the vegetation is heavy. Ordinarily, in humid regions, the mat of vegetation protects the loose soil below. If this protective shield is broken, or if slopes are steepened, the balance between erosion and rejuvenation of the soil is disrupted, runoff is increased, and erosion takes place.

Although soil erosion occurs from natural causes, humans have accelerated the process. The accelerated erosion, if unchecked, will destroy increasingly large areas of agricultural land, impair the use of land for settlement and urbanization, and contribute to sedimentation problems.

Sheet Erosion. Tiny trickles from accumulated rainfall can move grains of soil downslope a fraction of an inch or a few inches before the rill seeps into the ground. The total effect is *sheet erosion*, the imperceptible downslope transfer of the surface of the soil without conspicuous channeling (Figure 3-19). It is sometimes recognized by local thinning of the soil, by the appearance of bare spots, and by the accumulation of sediment downslope and in drainageways. Sheet erosion may take place on

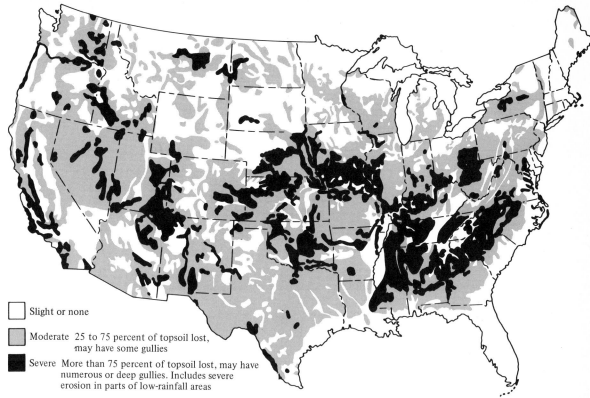

Slight or none

Moderate 25 to 75 percent of topsoil lost, may have some gullies

Severe More than 75 percent of topsoil lost, may have numerous or deep gullies. Includes severe erosion in parts of low-rainfall areas

FIGURE 3-18. **Soil erosion in the United States.** (*After U.S. Dept. Agriculture.*)

unprotected granular soils with surface slopes of only a few degrees, or it may require slopes exceeding 15 to 25° for firm soils.

Rill Erosion. *Rill erosion* is the removal of soil along small but visible channels. Because rills are easily obliterated in tilling, this type of erosion is commonly included under sheet erosion in soil-erosion studies.

Gully Erosion. *Gullies* are small valleys that are too deep to be obliterated in normal tillage. Although often spectacular, they generally affect smaller total areas of arable soil and do less regional damage than sheet and rill erosion. Locally, of course, the damage may be great. Gullies may start in the channels of natural drainageways or other depressions, in plow furrows, animal trails, vehicle ruts, and between crop rows trending up and down slopes.

Causes

Soil erosion has been going on as long as there has been soil, long before humans appeared. The causes may be natural or artificial.

As noted above, any factor that weakens or destroys the vegetative cover, or that over-steepens slopes, may accelerate runoff and erosion. Droughts, disease, insect infestations, fires induced by lightning or spontaneous combustion, or suffocation by volcanic dust or gases are natural factors that may destroy the effectiveness of the vegetative cover. Over-steepening of slopes may result naturally from erosion, subsidence, landslides, or uplifts due to faulting.

Deforestation for agriculture was the first cause of soil erosion in many regions. In the days of abundantly available land, many farmers locally exhausted the land and moved on. Uncontrolled deforestation by the lumbering industry also led to serious soil erosion. More

FIGURE 3-19. **Overgrazed (left) and properly managed (right) native pastures. Sheet erosion is an active process on the overgrazed pasture and, if overgrazing continues, wind erosion will become a factor. The field on the right has been rested for 1 year. (*U.S. Dept. Agriculture.*)**

judicious cutting and carefully planned reforestation are proving effective remedies.

Many modern *farming practices*, some unavoidable, contribute to soil erosion. During the preparation of the soil for crops, and for a considerable time after seeding or planting, the soil is bare and exposed to erosion. While some erosion at these times may be unavoidable, bad farming practices add to the damage. Plowing and harrowing up and down slope rather than laterally, provide relatively unobstructed channels for runoff and accelerated erosion. Overgrazing leads to soil erosion by destroying the grass cover.

Construction practices contribute to soil erosion and trigger landslides and mudflows. The oversteepening of slopes is a common cause. The excavation of steep road and railroad cuts and of steep cuts at the backs of building pads encourages gullying and slope failure.

A common cause of soil erosion is poor planning at development sites. Each year more than a million acres of agricultural land in the United States is converted to urban use, in addition to the forested land that is taken over. It is estimated that erosion of land being prepared for urbanization is 10 times greater than that of properly cultivated land, 200 times greater than that of pastureland, and 2000 times greater than that of forested land.

Much of the erosion at construction sites takes place during the construction period, but adjacent areas may be affected later because of accelerated runoff from paved areas or compacted soil. The magnitude of erosion at construction sites varies with the soil, the steepness of slopes, the intensity of rainfall, and the construction methods (see Figure 3-6).

Clogging of stream channels with sediment from soil erosion reduces their capacity to transmit water, making them more subject to floods. The sediment reduces storage capacity of reservoirs downstream and leads to rapid filling of small ponds and lakes. The amount of sediment contributed by urbanization is disproportionately high relative to that from the much larger acreage of agricultural land.

Preventive and Remedial Measures

Preventive Measures. The *prevention of soil erosion* involves control of surface runoff. This requires careful planning and neighborly cooperation. For example, gullies resulting from careless practices on one property may expand

and damage a neighbor's property upslope. Or, sediment from erosion upslope may inundate property below.

Factors to be considered in prevention of soil erosion are erodibility of the soil, slope declivity, amount and intensity of rainfall, and land use. Good management is especially important in drought-susceptible areas, where wind erosion may lower the surface of fields as much as 1 m (3 ft) in a single dry year and bury distant fields under the wind-transported sediment.

Preventive measures in agricultural areas include:

1. Plowing, planting, cultivating, and harvesting crops in rows that contour the slopes rather than go up and down slope (Figure 3-20). Each furrow traps runoff and sediment washed into it.

2. Minimum tillage. Rough plowing and simultaneous planting without smoothing or other treatment of the soil help where the rains come soon after planting, when the ground is bare.

3. Strip cropping (Figure 3-21). The alternation of dense erosion-resistant crops such as the small grains, clover, or alfalfa with open-spaced erosion-susceptible crops such as corn

FIGURE 3-20. **Contour furrowing retards runoff and erosion. (*U.S. Dept. Agriculture.*)**

is a low-cost method of erosion control. The erosion-resistant crops protect the ground immediately below from raindrop impact and runoff, retard the runoff from erosion-susceptible strips upslope, and trap the soil which washes down.

4. Crop rotation. Cultivated land is covered with soil-holding and soil-building crops when not producing cash crops. Rye grass, sweetclover, and winter barley are common covers. The cover crops may be used for pasturage, and legume crops can add nitrogen to the soil. Best soil-erosion control results if the crops are left on the land rather than harvested.

5. Terracing. Terraces check the speed of runoff and reduce soil loss.

6. Damming drainage lines or excavating basins for farm ponds to retard rapid runoff of storm waters.

7. Planting trees and shrubs. Much easily erodible land is better left in natural vegetation but, if already cleared and eroding, should be protected by judiciously planted trees and shrubs. Trees protect the soil and may produce profitable tree crops. In places, the black locust tree is used because it provides good fence posts.

8. Judicious management of pasture lands. A grass cover protects pasture land from erosion. Overgrazing should be prevented. Supplementary pastures may be used for grazing to permit permanent pastures to rehabilitate themselves. Cattle and sheep are the principal culprits in destroying ground cover, but hogs are locally more destructive because of their rooting habits.

Preventive measures against soil erosion at construction sites include:

1. Selecting sites where soils, topography, and drainage inhibit erosion. Thus, in routing roads or railroads, it may be less costly in the long run to deviate from the shortest route to avoid erosion-prone soils and slopes.

Bankers are increasingly using soil surveys in land appraisal for making loans. Land assessors use soil surveys in determining the tax base. Soil surveys have been completed for more than 2000 counties and miscellaneous areas in the United States. These surveys in-

FIGURE 3-21. **Contour strip cropping protects fields where clean-cultivated and close-growing crops are alternated. (*U.S. Dept. Agriculture.*)**

clude soil texture, soil depth, slope declivities, erodibility, permeability, degree of wetness, presence of impervious or permeable layers, and other data useful in construction.

2. Planning the land development for erosion control. Slopes subject to erosion should be covered with mulch or cover crops during construction. Terracing may be employed. Erosion of slopes may be inhibited by ditches at their crests which lead runoff away from the slopes. It may be necessary to line ditches and channels to prevent their erosion. Permanent vegetation and erosion control devices should be established as soon as possible.

3. Setting aside for open space and recreation those areas unsuitable for urban development.

4. Developing large tracts progressively so that sites are left denuded for as short a time as possible.

5. Disturbing the natural slopes as little as possible and preserving trees and other vegetation wherever possible.

6. Providing for safe offsite disposal of runoff and sediment.

Remedial Measures. The above measures prevent or decrease erosion of slopes. The stabilization and control of gullies, however, requires additional procedures.

The first step in gully control is to eliminate the causes of the gullying. The methods outlined above for control of runoff should be undertaken immediately. Because gullies expand headward along runoff lines, it is essential

first to divert the runoff from the head of the gully. This is done by excavating a shallow, arcuate diversion ditch around the gully head (Figure 3-22). On grazing land, it is also essential that the gully be fenced to prevent grazing animals from further destruction of the vegetation within the small area draining toward the gully. In response to these two simple measures, gullies may often be rapidly stabilized. Figure 3-23 shows a gully 6.5 to 8 m (20 to 25 ft) deep before and after being fenced in. It is stabilized and repairing itself by growth of new vegetation and entrapment of sediment.

FIGURE 3-22. **A diversion ditch above the head of this gully has stopped its growth. (*U.S. Dept. Agriculture.*)**

(a)

(b)

FIGURE 3-23 **A large gully** *(a)* **before and** *(b)* **after being fenced in and protected from grazing. (***U.S. Dept. Agriculture.***)**

Small check dams are commonly emplaced in gullies to retard erosion and trap sediment moving down the gully (Figure 3-24). To speed the healing process, new vegetation may be planted.

Policies on Soil Erosion and Sedimentation

The hub of the conservation program in urban development is the county, town, or other local governments in concert with the local conservation district. Conservation districts are "local public bodies responsible under state law to promote the conservation of soil and water and related resources."

The local government has as responsibilities establishing land development policy, community planning, zoning, approving site plans, issuing permits, and inspecting construction. An increasing number of local governments are establishing ordinances, codes, and regulations to provide criteria and guides for erosion and sediment control. Most communities, however, do not have codes, and some of the codes already established are so complicated or vague as to be difficult to adhere to.

The conservation districts and Soil Conservation Service are the technical arms. Districts represent the public interests and the Soil Conservation Service provides the technical information. The latter also makes soil surveys and administers small watershed programs.

As for the general problem of water, planning for utilization of surface and subsurface waters should be based on entire river or ground-water basins rather than segments thereof. Each basin, ideally, should have its own master plan into which all lesser plans would be fitted. However, it is unreasonable to expect all development to cease until a master plan is finalized. The realities of life lead to the piecemeal development of large basins even as consideration of master plans progresses. Those in positions of responsibility, however, must ensure that local plans are thoughtfully conceived and unlikely to prove harmful to other parts of the basin. Small watersheds, on the other hand, are the primary units of resource development, management, and conservation. Recognition of this is indicated by the large number of concerned watershed groups, associations, and legal organizations. Nearly every important national organization in industry, labor, agriculture, and conservation has endorsed or sponsored the local-state-federal–partnership approach to planning and implementation of programs within the framework of natural watersheds.

FIGURE 3-24. **Runoff from 18,200 ha (45,000 acres) had cut a gully** *(a)* **11 m (35 ft) deep in northwestern New Mexico. In 1937, an earth-check dam was built. By 1940, deposition of sediment had filled the dam-blocked gully** *(b)***. (*U.S. Dept. Agriculture.*)**

(a)

(b)

ADDITIONAL READINGS

Bailey, J. F., J. L. Patterson, and J. L. H. Paulhus: "Hurricane Agnes Rainfall and Floods, June–July 1972," U.S. Geological Survey, Professional Paper 924, 1975.

Chin, E. H., J. Skelton, and H. P. Guy: The 1973 Mississippi River Basin Flood: *U.S. Geological Survey and Nat. Oceanic and Atmospheric Ad-*

ministration, U.S. Geological Survey Professional Paper 937, 1975.

Davis, S. N., and R. J. M. DeWiest: "Hydrogeology," John Wiley & Sons, Inc., New York, 1966.

Leopold, L. B., K. S. Davis, and the Editors of *Life*: "Water," Time, Inc., New York, 1966.

Schwartz, F. K., L. A. Hughes, E. M. Hansen, M. S. Peterson, and D. B. Kelley: "The Black Hills—Rapid City Flood of June 9–10, 1972," *U.S. Geological Survey*, Professional Paper 877, 1975.

Thomas, H. A.: "The Conservation of Ground Water," McGraw-Hill Book Company, New York, 1951.

U.S. Dept. Agriculture: "What Is Soil Erosion?" Misc. Publ. No. 286, 1938.

———: "Water," The Yearbook of Agriculture, 1955.

———: "Soil," The Yearbook of Agriculture, 1957.

———: Controlling Erosion on Construction Sites, *Agriculture Information Bulletin 347*, 1970.

U.S. Dept. Housing and Urban Development: "National Flood Insurance Program," From Senate Rept. No. 93-583 (Comm. on Banking, Housing, and Urban Affairs), 1974.

U.S. Geological Survey: "Hydrology for Urban Land Planning—A Guidebook on the Hydrologic Effects of Urban Land Use," Circular 554, 1968.

———: "Water for the Cities—The Outlook," Circular 601-A, 1969.

———: "Urban Sprawl and Flooding in Southern California," Circular 601-B, 1970.

———: "Flood-Hazard Mapping in Metropolitan Chicago," Circular 601-C, 1970.

———: "Water as an Urban Resource and Nuisance," Circular 601-D, 1970.

———: "Sediment Problems in Urban Areas," Circular 601-E, 1970.

———: "Real-Estate Lakes," Circular 601-G, 1971.

———: "Extent and Development of Urban Flood Plains," Circular 601-J, 1971.

CHAPTER FOUR

coastal environments

INTRODUCTION

The shore is a relatively hostile environment, constantly changing. Among the important factors that influence its behavior are the state of the sea, configuration of the coast, terrestrial and submarine slopes, magnitude of inshore and offshore currents, local geology, and sediment brought into the sea by rivers. The interreaction of these factors imparts distinctive characteristics and behavioral patterns to the shore. The shore may remain essentially stable for decades, imparting a false sense of security to the inhabitants. Suddenly, however, a storm of hitherto unrecorded magnitude may devastate the coast, causing heavy loss of life and property. Such was the situation at Galveston, Texas, in 1900 and along the New England coast in 1938.

Human encroachment on the shore often compounds the natural dangers. Tampering with any part of a shore may have damaging effects not only in the immediate area but for considerable distances on either side. Even the damming of a valley inland may lead to serious coastal erosion by depriving the shore of its usual quota of protective sediment.

In this chapter, we consider the dominant coastal environments from the viewpoint of human occupation, the natural processes which affect these environments, the hazards they present, preventive and remedial measures, and planning recommendations. The long-term effects of slow worldwide changes in sea level are primarily matters of concern for future generations.

ENVIRONMENTS

Coastal environments may be divided into the beach zone, the perched zones on steep slopes above the sea, and estuaries and deltas. Each presents unique problems.

Beach Zone

The *beach zone*, or *shore*, is the zone of loose sediment extending from ordinary low-water level to the sea cliff, if present, or to the uppermost level reached by storm waves, surf, or swash (uprush) as indicated by permanent vegetation or other evidence of wave limit.

Beaches may form continuous aprons along straight, cliffed coasts. Along irregular steep coasts, the beaches may be restricted to shel-

tered coves between promontories as *pocket beaches* (Figure 4-1); along other irregular coasts they may also appear as sand spits and bars[1] (Figure 4-2). Along low plains coasts (Figure 4-3), beaches appear locally as fringes along the mainland shore but primarily along the seaward side of barrier islands.

Most beaches consist of sand. Because of turbulence shoreward of the breakers, smaller sedimentary particles are kept in suspension and eventually come to rest as mud in the deeper, quieter water offshore. These particles may be moved long distances by coastal currents before finally settling out.

Beaches at the foot of sea cliffs may include large rock fragments battered from the cliffs

---

[1]The term "bar" is restricted by some to permanently submerged ridges of sediment. We shall use the term in a general sense regardless of the duration of exposure above water. If entirely submarine, it shall be so stated.

FIGURE 4-1. **Pocket beaches. Indented cliffed coast. Santa Lucia Range, California. (*Arthur D. Howard.*)**

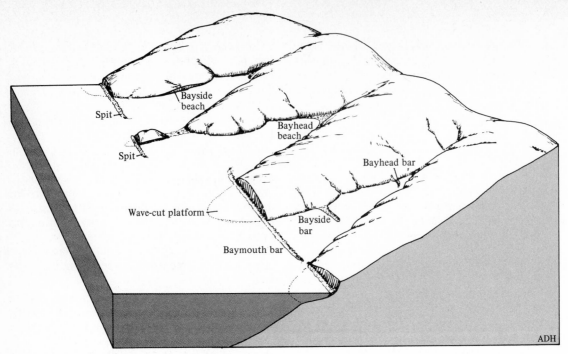

FIGURE 4-2. **Multiple beach sites: embayed coasts. The wave-cut platforms are the submerged wave-truncated ends of the headlands.**

FIGURE 4-3. **Beach sites: low plains coast. The seaward side of the barrier island is a continuous beach. The ragged, lagoon side of the barrier island represents sand washed over during exceptional storms. Although this fringe rapidly reverts to marsh, scattered beaches may be present. Marshes are more extensive on the shallow mainland side of the lagoon and beaches are rare. As the barrier island is driven landward, the lagoon becomes narrower and finally disappears.**

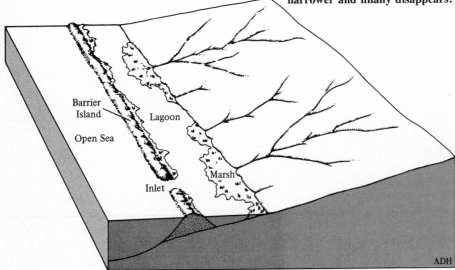

and worn smooth by wave attack. Beaches that consist almost entirely of pebbles or cobbles are *shingle beaches*. Other shores, in quiet waters, may consist of mud. These rapidly convert to marshes as grasses accumulate.

Figure 4-4 is a diagrammatic view of a *cliffed coastal zone*. This zone extends from seaward of the breakers to a point inland from the sea cliff where coastal features give way to a different type of terrain. Thus, the coastal zone includes both land and sea areas. The term "coast" is generally limited to the portion inland from the base of the sea cliff.

The *inshore zone* extends from seaward of the breakers to the low-water shoreline. Outside this is the *offshore zone*. The *beach* or *shore* extends from the low water shoreline to the sea cliff. It consists of the *foreshore*, or sloping front of the beach, and the *backshore* extending to the sea cliff. The backshore often displays one or more terracelike flats, or *berms*, resulting from storm-wave erosion (Figure 4-5).

Perched Zones

Perched zones (See Figures 2-15 and 2-17) are parts of the coastal zone above beach level and (1) at the crest or close to the edge of an actively eroding scarp, (2) on the scarp itself, or (3) in valleys or ravines dissecting the scarp. Scarps vary from moderately steep slopes to vertical cliffs.

Some coastal zones consist of a succession of terraces separated by scarps representing old sea cliffs. The foot of each scarp marks a former stand of the sea. Some scarps are solid rock; others consist of unconsolidated or semiconsolidated sediments; and still others are of rock mantled by unconsolidated or semiconsolidated sediments. The mantle sediments may be marine, that is, deposited originally below sea level; terrestrial, deposited above sea level; or a combination of these. Clearly, rocky shorelines are less susceptible to rapid and unexpected erosion than those composed of unconsolidated or semiconsolidated deposits.

Estuaries and Deltas

An *estuary* is that portion of the lower course of a river system that experiences tides. Estuarine sediments consist of the tiniest particles of sediment brought down into the estuary and shifted about by tidal currents until they settle to the bottom. Deposition of the sediments may take place far from their point of entry.

Estuaries with their associated marshes are drowned valleys. Tidal marshes, while common to estuaries, are not confined to them. They may develop in any sheltered area where silting occurs. Once the vegetation takes hold, it encourages further silting by acting as a sediment trap. Other marshes accumulate behind sand

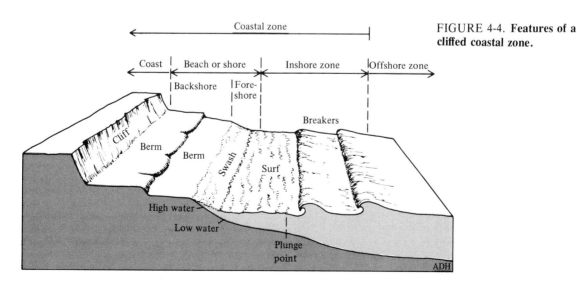

FIGURE 4-4. **Features of a cliffed coastal zone.**

FIGURE 4-5. Berm. Fire Island State Park, New York, 1937. (*Long Island State Park Commission.*)

spits or bars or at the heads of bays lacking large inflowing rivers.

Marsh grasses, principally *Spartina* (cord grass) and *Salicornia* (pickleweed), spread rapidly over the soft muds of estuaries and coastal flats. In tropical areas, including southern Florida and Hawaii, mangrove takes the place of marsh grass. Mangrove is primarily but not exclusively a saltwater plant, and its roots must be alternately submerged and exposed if it is to flourish.

Estuaries and tidal marshes are unique ecological preserves and are important stopovers for migrating birds. The San Francisco marshlands, for example, are a stopover on the flyway between Alaska and Central America. Reserves have been set aside on both sides of the bay, and permit access to otherwise inaccessible marshlands (Figure 4-6).

The life span of estuaries and bays is limited because of filling by sediment and growth of tidal vegetation. Many small bays created during the last major rise of the sea a few thousand years ago are already choked with sediment. Others are almost filled, with only tidal channels remaining. Large embayments such as San Francisco Bay (Figure 4-7) still have large expanses of open water, but as the marshes encroach, only a network of channels main-

FIGURE 4-6. Marshland reserve at Palo Alto, west side of San Francisco Bay, California. Boardwalks provide easy access to many parts of the tidal marshes. Tall plants—Spartina, or cord grass; lower plants—Salicornia, or pickleweed. (*Arthur D. Howard.*)

FIGURE 4-7. **Mud flats and tidal marshes in southeastern portion of San Francisco Bay, California. Salt pans in foreground. (*U.S. Geological Survey.*)**

tained by strong tidal currents or by human effort will remain. Legislation can delay the inevitable fate by restricting haphazard filling or hinterland activities that increase the volume of sediment carried into the bay.

The rate at which people are destroying many bays and estuaries is alarming. San Francisco Bay in 1850 had an area of 1760 sq km (680 sq mi). The area now is 1090 sq km (420 sq mi), or 40 percent less. About 670 sq km (260 sq mi) of marshes and shallow water have been "reclaimed" for agriculture and salt ponds. Hence, the San Francisco Bay Conservation and Development Commission was created to control private and public activities and to provide regional plans to ensure the best use of the bay shoreline for residential, recreational, and industrial purposes with sufficient area left for natural preserves. A "save our coast" proposition was passed overwhelmingly in California by popular vote on November 7, 1972.

Deltaic environments differ from estuaries in several ways. Whereas the bulk of the sedi-

ment of estuarine marshlands is very fine and settles from suspension in tidal currents, the bulk of the sediment of deltas is deposited directly at the point of entry of a river. Whether the delta consists largely of fine or coarse sediment depends on the size of the sedimentary particles being moved by the river. Another difference is that deltas are not restricted to embayments but may be built far out from mainland coasts. Deltas such as the Mississippi and the Nile extend seaward for scores of kilometers.

The growth of many deltas is rapid. During floods, the channels of delta streams are subject to change: old channels may be abandoned and new ones formed. Former river and seaports are now isolated from the sea, and new ports have taken their places. To prevent abandonment of important ports, permanent channels are maintained, often at considerable cost.

SHORE PROCESSES

The shore at any one instant represents a momentary stage in the struggle of land and sea. The sea effects change through storm waves, storm surge, and certain currents, while minor factors are the standing waves known as seiche, coastal winds, and sea or lake ice.

Waves

There are two principal types of waves: waves of oscillation and waves of translation (solitary waves). The *wave of oscillation* (Figure 4-8) is found outside the breaker zone. The water particles follow a more or less circular orbit with little or no permanent change in position. Only the wave form advances, like the waves moving across a field of wheat.

Waves of oscillation are termed *forced waves* if formed by local winds, and *swell* if formed by winds in some distant storm area.

Waves of translation have only a crest without a complementary trough. The wave that is pushed ahead of a ship over the calmer water in front is an example. The wave particles are displaced forward without a compensating backward movement. Surf, formed when waves of oscillation reach the plunge point and break (see Figure 4-4), is a wave of translation. Objects are moved directly ahead, not in an orbital motion. Other waves of translation are formed from waves of oscillation in which a sudden shallowing, due to a near-shore reef, disrupts the oscillatory motion permitting only the upper part of the wave to progress shoreward.

The orbit of the wave of oscillation becomes smaller with depth because of internal friction and is negligible below a depth equal to one-half the wavelength. Thus, most of the energy of a wave with a length of 30 m (100 ft) is confined to a depth of about 15 m (50 ft). On approaching the shore, the orbital energy is restricted to a smaller and smaller depth of water, resulting in increased orbital velocity and height of waves. At the same time, the shallowing bottom retards the advancing waves so that those farther from shore crowd on the heels of those ahead (Figure 4-9). Waves generally break when the depth of water below the trough is a little greater than half the height of the wave.

The breaking of a wave, with the resulting turbulence of the surf zone and the churning of the bottom, dissipates the wave energy. The surf rushes forward and sweeps up the beach as *swash*, which in turn recedes as *backwash*.

Short, steep waves generated by storms near the coast tend to tear a beach down by removing material from just above still-water level and carrying it into deeper water. Long, low waves, such as swell from distant storms, tend to rebuild beaches by bringing material in from shallower water and depositing it just above the still-water level. Closely spaced violent storms result in severe erosion because of insufficient time for restoration of the beach in quieter periods between storms. Because storms are most frequent in winter, beaches tend to be destroyed in winter and rebuilt in summer.

On an irregular coast, headlands are subject

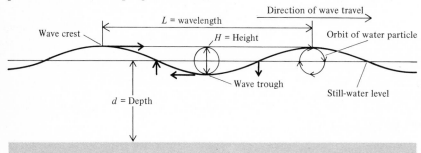

FIGURE 4-8. **Waves of oscillation. Heavy arrows indicate direction of movement of water particles in different parts of the wave. (*Modified from U.S. Army Corps of Engineers.*)**

FIGURE 4-9. **Crowding of waves approaching shore. Note also the progressive narrowing and sharpening of wave crests. These characteristics are accompanied by an increase in height. Porthkerry Beach, near Barry, South Wales, British Isles.** (*Courtesy W. W. Williams.*)

to more concentrated wave attack than intervening embayments because of wave refraction (Figure 4-10). Waves slow down as they enter shallow water, and because they first encounter shallow water off the headlands, they slow down there first. The remainder of the wave, running more freely on either side, thus envelops the headland. As the wave breaks, first at the headland and progressing around the sides, the wave energy is concentrated inward on the headland. In the intervening embayment, the wave form is deployed radially outward, resulting in dissipation of energy and weaker wave attack. Erosion is also more rapid at the headlands because the waves break nearer to shore owing to a steeper bottom profile than in the embayments.

Many cliffed, rocky shorelines are highly irregular, with numerous small promontories separated by coves. These small irregularities indicate local weaknesses in the rock. The weaknesses may be due to intense fracturing or

to lateral changes in composition and structure of the rocks. In these situations, the coves rather than the promontories are eroded more rapidly.

Along many rocky cliffed coasts, selective wave attack at the base of a cliff may excavate caves, some of which extend scores of meters into the cliff. Enlargement of a cave may so thin the roof that it collapses.

Storm Surge

The pressure of the wind may raise or lower sea level abnormally. A steady onshore wind raises sea level along the coast, whereas an offshore wind lowers it. The wind-induced rise of the sea against the land is *storm surge*.

Storm surges may reach considerable heights. The Galveston hurricane of 1900 raised

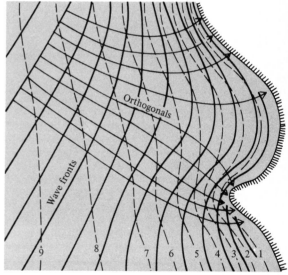

FIGURE 4-10. **Wave refraction. The dashed lines are bottom contours at intervals of 1 fathom (6 ft, approximately 2 m). The incoming waves (heavy lines) reach shallow water first off the headlands and are retarded, but still run freely ahead in the bays. The wave fronts thus tend to assume the configuration of the shore. The lines normal to the advancing wave fronts indicate the paths along which energy is transmitted to the shore. Note that because of refraction, wave energy is concentrated on the headlands and dispersed in the bays.** (*Arthur D. Howard.*)

water levels 5 m (15 ft) above mean sea level. Similar levels were reached by the hurricane which devastated New England in September 1938. Such storm surges provide elevated platforms from which the storm waves can attack areas normally out of reach. Storm surges are greater in shallow water and in funnel-shaped bays.

The effects of wave battering from the storm surge of the New England hurricane of September 1938 are illustrated in Figure 4-11.

Photo (a) is representative of the destruction of large areas formerly thought to be safe because of elevation and/or dune protection. Photo (b) shows one of the nine new inlets broken through the barrier beach. The resulting in-

FIGURE 4-11. **Effects of hurricane of 1938, south shore Long Island, N.Y.** (a) Destruction in areas formerly thought to be safe because of either elevation or dune protection. (*Photographer unknown.*) (b) One of nine new inlets in the barrier beach. (*Arthur D. Howard.*)

(a)

(b)

creased tides in the bay behind the barrier beach submerged property on the mainland shore. Some inlets were closed naturally by sediment drifted along the shore. Others required weeks of sustained effort to close artificially.

Seiche

Water levels are influenced not only by the tides but by the weather and earthquakes. Changes in atmospheric pressure, particularly over smaller bodies of water such as bays and lakes, may elevate or depress the water surface, initiating vertically oscillating standing waves, or *seiche* (pronounced say-sh). Or, persistent winds may elevate the surface at one end of a lake and lower it at the other. When the wind dies down, the water surface will slowly teeter-totter until stability is regained. The seiche period is determined by the strength and duration of the wind; the length, breadth, and depth of the water body; the configuration of the basin; and its orientation relative to strong wind directions. In the Great Lakes, seiche periods range from a few hours to a few days. Because of the characteristics of Lake Erie, only a few hours of strong winds are needed to raise the water level as much as 2.6 m (8.4 ft). Especially violent windstorms may create even greater seiches. Thus, on June 26, 1954, a violent windstorm raised the level of Lake Michigan 3 m (10 ft) in the Chicago area, with a loss of seven lives.

Seismic Sea Waves and Physically Similar Types

An extreme hazard results from earthquake-associated sea waves known as *seismic sea waves*, or *tsunamis* (tsoo-nah´-mees). Although these are sometimes called tidal waves, they have nothing to do with the tides. They result from faulting of the sea bottom or adjacent coast. Because seismic sea waves of large magnitude occur at relatively rare intervals and each may be destructive in only a small area, they have only a minor role in shaping the coasts of the world. Where such waves reach a coast in force, however, destruction is devastating.

Seismic sea waves originate as waves of translation advancing over undisturbed water in front. They are generally only a fraction of a meter in height, but may reach lengths of almost 200 km (120 mi) from crest to crest. Velocities up to 800 km (500 mi) per hour have been recorded. When these waves enter shallow water, they rise to incredible heights and produce gigantic surf (Figure 4-12). Observers cite seismic sea waves exceeding 25 m (80 ft) in height on low-lying coasts and rushing inland almost 2 km (more than 1 mi). The Hawaiian Islands are especially susceptible to tsunamis originating in Alaska and the Aleutian Islands. Depending on whether the original sea bottom was displaced upward or downward, the coast may experience an initial transgression or withdrawal of the sea.

Submarine volcanic explosions create similar catastrophic waves. The eruption of Krakatoa in the East Indies in 1883 created waves at least one of which was 21 m (70 ft) high and carried a gunboat 3.2 km (2 mi) inland, stranding it 9 m (30 ft) above sea level.

Other catastrophic waves are created by landslides into water bodies. If the water bodies are deep and narrow, the resulting waves may be enormously destructive (Figure 4-13).

Currents and Drift

Longshore (Littoral) Currents. The great ocean currents have little direct effect on coasts. However, *longshore (littoral) currents*, those following the shore, and to a lesser extent tidal currents, are important environmental factors. Littoral currents are caused by obliquely approaching waves and may transport enormous quantities of sediment along the shore (Figure 4-14). This littoral transport takes place on the beach, in the surf zone, and in a zone beyond the breakers down to depths of 30 m (100 ft). The greater part of the littoral drift is within a depth of 2 m (6 ft) and confined to a zone about 180 m (600 ft) wide. Thus, it has an important bearing on the design and dimensions of shore structures extending into the sea.

Beach drift is the movement of sediment along the beach (Figure 4-14). When an obliquely approaching wave breaks, the surf is thrust obliquely toward the shore, and the swash rushes obliquely up the beach. The

FIGURE 4-12. One of the early waves of the tsunami of April 1, 1946, breaking over Pier 1 at Hilo, Hawaii. The man in the figure was swept to his death. (*Press Assoc. Photo from F. P. Shepard et al., Bull. Scripps Inst. Oceanography, vol. 5, pl. 23, 1950.*)

swash returns down the beach slope in response to gravity, as backwash. The momentum of the swash and backwash impacts a parabolic zigzag path to the particles. The exact configuration depends on the size of the beach particles and the strength of the swash and backwash. These, in turn, depend largely on the longshore component of wave energy and the steepness of the beach.

Any natural or artificial obstruction to drift causes deposition on one side (updrift side) and erosion on the other side (downdrift or lee side). Prior to the obstruction, the ravages of erosion on the downdrift side were healed by the sand moving in along the beach.

Longshore drift occurs also in the *surf zone*, where bottom sediment and material placed in suspension by the breaking waves are moved laterally parallel to the shore. The zigzag paths are relatively sharp and compressed.

Outside the breaker line, bottom material (bed load) moves in zigzag fashion with the oscillatory movements of the oblique waves. It moves forward with the crest, backward with the trough.

In brief, *littoral*, or *longshore, transport* includes beach drift, movement in the surf zone, and movement beyond the breaker line.

Detection of Longshore Sand Transport. The longshore movement of sand may be determined by several methods. Distinctively colored particles such as crushed brick or glass may be spread on the beach and their migration observed. In another procedure, radioactive grains are added to the beach, or sand grains are coated with a radioactive substance, and the migration is detected with a radiation counter. The most common procedure involves dying the beach sand and observing its movement by color. Other dyes are fluorescent, and the sand movement may be detected with an ultraviolet lamp.

The sediment moved along some shores exceeds 1.5 million cu m (2 million cu yd) per year. Even in the Great Lakes, the amounts at some localities exceed 110,000 cu m (150,000 cu yd) annually. The rates tend to remain constant unless interfered with by humans.

Tidal Currents. The tides, caused by the gravitational attraction of the moon and to a lesser

FIGURE 4-13. **Effects of landslide, Lituya Bay, Gulf of Alaska. Lituya Bay occupies the site of a former valley glacier whose limits are marked by the encircling ridge of glacial debris, R. On July 9, 1958, a rockslide from the slope at S, created a giant wave which swept down the bay, trimming away the forest in the light areas. At T, the uprushing waters reached an altitude of 525 m (1720 ft). A fishing boat in the bay at B was carried over the sandspit at an estimated height of 24 m (80 ft) and wrecked outside.** (*Photo by D. J. Miller, U. S. Geological Survey.*)

extent the sun, are regular and predictable at any locality. They present no serious unexpected hazards.

The range of tide, that is, the vertical distance between high and low water, varies from place to place. The range is greatest when the sun, moon, and earth are in line, with the moon either between the sun and earth (new moon) or on the far side of the earth from the sun (full moon). Because the moon revolves around the earth in about 4 weeks, it is aligned with the other two bodies twice a month. These high-range tides, with exceptionally high tides and exceptionally low tides, are *spring*, or *flood*, *tides*.

When the moon is at right angles to a line connecting sun and earth, tidal ranges are low. These occur at half-moon stages, and the low-range tides are *neap tides*. Thus, there are two spring and two neap tides each month.

Tides are also influenced by latitude, distribution and configuration of the continents, and weather. Tides are especially high where the tidal bulge enters funnel-shaped bays. Thus, the tide in the funnel-shaped Bay of Fundy in Nova Scotia often exceeds 12 m (40 ft). An especially high tide may enter a narrow embayment or river mouth as a wall of water known as a *bore*. It may be quite destructive.

Tidal currents result either from the movement of large volumes of water in and out of funnel-shaped tidal estuaries or where the tide is constricted in narrow straits or in inlets through barrier beaches. Velocities exceeding 19 km (12 mi) per hour have been recorded. Erosion and transportation by such currents extend to depths of 100 m (330 ft) or more in coastal waters.

Rip Currents. During storms, erosion of the beach and removal of the material seaward may result in construction of a series of submarine bars. Beaches with slopes of less than 1 in 50 may have two or more bars. During the calmer summer months, the sediment of the bars is driven shoreward to enlarge the beach. The

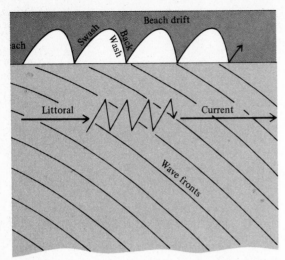

FIGURE 4-14. The littoral current running parallel to the beach is set up when waves move toward the beach at an oblique angle. Suspended sediment and bottom particles move in zigzag fashion with the littoral current. On the beach face, particles are driven obliquely upslope with the swash (uprush) and return seaward in the backwash. The rounded asymmetric path results.

higher part of the beach at Carmel, California, averages 100 m (300 ft) wider in September, after the quiet season, than in April, after the stormy season.

When high waves in rapid succession raise water levels inside a bar, the water rushes back with enough vigor to breach the bar at low places, eroding *rip channels*. Excess water shoreward of the bar moves laterally until it can escape seaward at these channels, producing a *rip current*. Velocities of these narrow, shallow currents may exceed 6 km (4 mi) per hour. Swimmers trapped in these currents are advised not to oppose them, but to escape laterally. Because rip channels slowly migrate, the currents likewise shift position. Rip currents are not significant factors as far as coastal construction is concerned.

Wind Deposition

Because most beaches consist of sand, coastal winds commonly heap the sand into dunes. The beach is nature's outer defense line against destructive attack by the sea; sand dunes are the next line of defense. The destruction of sand dunes for real estate developments, for commercial sources of sand, for unobstructed vistas of the sea, or for easy access to the beach often leads to dire results.

HAZARDS: PREVENTION AND REMEDIES

Beach Zone

Causes of Damage. *Natural Causes.* Even though seemingly fixed in position over the long term, a beach is subject to repeated erosion and rebuilding (Figure 4-15). Broad beaches have been obliterated in a single exceptional storm and restored in the succeeding quiet interval. The shoreline of the south coast of Long Island, New York, has moved back and forth 60 to 90 m (200 to 300 ft). Most shorelines, whether on barrier beaches or along cliffed coasts, are in a state of long-term retreat, perhaps due to slow rise in sea level. Some investigators suggest a rate of rise of about 1 mm (0.04 in.) per year. If true, human coastal defenses, viewed over the long term, are futile attempts to impede the natural flow of events.

Beaches, even over the short term, are continually changing shape and dimensions and sometimes disappear completely. If beaches were permanent, their presence at the foot of a sea cliff would forever protect the cliff from erosion. Yet, beach or no beach, sea cliffs continue to be eroded back, in some places at alarming rates.

Prediction of beach changes is difficult because of the variety of forces involved and their often unpredictable changes in relative importance. Changes from oscillatory to translational waves, changes in wave height, length, and direction, the state of the tide, the changing frequency of storms, changes in offshore currents, variations in amount of sediment introduced by rivers, and changes in bottom configuration all combine to make the prediction of beach behavior hazardous. Consider, for example, the possible effects of a change in bottom configuration caused by deposition of sediment. The wave-refraction pattern may be changed sufficiently for waves approaching from a certain direction to focus their energies at one

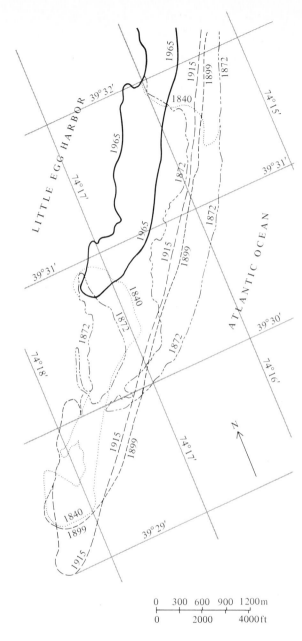

0   300 600 900 1200m
├────┼────┼────┼────┤
0   2000   4000ft

FIGURE 4-15. **Shoreline changes along a relatively uninhabited and unprotected stretch of the New Jersey coast. The few surveyed shoreline positions give only an inkling of the endless changes to which such coasts are subject. Part of the Holgate Wildlife Preserve at the south end of Long Beach Island.** (*Condensed from part of chart prepared by Philadelphia District, Corps of Engineers.*)

particular spot along the shore. This may cause rare and localized damage to a beach that appears to have been stable for a long period of time.

We cannot permanently prevent coastal change. It is impractical for public funds to be spent to restore or protect almost uninhabited shorelines. If a beach is hazardous, individuals and developers should be discouraged from building, or legislation should make it clear that the risks and costs are theirs.

In August 1972, a new inlet was eroded through Bogue Banks, just to the east of Bogue Inlet, North Carolina. The cause of the break is still in dispute, whether natural or due to dredging and land development. Erosion at the new inlet has washed away the end of a road built by developers, and the eastward migration of the inlet is threatening the small community of Emerald Isle. Property owners and developers are seeking public funds for stabilization of the new inlet. Considering possible costs against the sparse population and the probability of similar events in the future, neither the North Carolina Department of Natural and Economic Resources nor the U.S. Corps of Engineers is optimistic over the funding of remedial measures. Historic records are replete with examples of coastal change in the area, and it is unfortunate that the hazardous nature of these shores was not publicized before property owners became committed. However, it is questionable whether public funds should be expended for costly stabilization projects to protect the relatively few homes in the area. It might be more logical and financially feasible to buy out the inhabitants and convert the area to a national seashore or a state park.

The harsh measures discussed above cannot be applied to communities that have grown to considerable size and contribute considerable taxes. Communities such as Atlantic City, New Jersey, and Galveston, Texas, must receive consideration in times of disaster or threatened disaster. However, the practice of haphazard development followed by demands for relief requiring expenditure of tax money is objected to by many. Too often, thoughtless modification of the shoreline during development may provide the seed for later difficulties, not only at the development site but up and down the coast. Thus, the leveling or breaching of dunes, the dredging of channels, the widening of beaches, the building up of shoal areas, the opening of new inlets in bars and barrier islands, and the ill-conceived construction of protective devices may eventually prove detrimental.

One difficulty in protecting against natural hazards in many areas is the lack of long-term records of the interactions of sea and shore. Hurricanes, for example, have recently reached some coasts previously thought to be outside their range. Rare storms of lesser intensity, but coinciding with an especially high tide, may also render short-term records inadequate.

Even if the occurrence of rare storms were known, fiscal restraints might make protection impractical. Suppose, for example, that at rare intervals hurricanes along a shore raise sea level 5 m (15 ft) by storm surge. Even neglecting the stage of the tide, the construction of a sea wall large enough to protect against such possible hurricane damage of a sparsely settled shore should have low priority. Impossibly high costs would accompany any attempt to protect not only against rare hurricanes but also against the possible rare combination of hurricanes and seismic sea waves. Fiscal constraints may restrict safety measures to warning systems. Even the costs of repairing damage to structures in sparsely settled areas from the rare unusually severe storm would presumably be less than the costs of building and maintaining long sea walls of exceptional height.

Possible loss of life in the absence of maximum shore protection must be considered as well as property damage. However, it may be unreasonable to spend hundreds of millions of dollars to ensure the safety of a few who might be unaware of or disregard evacuation warnings when the money might be spent for medical research, highway safety, or other projects affecting a far greater number of people.

It sometimes happens that great swells from distant storms focus on a stretch of coast and create serious hazards. A classic example is the Atlantic coast of Morocco. Here, the coast is periodically subject to great swells that severely damage vessels at anchor or alongside wharves. In January 1913, the port of Casablanca was paralyzed by damage and destruction of vessels by a series of these large swells. Subsequent studies revealed that when an atmospheric low-pressure system passed between Iceland and Ireland, 2600 km (1600 mi) away, the giant swells would appear on the Moroccan coast 2 to 4 days after passage of the depression. The larger and deeper the low-pressure system, the larger the swells that reach the Moroccan coast, some of them reaching heights of 5 m (15 ft). Daily analysis of northeastern Atlantic weather at a central station in Rabat, Morocco, now permits swell predictions that are 80 percent accurate for the following day at Moroccan ports.

Other natural causes of shoreline alteration are changes in the level of the coast due to faulting, coastal and submarine landsliding, deposition of organic growths, or variations in supply of sand from streams. The effects of faulting were demonstrated during the Alaska earthquake of March 27, 1964, when some parts of the coast were elevated and other parts depressed.

Landslides are common along cliffed coasts. If the landslide deposits interfere with movement of sediment along the beach, the sediment may accumulate, resulting in an expanding beach. On the other side of the obstruction, the beach may undergo erosion because of sediment loss. Submarine landsliding, by changing the bottom configuration, affects the size of waves reaching the shore and modifies the refraction pattern and the pattern of longshore currents. If submarine sliding provides deeper water immediately offshore, larger waves are able to reach the shore, and erosion is accelerated.

Similar modifications of shore processes result when the bottom is changed by organic growths such as coral reefs or by accumulation of sediment. Variations in the amount of sediment brought to the sea by streams may lead to either expansion (*prograding*) of the beach or contraction (*retrograding*). Many beaches shrink in times of drought when sediment-bearing streams dry up.

The processes mentioned above—wave and current activity, faulting, landsliding, and the accumulation of organic or sedimentary materials—produce effects on the beach that are normal and expectable in shore environments. The rare hurricanes, seismic sea waves, and unusual swell may locally be regarded as abnormal.

Artificially Induced Causes. Too often, developers are unaware of possible consequences of coastal projects and may instigate rapid and disastrous shoreline changes. Some years ago, a real estate development was opened on the west side of one of the fingerlike bays that indent the north shore of Long Island, New York. A strong selling point was a beautiful sandy beach. Across the bay was a bayside bar projecting fingerlike toward the development beach. The bar was submerged at high tide but exposed at low tide, leaving a channel several hundred feet wide near the development beach. To convert it into a public beach, the county had the bar built up above water level at all stages of the tide. This concentrated all flood and ebb waters in the narrow channel next to the development beach, causing deepening of the channel. The accelerated channel erosion steepened the beach profile and imparted additional vigor to the backwash and undertow. In a short time, all sand and finer materials were winnowed from the beach, leaving a pebbly concentrate. Legal settlement involved restoration of the bayside bar to its original condition.

Preventive and Remedial Measures. Almost any human activity at or near the shore can prove harmful somewhere else. To protect a beach from erosion, or to restore or expand it, a number of different types of structures may be built. They include (1) revetments, sea walls, and bulkheads, (2) groins, and (3) jetties and breakwaters.

Revetments, Sea Walls, and Bulkheads. Revetments, sea walls, and bulkheads are parallel to the shore at the line of demarcation of land and sea. The primary purpose of the revetment and sea wall is to protect the land against wave attack and incidentally to serve as retaining walls. The bulkhead is primarily a retaining wall against landsliding or creep and secondarily a protection against wave attack in storms.

    A *revetment* is a sloping apron of poured concrete, blocks of concrete, or rock fragments at the foot of a coastal bluff (Figure 4-16). The design of the revetment, including materials, size of blocks, whether blocks are to be fitted together or dumped haphazardly, slope of the revetment, its width normal to the beach, and

(a)

(b)

FIGURE 4-16. Revetments. *(a)* Interlocking concrete-block revetment. Jupiter Island, Florida. *(U.S. Army Corps of Engineers, Coastal Engineering Research Center, Tech. Rept. no. 4, 3d ed., June 1966.) (b)* West Cliff Drive in Santa Cruz, California, along with many dwellings, would be part of the Pacific Ocean floor were it not for the shore protection. *(U.S. Army Corps of Engineers, San Francisco District.)*

the foundation, is dictated by local shore conditions.

Sea walls (Figure 4-17) are massive structures designed to withstand the largest storm waves. Vertical sea walls do not protect the beach at their bases because of the downward forces exerted by the striking waves. Thus, a revetment is commonly provided at the foot of the wall. At Galveston, Texas, and San Francisco, California, the sea walls are curved to form sloping aprons in their lower portions. The base of the structure at Galveston is armored with a revetment of huge rocks with their flat sides uppermost. At San Francisco, the basal curve of the concrete sea wall is continued outward as a stepped concrete apron. Vertical sheet piling under the outer parts of both structures extends below low-water level to prevent undermining. The height of the Galveston wall above the beach is 5 m (16 ft); that at San Francisco, more than 6 m (20 ft).

Bulkheads (Figure 4-18), designed to prevent sliding of the land or fill behind them, may consist of steel or timber sheet piling or of concrete. If waves are expected to strike against the bulkhead, causing scour at the base, a basal revetment will be necessary.

The type of structure depends on the magnitude of the forces, the foundation conditions, the availability of materials, and costs. Availability of materials and cost are related. For

(a)

FIGURE 4-17. **Sea walls.** *(a)* **Galveston, Texas.** *(b)* **San Francisco, California.** (**U.S. Army, Corps of Engineers, Coastal Engineering Research Center, Tech. Rept. no. 4, 3d ed., June 1966.**)

(b)

FIGURE 4-18. **Timber sheet pile bulkhead, Avalon, New Jersey, 1962.** (*U.S. Army Corps of Engineers, Coastal Engineering Research Center, Tech. Rept. no. 4, 3d ed., June 1966.*)

example, hard rock suitable for *riprap* (rock fragments used for protective armor) is abundant and readily accessible along much of the West Coast of the United States but rare along much of the East Coast south of New York. Transportation costs for riprap along much of the East Coast, therefore, may be excessive.

Groins. Groins (Figure 4-19) are protective structures built out more or less perpendicular to the shoreline to retard erosion by impeding oblique waves or to trap sediment for restoration or expansion of beaches. Groins are narrow, and the length varies from less than 30 m (100 ft) to several hundred meters. Some are extended out to a depth of 2 m (6 ft) below low water on the grounds that 80 percent of the nearshore sediment drift is within this depth range. Groins may be permeable, permitting passage of part of the littoral drift, or impermeable, blocking all drift. In the latter case, drift will resume when the beach has been built high enough to overtop the groins.

Groins may be constructed of timber, steel, concrete, rubble, or less commonly, asphalt.

FIGURE 4-19. **Groin system, Willoughby Spit, Virginia.** (*U.S. Army Corps of Engineers, Coastal Engineering Research Center, Tech. Rept. no. 4, 3d ed., June 1966.*)

Groins of brush contained by chicken wire have proved effective in temporary restoration of severely eroded beaches (Figure 4-20).

Indiscriminate groin construction robs downdrift areas of the sediment needed for their preservation. The top profiles of groins should be no higher than those of beaches of reasonable dimensions so that drift can resume once these dimensions have been reached. Permeable groins are preferred so as not to deprive downdrift areas of all sustenance during repair of updrift areas.

FIGURE 4-20. **Chicken wire and brush groins. Same beach at three stages of restoration. (*Works Progress Administration, N.Y. State.*)**

Jetties and Breakwaters. Jetties resemble groins but are larger and longer. They are generally constructed at right angles to the shore to confine a stream or tidal current to a channel and thereby prevent or reduce deposition of sediment and shoaling (Figure 4-21). Jetties also protect channel entrances from storm waves and shifting currents and help to stabilize inlets through barrier beaches. They are thus primarily aids to navigation.

Note in Figure 4-21 that jetties, like groins, retard longshore drift, causing deposition of sediment on the upcurrent side. To be effective aids to navigation, they must block the longshore drift out to a depth equal to the depth of the channel. Like groins, they deprive downdrift areas of sediment, causing beach erosion. To compensate, sand must be (1) imported from other coastal or inland sources, (2) dredged from offshore, with the probability that this interference with natural bottom conditions may lead to other problems, or (3) artificially piped across the inlet or harbor entrance.

Breakwaters are structures to break the force of waves or currents, mainly for protection of navigation. They may be connected to the shore (Figure 4-22) or lie offshore. Like jetties, they are generally constructed of rock rubble, of huge, four-legged concrete forms of massive concrete, of steel or timber piling, or of combinations of these. The rubble type is adaptable to any depth of water or foundation and is the most common type in the United States.

Offshore breakwaters may be designed to protect the shore in its lee from erosion, to provide an area of calm water for anchorage, or both. A common undesirable effect of the sheltering of the shore from impinging oblique waves is the weakening of the longshore currents. This causes the drifting sediment in the sheltered area to accumulate, the shore to expand seaward, and the beaches downdrift to become impoverished and erode.

Importation of Sand (Nourishment). Many beaches are being expanded by *importation of sand (nourishment).* This can create other problems. A shoreline may be adjusted to the size of the sand grains available to it. If a smaller grain size is imported, it may be rejected and win-

FIGURE 4-21. **Santa Cruz, California, with Santa Cruz Harbor in the foreground, San Lorenzo River flood-control project in the background, and beach-protection work along the shoreline. The jetties at the mouth of the harbor confine the tidal currents and increase their effectiveness in preventing shoaling.** (*U.S. Army Corps of Engineers, San Francisco District.*)

FIGURE 4-22. **Breakwater and mooring facilities. Redondo Beach, California.** (*U. S. Army Corps of Engineers, San Francisco District.*)

nowed out to deeper water. If the sand is too coarse, its rate of migration along the beach may be impeded, or it may form an armor over the sand beneath, impeding longshore drift. Imported sand should be as nearly identical to the natural beach sand as possible.

An elaborate importation scheme was involved in the expansion of Copacabana Beach in Rio de Janeiro. The source of the sand was the bottom of Botofogo Bay, 2.5 km (1.5 mi) north of Copacabana Beach on the far side of a high rocky peninsula. The sand was transported to Copacabana by barge, from which it was dumped offshore, and by overland pipeline, from which it was splayed out over the beach. The artificially expanded beach, however, was severely eroded by storm waves during the operations. Groins and/or periodic importation of sand may be necessary to maintain the new beach configuration. What the future holds for the beach at Botofogo as a result of the off-shore dredging remains to be seen.

It is not necessary that imported sand be distributed uniformly over a beach. It may be dumped at the updrift end and left for distribution by beach drifting (Figure 4-23).

By-passing (Figure 4-24) involves dredging sand from the updrift side of a channel and piping it across the opening to the downdrift side. This practice is more expensive than maintaining the channel by dredging, but it contributes directly to preservation of beaches downdrift.

On the California coast, much sand drifts into the heads of submarine canyons and is lost to deep water. Facilities have been established to trap the sand on one side and transport it to the other side of the canyon, where it can continue to nourish beaches during its long-shore drift.

Sand Dunes. Sand dunes serve as an outer line of defense against the sea. In some areas, shorelines are robbed of this line of defense by commercial exploitation of sand, by leveling for real estate developments, or by breaching for access to the beach or unobstructed ocean vistas. In other areas, artificial dunes are created to provide this defense.

Shore dunes do not offer permanent defense against the sea. During the hurricane of

FIGURE 4-23. **Sand dumped on a beach is distributed by wave action at high water. Shark River Inlet, New Jersey. (*U.S. Army Corps of Engineers, Coastal Engineering Research Center, Tech. Rept. no. 4, 3d ed., June 1966.*)**

FIGURE 4-24. **Sand-bypassing installation at the entrance to Palm Beach harbor, Florida. The pumping plant transfers sand brought from the right by littoral drift across the harbor entrance to the left. (*U.S. Army Corps of Engineers, Jacksonville District.*)**

September 1938, protective sand dunes were washed away from long stretches of the barrier islands off the south shore of Long Island, New York, exposing the homes behind the dune line to destruction. In other places, the sea failed to breach the dune line, and many structures were spared. It is essential that natural dunes be maintained and stabilized and that the creation of artificial dunes be considered where natural dunes do not exist.

Artificial stimulation of dune growth requires that obstructions be placed across the path of the wind. Common obstructions include sand fences, brush barriers, and vegetation. In contrast to slow stimulation of dune growth, large artificial dunes may be built rapidly by mechanical or hydraulic means.

The type of fence most commonly used is the cheap, prefashioned, slat-type snow fence, although brush fences are also employed (Figure 4-25). Four-foot-high snow fences are best used on flat beaches and will result in dunes about 4 ft (1.2 m) high, often in less than a year.

Dune growth may also be stimulated by planting beach grasses that are rapid growers

FIGURE 4-25. **Sand fencing. *(a)* Commerical snow-type straight fencing with side spurs. *(b)* Brush-type straight fencing with side spurs. (*U.S. Army Corps of Engineers, Coastal Engineering Research Center, Tech. Rept. no. 4, 3d ed., June 1966.*)**

and have the capacity to continue growing through accumulating sand (Figures 4-26). The grasses trap the drifting sand much as do sand fences. Once formed, the dunes must be assured of a stabilizing cover of vegetation. Dune stabilization by vegetation is discussed further in Chapter 5.

In some circumstances, the creation of artificial dunes may actually lead to progressive narrowing and eventual destruction of beaches. A barrier island complex is normally driven slowly landward under wave attack. The sediment eroded from the seaward shore in storms is swept across the island and added to the lagoon side. Thus, the beach, the dunes, and the sedimentary fringe on the lagoon side are driven as a unit toward the mainland. If natural or artificial dunes are stabilized, however, the landward side of the beach will remain fixed, while the seaward side continues to be eroded. The beach may thus become progressively narrower until it is no more than a fringe at the foot of the dunes. As an illustration, the unmodified Core Banks off the North Carolina coast still display broad beaches, whereas the beaches along the Cape Hatteras shoreline, with its rampart of artificial dunes, are reduced to narrow remnants.

Perched Zone

Human invasion of the coast includes the terrain back of sea cliffs, the crests of sea cliffs, and even niches or reentrants in cliffs. These sites make up the perched zone and may present unique problems.

Wave Erosion at the Base of Steep Slopes. Wave erosion at the base of steep slopes may cause landsliding, earthflows, or creep. Too often, developers fail to recognize old coastal landslides, and their activities may stimulate renewed movements. Such was the case at Pacific Palisades described in Chapter 2.

To prevent the removal of support by wave erosion, bulkheads are often constructed. If the costs of bulkheads are prohibitive, the hazardous area might better be set aside for such low-population-density uses as golf courses, recreation fields, and parks. Whenever bulkheading is decided upon, account must be taken of the strength of the slope materials, lateral changes in these materials, the possible influence of impeded surface water and ground water, and the nature and configuration of the surface below the possible slide mass. To reduce the potential harmful effects of surface and subsurface water, drainage may be necessary.

Many sea cliffs consist of rock in their

FIGURE 4-26. **Stabilization of dunes with sand fences and sea oats. Tar Landing, Atlantic Beach, Outer Banks, North Carolina. (*Arthur D. Howard.*)**

lower part and an overlying mantle of unconsolidated sediment. The mantle may be quite permeable relative to the underlying rock. Thus, rainwater seeping into the ground may be blocked at the rock surface and escape by way of seeps below the sediment. This impeded ground water may encourage landslides or earthflows around the borders of the sedimentary mantle.

The edges of rocky sea cliffs are reasonably safe homesites, whereas comparable cliffs in unconsolidated sediments may be extremely dangerous. Even in solid rock, account must be taken of the degree of fracturing and the inclination of the fractures. An intensely fractured rock may be as weak as unconsolidated deposits, and fractures inclined seaward are more conducive to landsliding than those inclined landward. In brief, all the factors discussed relative to landsliding must be considered in the occupation of cliffside or clifftop sites.

Construction at the edge of sea cliffs is particularly hazardous in earthquake areas, as seen in the devastation at Turnagain Heights in Anchorage, Alaska, during the earthquake of 1964.

Collapse of Sea Caves. The collapse of sea caves generally occurs piecemeal because caves are usually eroded into firm rock and expand relatively slowly. The collapse sometimes begins as a small hole from which a geyserlike spurt of water erupts with each large wave that enters the cave. This hole may expand slowly. At other places, however, waves have driven deep caves into high cliffs at the edge of marine terraces (Figure 4-27). The caves continue to expand under the prying action of compressed air and water as waves rush into the cave. Eventually, the ceiling may collapse like the roof of a limestone cavern (Figure 4-28).

Every construction site near the seaward edge of a terrace must be examined carefully. The presence of caves at the base of the cliff is cause for detailed investigation to determine the probability of collapse in the foreseeable future. The fact that caves extend only part way up toward ground level does not reduce the hazard if its roof is close to the contact with overlying weak sediments.

FIGURE 4-27. **Sea caves, California coast, 48 km (30 mi) south of San Francisco. (*Arthur D. Howard.*)**

Accelerated Gullying. *Accelerated gullying* is a hazard on terrace flats above sea cliffs. Wave erosion at the base of the cliff steepens gully gradients and accelerates erosion of the terrace flat. Valuable farmland has been destroyed and numerous dwellings endangered by expansion of such gullies. Measures for gully control were considered in Chapter 3.

An interesting situation arises when gullies erode into a terrace consisting of unconsolidated sediments over solid rock. We have noted that ground water may be impeded at the rock contact and appear as seeps. The escaping ground water may undermine the sediment or lubricate the contact to cause large-scale slumping. The loosened material may rush down the gully as an earthflow and come to rest on the beach below. The debris flow in Figure 2-23 carried part of a coastal highway with it. The heads of gullies in unconsolidated sediment must be examined for seeps and for evidence of incipient failure such as cracked or settling ground.

PLANNING

The Dilemma

People may build expensive protective works against the greatest storms known, only to find that some unusual combination of meterological, hydrographic, and geologic factors creates a devastating storm exceeding all records. Thus, hurricane invasion of an area previously thought to be safe from hurricanes, or a seismic sea wave of historically unknown magnitude, may render all previous protective measures in a given area impotent.

Coastal problems are especially serious because of the indications that sea level is slowly rising as a result of melting of the earth's glaciers. What possible measures can protect the earth's vast, low-lying coastal regions and the great cities located therein?

FIGURE 4-28. **Collapsed caves, Sunset Cliffs, San Diego, California. (*M.P. Kennedy, California Division of Mines and Geology.*)**

Even though we restrict our concern to present sea level, total protection of thousands of miles of coasts against all conceivable hazards is impossible. Extensive damage from rare events must be expected. The costs of repairing damage from these rare events, however, may be far less than the costs of massive, far-flung structures built to withstand the maximum catastrophe conceivable.

Unsettled or sparsely settled areas offer the best opportunities for successful planning, whereas already densely occupied areas limit the freedom of action. The accumulation of adequate data may require years. The investigation might include (1) topographic and hydrographic surveys, (2) study of historical coastal changes, (3) investigation of wind, waves, tides, and currents, (4) study of longshore drift, (5) sampling of materials, and (6) investigation of coastal geology. In this phase of the program, developers and planners may be at odds. For financial reasons, developers initiate construction as soon as possible after acquiring the land. Such construction before the hazards have been investigated may prove devastating in terms of possible repair costs and law suits. However, no combination of protective measures will guarantee against all possible damage. Properly constituted agencies can only advise on and consent to procedures to reduce hazards to a minimum.

An outstanding example of presentation and analysis of data bearing on the geologic environment of a vast coastal belt is the seven-volume "Environmental Geologic Atlas of the Texas Coastal Zone" prepared by the Bureau of Economic Geology of the University of Texas. Each volume includes an environmental geology map and a series of special-use environmental maps that include physical properties (soils, landforms, geography, etc.), biologic assemblages, current land use, mineral and energy resources, active processes, man-made features, rainfall–stream discharge–surface salinity, and topography and bathymetry.

Individual and Community Planning

No part of a shoreline can be tampered with without the effects being transmitted in both directions. The unregulated construction of a jetty or the excavation of a boat shelter by a private owner or community is, therefore, not permissible. The construction of a jetty or harbor entrance in one part of a community may lead to destruction by erosion elsewhere in the same community.

Some common practices and their consequences are:

1. Construction of groins and jetties to protect the shore from wave erosion, to trap sand for expansion of a beach, or both. The entrapment of sand prevents normal longshore drift of sand and leads to erosion of the beach elsewhere.

2. Breaching a dune ridge for access to the beach or an unobstructed sea vista. During storms, overspill through these low places may erode a channel whose expansion often results in dune destruction and damage of areas behind the dunes.

3. Local widening of a beach. This may increase longshore drift and widen beaches downdrift, stranding piers and other structures.

4. The construction of a short revetment or sea wall to protect an owner's property. The general result is continued erosion at the ends of the structure, with damage to the owner's property and that of any neighbor.

5. The deepening of channels in lagoons for access to developments. This may result in stronger currents, leading to beach erosion, steepening of the beach profile, and removal of sand.

6. The excavation of small harbors. Beach drift swept into the harbor by tidal currents will be lost to beaches downdrift.

7. The construction of artificial lakes. This may deprive the coast of the sediment required for stabilization. Beach erosion may result.

8. Real estate development without planning for the effects of increased runoff from the covered ground may accelerate gully erosion and the dumping of sediment on the beach. This may cause beach expansion or, by changing the bottom configuration, may shift the locus of wave erosion.

9. Clearing vegetation for real estate development. This may lead to accelerated soil erosion. The eroded sediment is dumped on the beach with the side effects already mentioned.

10. Constructing a home at the seaward edge of a dune belt. This is one of the most ill-advised procedures, usually subjecting the property to potential storm-wave damage.

11. Constructing a home at the edge of a sea cliff composed in whole or part of unconsolidated sediments. Landslides and gully erosion are expectable.

Local communities are normally permitted by state zoning laws to regulate the use of land within their jurisdictions. No construction should be allowed in the fluctuating shore zone without planning and supervision. Regulations should require preservation of dunes and adherence to building standards to minimize storm damage. Projects should be coordinated to minimize deleterious effects in intervening areas.

Regional planning, with the coastline, not the individual, as the major consideration, seems essential. While projects may be well conceived to minimize harmful effects within the bounds of a community, they may have adverse effects on neighboring communities. For example, a jetty or harbor entrance in one community may impoverish beaches for many miles downdrift. When an artificial inlet was opened at Palm Beach, Florida, in 1945, the entrapment of sand against the north jetty resulted in shrinkage of beaches for more than 24 km (15 mi) to the south.

The State Role

Some states engage in regional shoreline supervision. State permits are required for construction in coastal waters, estuarine waters, tidelands, marshlands, and state-owned lakes. In navigable waters, separate permits must also be secured from the U.S. Corps of Engineers. Before granting a permit, the state must give notice of the project to adjoining riparian owners, who have 30 days to file written objections.

On November 7, 1972, the people of California passed Proposition 20, the Coastal Zone Conservation Act Initiative. This act, subsequently passed by the legislature, set up the State Coastal Zone Conservation Commission and six regional commissions to prepare the Coastal Zone Conservation Plan for "balanced utilization and preservation of coastal resources." The plan, with modifications, was passed by the state legislature in 1976.

The Federal Role

Congress, in 1930, established the Beach Erosion Board of the Corps of Engineers. Its functions were to assemble the scattered data on shore processes, to synthesize engineering experience, and to undertake investigations in cooperation with state agencies to protect coastal and lake shores from erosion by waves and currents. In 1936, the Beach Erosion Board was supplanted by the Coastal Engineering Research Center with elaborate experimental facilities. The U.S. Army Engineer Waterways Experiment Station at Vicksburg, Mississippi, also contributes to coastal knowledge in phases other than beach protection. These organizations have published hundreds of valuable reports.

The Corps of Engineers Lake Survey District in Detroit, Michigan, and its Coastal Engineering Research Center conduct research on coastal aspects of the Great Lakes.

Federal assistance in beach-erosion control applies to publicly owned shores and to private shores where public benefits would result. Projects involving less than $400,000 of federal funds, normally 50 percent of costs, may be approved by the Secretary of the Army; otherwise, Congress must authorize the federal aid. Nonfederal publicly owned shores involving parks and conservation areas as well as permanent habitations may be supported up to 70 percent by federal funds. The maintenance of shore protection works is not eligible for federal aid. The Corps of Engineers assists local authorities in applying for federal aid.

Projects in navigable waters for which permission must be obtained include dams, dikes, dredging, wharves, piers, booms, weirs, breakwaters, and jetties. The work must be approved by the Chief of Engineers and authorized by the Secretary of the Army. A federal permit does not eliminate the necessity of obtaining a state permit.

Until 1972, the Corps of Engineers' jurisdiction extended only to the mean higher high-water line. For areas subject to especially high tides twice a month, this is the average of these higher high tides. Now, salt ponds, tidal marshes, and unfilled diked areas fall within the jurisdiction of the Corps. The new definition of "navigable waters" may add up to 500 sq km (190 sq mi) to the Corps' area of jurisdiction in San Francisco Bay alone and will have far-reaching effects nationwide.

ADDITIONAL READINGS

Bascom, W.: "Waves and Beaches," Doubleday & Company, Inc., New York, 1964.

Coates, D. R., ed.: Coastal Geomorphology, *Publications in Geomorphology*, State University of New York, Binghamton, 1973.

Dolan, R., P. J. Godfrey, and W. E. Odum: Man's Impact on the Barrier Islands of North Carolina, *American Scientist*, vol. 61, 1973, pp. 152–162.

Guilcher, A.: "Coastal and Submarine Morphology," Translated by B. W. Sparks and R. H. W. Kneese, John Wiley & Sons, Inc., New York, 1958.

Johnson, D. W.: "Shore Processes and Shoreline Development," John Wiley & Sons, Inc., New York, 1919.

Kuenen, Ph. H.: "Marine Geology," John Wiley & Sons, Inc., New York, 1950.

Paige, S., Chairman: "Application of Geology to Engineering Practice," Berkey Volume, *Geological Society of America*, 1950.

Shepard, F. P., and H. R. Wanless: "Our Changing Coastlines," McGraw-Hill Book Company, New York, 1970.

U.S. Army Coastal Engineering Research Center: "Land Against the Sea," *Corps of Engineers*, Misc. Paper No. 4-64, 1964.

———: "Shore Protection, Planning, and Design," 3rd ed., *Corps of Engineers*, Tech. Rept. No. 4, 1966.

University of Texas, Bureau of Economic Geology: "Environmental Geologic Atlas of the Texas Coastal Zone," 7 volumes, 1972.

Woodhouse, W. W., Jr., and R. E. Hanes: "Dune Stabilization with Vegetation on the Outer Banks of North Carolina," U.S. Army Coastal Engineering Research Center, *Corps of Engineers*, Tech. Memo. No. 22, 1967.

CHAPTER FIVE

dry
environments

INTRODUCTION

Almost one-third of the earth's land area is arid or semiarid (Figure 5-1). About the same proportion of the coterminous United States (the 48 contiguous states) is so classified (Figure 5-2).

Unless interrupted by lengthy droughts, human invasion of the arid lands of the United States will continue. Prosperous oases such as Salt Lake City, Las Vegas, and Tucson will continue to grow, and new ones will arise. The principal limiting factor will be the availability of water.

Part of the water problem in the arid lands of the United States is due to the insistence of newcomers on importing the conveniences and comforts to which they have become accustomed in humid climates. Large amounts of water are consumed in watering lawns, washing cars, filling swimming pools, and other wasteful uses. Part of the blame rests with local governments, which charge about the same for water in arid regions as in humid regions. This hardly induces conservation.

In this chapter, we consider the characteristics of dry environments, the natural processes which operate there, the problem of water in all its ramifications, and the question of regional planning.

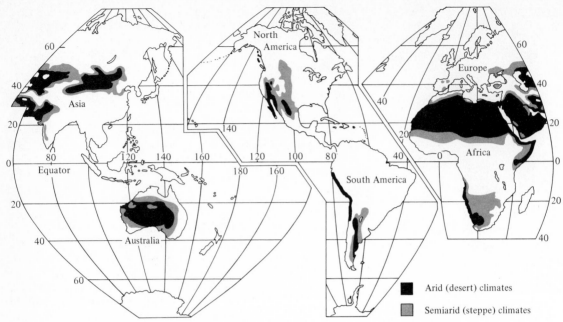

FIGURE 5-1. **Arid and semiarid climates of the world.** (*After R. F. Flint and B. J. Skinner, "Physical Geology," John Wiley & Sons, Inc., 1974.*)

■ Arid (desert) climates

▨ Semiarid (steppe) climates

CHARACTERISTICS OF DRY REGIONS

The "dry" regions are generally divided into semiarid and arid. Classifications differ in the criteria used for differentiation. A classification based on amount of rainfall regards regions with more than 50 cm (20 in.) of rainfall annually as humid; those with 25 to 50 cm (10 to 20 in.) as semiarid; and those with less than 25 cm (10 in.) as arid. These rainfall limits correspond approximately to changes in vegetation, soil, and drainage. Amount of rainfall alone, however, is not a satisfactory criterion, because regional differences in temperature may determine how much rainfall is preserved on the surface or in the soil. High temperatures result in high rates of evapotranspiration, and regions may be parched in spite of rainfall exceeding the above limits. A more satisfactory definition regards as arid those lands in which evapotranspiration exceeds precipitation so that the region is water-deficient.

High temperature is not a requisite for aridity. Cold air has less capacity for moisture than does warm air. Hence, in very cold regions, even though the air is saturated, the actual amount of moisture may be so small that the precipitation potential is very low. Thus, Antarctica is a cold desert. Tropical rain forests also demonstrate that high temperature alone does not create deserts.

Modern climatic classifications give greater attention to seasonal variations in rainfall and temperature. If the bulk of the rainfall in arid lands comes in a cool winter season, evapotranspiration will be less than if it occurs during a hot summer. More water is available for agriculture in the former situation than in the latter. And in some dry areas, dew and frost contribute to the water budget.

Some investigators use vegetation as an index of aridity, defining the zonal boundaries by the distribution of drought-resistant plants. One vegetative adaptation to drought consists of deep roots to tap water at depth, a characteristic of the plants known as *phreatophytes*. Examples are the date palm, salt cedar or tamarisk, and mesquite. Most desert plants, however, are *xerophytes*, which resist transpiration by their tough leathery skins, by an absence of leaves or a loss of leaves during the dry season, by the small size and waxy surfaces of leaves, by the shallow, widely branching root system designed to obtain as much moisture as possible in a short time, and by the ability to store water for long periods of time. Examples of the last are the succulents, including cacti. Examples of other xerophytes are creosote and burrobush.

Other investigators classify the arid lands on the basis of whether drainage escapes from the region. With rare exceptions such as the Nile, Niger, and Colorado, no drainage escapes from arid regions. Streams fade away by evaporation and seepage underground. Those that

persist start with sufficient volume to persevere. Thus, the Niger rises in the rainy tropics of West Africa, wanders northeast into the Sahara Desert, and southeast through the rainy tropics of Nigeria. It shrinks so low in the desert that cattle are pastured in the river bed.

Still other investigators classify arid lands on the basis of soils. The common soils of dry regions are relatively rich in soluble carbonates and indicate that evapotranspiration exceeds precipitation. If precipitation exceeds evapotranspiration, soluble carbonates are leached out of the soils.

Regardless of the classification used, the arid lands are water-deficient. We will use the terms "desert" and "arid lands" as synonyms, applicable in a rough way to regions having less than 25 cm (10 in.) of rainfall per year. The semiarid lands, also known as *savanna* or *steppe*, are transitional between arid and humid regions; annual rainfall approximates 25 to 50 cm (10 to 20 in.) and the vegetation consists primarily of grass and brush. There may be scattered trees or even dense groves in valley bottoms. In contrast to deserts, the semiarid lands are already more or less fully devoted to agriculture.

The boundaries of the arid and semiarid zones shift with changing rainfall patterns. Between 1944 and 1947, the desert margin in Tunisia in North Africa migrated 270 km (170 mi) farther north than its position during the moist period 1931–1934. During the great drought of the early thirties in the United States, the arid border moved eastward into the Great Plains, and the semiarid boundary moved practically to the Mississippi Valley. Much of the Great Plains became unfit for agriculture. Thousands became homeless, and agricultural losses were staggering.

In any one year, the semiarid boundary in the United States may move as far east as Wisconsin, Iowa, and Louisiana. In the southern Sahara, the semiarid grasslands expand 1600 km (1000 mi) northward as the rainfall zone expands. In dry years, they shrink back. At present, the semiarid region known as the Sahel is becoming a desert, with incalculable loss of life, capital, and hope for the future. A similar situation exists in northwest China. During the wet periods, the Chinese farmers spread

FIGURE 5-2. **Dry regions of the United States.** (*Modified slightly from C. Hodge and P. Duisberg, eds., 1963, "Aridity and Man," American Association for the Advancement of Science, Publ. no. 74. Copyright 1963 by the American Association for the Advancement of Science.*)

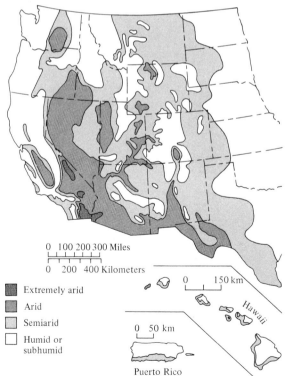

0 100 200 300 Miles

0   200   400 Kilometers

0   150 km

Hawaii

0   50 km

Puerto Rico

■ Extremely arid
▨ Arid
▧ Semiarid
□ Humid or subhumid

out to the north and west, crowding out the Mongol herdsmen; during dry periods, they retreat, and the herdsmen take over. Exploitation of the borders of semiarid lands is a hazardous undertaking.

DISTRIBUTION AND ORIGIN

Deserts (Arid Lands)

Deserts are present in all parts of the world. Their distribution would be relatively simple if the sole control was the global pattern of atmospheric circulation (Figure 5-3). Note that there are belts of descending air at roughly 30° north and south of the equator, the subtropical high-pressure belts. As this air descends and is compressed, it becomes warmer and exerts a drying influence on the ground. That part of the air which spreads equatorward at ground level is heated up further as it moves into warmer latitudes, again drying the ground in its path. In contrast, where air rises, or moves poleward, it cools and is forced to relinquish some of its moisture as precipitation. This is true at the equator and in the zone of westerly winds.

The distribution of deserts is far more variable, however, than suggested by the pattern of atmospheric circulation. Additional influences are the distribution of continents and oceans, the size and shape of land and water masses, the topography of the lands, the influence of warm and cold ocean currents, the seasonal shifting of the climatic belts, and the seasonal presence of localized high- and low-pressure cells. Thus, there are several kinds of *midlatitude deserts*. There are interior deserts, such as the Turkestan Desert, far from oceanic

FIGURE 5-3. **The general scheme of atmospheric circulation. Deserts due principally to this circulation pattern occur under the subtropical belts of descending, warming air and in the trade wind zones where winds blow equatorward. Polar deserts are due to the low moisture content of the cold air.** (*After A. N. Strahler, "Introduction to Physical Geography," John Wiley & Sons, Inc., 1965.*)

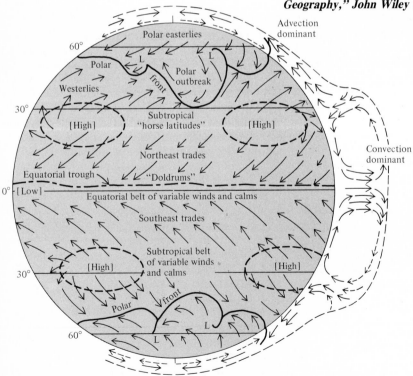

supplies of moisture. There are mountain barrier deserts in the lee of high mountains which rob the winds of their moisture. The deserts east of the Sierra Nevada in the United States and east of the Andes in southern Argentina are examples. And, there are combinations of the above as exemplified by the Gobi Desert in the interior of Asia beyond the barrier of the Himalayas.

Low-latitude deserts include those below the high-pressure belts of descending air and those in the path of the trade winds moving equatorward. The two commonly merge. Examples are the Sahara Desert of North Africa and the Sonoran Desert of northern Mexico and southern United States.

Some low-latitude deserts lie along the western margins of continents where the drying trade winds blow seaward. These deserts are cool and foggy because of the cold ocean currents offshore. During the day, fog is drawn inland. The coastal deserts of Peru and northern Chile are examples.

In environmental investigations, microclimates, such as may prevail on opposite sides of a ridge, may be more important than the characteristics of the broad climatic zone in which the area is located.

Semiarid Lands

In arid regions, the climatic limitations are sharply defined, and there is little justification for blindly initiating agricultural or ranching ventures. The semiarid transitional lands, on the other hand, are hazardous because of their climatic vagaries. A shift in the arid-semiarid boundary may parch lands that were already marginal for agriculture. On the other hand, the boundary may shift favorably. In 1905, one of the rainiest years in the Great Plains, the semiarid climate was supplanted by a subhumid climate. Apparently, the only consistent factor in the semiarid climate is its inconsistency.

The Great Plains, stretching eastward hundreds of miles from the Rocky Mountains, is the largest area of savanna or steppe in the United States. Precipitation is unreliable throughout the Great Plains. Frequently, the rainfall surpasses the expectable, encouraging agriculture, but droughts lasting 35 or more

consecutive days may occur annually and droughts of 60 to 70 days once a decade. More rarely, droughts lasting 3 to 4 months may occur for several consecutive years. Although the great drought of the 1930s is freshest in our minds, there have been other calamitous droughts since agriculture supplanted cattle raising on the plains after the Civil War. In the 1890s, a severe drought resulted in wholesale exodus of farmers. Another serious drought occurred in the 1920s.

The suitability of the semiarid lands for cultivation and grazing also depends on the season in which the rainfall occurs. The steppes north of the Sahara Desert, for example, have winter precipitation, some of which is in the form of snow. Hence, moisture losses by evaporation are reduced. The southern steppes, on the other hand, are more arid because the rains come during the hot summer.

PROCESSES OF LANDSCAPE SCULPTURE

The processes of weathering, erosion, and sedimentation in arid climates are similar to those of humid climates but differ in relative importance, intensity, and superficial effects. Weathering is widespread but, in the absence of abundant moisture, is dominantly physical. Streams are the principal agents of erosion because of the intense runoff from the barren slopes. Even in the Death Valley region, the efficacy of stream erosion is clearly evident in the profusion of typical stream-eroded valleys on all but fresh depositional slopes (Figure 5-4). Part of the erosion, however, particularly at lower elevations, probably took place in former, more humid climatic episodes.

Most desert streams are intermittent and short but have enormous flows in time of flood. In one Arizona flash flood, the flow reached 1300 cu m (46,000 cu ft) per sec. In contrast, the mighty Colorado River at Yuma, Arizona, in 1930, before construction of the dams upstream, had an average flow of only 850 cu m (30,000 cu ft) per sec.

Wind is a far less effective agent of erosion than water. It is primarily a transporting medium, creating dust and sand storms of the materials broken up by other processes. The re-

FIGURE 5-4. **Stream-eroded landscape bordering Death Valley, California. This mountain-encircled basin is a bolson. The dry floor is a playa. The slopes extending away from the foot of the mountains consist of coalescing alluvial fans. Seeps are common around the bases of many fans where groundwater flow is impeded by fine-textured impermeable sediments. View from Dante's View in the Funeral Mountains. (*Photo by Mary Hill.*)**

moval of sediment by the wind is *deflation.* Where wind-borne sand is swept continuously across an area, however, it may sandblast even hard rock.

In cold deserts, frost shattering is effective, and glaciers erode and deposit as elsewhere. Some of the processes of landscape modification in cold, dry, ice-free environments are unique, owing to the presence of permanently frozen ground (see Chapter 7).

Mass movements, including slump, landslides, rockfalls, and mudflows, are common, but creep and earthflows, which involve soil and other loose materials, are less common.

THE LANDSCAPE

General Considerations

The common conception of a desert is a vast sand-covered plain. Actually, desert regions may have very little sand and considerable relief. The Sahara is about one-ninth sand, much of it concentrated in vast sand seas (Figure 5-5). The remainder is bare rock, alluvi-

um, alluvium-veneered rock plains (pediments), and wind-blown dust (loess).

The major difference between arid and humid landscapes is that the former are dominantly angular and the latter are dominantly rounded (Figure 5-6).

In humid climates, chemical weathering produces a thick soil which together with vegetation serves as a sponge to reduce the amount and intensity of runoff. The soil is not easily washed away. Chemical weathering, because it attacks sharp crests and edges from two sides, rapidly rounds these off, and the waste accumulates. Through the process of creep, the soil moves slowly downslope, resulting in smooth flowing slopes that are convex at the crest and concave below. In deserts, on the other hand, soil and vegetation are generally absent, and the uncontrolled runoff sweeps away debris that would otherwise collect on and at the foot of slopes. Physical weathering contributes to the angularity by prying fracture-bounded rock fragments from the steep slopes, thus perpetuating their steepness and sharp crests.

Landforms

Desert landscapes are varied. They include mountains, plateaus, basins, plains, sand fields, and minor landforms.

Mountain ranges in arid regions may be arid from base to crest or may have humid summit areas. Those that are totally arid are too low to generate their own rainstorms or too far removed from sources of moisture. Those that have humid uplands include the higher ranges of our Western deserts and those nearer the windward side of the deserts. Some are forested. If high enough, they may have both an upper *tree line*, above which the cold discourages tree growth, and a lower tree line determined by aridity. The intermediate zone, offering welcome respite from the heat of the deserts below, is being invaded by people in increasing numbers.

Many deserts include mountain-encircled basins known as *bolsons*. Death Valley is an example (Figure 5-4). They have no outlet for the waters which on rare occasions flood their floors. The shallow temporary lakes so formed are *playa lakes* and may last a few days or a few weeks. The dry lake floors are termed *playas*. They may appear white from precipitated salts or gray if the surface consists of clay.

Alluvial fans are common at the mouths of mountain valleys opening onto lowland plains. Fans built out from adjacent valleys may coalesce to form an undulating sloping plain (Figure 5-4).

Still farther from the mountains, beyond the periphery of the alluvial fans, there is generally a smooth alluvial plain. In intermontane basins, this merges into a playa in the lowest part of the basin. In some western basins, the thickness of alluvium and playa sediments exceeds 1.5 km (1 mi).

FIGURE 5-5. **View in the sand sea known as the Great Eastern Erg, Sahara Desert. (*Arthur D. Howard.*)**

(a)

FIGURE 5-6. **Contrasts in arid** *(a)* **and humid** *(b)* **topography.** *(a)* **Colorado Plateau in northern Arizona.** (*Arthur D. Howard.*) *(b)* **Appalachian Plateau, West Virginia.** (*John Rich, Courtesy Amer. Geogr. Soc.*)

(b)

Where the mountain front has receded by weathering and erosion, a smooth rock surface, or *pediment* (Figure 5-7), is left behind. Pediments may have a thin, discontinuous cover of alluvium or a continuous veneer up to 16 m (50 ft) thick.

The erosional pediment may be distinguished from a depositional piedmont plain wherever ravines have cut through the thin cover to reveal the rock surface below. Without dissection, the distinction can be made for some surfaces by the degree of slope. Pediment slopes do not exceed 5 or 6°, whereas the slopes of many fans, as on the east side of Death Valley, exceed 20°. Because of their influence on ground-water supplies, the distinction be-

tween thinly veneered pediments and thick alluvial fans is important.

Many pediments in the Southwest were apparently formed under moister climatic conditions than now, because they are either buried under thick alluvial fans or encroached upon by sand dunes.

As in any climate, the bottoms of broad valleys in desert regions have floodplains. The problems they present may be quite different from floodplains elsewhere because of the difference in their relation to the water table. In hilly terrain, the water table in dry regions is farther below the surface than in humid areas. In flat terrain, the water table in dry regions is highest under perennial streams because water seeps downward from the stream bed to build the water table upward. This is evident along permanent rivers like the Colorado and Nile in their passage through level plains; wells tap water close to the surface near the river but have to go to increasingly greater depths farther away.

Desert pavement is common in many deserts. It consists of a sheet of pebbles left behind

FIGURE 5-7. **Pediments, east side Turtle Mountains, Mojave Desert, California. As pediments encroach into a desert range, they ultimately dismember it. (*Stanford University collection.*)**

as the finer sediment is winnowed out of alluvium by the wind (Figure 5-8). The surfaces of the pebbles may be flattened and polished by sandblasting to provide an amazingly smooth surface. Desert pavement protects the materials below from further deflation.

Sand dunes and *sand sheets* are common. The sand is generally quartz, but at least one large dune field in the United States, the White Sands of New Mexico, consists of gypsum. In Bermuda, dunes are built of calcite derived from beaches underlain by limestone. Small dunes of salt occur on the exposed floor of Great Salt Lake in Utah.

Dunes occur in a variety of forms (Figure 5-9), including *transverse dunes*, wavelike forms normal to the wind direction; *barchans*, crescentic forms with the horns extending to leeward; *deflation dunes*, also crescentic but with the horns facing into the wind; *longitudinal dunes*, long ridges strung out parallel to the wind; and *dune complexes*, assemblages of dune types. The variation in type is due to the quantity of sand available, to velocity, direction, and persistence of the wind, and to the presence or absence of desert vegetation.

Transverse dunes are found where sand is abundant and winds moderate in strength. They are dismembered into barchans away from the source of supply as sand becomes less abundant. Deflation dunes are found where stabilizing vegetation has been breached, and the exposed sand is heaped up around the leeward side of the deflation basin. Longitudinal dunes

FIGURE 5-8. **Desert pavement, southern Nevada. (*Photo by Chester R. Longwell.*)**

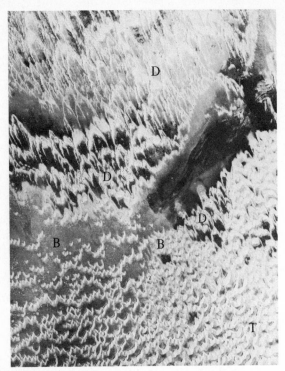

FIGURE 5-9. **Sand dune types. T—transverse; B—barchan; D—deflation. The prevailing wind is from the lower right. (U.S. Dept. Argiculture.)**

as Big Hollow in the Laramie Basin of Wyoming, 10 by 15 km (6 by 9 mi) across and 100 m (300 ft) deep, and the Qattara basin in western Egypt, more than 240 km (150 mi) long and up to 120 km (75 mi) wide, with its bottom locally 130 m (420 ft) below sea level. The size of these basins is enlarged by stream erosion during desert rains, but the depth does not diminish as it should if the sediments were trapped inside. The excess sediment is removed by the wind. Deflation is effective down to the water table.

ENVIRONMENTAL PROBLEMS AND REMEDIES

Many of the environmental problems of deserts are similar to those of humid regions but differ in relative importance. A few are unique.

Water Supply

Permanent streams are absent from most deserts, and, because of the small annual rainfall, the brief, high-intensity storms, the almost total runoff from hilly areas, and the loss by evapotranspiration, there is little recharge of ground water.

Hydrologists attempt to overcome water deficiencies by (1) preventing excessive runoff, (2) reducing evaporation, (3) reducing transpiration, (4) locating sources of ground water, including deep aquifers, (5) assisting in regional planning for efficient and fair use of available water, and (6) importing water from other drainage basins.

Retardation of Runoff. Runoff may be retarded and conserved by dams, detention reservoirs, small check dams, terracing, cisterns, and other devices.

As in humid climates, large dams are the conventional means of retarding runoff and storing water for human, agricultural, and industrial consumption and for flood prevention and generation of power. Small reservoirs, particularly farm and ranch ponds, are primarily for local water needs, recreation, and aesthetics.

In desert areas, the scarcity of permanent rivers reduces the prospects for large dams. In

occur where strong, unidirectional winds are unobstructed, as on plains and plateaus. In parts of the Sahara, where sand forms only a thin sheet, it may be swept up into linear complexes separated by corridors of bare rock. These corridors are used by camel caravans but shunned by motorcars because of the rough rock.

In places in sandy deserts, the sand is so thin in interdune depressions that water is obtainable at shallow depths. After rains, temporary lakes may appear. Active dunes, those presently migrating as sand is driven up their gentle windward slopes and over their crests, locally create serious problems.

Deflation gives rise to *deflation basins*. While many are small, a few meters across and a few centimeters deep, thousands of others, as in the High Plains of the American West, are 1.5 km (1 mi) or more across and tens of meters deep. Some are major landscape features, such the southwestern deserts of the United States, only the Colorado and Rio Grande rivers and some of their larger tributaries are perennial. In North Africa, only the Nile and Niger rivers cross the Sahara.

Detention reservoirs (see Figure 3-16) are constructed on alluvial fans at the foot of desert mountains. The water may be used for agriculture, recreation, or other uses, and contributes to ground-water supplies by seepage.

Desert dwellers commonly preserve moisture and create flat areas for agriculture by building a succession of small check dams across the beds of steep ephemeral streams. The area in back of each check dam traps the sediments washed into it, and these sediments retain moisture long enough for the growth of suitable crops. The procedure dates back to antiquity (Figure 5-10).

The same effect is achieved on sloping surfaces by terracing. Low walls are built of stones or earth. Runoff washes sediment into the space in back of each wall to create a strip of fertile moisture-retentive ground. This method was widely used by the Indians of the American Southwest.

In many desert areas, the use of *cisterns*, artificially enclosed reservoirs or tanks, is common. They are invariably roofed to eliminate evaporation.

Large cisterns may have roofs shaped to catch rainwater and lead it below. A comparable practice is used with individual buildings, the rainwater being piped into small cisterns, hogsheads, or other receptacles. Livestock ponds, also known as tanks, are constructed to trap occasional rains.

Evaporation Control. Water losses by evaporation along watercourses alone in the 17 Western states is estimated at 24 million acre ft per year, or nearly 30 billion cu m. The annual evaporation from Lake Mead behind Hoover Dam is 850,000 acre ft, or well over 1 billion cu m. An example of the rate of evaporation was provided when the Colorado River flooded the Imperial Valley in southern California in 1904–1907 to create Salton Sea. After the river was brought under control, the level of Salton Sea decreased 150 cm (60 in.) per year by evaporation before equilibrium was attained.

FIGURE 5-10. **Reconstructed prehistoric check dams at Mesa Verde, Colorado, held back soil and conserved moisture for crops. (*R. B. Woodbury, in C. Hodge and P. Duisberg, eds., 1963, "Aridity and Man," American Association for the Advancement of Science, Publ. no. 74, Copyright 1963 by the American Association for the Advancement of Science.*)**

A few measures to control evaporation have been noted in Chapter 3. Where options in the location of dams are available, they should be located where the ratio of surface area to depth is small.

Another approach toward reduction of evaporation is the use of chemicals to form a protective film shielding the water below. The films, however, are disrupted by wind and waves, biological processes, and ultraviolet radiation. Hence, large reservoirs, because of turbulence during wind storms, are not suitable for use of chemicals. On the other hand, chemical films may be effective where conditions are favorable. Decreases of 9 to 22 percent have been achieved on lakes in Oklahoma, Arizona, and California. The degree of success depends on the concentration of the chemical, weather, size of the lake, and frequency of application.

Evaporation control is furthered by the use of underground reservoirs. To date these are small because of costs relative to surface reservoirs. Many deep abandoned mines in the American West are flooded and constitute ready-made underground reservoirs. One at Eureka, Nevada, has so great an influx of water at depth as to defy economically feasible dewatering. There is insufficient demand in the immediate area to justify the cost of pumping and transmission. At least one small town in the western foothills of the Sierra Nevada draws most of its water from an abandoned mine.

Evaporation may be reduced by the substitution of pipelines for open canals wherever possible. This also reduces seepage losses. Most of Israel's conveyance system is by buried pipeline.

Finally, evaporation may be reduced by trickle irrigation, whereby water trickling from perforated pipes nourishes only the plants, not the unplanted areas between (Figure 5-11).

Reduction of Transpiration Losses. Transpiration, the escape of water from the aboveground parts of plants, is quite variable in desert vegetation, depending on the type of plant, depth to water table, and local climatic factors such as length of growing season, length of day, temperature, and humidity. The phreatophytes, which occupy 6 million ha (23,500 sq mi, or 15 million acres) of land in the United

FIGURE 5-11. **Trickle irrigation. Young pepper plants under drip irrigation in the southern Arava, Israel. The researcher is reading a tensiometer. The tensiometer determines the availability of soil moisture to the plants. Fences in background prevent wind erosion.** (*Courtesy Aaron Wiener and American Scientist.*)

States, and are represented by such forms as salt cedar, greasewood, cottonwood, and mesquite (Figures 5-12, 5-13, and 5-14), use astonishing quantities of water. They are found in valley bottoms and along streams. The xerophytes, such as the cacti, rooted well above the water table, are independent of ground water and use little water.

The high consumption of water by phreatophytes is a serious problem. Transpiration exceeds evaporation because plant roots draw water from depths below the reach of evaporation. In the 12 months ending September 30, 1944, consumption of water by phreatophytes over an area of 3750 ha (9300 acres) along the Gila River in Arizona was 35 million cu m (28,000 acre ft), an average of 1500 cu m per ha (3 acre ft per acre).

As a general rule, the shallower the water table, the higher the water consumption by phreatophytes. Salt grass prefers a water table between 2 and 4 m (6 and 12 ft) below the

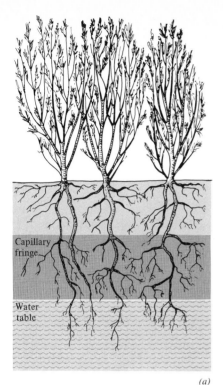

FIGURE 5-12. **Distinction between phreatophytes** *(a)* **and xerophytes** *(b)* **shown by their occurrence in relation to the water table. Phreatophytes will also grow wherever water is supplied artificially, as along ditches and canals, even though their roots do not reach the water table. Alfalfa is a useful phreatophyte grown by irrigation regardless of the depth to the water table. (*After U.S. Geological Survey.*)**

surface. Salt cedar, willow, and cottonwood thrive where the water-table depth is between 3.3 and 6.6 m (10 and 20 ft). Greasewood grows best where the water table is no farther than 5 m (15 ft) from the surface. Mesquite can send its roots to exceptional depths of 30 m (100 ft). These plants are rough guides to the depth to the water table.

Phreatophytes can be removed temporarily by mechanical equipment, burning, or chemical sprays. For permanent removal, the plants must be denied water by cutting off the supply from upstream, piping the water, or lowering the water table. In replacing phreatophytes by other vegetation, it is necessary to understand the requirements of the substitute plants to be sure that they will thrive in that environment. Forage crops such as alfalfa and grasses adapt well.

Phreatophytes grow densely on the delta plains at the heads of large reservoirs. They use large quantities of water that would otherwise contribute to reservoir storage. For example, if the delta of the McMillan Reservoir on the Pecos River in New Mexico were cleared of phreatophytes, it would free 48 million cu m (39,000 acre ft) of water to the Pecos River annually. The Bureau of Reclamation has channelized the Rio Grande for 56 km (32 mi) above Elephant Butte Reservoir, also in New Mexico. Whereas almost no water reached the reservoir prior to channelization, 123 million cu m (100,000 acre ft) per year now reach the reservoir.

Phreatophyte elimination would save Nevada annually 1.8 billion cu m (1.5 million acre ft) of water, of which 25 percent would be salvageable.

On the other hand, phreatophytes serve as

FIGURE 5-13. Dense saltcedar growth along Gila River, 65 km (40 mi) east of Phoenix, Arizona, near Powers Butte. (*U.S. Geological Survey.*)

screens, blocking much sediment from reaching the central portions of reservoirs, and provide havens for birds and other wildlife. There have been proposals to substitute useful plant crops at the heads of reservoirs not only for their value but to impede reservoir filling by sedimentation. In other places, grasses and low-growing shrubs, such as the wild rose, sand cherry, hackberry, and olive, might be planted to provide food for wildlife. Individual land-owners have objected to removal of phreato-phytes, which in many areas provide the only greenery in an otherwise barren landscape.

Outside the channels and deltas, the dominant plants are the xerophytes, which are truly

FIGURE 5-14. Willows and cottonwood (tall trees) line the banks and cover the adjacent lowland of the Carson River in Carson Valley, Nevada. (*U.S. Geological Survey.*)

adapted to dry conditions (Figure 5-12). Lacking deep roots, they expand their roots laterally and draw on the moisture in the zone below the evaporation depth. Because the quantity of moisture in this zone is small, the roots must cover a large area to sustain the plant. Hence, xerophytes occur as scattered individuals. Transpiration from these plants is low.

Exploitation of Ground Water. The discovery of sources of deep ground water, much of it artesian, has changed many desert areas. In North Africa, an aquifer of vast extent has been discovered south of the Atlas Mountains from Libya on the east to Morocco on the west. It extends hundreds of miles to the south. The first borehole was drilled at Ghardaia, Algeria, in 1950. The aquifer was tapped at a depth of 1200 m (4000 ft), and the water gushed up in a fountain 180 m (600 ft) high.

Prior to the discovery of this prolific aquifer, the oases of Ouargla and El Ouad in the Algerian Sahara were threatened because of insufficient water and drifting sand. The limited application of available irrigation waters brought salts to the surface, and the date palmeries were being destroyed. With the abundant water now available, the salts have been flushed out of the soil and the dunes controlled by vegetation and other means. However, in other arid areas, ground water is a dwindling resource.

The rate of water movement in confined aquifers is slow, from a fraction of a centimeter to a meter or so per day, with exceptional daily rates up to 100 m (several hundred feet). Thus, water pumped from the center of a large basin may have been in transit for scores of years or centuries. Clearly then, if water is extracted wastefully from these aquifers, they cannot be naturally replenished at the wasteful rate. Water levels in wells tapping the Dakota Sandstone, the principal aquifer under parts of the semiarid Great Plains, have been lowering steadily since these reserves were discovered.

The sediment-filled basins of the Basin-Range province of the Western United States are ideal for the entrapment of ground water. Not only are there extensive artesian aquifers within the basin sediments, but because of vertical and horizontal changes in the permeability of the sediments, perched water tables and complicated interconnections between aquifers are common. Because the finest textured and least permeable sediments are farthest from the mountain sources, the central areas of basins may be less productive than marginal areas, which consist of coalescing alluvial fans. Furthermore, the waters of the central areas are poorer in quality because of mineral matter dissolved in their long transit.

Water that seeps into the coarse sediments at the head of a fan follows the permeable strata downslope. The decreased permeability toward the foot of the fan acts as a barrier to water movement, and the water appears locally as seeps, often arranged in a contourlike arc (see Figure 14-18). Thus, successful wells are more likely farther up the fan.

In view of rapidly expanding water use, we can no longer afford uncontrolled drilling and waste of ground water. In irrigation for tree crops, for example, it may be necessary, following ancient Indian and modern Israeli practices, to pipe water directly to each plant. Even where entire fields are watered, no more water than necessary should be applied. Actually, agriculture uses far more water than do municipalities and most industries. Perhaps industry should receive preference over farming in dry regions.

Importation of Water. The importation of surface water into dry regions may reduce the demands on ground water. It may be used to recharge ground water where withdrawal exceeds natural replenishment, or it may freshen or substitute for irrigation waters that are already so saline as to affect crops.

Some cities in dry regions import water from great distances. Los Angeles derives much of its water from Owens Valley, 390 km (240 mi) away; Mono Lake, 560 km (350 mi) away; the Colorado River, 730 km (450 mi) away; and northern California, 970 km (600 mi) away. Phoenix derives its water from Roosevelt Dam, 188 km (117 mi) to the east on the Salt River. Unfortunately, most transmission is by open canal; hence, high evaporation losses in transmission are added to evaporation losses from the reservoirs in the systems. Furthermore, such diversions may deprive the source areas of needed water. Diversion of water to Los Angeles has led to actual conflict with the ranchers of Owens Valley.

New water supplies will be needed as demands increase. Recycling of waste waters and desalinization of seawater will receive greater attention. In Israel, sewage formerly discharged into the Mediterranean is now treated in shallow, impervious open-air basins, from which it is led into sandy spreading areas upslope from recovery wells.

Desalinization and Cloud Seeding. *Desalinization* involves the removal of salts from saline waters. Some methods convert the water to vapor which is then condensed to fresh water. A more costly process suitable for small operations involves the chemical substitution of other elements for those that cause water hardness. This is the system used in ordinary water softeners. Finally, water-hardening substances may be removed by an electrically aided filtering system.

Desalinization is an expensive and energy-consuming process. Costs are not competitive with other means of supplying fresh water, except at special sites where development is essential despite the cost. At present, desalinized water is 10 times more expensive than irrigation water and 4 times as expensive as the water provided in 400 American cities. The extraction of useful minerals from seawater may eventually lower costs.

The goal of a major desalinization project planned by the Orange County Water Department in southern California is 11 million liters (3 million U.S. gallons) of fresh water daily to be pumped into the ground to replenish ground-water supplies. Costs of the desalinization alone are approaching $1 per 3800 liters (1000 gal), or about double that of delivered natural waters of comparable quality. The disposition of the enormous volumes of liberated salts presents a problem: 3800 liters (1000 gal) of seawater contains 90 kg (200 lb) of salt.

Cloud seeding, in which silver iodide crystals are dispersed in clouds to act as nuclei for water droplets, will not change an arid climate to humid because clouds are too spotty in occurrence and distribution. The number of rainstorms may be increased, but they will remain localized. The increased total precipitation, however, could increase the flow of the few permanent streams, could increase the surface life of ephemeral streams, and could increase ground-water storage. Some maintain that the rain attributed to cloud seeding would have fallen anyway a short distance away so that the precipitation total of the region is not increased. If, on the other hand, cloud seeding causes precipitation from air masses that would have otherwise passed uninterruptedly over the lands to the sea, then the hydrologic cycle will be altered. It is uncertain whether cloud seeding can be quantitatively satisfying except for small areas at generally unpredictable points in time. Recent experiments in California indicate that cloud seeding may locally increase precipitation about 5 percent.

Human Restraints and Recycling. Part of the excessive water demands in growing desert communities are due to water-wasteful practices. Farmers use more water in irrigation than needed. Experiments indicate the possibility of reducing evaporation of soil moisture and decreasing transpiration of crops by using plastic covers. Another experimental procedure introduces evaporation-retardant chemicals into the soil, from which they are extracted by the crops with reduction in transpiration.

An increasing effort is being expended in water-recycling investigation and implementation. The steel mill at Fontana, California, recycles its water so many times that the water demand is only about 2 percent of that normal in the industry. The city of Amarillo, Texas, attracted a new oil refinery by offering it treated sewage water.

Many simple water-conserving measures may be taken by individuals. These include the repair of leaks, washing from a partially filled basin rather than a running spigot, using smaller capacity toilet tanks, avoiding wasteful practices in washing cars, using air conditioners that are not wasteful of water, using drought-resistant vegetation in place of lawns, watering plants in the cool of the evening when evaporation is less, avoiding runoff into the streets, replacing phreatophytes with plants more frugal of water, designing desert houses with maximum natural circulation of air to minimize need for air conditioning, foregoing individual swimming pools, and—above all—adapting to the desert environment rather than trying to supplant it. On a community basis, ditches and canals used for transport of water should be lined to prevent leakage, pipes should be substituted for ditches wherever feasible, tanks and reservoirs should be covered, water should be recycled, and priority should be given to high-return activities such as domestic and industrial uses rather than agricultural uses.

It is doubtful that much water can be saved on a voluntary basis while water is supplied to users below cost. The National Water Commission has recommended that the costs of irrigation projects and other projects benefitting special groups be repaid in full by the users. Thus, the answer to water wastage may lie in treating water as any other natural resource which is in short supply and pricing it accordingly. It can be argued that this will be hardest on the poor and will not deter the wealthy. In that case, costs should be accelerated above a per capita minimum deemed adequate for a comfortable existence.

Next to irrigation, industry is the biggest user of water. There is little use in attempting to attract industry to increase job opportunities if there is insufficient water or if it is too costly. One must bear in mind that 18 barrels of water are required to refine a barrel of oil, 10 liters to refine a liter of beer, and that a large paper mill uses more water than a city of 50,000. Thus, a simple increase in water rates based on consumption alone is not the sole answer. Costs should be sufficiently high to encourage large users to recycle without driving them out of business.

Planning. It is now accepted in the United States that watersheds of a few thousand to a few hundred thousand acres are practical management units. Too little is known about many ground-water watersheds to justify defining an undergound management unit at this time. The watershed may be treated as a socioeconomic unit involving not only water resources but soils, minerals, and faunal and floral communities.

The Watershed Protection and Flood Prevention Act of 1954 places responsibility for watershed management in state and local communities. Federal assistance is limited to aiding local organizations in undertaking necessary work. The act deals with measures beyond the capabilities of individual farmers or ranchers but emphasizes the need to relate individual measures to those involving the entire watershed.

General measures for conservation of water are:

1. Water laws to control use among competing needs, to protect water rights of users, to maintain control over the amount of available water, and to prevent misuse and wastage

2. Planning on a watershed basis

3. Reclamation of waste water

4. More efficient reuse of industrial water

5. Improved land-use practices to control runoff

6. More efficient use of irrigation water

7. Reduction of evaporation losses

8. Reduction of transpiration losses

9. More realistic charges for water

It is interesting that the federal commission which was set up after the Civil War under John Wesley Powell recommended a minimum homestead of 4 sq mi (10 sq km), or about 2560 acres (1600 ha), per family in the arid west instead of the 160 acres (64 ha) which was considered adequate in the humid east. It recommended that, instead of rectangular plots, the allocations consider the configuration and other characteristics of the drainage network. Many of the dry-farming failures in the arid West are results of not following such recommendations.

One of the most serious water problems facing many nations is the increased mineralization and salinity of their waters. Enormous amounts of water leave industrial plants often heavily charged with waste mineral and chemical products. These are often emptied into surface streams, and much seeps into the ground to mingle with the ground water. Other minerals are contributed by irrigation waters which leach substances from the soil. Evaporation from reservoirs also concentrates salts. The result is that the waters of many rivers and aquifers are so laden with salts and other mineral matter as to be useless for agriculture. The

Colorado River is so saline in its lower reaches that millions of acres of once-productive land are rapidly becoming barren. Protests have been registered by states bordering the lower Colorado and by Mexico. In August 1973, the United States announced a settlement of the dispute with the construction of the world's largest desalting plant in southern Arizona. The overall costs, including drainage works in Arizona and a concrete-lined canal to discharge saline waters into the Gulf of California, total $115 million. The plant is designed to produce 400 million liters (100 million gal) of desalted water daily starting in 1978.

Many multipurpose reservoirs have been built along permanent desert rivers such as the Colorado and Nile. One of their important contributions is hydroelectric power. Egypt is considering a novel use of the vast Qattara basin, with an area of about 13,000 sq km (5000 sq mi) and its bottom 130 m (420 ft) below sea level. The north edge of the basin is only 68 km (42 mi) from the Mediterranean Sea. To increase its energy resources, Egypt is considering blasting a canal through the intervening mountains to lead water from the Mediterranean to the lip of the depression where it would plunge down, turning huge generators. The depression would be filled to a depth of 30 to 60 m (100 to 200 ft), and the inflow would be controlled so that it balanced the amount removed by evaporation. The problem of salt disposal would have to be resolved.

Floods

The arid lands are unique in that devastating floods may result from storms of even brief duration. This is due to the high rate of runoff from the barren slopes. Whereas in many humid regions an annual flood is the rule, and whereas in hurricane-prone areas there is ordinarily considerable warning time, there is no such regularity or lengthy warning time for desert floods. In dry regions, except in the mountains, the storms are spotty due to convectional ascent of locally heated air to colder reaches of the atmosphere. Hence, the vagaries of the weather make them more or less unpredictable. However, progress has been made in establishing a prediction system for flash floods.

Desert rainfall is erratic spatially and chronologically. In any one year, the annual rainfall may exceed the long-term average by 4 times or more and may occur in a single storm. Add to this the high rate of runoff, and it is clear why desert storms are so devastating. The destruction of grass cover in semiarid areas by overgrazing and the blanketing of ground in urban development have contributed to the flood hazard. Floods are especially severe in mountain valleys and in the lowlands immediately beyond.

The National Weather Service (formerly the Weather Bureau) has three programs contributing toward flash-flood warnings. A severe-storm unit forecasts severe weather for the entire United States. A quantitative precipitation unit forecasts precipitation expected in the next 24 hours over the country and delineates areas of expected heavy rainfall to alert local flash-flood-warning networks. The third program involves radar surveillance, which is especially useful in spotting scattered thunderstorms. Current and repetitive data from earth-orbiting satellites are proving especially valuable.

Avoidance of flood-prone areas is much more important in arid regions than elsewhere because of the greater unpredictability of flash floods. Occupation of flood-prone areas should be discouraged, and public tax money should not be used to recompense those who occupy such areas in spite of warnings. Areas already heavily settled, of course, must be protected.

The possibility should be considered of permitting agricultural development on arid floodplains only if dwellings and communities are located on higher ground elsewhere. While this would result in savings in life and property, the inconvenience would probably result in strong objections from those involved.

Erosion and Sedimentation

The effectiveness of slope erosion in dry lands is due to uninhibited runoff from the largely barren hillsides. But, by denuding bottom lands of their grassy cover, people have contributed to erosion that has rendered valueless millions of acres of once useful land.

The San Simon River in southeastern Arizona drains 500,000 ha (1¼ million acres) of public lands known as the Safford Grazing District. Immigrants in the 1870s described this as "a promised land" with a thick sea of perennial grass and wild flowers reaching to stirrup level. Antelope abounded. A clear stream lined with cottonwood trees ran down its center. The grass formed so dense a foliage and root mat as to discourage penetration by other plants. By the 1880s, 50,000 head of cattle were consuming 1000 short tons of grass a day. In a dozen years, the district had been grazed to the ground. Pounding by the hoofs of cattle compressed the roots, damaged the blades, and prevented seeding. Bare spots spread and sagebrush and cactus took over. The rains, no longer hindered by a mat of vegetation, carved great gullies. By 1900, the district was largely a wasteland. Only in 1934 did the government step in to stop the damage. Yet today, the area is still being overgrazed. Many pastures are 90 percent bare ground. If grazing is restricted to reasonable limits, the district can be restored to its original condition in 15 to 20 years.

Restoration of the overgrazed lands requires reduction of herds, building of fences, and scattering of water holes to disperse the cattle. This permits the grass to reseed itself. Rotation of cattle from pasture to pasture affords time for vigorous root-and-seed-producing growth. Mule Park, a 3.5 sq km (2 sq mi) meadow in Colorado's Gunnison National Forest, was rehabilitated from largely bare dirt to a rich sea of perennial grass 1 m (3 ft) high in a dozen years. The Forest Service in 1950 required the rancher to reduce his herd from 500 to 291 animals. In 1963, he was allowed to increase it to 490, just 10 below the original number. Conservation measures are expected to maintain the range for this number of cattle.

Sites of sedimentation in dry climates are essentially the same as those in humid climates, with a few important exceptions. Most streams in deserts die out within a short distance of their mountain sources. If they are heavily debris-laden, their rapidly decreasing vigor causes them to dump the sediment in alluvial fans, whereas in humid climates, most of this debris would be carried down the drainage system and ultimately into the sea. A second difference is that closed basins in humid regions are occu-

pied by lakes which only interrupt the journey of water and sediment to the sea, whereas the dry basins of arid lands are long-term repositories of the sediment shed from the barren slopes. Because of the absence of long-distance transport in most arid regions, sediment accumulates and submerges the lower parts of the landscape.

Permanent desert streams, however, do move large quantities of sediment downvalley. Where these streams enter reservoirs, they fill them rapidly because of the quantity of sediment swept from the barren slopes. The deltas built into the reservoirs unfortunately provide an excellent environment for the growth of water-wasteful phreatophytes.

In the desert, areas of fluvial sedimentation receive first priority for settlement because of the moisture that is commonly conserved in them and because they are relatively flat. Yet many of these sites, such as floodplains and alluvial fans, present the worst flood hazards.

Sewage sediment and other wastes, treated or untreated, are discharged into nearby channels in many desert areas. These wastes, which would be carried off by the permanent streams of humid climates, seep into the ground and cause ground-water pollution. Proper processing of waste water will be an increasingly important economic factor in desert regions.

Subsidence: Withdrawal, Hydroconsolidation, Piping

Excessive extraction of ground water can lead to compaction of buried sediments and subsidence of the ground. As an example, extraction of water from under the Tucson, Arizona, area has led to subsidence of 254 mm (10 in.) and to widespread ground fissuring.

As noted in Chapter 2, subsidence may also result from addition of water in the process of hydroconsolidation. If the sediments include considerable clay, the water may be present only as thin films around the particles but be sufficient to bind the particles together by surface tension at points of contact. When irrigation or other water floods the sediment, the water films disappear and the bond is destroyed. The weight of the overburden is now able to compact these layers at the expense of remaining pore space. Most of this hydroconsolidation takes place at shallow depths.

Also in Chapter 2, we mentioned piping subsidence due to selective removal of fine sediment by subsurface water. The phenomenon occurs in clays or clayey silts, which expand greatly when wet and shrink when dry. Because shrinkage contributes to piping, dry climates in which prolonged dessication is common provide the most favorable environment.

The piping usually extends down to the level of the nearest drainage channel, generally a few meters. However, in areas where the susceptible deposits are thick and the descent to drainage channels is great, it may be 15 m (45 ft) or more. This is the situation in parts of the great loess fields of northwest China, where depressions and natural bridges of this magnitude have been described.

The undermining and collapse associated with piping has damaged and imperiled bridges, roads, railroads, dams, canals, and other structures throughout the arid West. Bridge abutments, and their side wings (wingwalls), are especially vulnerable. Drainage should never be allowed to spill directly onto these slopes or down through joints in the bridge deck. Drainage should be piped down to the channel bed. Runoff should also be diverted from all slopes adjacent to structures.

Piping affects clay dams, leading to leakage and failure as in the case of the 16-m- (50-ft-) high dam on Caney Coon Creek in Oklahoma. The dam failed 2 days after the reservoir was filled. Piping may start on a slope below a concrete-lined canal and may work headward and undermine and damage the canal itself.

Wind Hazards

The impact of wind on human activities in dry environments is locally important. The environmental impacts are due to erosion, transportation, and deposition.

Erosion. During the drought of the middle 1930s, when crops failed over much of the western plains leaving dry ground exposed, the fertile topsoil was removed to depths of several feet by deflation. Figure 5-15 shows the kind of topographic evidence used to determine the depth of deflation; small remnants of the origi-

FIGURE 5-15. **Evidence of deflation, Danby Playa, southeastern California. The flat tops of the small hillocks once formed an unbroken surface, the bed of an ancient lake. Deflation by the wind carried the fine sediment out of the basin. The depth of deflation is as much as 4$^{1}/_{4}$ m (14 ft). (*Eliot Blackwelder.*)**

nal surface may stand like islands above the deflated surface.

Around the Mediterranean, with its long, dry summers, overuse of agricultural lands has caused a continuous shrinkage of the agricultural belt, especially in Spain and Algeria. The U.S.S.R. has encountered similar difficulties, culminating in recent years in great dust storms east of the Caspian and Aral seas. These events point out the need, especially in agriculturally marginal lands, for restraints in the use of land as described in Chapter 3.

Deflation has at least one beneficial aspect. In some desert areas, deflation basins have locally been deepened to the level of the water table. Some of the fertile oases west and southwest of Cairo originated in this manner.

Because farmlands must be denuded of vegetation during certain phases of cultivation, they are particularly susceptible to erosion by deflation. The wind winnows out the clay, silt, and organic matter, and leaves the sand behind. The residual sand is infertile and subject to migration. Measures to protect farmland against this winnowing are the maintenance of a plant or mulch cover, roughening the surface, producing stable soil aggregates, and constructing wind barriers.

The maintenance of a continuous plant cover is not feasible in most kinds of agriculture. In many dry areas, fallowing is a common practice during which the ground is churned up to provide as much surface as possible to entrap and conserve moisture. The conservation of moisture requires that the ground be free of transpiring plant growth. In the colder regions of the United States, this may mean as much as 20 months during which the ground is exposed. In southern areas, wheat is generally sown in autumn, and if growth is favorable, it protects against the strong winds of spring. If growth is delayed or poor, wind erosion may ensue. Some soils are so susceptible to wind erosion that they should be kept continuously in grass or other natural vegetation.

Crops offer different degrees of protection. Some, like wheat and soybean, provide a dense cover, while cotton, tobacco, potatoes, and others offer sparse cover. In addition, the frequent tillage necessary to control weeds in the open crops exposes the soil to further erosion. The danger, however, is greatest when no cover is present. Then, measures must be taken to reduce the surface wind velocity. This may be effected by ridges of stubble or earth. Tillage, in which the crop residues are mixed into the soil, is helpful but not fully satisfactory, because mulch should lie on the surface.

The formation by tillage of a soil structure that will protect against wind erosion is often beyond the farmer's control. Many soils have a tendency to clod, the size of the clods being determined by the characteristics of the soil. Clods larger than wheat kernels are generally secure against removal by the wind. Large clods, however, do not form a continuous cover.

The construction of barriers to the wind is a widespread practice. In strip cultivation, bands of dense, wind-resistant crops are alternated with more susceptible types. Or, planted strips may be alternated with fallowed strips. Another procedure is to plant lines of shrubs or trees, or to construct fences, to protect relatively large areas to leeward.

Wind abrasion, a form of natural sandblasting, results from the sweep of sand-laden air against obstructions. Rock exposures in the desert may be worn smooth, or—if composed of intermixed weak and resistant materials—may be etched into intricate detail. Exposed window panes may be frosted and paint scoured off exposed surfaces in a single sandstorm. Wooden telephone poles have been undercut in a single season; others are shredded by the crystallization of brine (Figure 5-16).

FIGURE 5-16. **Utility poles in the desert.** *(a)* **Utility poles near Palm Springs, California. The poles are sheathed in metal to protect them from sand erosion.** (*Southern California Edison Company.*) *(b)* **Telephone pole shredded by crystallization of brine soaked up from below. Great Salt Desert, Utah.** (*Eliot Blackwelder.*)

(a)

(b)

Structures should be protected by restricting the moving sand or by using only durable construction materials. Metal and brick are preferable to wood for structures exposed to the wind, and metal shutters for windows are advisable. The restrictive and protective measures are necessary only within the few feet of the ground to which sand movement is confined. This explains why wooden poles are sheared off near the ground rather than higher up. Dust, on the other hand, although lifted to great heights, is not effective in abrasion.

Transportation and Deposition. Sand grains are relatively large and heavy and are not transported very far by the wind. Dust, however, may be carried to heights of 3.2 km (2 mi) or more and drift around the earth. Photos of so-called sandstorms extending to heights of 1500 to 3000 m (5000 to 10,000 ft) actually portray dust storms. Dust may bury fertile agricultural land (Figure 5-17) under a cover that is at least temporarily infertile. In time, however, it weathers to fertile soil. Vast areas are covered by ancient, hardened dust deposits known as "loess" (pronounced "luss") (Figure 5-18).

Figure 5-19 shows the distribution of loess in central North America. In some areas, the dust was whipped from river floodplains. Because of prevailing westerly winds, a large area of loess lies east of the Mississippi floodplain. Close to the valley, its thickness locally exceeds 30 m (100 ft). In Alaska, thicknesses of up to 60 m (200 ft) have been encountered. Other sources of loess in North America are glacial sediments and ancient alluvial plains. Extensive areas of loess are also found in northern Europe and in Asia.

Dust suspended in the air can be carried great distances and settle everywhere over the topography. Sand, moving almost entirely along the ground, is more amenable to control.

Methods of protection against drifting sand are applicable wherever sand occurs, whether it be along the shores of seas and lakes, on the lee side of sandy floodplains, or leeward of sand deposits spread out by ancient glaciers. They are used to protect highways, railroads, buildings, fields, and desert oases (Figure 5-20).

The rate of movement of sand dunes depends on the velocity and persistence of the winds, the constancy of wind direction, the presence and density of vegetation, the size of the dunes, and the size of the sand grains. Large dunes may move up to 6 m (20 ft) per year; small ones may exceed 23 m (75 ft) per year.

Sand may be unstable because a permanent plant cover has not yet been established or because an already established cover has been destroyed by cultivation, overgrazing, fire, or

FIGURE 5-17. **Drifting top soil has partially buried these farm buildings in Beadle County, South Dakota. During the "dust bowl" days of the 1930s, roads were blocked, crops and farm machinery were buried, and fields were converted to arrays of hummocks.** (*U.S. Dept. Agriculture.*)

FIGURE 5-18. **Ancient wind-deposited dust (loess), Shensi Province, China. Loess is cohesive and can stand in vertical cliffs. Because of its insulating qualities, dwellings in loess are cool in summer and comfortable in winter. Walls may be trimmed smooth and whitewashed. During earthquakes, the bonds between the dust grains are loosened and the material moves in rapid fluid flow. During the earthquake of 1920 in adjacent Kansu Province, loess flows inundated villages in valley bottoms and snuffed out 180,000 lives.** (*Arthur D. Howard.*)

careless construction. Stabilization involves halting the sand by establishing permanent vegetation on it. The sand can be stilled temporarily by other means, such as artificial barriers.

Figure 5-21 shows the use of grass to stabilize dunes in Libya. The choice of vegetation for a particular locality is a matter for botanists. Sand-stilling vegetation must be able to thrive in moving sand and either survive temporary burial or keep pace with deposition. The commonly used plants have coarse, stiff stems to resist sandblasting, are unpalatable to livestock, and are fast-growing on a starvation diet. These plants seldom provide a complete cover. The reader is referred to the U.S. Corps of Engineers publication "Dune Formation and Stabilization by Vegetation and Plantings" for a discussion of plants suitable for beaches in the United States.

Artificial barriers and restraints to still the movement of sand that cannot be planted effectively include application of asphalt and oil, covers of clay or gravel, fences, brush matting, and surface mulches.

Emulsfied asphalt cut with 75 percent water has been used in North Carolina in emergencies. The applications penetrate 2.5 cm (1 in.)

into the sand and do not crust. Thus, seedlings are able to germinate and grow.

Hot crude oil has been sprayed on sand in arid and semiarid dune areas. Heavy oil has also been used to stabilize cuts in sand and to protect a railroad in West Africa.

Clay as a protective cover is expensive. A thickness of 10 to 15 cm (4 to 6 in.) will permit permanent seeding or planting.

Gravel or crushed rock has been used as a veneer to protect railroad cuts in sand in Oregon and Washington. Coverage must be complete, with fragments too large to be moved by the wind. A 5-cm (2-in.) thickness was used on 18° slopes and a 10-cm (4-in.) thickness on steeper slopes.

Fences are used either to still sand, in

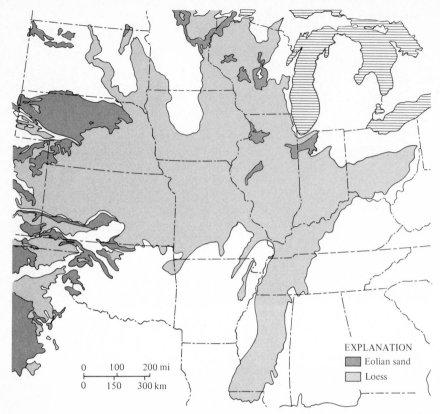

FIGURE 5-19. Distribution of windblown sand and loess in part of central North America. (*After R. F. Flint, "Glacial and Quaternary Geology," John Wiley & Sons, Inc., 1971.*)

FIGURE 5-20. Sand submerging part of oasis of Ouargla, Algerian Sahara, 1952. (*Arthur D. Howard.*)

FIGURE 5-21. **Use of grass in dune stabilization in Libya. This is a stage preliminary to forestation.** (*A Shell Photo.*)

which case they are arranged approximately at right angles to the prevailing winds, or to direct the wind elsewhere, in which case they are oblique to the prevailing winds. Picket fences are common (see Figure 4-26). Commercial snow fencing is commonly employed. In some places fences of palm leaves or brush are used (Figures 5-22 and 5-23).

Brush matting, as a surface mulch, temporarily checks sand movement, particularly in deflation hollows. The brush is laid in rows, with butts of the plants pointing into the wind. Succeeding rows overlap like shingles. Perma-

nent plantings of trees may be set into the brush mat. Brush mats are generally impractical where wind velocities exceed 65 km (40 mi) per hour. Fire is always a hazard.

FIGURE 5-22. **The sand dunes of Tit, west of In Salah in the Sahara. The crests of these dunes are "planted" with palm leaves to prevent them from encroaching on the oasis.** (*A. W. Pond, from "The Desert World," with permission of the reprint publishers, Greenwood Press, a division of Williamhouse-Regency, Inc.*)

FIGURE 5-23. **Brush barriers impeding movement of sand dunes along lakes shore, Muskegon County, Michigan. In the background is a stabilized area on which vegetation is well established. The foreground is a large blowout or deflation area caused by a break in the vegetation cover. (*U.S. Dept. Agriculture.*)**

Hay mulches may be used as a temporary measure in deflation hollows or in small sand areas. The mulch deteriorates rapidly, and so permanent seedings or plantings must be made as soon as weather permits. Fire is also a hazard.

Sand-stilling vegetation and artificial constraints are only temporary measures in dune stabilization; the final objective is the establishment of permanent vegetation, the native vegetation of the area. This may be grasses, shrubs, trees, or combinations. The permanent vegetation is established after the sand has been stilled.

Pines are commonly planted on dunes because of their preference for sandy soils. Mixed plantings of shrubs and trees are recommended, however, to ensure that the cover is not completely lost if the trees are killed by disease or insects. The most commonly used shrub along the West Coast is Scotch broom, which provides good protection, is aesthetically pleasing, and supplies nitrogen to the soil.

Where stabilization is impractical, sand may be *bypassed* around threatened areas. In 1952, burial of the date palms and part of the village of Ouargla, Algeria, by encroaching sand (Figure 5-20) was being thwarted by transportation of sand by truck from the windward to the leeward side of the oasis. To protect a railroad in Southwest Africa, solid fences leaning toward the roadbed were constructed in the hope that, under strong winds, the sand swept up these surfaces would leapfrog the tracks. The method was successful at some times, unsuccessful at others.

MANAGEMENT OF THE DRY LANDS

General Problems

Because fresh water is the determining factor in successful occupation of dry regions, because it is in limited supply, because it must commonly be imported from great distances, because it is not often used wisely, and because its withdrawal in one place may affect areas nearby or far-removed, water resources must be managed on a national or even continental scale. A complete national water plan for the United States will not be achieved in the near future because the country is so vast and consists of so many large watersheds. A preliminary national plan should be set up as soon as possible, however, and local projects should be planned to fit into this overall plan. The national plan should not be restricted to water resources; it should include all vital resources.

Wise management of dry lands requires many difficult decisions. For example, how much of the limited water supplies should be devoted to dry region agriculture which, in spite of the great quantities of water used, provides limited employment and returns in comparison with industry? Should industry be given preference to agriculture in these regions? If so, what can be done about air pollution where there is little rain to wash the air clean and frequent temperature inversions inhibit natural convection? Can we continue to expand irrigation without solving the problem of increasing salinity as the salt-enriched waters return to the system? What quantities of water should be imported from regions with surpluses? What might the climatic and other long-term effects of such tampering with the natural environment be? Note, for example, the increasing incidence of fog in heavily irrigated areas. To what extent can water supplies be stretched by cloud seeding, desalinization, and recycling? How shall the recycling wastes be disposed of? Should population be restricted in the dry regions? These and many other questions require careful study.

Piecemeal steps toward conservation and protection of lands and waters have, of course, been taken. In 1933, Congress created the National Resources Planning Board, the Tennessee Valley Authority, and the Soil Erosion Service. In 1934, the soil erosion service was enlarged into the Soil Conservation Service. Also in 1934, a timber shelterbelt program was initiated, and the Taylor Grazing Act was enacted to prevent overgrazing and soil deterioration of the public range. In 1936, the Department of Agriculture was authorized to construct reservoirs, ponds, wells, and other facilities in arid regions. Also in 1936, the Flood Control Act was passed, encouraging states to enter compacts with the federal government on flood-control measures. These and other measures have been enacted as needs became evident, but not as part of a national plan. The National Environmental Policy Act of 1969, considered in Chapter 15, was a major step toward a national policy.

A start toward coordinated management of arid lands was the provision for cooperation between the 11 Western states and Alaska and Hawaii. All these states have dry areas. The coordinating links are a governors' council, the Western States Council for Economic Development, and the Western Interstate Commission for Higher Education. Unfortunately, the scope is too narrow and actual operation too slow.

Management of Dunes

There remains the problem of our extensive sand dune areas. The sand dune area of Nebraska alone is larger than the state of New Hampshire. The U.S. Department of Agriculture lists seven general uses for dune areas and the precautions to be taken.

1. *Recreation.* Trails and roads should be covered with gravel, crushed rock, or asphalt to prevent destruction of the sod or other protective ground cover. Elevated boardwalks which involve little contact with the fragile surface are even better. Similar attention must be given to parking, camping, bathing, and other facilities within the dune area. Precautions must be taken against fire.

2. *Wildlife.* An adequate cover is needed for shelter and food. Fire precautions are essential.

3. *Homes.* Homes should be prohibited in areas of unstable sand. Where built among dunes, their construction should not disturb the natural cover unnecessarily. Roads and trails should be promptly surfaced to prevent deflation. Breaks in the sod should be repaired at once.

4. *Woodlands.* Woodlands on dunes provide fuel, posts, and pulpwood, and the under-

growth furnishes Christmas greenery, including salal, sword fern, cascara, huckleberry, and bearberry. Seed trees of native species should be included in the woodlands to ensure permanence. The cutting of trees should be regulated.

5. *Watershed cover and storage.* Because of their permeability, stabilized dunes act as ground-water reservoirs and regulate streamflow, as in the Sandhills of Nebraska. A good cover prevents clogging of channels with sand, retards runoff, and increases infiltration to ground water.

6. *Special crops.* Soils suitable for crops may exist in the troughs between dunes or in the flat areas between complexes of stabilized dunes. Some of these sites are used for such crops as bulbs, blueberries, and cranberries. Only enough cultivation to control weeds should be undertaken, windbreaks should be established, and crops should be rotated to restore productivity of the soil.

7. *Grazing.* Grazing on dunes must be carefully controlled to ensure an adequate cover at all times. Measures include careful consideration of stocking rates, rotation and deferred grazing, and dispersal of salt sites and watering places to encourage dispersal of cattle. Deflation hollows should be repaired immediately with hay mulch covered by brush to keep cattle out.

ADDITIONAL READINGS

American Scientist The Development of Israel's Water Resources, vol. 60, 1972, pp. 466–473.

Cooke, R. V., and A. Warren: "Geomorphology in Deserts," University of California Press, Berkeley, 1973.

Davis, S. N., and R. J. M. DeWiest: "Hydrogeology, " John Wiley & Sons, Inc., New York, 1966.

Dregne, H. E., ed.: "Arid Lands in Transition," *American Association for the Advancement of Science*, 1970.

Hodge, C., and P. Duisberg, eds.: "Aridity and Man," *American Association for the Advancement of Science*, Publ. No. 74, 1963.

Hornby, W. F., and P. Newton: "Africa," University Tutorial Press, Ltd., London, 1974.

Leopold, L. B., M. G. Wolman, and J. P. Miller: "Fluvial Processes in Geomorphology," W. H. Freeman and Company, San Francisco, 1964.

Pond, A. W.: "The Desert World," Thomas Nelson & Sons, New York, 1962.

Thomas, H. E.: "The Conservation of Ground Water," McGraw-Hill Book Company, New York, 1951.

U.S. Corps of Engineers: "Dune Formation and Stabilization by Vegetation and Plantings," *Beach Erosion Board*, Tech. Memo. No. 101, 1957.

U.S. Dept. of Agriculture: "Water," *The Yearbook of Agriculture*, Washington, D.C., 1955.

Walton, K.: "The Arid Zones," Aldine Publishing Company, Chicago, 1969. -

CHAPTER SIX

tropical rain forest environments

INTRODUCTION

Distribution of Forests

Forests cover one-third of the land area of the earth. They form broad latitudinal zones, each zone having characteristics determined by global climates (Figure 6-1).

The polar regions are mapped in Figure 6-1 as "tundra and ice." Outside the tundra zone is the taiga, or coniferous forest, in which the common trees are spruce, pine, fir, hemlock, cedar, and, in a narrow coastal strip in California, the giant redwoods. Beyond the coniferous zone is the temperate zone forest, including relatively small areas of temperate rain forest and large areas of deciduous forest with trees such as oak, beech, and maple. Finally, in the equatorial zone, are the tropical forests, consisting of the always-humid rain forest and the winter forest with its short, dry winter season. The perennially humid rain forests, in which teak, mahogany, ebony, and other broad-leafed trees are common, remain evergreen because of the year-round growing season. The trees of the winter forest outside the perennially humid areas lose their leaves at the beginning of the dry winter season. Figure 6-1 combines tropical rain forest and winter forest under "tropical rain forest" and shows other vegetation zones in addition to those considered above. The grasslands of

153

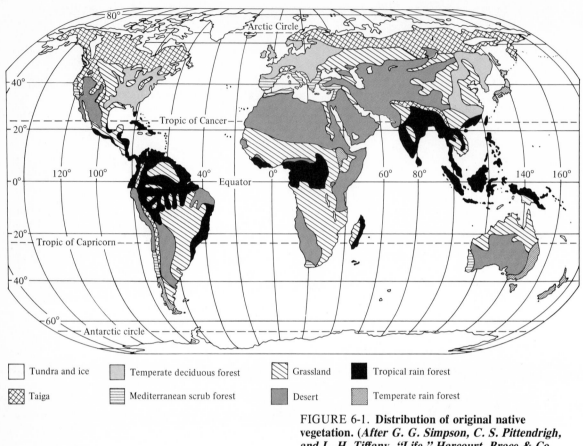

Tundra and ice

Taiga

Temperate deciduous forest

Mediterranean scrub forest

Grassland

Desert

Tropical rain forest

Temperate rain forest

FIGURE 6-1. **Distribution of original native vegetation. (*After G. G. Simpson, C. S. Pittendrigh, and L. H. Tiffany, "Life," Harcourt, Brace & Co., 1957.*)**

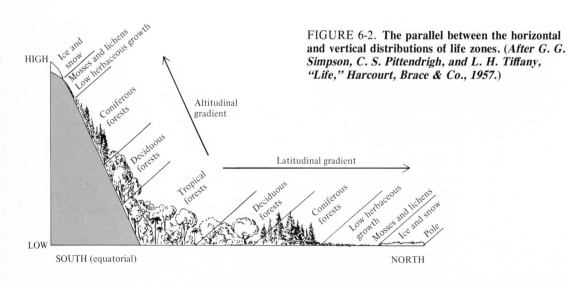

FIGURE 6-2. **The parallel between the horizontal and vertical distributions of life zones. (*After G. G. Simpson, C. S. Pittendrigh, and L. H. Tiffany, "Life," Harcourt, Brace & Co., 1957.*)**

the tropical and subtropical zones are the savannas.

Forests are zoned not only according to latitude but also according to elevation (Figure 6-2). In the tropics, one need ascend only a few kilometers in high mountains to encounter climatic and vegetative zones that occur thousands of kilometers farther north or south in the lowlands.

The boundaries of the zonal forests are highly irregular. Certain forest types grow far north of their usual limits because of the influence of relatively warm ocean currents. Examples are the tropical forests in Florida and Bermuda, well outside the tropics. Elevation, too, causes deviations in pattern. Thus, the northern coniferous forest has three tentacles extending far to the south in North America. One extends down the mountains of the Far West into California, a second extends down the length of the Rocky Mountains, and a third extends down the Appalachian chain.

Human Invasion of the Tropical Rain Forests

The tropical rain forests are being invaded at an increasing rate due to expansion of agriculture, lumbering, and mining. Exploitation of the resources of the Amazon and Congo basins has involved the building of roads and railroads, dams, processing facilities, and complete towns. The accelerating impact of people on this unique environment is a cause for concern.

TROPICAL RAIN FORESTS

The Tropical Climatic Zone

The tropical zone owes its characteristics to global location and global atmospheric circulation (Figure 5-3). The location, straddling the equator, makes it the warmest zone on earth. Because the sun shifts only $23^1/_2°$ north and south of the equator, average monthly temperatures vary only 5.5°C (10°F) from the annual average of 27°C (80°F), although in places the daily changes exceed this.

At the equator, the heated air holds large quantities of moisture, and the hot, moist air forms a low-pressure belt known as the *doldrums*. Rainfall generally exceeds 200 cm (80 in.).

The doldrum belt has a width of about 10° in latitude. Because global pressure belts shift north and south with the sun, areas marginal to the doldrums periodically come under the influence of the trade winds, or locally of monsoon winds. Variations in this simple picture are due to distance from sources of moisture, elevation, and size and orientation of mountain ranges. Highlands have lower temperatures, particularly at night; they generate orographic rainfall and create rain shadows in their lee. Thus, dry areas are found in the lee of some mountain barriers, as well as in marginal parts of the tropical zone.

The largest rain forest regions are the Amazon basin, the Congo basin, and Southeast Asia (Figure 6-1). In these regions, rainfall occurs almost daily throughout the year. Torrential rains occur in the afternoon when the ground and air are hottest. The expanded warm air rises and chills at high levels to liberate moisture. Cumulo-nimbus clouds up to 3.2 km (2 mi) thick may develop within a half hour, and the vertical air currents may reach 160 km (100 mi) per hour.

Sunshine incidence at ground level is lowest at the equator because of the high humidity which filters out the shorter wavelengths. It increases to the north and south. There is a diurnal rhythm of sunshine; mornings start out clear, afternoons are cloudy and rainy, and the skies clear toward sunset.

Humidity varies in the doldrum belt. Coastal areas are more humid than inland areas, and highlands are more humid than plains. Tropical areas with relatively dry seasons enjoy periods of low humidity.

Rain Forest, Winter Forest, Jungle, and Savanna

The perennially humid *rain forest* is restricted to areas with a warm, moist climate throughout the year, whereas *winter forest* is largely restricted to the margins of the tropical belt where there is a dry winter season.

In the rain forest, the trees remain evergreen because of the uninterrupted growing season. Leaves are shed, but they are not shed synchronously. The canopy is so thick as to create almost perpetual twilight. Undergrowth is suppressed so that the forest floor is relatively open. *Jungle* is rain forest in which the tall trees are less densely packed and the ground is covered by an almost impenetrable jumble of shrubs and other small plants. Jungle generally borders rivers and other areas subject to periodic inundation (Figure 6-3).

In the rain forest are many epiphytes, plants including orchids, ferns, and wild figs, whose seeds take root in debris trapped in crannies high above the ground. Some, like the strangler fig, send cablelike roots toward the ground. Once contact is made, the food supply increases enormously, and the epiphyte grows rapidly and sends additional roots downward. Some of these encase the original tree, literally squeezing it to death. The host tree finally rots away, leaving what appears to be a hollow tree. In addition to the epiphytes, hordes of climbers (lianas) use the trees as supports (Figure 6-4). These, too, may eventually kill the host tree.

FIGURE 6-4. **Amazon rain forest at Belém. The climbing plants encasing the bole of the large tree may eventually kill it.** (*Arthur D. Howard.*)

FIGURE 6-3. **Amazon jungle near Belém. The dense undergrowth in these periodically flooded areas makes penetration difficult.** (*Arthur D. Howard.*)

Locally, within the rain forest, but largely beyond its margins, there are areas of grassland known as *savannas*. Some of these within the rain forest are natural; others are the result of destruction of the forest by humans. Some may have formed in earlier, drier climates.

In the winter forest, leaves are shed at the beginning of the dry winter season. This synchronous shedding of leaves rids the broadleafed trees of moisture-using appendages.

Topography

Rain forests are not restricted to lowland plains or to regions of low relief. They are found above 3000 m (10,000 ft) in New Guinea and in the Andean headwaters of the Amazon. Figure 6-5 shows the rain forest in the high mountains of Venezuela.

FIGURE 6-5. **Rain forest in the mountains of Venezuela protects the soil from erosion and holds it in place on steep slopes. The rapid downward stream erosion of the valley bottoms oversteepens slopes, with resulting landslides, avalanches, and soil creep. (*U.S. Air Force.*)**

The presence of dense forest on steep slopes is common. The preservation of soil on such slopes is due to the forest canopy which impedes the rain on its way to the ground, to the forest litter which acts as a sponge to retard runoff, and to the webbing of roots.

The rain forest changes gradually with altitude. In Southeast Asia, for example, such familiar temperate-climate trees as the oak and chestnut become more common with increasing elevation. At still higher elevations, the trees diminish in size, and the long, straight-bole trees of the lowlands become twisted and gnarled. Above 1500 m (5000 ft), the forest is draped in pendants and festoons of moss. Above 2000 m (6500 ft), grasses and shrubs blanket increasingly large areas of open and rolling landscapes. Above this is the relatively barren alpine environment.

Rain Forest Ecology

Sunlight and Soil Factors. Part of the complexity of the rain forest association may be traced to the *sunlight factor*. Some plants require exposure to sunlight throughout their lives, others require sunlight only in the germination period, while still others do not grow at all in sunlight. In the rain forest, the tall trees struggle for sunlight. Most form a dense canopy, the roof of the forest, which varies in height in different regions. In the Amazon it is 30 to 35 m (100 to 120 ft) high. Above this protrude scattered emergent giants.

Soil, in the agricultural sense, is thin and deficient in nutrients. Because of the warm temperatures and abundant rainfall, vegetable matter is rapidly decomposed, rendering the percolating waters acidic. These waters convert some of the rock-forming minerals to soluble substances which are rapidly leached out of the soil, leaving less soluble materials, including iron oxides. The soil is thus impoverished. Plant nutrients, however, are continuously introduced into the topsoil by decay of the humus. These are primarily responsible for the continued existence of the forest. The forest feeds on itself. If the forest is cleared so that there is no longer accumulation of litter, the topsoil rapidly loses its nutrients and becomes infertile.

Root Systems and Buttresses. Because only the thin upper layer of soil in the rain forest is fertile, there is no reason for deep roots. Instead, even the tallest trees have shallow roots with tentacles probing widely through the fertile layer (Figure 6-6).

The lack of deep roots explains many of the novel supporting structures at the base of the trunks. These generally consist of exposed roots in one form or another. Some trees, like the strangler fig (Figure 6-7), develop a radial

FIGURE 6-6. **Shallow, widely spreading roots in thin tropical soil. Near Telok Chempedak, east coast of Malay Peninsula.** (*Cathy Conner.*)

FIGURE 6-7. **Buttressed finlike roots of the strangler fig.** (*Arthur D. Howard.*)

set of finlike buttresses. Others, like the pandanus tree (Figure 6-8), have a tepeelike arrangement of stiltlike roots. The banyan tree, like the strangler fig, starts from epiphytes in the limbs of a host tree. It sends out a network of roots which hang from the branches and grow until they become rooted in the ground. The pendant roots become thicker and ramify. The host is ultimately killed, and what appears to be a grove of trees remains. Other trees have ribbed or fluted trunks for added strength. Some trees dangle roots down, not for support, but to tap additional soil and water.

Trees with broad, high buttressed trunks can be difficult to fell. Scaffolding is generally erected to permit workers to operate above the buttress level. The hardness of the tropical wood presents another difficulty. Still another

disadvantage is that the fall of one tree may carry with it neighboring trees due to the intimate lacework of strong creepers and vines.

Abundance of Species. Tropical rain forests differ from temperate zone forests in one very important respect. In temperate zone forests, there are extensive stands of individual tree types. A pure stand means that the temperature-moisture-soil conditions are ecologically favorable for that tree type, which therefore enjoys an advantage over its competitors. In an adjacent area, the ecological factors may be sufficiently different to give some other tree type the competitive edge. In the rain forest, however, the ecological factors are amazingly constant except in restricted local environments. These local environments provide stands of mangrove, Nipa palm, and sago palm.

FIGURE 6-8. **The pandanus tree. The stiltlike roots help support this top-heavy tree.** (*Arthur D. Howard.*)

For the most part, however, the uniform temperature, moisture, and soil conditions provide little advantage for one tree type over its neighbors. Thus, there is a thorough mixture of species. In Southeast Asia, as many as 30 tree species have been reported in a plot 100 m (330 ft) square. In Guyana, 86 species more than 5 m (15 ft) high have been reported in an area of 2 ha (5 acres). In the Amazon rain forest, there may be as many as 3000 species of plants per sq mi (2.6 sq km).

Processes and Landforms

The geological processes and landforms of the tropical rain forest are fundamentally the same as those of the temperate zone. The processes, however, have different relative importance.

Mechanical disruption of rocks is negligible compared to chemical alteration, fluvial erosion is retarded by the dense forest canopy and the mat of forest litter, and coastal erosion is affected by stands of mangrove and by beach rock and coral reefs.

Drastic climatic changes have taken place in the rain forests during the last 2 million years, the Quaternary Period. Forests have spread and contracted, as have the grasslands. Thus, some of the landscape features are residual and owe their characteristics more to past climates and processes than to those of the present.

Weathering. Physical weathering is unimportant at low elevations in the tropical rain forest. Frost, however, may occur at high elevations. During cloudy days, the ground is shielded from solar radiation, and if clear, calm nights follow, frost is probable in areas of light plant cover. If the area is heavily forested, warm air is preserved under the canopy and discourages freezing. Because the temperature changes are largely diurnal, they do not penetrate deeply. Hence, frost disruption of rocks is superficial, and only fine debris is produced. To protect tea gardens in Java on frost-susceptible slopes, narrow, deep ditches are dug trending downslope to drain off the heavy cold air.

The intense chemical weathering, with removal of all readily soluble materials, leaves an infertile reddish soil consisting largely of clay and iron oxides. We shall refer to these reddish leached soils as *lateritic soils*. Under certain conditions, hard lateritic crusts may develop. Their development seems to require exposure to drying conditions and partial dehydration. Drying and hardening to form crusts has resulted from the clearing of forests. The clearing removes the shade cover and exposes the ground to the sun. Dehydration may also result from erosion and lowering of the water table. Some of the laterite crusts in parts of the tropical rain forests, however, may be holdovers from a former, drier period.

Because of their resistance to erosion, laterite-crusted areas remain as tablelands, or hills and ridges. Laterite crusts are locally rich enough in iron, manganese, or aluminum to serve as ores. In Ghana, aluminum-rich laterites (bauxites) cap the highest tablelands, where they form crusts up to 12 m (40 ft) thick.

The formation of laterite crusts involves a reduction in volume. Some crusts display a shallow, broken topography, with subsidence depressions and low scarps. Landslides result when the laterite crust is undermined. Many slopes show jumbles of laterite blocks from earlier falls and slides.

Vegetation on laterite crusts is especially sparse and stunted. This is due to the infertility of the soil and to the high porosity and permeability. The porosity and permeability are due to the leaching of soluble materials, to the presence of root tubes, and to fractures. Rainwater sinks readily through the crust, and insufficient moisture is retained for normal forest growth.

Steep-head valleys are common. Water seeping down through the lateritic crust and on through the soil below is stopped when it encounters fresh rock at depth. The water escapes laterally as seeps, which, by undermining the slopes above, results in steep valley heads. Some of these steep-head valleys are kilometers in length and by their ramification may convert broad tablelands into a maze of steep narrow valleys separated by narrow winding divides. Many of the commercial and residential areas of Salvador, Brazil, are concentrated either on the relatively narrow but flat floors of these steep-head valleys or on the narrow divides between them. Landslides are common on the steep slopes, and torrential rains wash away poorly constructed dwellings and inundate the valley floors.

Mass Movements. Limestone underlies large areas of rain forests, as in the East Indies. As in any limestone region, the sudden collapse of the roof of an underground cavern is possible. Some tropical limestone regions, where solution is well advanced, are extremely rugged (Figure 6-9). The terrain appears as though diced with a cleaver, with a maze of steep-sided hills and ridges, a mosaic of undrained sinks, and an occasional stream which may pass through the area or disappear underground.

On all steep slopes in the rain forest, landslides must be expected. For the most part they involve only the deeply weathered soil. On hillsides, the contact between the weathered mantle and bedrock is often sharp rather than gradational (Figure 6-10). This indicates sliding of the weathered material over the bedrock.

An interesting variety of *soil flow* is common but not restricted to tropical rain forests. It takes place on gentle to moderate slopes under the mat of vegetation. As a result of subsurface soil flow, the ground surface, with its vegetation intact, may settle as much as 0.3 to 0.6 m (1 to 2 ft). The vegetative mat is bulged and often breached farther downslope.

Stream Erosion and Deposition. Erosion under a tropical rain forest canopy, except in the mountains, differs considerably from that in unforested areas. The dense canopy prevents part of the precipitation from reaching the ground. Instead, this water evaporates after the rain ceases. Much of the precipitation that does reach the ground is transpired and also plays no role in erosion. The remainder may contribute to erosion, depending on the quantity, the amount of forest litter, and the slope.

There is no immediate concentration of runoff in channels of rain that falls in forested areas. The raindrops seep into the mantle of forest litter. If the rain is intense enough, the litter becomes saturated and the ground below becomes soaked. Runoff may then take place as sheet wash and rills *under the litter*, invisible at the surface. This runoff, as it gathers into streams, may eventually become effective in erosion.

Slope influences runoff as in other environments. Although retarded by the dense forest cover, gullying may take place wherever slopes are oversteepened by undercutting or landsliding, or where vegetation is destroyed by natural events such as avalanching, or by human activities. The deeply weathered soil is especially susceptible to gullying, and gullies of great size and depth are common.

In the mountains, the heavily vegetated slopes remain steep as they recede under vigorous erosion at their bases, and landslides and other mass movement above. Opposing slopes, therefore, meet in sharp, steep-sided ridges in contrast to the smooth, rounded topography of humid climates elsewhere. In the mountains, the rate of erosion may be enormous. The efficacy of erosion is due to the steep slopes, torrential rains, and deeply weathered soils.

FIGURE 6-9. **Rugged karst topography (an advanced stage of solution in fractured limestone.) Dry Harbour Mountains, northcentral Jamaica, West Indies.** (*Courtesy Raymond Wright.*)

Away from the mountains, stream gradients are flatter than those in temperate humid climates. Because of the thorough weathering, the sediment carried over the lowlands by rain forest rivers is fine, ranging from clay to fine sand. Such material can be transported on very low gradients. In the reaches between rapids or falls, the gradient may be so low that there seems to be little if any current at times of low water. Even at rapids, erosion of the resistant rock in the riverbed is unusually slow because of the fineness of the sediment available for erosion.

FIGURE 6-10. **Red lateritic soil (upslope) resting on gneiss. The slope between the soil scarp above and the rock scarp below has been bared by an earthslide. Slides are commonly indicated by sharp contacts between soil and bedrock. Salvador, Brazil.** (*Arthur D. Howard.*)

The great lowland rivers, such as the Amazon, Congo, and their larger tributaries, engage in geological activities similar to those of large rivers elsewhere, but with a few differences. We will use the Amazon as an example.

The Amazon River, 6300 km (3900 mi) long, and up to 16 km (10 mi) wide exclusive of the enormous width at its mouth, drains 6 million sq km (2.3 million sq mi). (Figure 6-11). Its average discharge is 10 times that of the Mississippi River and accounts for 15 percent of all the fresh water discharged into the oceans. Its basin has been leached of soluble materials to such a degree that the waters are almost entirely free of chemical impurities. Whereas the Amazon and some of its large tributaries that rise in the Andes are turbid with fine sediment, the sediment consists of insoluble particles of clay and silt, which do not affect the chemistry of the waters.

At low water, the Amazon is braided in many places. That is, it flows in a number of channels that divide and recombine continuously. It wanders across a broad, low floodplain known as the *várzea* (vár-zee-uh). This low-water plain, with relief of a few meters, consists of silty alluvium and, because of annual flood replenishment, remains fertile. The várzea is

FIGURE 6-11. **The Amazon drainage basin, showing the principal tributaries. The greater part of the basin is south of the equator. When the doldrums (see Figure 5-3) are south of the equator, floods in the southern tributaries contribute to high flood levels in the Amazon. When the doldrums are north of the equator, floods from the smaller streams do not contribute as heavily to the Amazon discharge. This factor, plus the seasonal snow melt in the Andes, explains the great seasonal variation in flow of the Amazon as compared to the Congo. (After B. J. Meggers, E. S. Ayensu, and W. D. Duckworth, eds., "Tropical Forest Ecosystems in Africa and South America: A Comparative Review," Smithsonian Inst. Press, 1973.)**

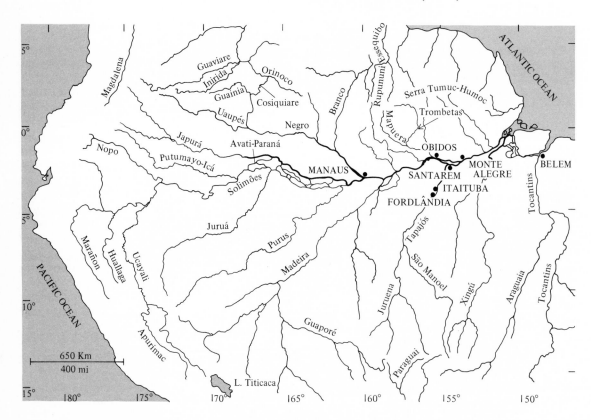

largely covered by high grass and patches of trees and brush and is generally good agricultural land. Along many stretches, the channels are flanked by forested natural levees. During the annual floods, the várzea is largely inundated. The higher ground against which it abuts, the *terra firma*, is out of reach of the floods.

Along some tributaries, there is no várzea; the river flows between slopes of terra firma, and the rain forest reaches the river. Elsewhere, as along the Rio Negro, the river channel is flanked by strips of low plain several kilometers wide which are inundated during much of the year. Here is dense forest of smaller trees, with roots almost always under water.

The várzea has extensive backswamps between the levees and the terra firma and between the levees of the various channels and tributaries. It also contains numerous lakes, most of which are abandoned channels. The annual inundation adds new alluvium to the várzea. At the same time, erosion takes place where suitable conditions prevail, so that there is a redistribution of sediment, with creation of new bars and islands and destruction of others.

The Amazon and its tributaries erode their banks as do temperate zone rivers. A unique type of bank erosion occurs when flood waters start to rise and soak the lower parts of the banks. These lower parts fail and the upper part breaks away, forest and all. The failure may propagate upstream and downstream for kilometers. Often, slumped sections form extensive temporary terraces. Entire stretches of forest, often with huts, people, and animals, are torn away.

Although the Amazon and many of its large tributaries are braided for considerable distances, many of the tributaries have meandering (sinuous) reaches with successive arcuate bars indicating progressive growth of the meanders (see Figure 14-12, locality B, for a view of such bars).

From the mouth of the Amazon, 200 km (125 mi) wide, an enormous volume of fresh water, 175,000 cu m (about 6 million cu ft) per sec, flows far out to sea while being deflected to the northwest by ocean currents. Amazon sediments are thus spread northward as far as the French Guiana coast and contribute to the regularity of the shoreline. On the other hand,

for 500 km (over 300 mi) southeast of the Bay of Marajo, the easternmost mouth of the Amazon, the shoreline is highly indentate. Here, in the absence of drifting Amazon sediments, the shore retains the erosional irregularities inherited from the last low stand of the sea.

Although the Amazon is not building a clearly defined delta, sedimentation offshore is rapid and widespread. Sedimentation at the mouths of other tropical rivers has isolated coastal communities from the sea. Shallow flats are built up offshore, change constantly, and navigation becomes hazardous. Thus, in Vietnam, Ho Chi Min City (Saigon) is accessible to vessels of 9-m (27-ft) draught, but shifting bars are a constant danger.

Because of the vast shallows north of the main channel of the Amazon, a destructive tidal bore is created during the high "spring" tides. As the broad tidal wave enters the offshore shallows, the sudden restriction in depth causes it to pile up abruptly as a wall of water locally more than 4m (12 ft) high, covering the whole horizon. The impact of this tidal bore on the shallows, the islands, and the low-lying coast is devastating. Bottom topography is radically changed with new patterns of shoals and channels. Small islands have their dense forest cover sheared off and may disappear. Other islands are severely eroded. Forest debris is scattered everywhere. In other places, however, sedimentation creates new bars, and old islands are enlarged.

The Congo drainage basin (Figure 6-12), unlike the Amazon, is hemmed in on all sides, including the ocean side, by highlands. The central, lower part of the watershed, the so-called Congo basin with an average elevation of about 400 m (1300 ft), is the site of much of the rain forest. This shallow topographic saucer embraces 1,040,000 sq km (400,000 sq mi), or about 25 percent of the total area of the watershed.

The Congo River, 4650 km (2900 mi) long, escapes to the sea through a deep, winding, 350-km (217-mi) gap in the Crystal Mountains which form the western border of the basin. Falls and rapids in this gap render navigation impossible. In contrast to the Amazon, which is navigable for oceangoing vessels for 3700 km (2300 mi) from its mouth to Iquitos, Peru, only

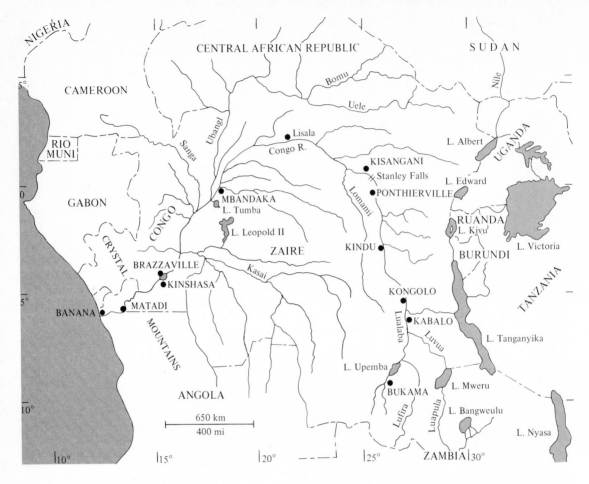

FIGURE 6-12. **Congo watershed, showing the principal tributaries and lakes.** (*Prepared from base map by B. J. Meggers, E. S. Ayensu, and W. D. Duckworth, eds., "Tropical Forest Ecosystems in Africa and South America: A Comparative Review," Smithsonian Inst. Press, 1973.*)

the lower 136 km (85 mi) of the Congo are accessible from the sea. Inland from the Crystal Mountains, the river is again navigable for 1600 km (1000 mi) as far as Kisangani, formerly Stanleyville. Beyond here, navigable and impassable reaches alternate. Access to the interior is possible by a combination of river and rail transport, the railroads bypassing unnavigable stretches. The Congo system has 9600 km (6000 mi) of waterway navigable for vessels drawing less than 1.5 m (5 ft) of water. The Amazon drainage system, on the other hand, has 50,000 km (31,000 mi) of water navigable to vessels of deeper draft.

A second contrast with the Amazon concerns the flood regime of the Congo. Average flood flow of the Congo is only twice the average of low-water flow. In contrast, the ratio for the Amazon is about 5 to 1. The explanation is threefold. The Congo straddles the equator so that the rainy seasons to the north and south alternate and equalize the flow of the Congo. Nine-tenths of the Amazon basin, on the other hand, is south of the equator and lacks this equalization factor. Secondly, whereas the summer rainy season in the Amazon basin coincides with the melting of snows in the Andes to increase flood levels, no comparable lofty range borders the Congo. Finally, vast, natural marshes along the Congo serve as overflow basins and reduce the flood volume downvalley.

Rill and stream erosion under the Congo forest canopy is negligible, as in the Amazon and Southeast Asia. Wherever the forest is cleared, however, the land is subject to gullying and loss of topsoil. Large areas around many water holes are bare of vegetation because of trampling and overgrazing by animals. Many of these are scenes of destructive gullying. Overused game trails suffer the same fate. Where permanent rather than shifting agriculture is practiced, terracing is common to retard erosion.

The Amazon and Congo differ also in the conditions at their mouths. Whereas the mouth of the Amazon, exclusive of peripheral areas to the north and south, is deep and accessible to ocean vessels, the lower part of the Congo, from the Crystal Mountains to the sea, is smothered in silt and sand with shifting shoals and islands. Currents and whirlpools are dangerous for small craft. Navigation has been improved by deepening the main channel and maintaining it by dredging.

Coastal Erosion and Deposition. Coastal erosion in many parts of the tropical zone is influenced by several unique protective phenomena: tidal mangrove forests, resistant beach rock, and coral reefs.

Much fine sediment carried to the coast by tropical rivers is deposited in deltas, estuaries, and shoals. The shallow-water and fine sediment provide ideal growth conditions for *mangrove*. Mangrove is a tidal tree with stilted roots that forms dense, dark green fringes of forest from a few tens of meters to several kilometers wide (Figure 6-13). The forests are of uniform height, only a few meters high in some areas and up to 30 m (100 ft) high in others. Mangrove is an effective sediment trap and contributes to expansion and protection of the shore.

The uncontrolled expansion of mangrove into narrow inlets, estuaries, and bays may be undesirable. A marine wood-boring organism has destroyed many of the mangrove trees in

the Ten Thousand Islands region of southwestern Florida. While the destruction is considered a catastrophe by some, others believe that it will restore the former land-water ratio, old salinity regimes, and tidal-flushing patterns. Wave erosion of shores exposed by natural destruction of mangrove is regarded by many as part of nature's scheme to perpetuate the larger ecosystem.

Beach rock, or *beach sandstone* (Figure 6-14), is common along the tropical coasts of Brazil, South Africa, the West Indies, Florida, and elsewhere. It occurs in layers as thin as 0.3 m (1 ft) and as thick as 3 m (9 ft), depending on the tidal range. The crustlike layer slopes seaward. Beach rock is beach sand, including shells and pebbles, cemented by the mineral calcite dissolved from the upper layers of the beach and deposited below. Because of its resistance, it forms reefs in the intertidal zone. Its durability provides protection for beaches except in great storms.

Far more protective than either the mangrove forests or the beach rock are the *coral reefs*. These reefs, formed of algae (marine plants), corals, and other marine animals, form fringes attached to the coast (fringing reefs), offshore barriers beyond a lagoon (barrier reefs), or isolated circular or elliptical reefs (atolls). The reefs, which are restricted to warm, clear, shallow waters, may have widths of several kilometers and lengths of hundreds of kilometers. Because of their resistance to erosion, and the ability of the reef animals to repair storm damage, living reefs provide excellent coastal protection.

Passage through reefs is through channels, some of which are up to 2 km (1¼ mi) across and 5 m (15 ft) deep. Many are created by storm erosion, but others are due to local conditions which discourage growth of reef organisms. The unsuitable conditions include excessive turbidity and insufficient salinity opposite river mouths, as well as water temperatures too low for reef organisms. Any human activities that spread these unsuitable conditions inhibit reef growth and lead to coastal erosion.

Elevated reefs commonly form a succession of terraces along tropical coasts (Figure 6-15). Until weathering has provided soil, and erosion has reduced the steep slopes, they

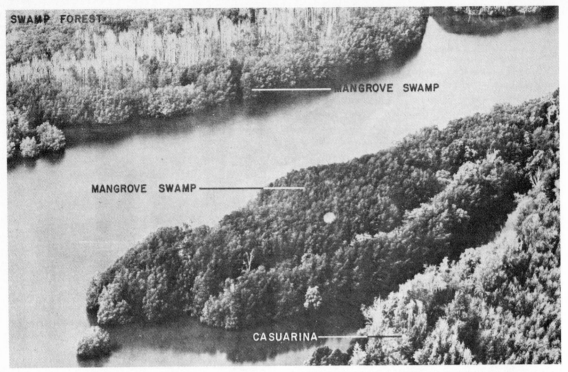

FIGURE 6-13. **Low-oblique air view of mangrove swamps. The casuarina trees indicate well-drained sandy soils above tide level. New Guinea. (*U.S. Navy.*)**

present agricultural and transportation problems.

Wind Activity. Wind erosion and deposition are rare in the humid tropics. Wind is practically nonexistent in the forests, and the heavy precipitation keeps the ground too wet for deflation. Wind activity is prominent, however, along stretches of coast where the rain forest is absent and where sandy sediment is available. The absence of forest at the shoreline may be due to natural or artificial causes. For example, mangrove is not found along steep exposed shores, and rain forest is inhibited by cool temperatures induced by ocean currents. Thus,

FIGURE 6-14. **Beach rock, Sandy Point, Great Abaco, Bahamas. (*Photo by C. Nicholas Raphael.*)**

FIGURE 6-15. **Coral terraces. Aguijan Island, Marianas, S.W. Pacific. The steep, rocky descents from one level to the next make movement between terraces difficult. Coral terraces remain unsuitable for agriculture for considerable periods because of the high permeability. Suitable conditions must await weathering and formation of a more moisture-retentive soil.** (*U.S. Air Force.*)

formed. This is true of much of the coast of northeast Brazil.

In places, waves and currents build sand spits and offshore bars, and here, too, the sand may be piled in dunes. In the sheltered lagoons behind offshore bars and sand spits, the muddy mainland shores are usually fringed with mangrove.

Activity of Soil Organisms. Warm, moist tropical soils abound in organisms. It is the abundance of aerobic (oxygen-requiring) bacteria that accounts for the rapid destruction of humus. This is true everywhere in the rain forest except in swampy areas, where water excludes air from the soil and humus accumulates. Only in such areas is the soil dark rather than reddish.

In some regions the only source of concentrated clay is in termite mounds, some up to 6 m (18 ft) high with volumes up to 600 cu m (780 cu yd). Beyond this, however, termite mounds hamper cultivation by their relatively close spacing.

HUMANS AND THE RAIN FOREST ENVIRONMENT

Agriculture

The long-established practice of shifting agriculture involves the clearing of forestland for cultivation by cutting and burning. Trees of special value, such as the palm oil tree, are generally left standing. During the trash burning, live standing trees are only scorched.

The land is cleared at the end of the season of heavy rains, and the trash is left to dry. The trash is burned before the next wet season, and the ash, with its potash and other nutrient substances, fertilizes the soil. Lightning during the incessant torrential storms creates nitrates from atmospheric nitrogen, and these are dissolved in the rain water and added to the soil, but not in adequate amounts. After one, two, or three crops, the soil is exhausted, the yields drop, and the field is abandoned to remain fallow. The fallow period may be as little as 1 year or as many as 20. As population increases and the availability of new land decreases, the fallow period is reduced, as in Nigeria and other

south of the mouth of the Congo, the cold north-flowing Benguela Current is responsible for coastal savannas in place of rain forests. In contrast, on the other side of the Atlantic, the warm, south-flowing Brazil Current extends rain forest conditions much farther south than is normally expectable. Humans have destroyed vast areas of rain forest for agriculture. Wherever the forest cover is destroyed, a greater proportion of the precipitation runs off the exposed surface. Dissection lowers the ground-water level, and the surface soil dries out. In the interior, shrub or savanna grasses soon take over so that even without the forest, wind erosion is negligible. Along exposed coasts, however, extensive sand dunes may be parts of west Africa. No comparable population pressures exist in most parts of the Congo rain forest or in the Amazon basin or much of Southeast Asia. However, the rush to achieve prosperity by the underdeveloped nations is accelerating destruction of the rain forest. In Brazil, the forests are rapidly disappearing in broad belts along the Belém-Brasilia and the Trans-Amazon highways (Figure 6-16). It is estimated that homesteading along these highways and feeder roads will convert one-third of the rain forest to farmland. There is some concern that widespread clearing of tropical rain forests may alter global climates.

In many rain forest regions, the first fields to be cultivated are those closest to the village. The zone of cultivation expands outward as soils become depleted, and nearby fields are abandoned. In time, if the distance to the fields becomes too great, the village may be relocated. In Burma, many villages are relocated every 4 or 5 years.

The clearing of fields on a large scale exposes the ground to rapid erosion, causes laterite crusts to develop, causes silting at lower topographic levels, and lowers the water table.

Slash-and-burn on steep slopes often results in catastrophic runoff. The vegetation that takes over these abandoned tracts may be useless even for pasturage. When we consider that in Burma alone 2.5 million people are involved in shifting cultivation, and that 565,000 ha (1.4 million acres) are exposed to erosion at any one time, the problem of erosion becomes obvious. About 103 million ha (250 million acres) are under shifting cultivation in Southeast Asia. Because of the environmental damage, shifting cultivation is now illegal in much of Southeast Asia but still persists. The United Nations Food and Agricultural Organization estimates that 5 to 10 million ha (12.5 to 25 million acres) are cleared of forest each year in Latin America.

Shifting cultivation is an indigenous technique, primarily for local self-sufficiency. To accommodate large populations and cities, a more permanent type of agriculture is needed, a type that will stabilize populations, permit orderly crop rotation, lend itself to mechanization, provide a surplus for export, and yet retain the worthwhile aspects of shifting agriculture.

FIGURE 6-16. **Large-scale clearing of forests along Belém-Brasilia highway. (*Arthur D. Howard.*)**

To meet these requirements, the Belgians introduced the "corridor" system of shifting agriculture in the Congo. The land is laid off in long corridors. Clearing, cultivation, and fallowing follow successively along the corridor. The cycle takes about 20 years.

Ground crops and tree crops are raised in the cleared areas of the rain forest. The former include cassava (manioc), cane, tobacco, jute, tea, and rice. Rice is grown in seasonally flooded areas. The tree crops are palms, coffee, cocoa, pepper, rubber, and banana. Jute was introduced to the várzea of Amazonia and pepper trees to the terra firma by Japanese colonists.

In contrast to procedures in temperate zones, clean weeding is unwise because of rapid erosion by torrential rains. This is especially true on steep, easily erodible, volcanic slopes, as in the East Indies. Although the retained weeds consume some of the scarce soil nutrients, their aid in impeding soil erosion is important. In Malaysia, undergrowth is allowed to grow in the rubber plantations, with paths cut from tree to tree. Terracing to prevent erosion is also common.

A minimum of 2000 mm (80 in.) of rainfall is believed necessary for development and expansion of virgin rain forest. Yet rain forest has persisted in areas such as Cuba, where the rainfall has dropped as low as 1100 mm (45 in.) and a marked dry season has intervened. The persistence of the forest is attributed to the retention of moisture under the rain forest canopy even during droughts. However, once such a forest is destroyed, it is unlikely to reappear.

Once the soil nutrients in a cleared field are used up, the field must be abandoned or new nutrients provided. Unfortunately, the mineral fertilizers do not contain mineral or organic colloids (ultrafine particle suspensions) which enable the fertilizers to be retained in the soil. Hence, the mineral fertilizers are washed out in 1 or 2 weeks. Repeated applications prove uneconomic. However, the manure of farm animals contains the necessary colloids. Thus, a combination of fodder, crops, and farm animals (chickens, pigs, cattle) has proved successful. This system is in wide use in the vast Zona Bragantina extending east from Belém, Brazil.

Considerable areas of rain forest have a relatively dry season when the doldrum belt shifts to the other side of the equator. The onset of the wet and dry seasons is generally unpredictable. If crops are planted after the first shower on the assumption that the rainy season has started, a dry spell may jeopardize the crop. If harvesting is delayed, a destructive dry spell may destroy the crops. If the crops are harvested too early, and rain follows after cutting but while the crop is still on the ground, the crop may rapidly rot. The difficulty of maintaining a consistent yield limits population growth.

In the Amazon basin, the várzea presents problems different from those of the terra firma. The main channel and many of the braided channels are bordered by levees. These are above the annual flood level and are commonly forest-covered. Elsewhere, the periodically inundated plain is largely scrub and savanna. The terra firma has poor, leached soils, whereas the fertility of the soils of the várzea is restored each year by the clays and silts washed down from the Andes. Because of the annual flooding and erosion, however, the várzea is suitable only for short-lived agriculture. To increase the cultivable areas, attempts are being made to fill the large várzea lakes with sediment. This is done by cutting channels through the levees and leading the sediment-laden waters into the lake basins. The terra firma, on the other hand, is more suitable for plantations, principally tree crops. Tree crops on sloping land are relatively immune to erosional damage, but combinations of tree and ground crops are even better.

Rain forest estuaries and deltas are the most fertile regions. The soil is replenished at times of flood overflow, and the high water table prevents significant leaching. In many deltas however, much of the terrain is unstable, with land and water boundaries constantly changing. Demarcation of property boundaries, and their subsequent retrieval, is commonly impractical or impossible.

Cattle Raising

Cattle raising, because of large pasturage requirements, is more destructive of rain forest than is the raising of tree or ground crops. Large areas of savanna, many with useless lateritic crusts, may be the result of clearing land for pasturage.

Along the Amazon, cattle are led from the floodplain to pastures on the terra firma during the flood season. Even so, extensive losses occur.

In the Congo, few tribes keep cattle, and these are mostly in the higher areas around the rim of the basin. The reason for this is the tsetse fly, which transmits sleeping sickness to humans and cattle. The tsetse fly is restricted to shady areas and does not spread above 1400 m (4600 ft). The larvae of the fly are destroyed by direct rays of the sun. Hence, ironically, the best countermeasure is destruction of the shade-providing forests. Goats, some sheep, and a dwarf variety of cattle, however, are immune to the disease.

In Southeast Asia, there is little interest in raising cattle other than water buffalo and small oxen for plowing. In addition to the climatic factors and a variety of pests, there is little economic incentive to raise animals for meat. The Buddhists and Hindus do not eat beef, Moslems do not eat pork, and other groups have similar inhibitions.

Lumbering and Tree Crops

Dense stands of uniform tree types so characteristic of the temperate zone are largely lacking in the rain forest. In spite of this and other difficulties, lumbering is increasing in many of the rain forest countries, as in Gabon, the Congo Republic, and Zaïre. Here, forest products are the major sources of income.

Until after World War II, the policy of "extractivism" prevailed in the Amazon. Products like rubber were extracted without regard for a lasting, productive plan for utilization of the resource. At the height of the rubber boom, trees were actually chopped down to secure all the latex in one wasteful operation.

In 1926, the Ford Company obtained a 10,000 sq km (3850 sq mi) concession in the Amazon for culture of the rubber tree. This project, "Fordlandia," and a later one, "Belterra," failed because of a fungus disease which destroyed the trees. Because individual tree types are widely scattered in the rain forest, a disease common to one type is not easily spread. But, when these same trees are artifically grown in close spacing, the disease is easily spread. Although the rubber tree disease was finally eliminated by a complicated multiple grafting procedure, the process proved uneconomic, particularly when the rubber boom declined after World War II.

There are presently six large rubber concessions in Liberia. Firestone has a 32,000-ha (80,000-acre) plantation at Harbel, employing over 20,000 workers. It has a smaller one of 4000 ha (10,000 acres) at Cavalla. Goodyear has a large plantation north of Monrovia.

Few attempts at reforestation have been made in the rain forests. In the Congo, reforestation involves combining lumber trees with other useful trees such as oil palm or bananas.

It seems doubtful, because of economic pressures, that the rain forest countries will benefit from the environmental mistakes of the temperate zones. Cutting goes on with little regard for side effects. A large corporation has purchased 1,600,000 ha (4 million acres) of forest south of the Amazon near the Tocantins River for tree farming, pulp, paper manufacturing, and agriculture. Another major corporation is cutting hardwoods in a concession farther east. As the world's needs for forest products increase, so will encroachment on the rain forest.

Engineering

Modern engineering projects, some of debatable benefit, disturb the rain forest environment to varying degrees. Within the last few years, two major highways have been constructed in the Amazon basin. Construction involved commitments to clear land within strips 20 km (12 $^1/_2$ mi) wide to encourage settlement in the new territories. Enormous areas of rain forest were destroyed in these two operations (see Figure 6-16).

Wherever highway cuts are made, landslides are common, because lateritic soils (exclusive of lateritic crusts) are thick and wet. Soil creep is rapid and requires constant removal from roadside ditches. Gullying starts early in steep road cuts and propagates rapidly headward into the deforested zone.

In the Congo, unnavigable stretches of river

are bypassed by stretches of railroad. A bypass canal is contemplated in Zaïre at Kisangani, on the upper Congo, to replace the present 125-km (78-mi) bypass railroad. Each such project involves additional loss of rain forest.

The construction of dams for hydroelectric power will increase. The Congo River system has the greatest electric potential of any drainage basin in the world. It is estimated at 125 million horsepower (at average minimum river flow), which is 18 percent of the world's total and almost half that of all Africa. The bulk of this potential is in Zaïre. The potential is especially great between Kinshasa and Matadi in the Crystal Mountains (Figure 6-12). Here, the river falls 270 m (900 ft) in a series of 32 rapids known as the Livingstone Falls. This stretch has a potential of 85 million horsepower. The small variation in flow of the Congo River ensures the dependability of the water-generated power.

Water-power resources are also available in the rain forest regions westward around the bulge of Africa from the Congo. The inconstant flows of the rivers, however, make them undependable power sources. A dam and reservoir have been constructed at Edéa, on the Sanaga River in Cameroon. The power supplies an aluminum refinery. The plant has been forced to close at times of low water on the river. A storage dam upriver to ensure a continuous flow of water is being considered.

Huge potentials of hydroelectric power also exist in the Amazon basin. Falls and rapids are common in almost all the tributaries of the Amazon in their descent to the floor of the basin. The falls and rapids farthest downstream are generally at the contact between the older, harder rocks of the flanks of the basin and the more recent, unconsolidated sediments of the basin floor.

Southeast Asia has a hydroelectric potential of 6½ million horsepower, less than one-quarter of which is being produced, mostly in Java. As in the Congo and Amazon basins, the sites of water-power availability are generally far from the centers of trade. Because of this, electric power in urban centers is obtained by burning imported coal. A consortium of Japanese and American firms may build a huge aluminum smelter on the east coast of northern

Sumatra and a 480,000-kilowatt hydroelectric station on the nearby Asahan River. Thus, in the Congo, Amazon, and Southeast Asia, dispersed industrialization at water-power sites may be part of the answer to utilization of the huge hydroelectric resources.

The mightiest hydroelectric project envisions damming the Amazon itself. The project, proposed by an American organization, would dam the river at Santarem (Figure 6-11). The vast lake thus formed would have a hydroelectric potential one-third that of the total United States output. A smaller project, proposed by a Brazilian engineer, would dam two converging Amazon tributaries near Belém to create a lake as large as Lake Superior in the United States.

Proposals to dam the Amazon are meeting opposition. Dams would submerge vast areas of the várzea, the best agricultural land in the Amazon basin. The unflooded portion below the dam, although protected from annual flooding, would be deprived of the annual gift of fertile silts. Above the dam, there would be the danger of further spread of the freshwater snail involved in the transmission of the disease schistosomiasis. The turbulent rapids along many of the tributaries now bar the spread of these calm-water snails. According to critics, the concept of an Amazon dam represents temperate zone thinking applied too loosely to the tropics.

One of the problems facing the rain forest nations bordering on the sea is the scarcity of natural harbors. Waters offshore are generally shallow and deep bays are rare. Shallow lagoons or swamps separate outer reefs from the mainland. Even river mouths generally offer poor ingress because of heavy deltaic deposition. Thus, costly transfer of cargo by lighter craft between ship and shore is common.

Mining

Mining in the rain forest regions presents the same environmental problems as in temperate zones (Figure 6-17). Because of the need to exploit resources, there is no serious attempt to solve environmental problems, particularly in the distant hinterlands.

Intense tropical weathering locally leaves laterites rich in iron, aluminum, manganese, and

FIGURE 6-17. **Mount Nimba iron mine in Liberia near joint borders with Ivory Coast and Guinea. The site is far inland, and the construction of a 250-km (160-mi) railroad to the sea has opened up the interior to forestry and agriculture. (***Courtesy the Grängesburg Company, Liberia Division, Stockholm, Sweden.***)**

even nickel. These are mined by first removing the vegetation and overburden down to the enriched layer. The overburden is dumped nearby, and extraction of the ore leaves great barren areas. These are revegetated but are susceptible to erosion while exposed. When the waste is permitted to wash into the streams, the overloaded streams silt their beds, their channels become shallow and subdivide, and swamps develop. Although economic laterites may be widely scattered, some are of such great size that their exploitation means considerable environmental degradation. One as yet unexploited bauxite deposit in Cameroon is estimated to contain 1 to 2 billion tons of ore having 43 percent alumina. Its exploitation will leave a huge barren area and result in unwelcome consequences downvalley.

Deep tropical weathering frees many important minerals from solid rock. These minerals include cassiterite, a source of tin and the most important mineral resource of Malaysia. The cassiterite is washed down into the valleys and incorporated in the alluvium. The extrac-

tion of the mineral by dredging, hydraulic mining, or strip mining creates the problems discussed in Chapter 11.

Zaïre contains more than 300 mines, an equal number of quarries, and 100 processing plants largely in the peripheral highlands. Waste products and liquid chemical pollutants will present a danger to the drainage basin unless appropriate measures are taken.

Similar situations exist in the Congo Republic, Gabon, and the new states along the African coast to the west. Gabon has one of the largest bodies of manganese ore in the world, deep within the rain forest in the southeastern part of the country. An open-pit mine, connected to a rail head by a 75-km (47-mi) cableway, is already part of the landscape. Gabon also contains enormous iron reserves, a large uranium deposit, and a large number of other mineral prospects, including the bauxite mentioned earlier. The opportunities for environmental degradation are great.

The Amazon basin is another storehouse of minerals, and the Brazilian government is stimulating exploration and exploitation. A huge deposit of high-grade iron ore has been discovered 500 km (300 mi) south-southwest of Belém in the state of Pará, and large deposits of manganese in the territory of Amapá and newly discovered deposits of bauxite in the state of Pará are being mined. In the alluvium of streams are gold, tin, and precious stones, including diamonds. As in the Congo, exploitation will increase the environmental impacts.

Water Supply

The heavy and relatively constant precipitation in much of the rain forest ensures an adequate water supply. The volume of water in tropical rivers is several times that of temperate latitude rivers draining watersheds of comparable areas.

In low areas of the rain forest, the water table is close to or at the surface, and its level varies with the level of nearby rivers. Depressions commonly become swamps at high-water stages. In spite of intuitive expectations, the water table may be lower in the rain forest than in undissected savannas because of the greater moisture loss by transpiration. Yet, the soil above the water table remains moister under the forest canopy than in the exposed savanna. Unfortunately, water wells are not successful in many areas of the rain forest, because the high clay content of the lateritic subsoil impedes seepage into the wells.

Water quality is a serious problem in some rain forest regions. The Amazon presents no problem because of the sparse and scattered population. Its waters can be drunk with safety everywhere except close to the few large communities. In the Congo and Southeast Asia, much of the surface water is polluted from unhygienic practices. Diseases contracted from polluted waters cause many deaths, especially in the Congo.

Large communities in rain forest regions now have modern water-supply systems. Many

purify and sterilize river waters; other tap ground water. Kisangani in Zaïre uses ground water from a neighboring plateau. Other cities, such as Lubumbashi (Elizabethville), in southeasternmost Zaïre, have deep wells. Inferior water quality accounts for the small populations toward the shores of deltas, where the tides make the ground water brackish and affect its suitability for agriculture.

Along low coasts fringed by mangrove, which grows in saline or brackish waters, fresh water may be detected by stands of Sago palm immediately inland. These are freshwater types, up to 10 m (30 ft) high, easily recognized by the flower stalks which project as rosettes high above the palm frond canopy (Figure 6-18).

THE FUTURE

The rate of destruction of the rain forests is accelerating. Timber interests in the United States, Japan, Great Britain, the Netherlands, and Scandinavia have enormous concessions in the tropical countries, including Malaysia, Indonesia, the Philippines, and Brazil. These represent only part of the expanding international interest in the tropical hardwoods. It is questionable whether sound forestry practices will be employed, because of the lack of sufficient knowledge of rain forest ecology. Even the removal of the average 25 commercial trees per hectare (10 per acre) involves the destruction of three-quarters of the neighboring trees because of the interlocking crowns and strong webbing of vines and climbers.

Even the mangrove, affording a measure of protection to low tropical coasts, is exploited for wood fiber necessary in the manufacture of rayon. Three large mangrove concessions in north Borneo have been sold to the Japanese. Coastal erosion may result, and unless measures are taken, the rain forests of Borneo may disappear in a generation.

Conservation measures may take the form of allowing the forest to rehabilitate itself, replanting native types, or substituting tree crops such as oil palm, rubber, and pepper. The pressing need is for sustained-yield forestry, which requires increased knowledge of tropical

rain forest ecology and recognition that temperate zone practices may not work in the tropics.

Improper practices may subject large areas, whether converted to agriculture or not, to irreversible damage, such as the development of lateritic crusts. Unchecked torrential runoff leads to gullying, removal of top soil, increased flood levels, and lowered low-water levels along rivers. Huge quantities of sediment may be added to streams, causing silting of channels, accelerated filling of reservoirs, and accelerated silting of river mouths, thereby aggravating navigational hazards and disturbing the shore processes. While there is time, long-range plans should be prepared to protect the rain forest environment from unnecessary damage.

FIGURE 6-18. **Sago palm, a freshwater indicator. New Guinea.** (*U.S. Navy.*) *(a)* Sago palm showing tall flower stalk. *(b)* Stereopair showing rosettes of Sago palm.

(a)

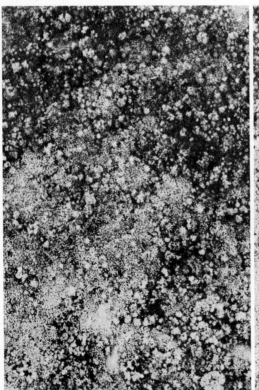

(b)

ADDITIONAL READINGS

Andrews, N.: Tropical Forestry: The Timber Industry Finds a New Last Stand, *Sierra Club Bulletin*, vol. 58, no. 4, 1973, pp. 4–9.

Dobby, E. H. G.: "Southeast Asia," John Wiley & Sons, Inc., New York, 1950.

Farb, P. and the Editors of *Life*: "The Forest," Time, Inc., New York, 1961.

Garlick, J. P., and R. W. J. Keay: Human Ecology in the Tropics, *Symposia of the Society for the Study of Human Biology*, vol. 9, Pergamon Press, 1970.

Hance, W. A.: "The Geography of Modern Africa," Columbia University Press, New York, 1964.

Hornby, W. F., and P. Newton: "Africa," *Advanced Level Geography Series*, Book Seven, Tutorial Press, Ltd., London, 1971.

Jarrett, H. R.: "Africa," MacDonald and Evans, Ltd., London, 1974.

Meggers, B. J., E. S. Ayensu, and W. D. Duckworth, eds.: "Tropical Forest Ecosystems in Africa and South America: A Comparative Review," Smithsonian Institution Press, Washington, D.C., 1973.

Mohr, E. C. J., and F. A. Van Baren: "Tropical Soils," N. V. Uitgeverij W. Van Hoeve, The Hague, and Interscience Publishers, Inc., New York, 1954.

Thomas, M. F.: "Tropical Geomorphology," John Wiley & Sons, Inc., New York, 1974.

CHAPTER SEVEN

cold environments

INTRODUCTION

The Cold Environments

Cold environments are those in which below-freezing temperatures are common and much of the precipitation is in the form of snow. The environments may include active glaciers, where snow supply exceeds loss, or may be glacier-free, where summer wastage exceeds winter's supply.

Even though free of glacial ice, the ground in large parts of the cold regions is frozen to great depths below a shallow surface zone which freezes and thaws with the seasons. This permanently frozen ground, or *permafrost*, is largely a holdover from the more severe climates of the last glacial age.

We shall distinguish between *glacial environments*, those areas presently occupied by glacial ice or so close to glaciers that glacial activity presents a direct hazard, and *glaciated environments*, areas farther afield, formerly occupied by ice and presenting indirectly related hazards. *Periglacial environments* are the cold regions around the periphery of modern and ancient glaciers and the comparable environments in high mountains in low latitudes.

During the Great Ice Age which began about 2 million years ago, glaciers accumulated in regions having surplus snowfall, and permafrost developed in the periglacial zones marginal to the ancient ice sheets. Since withdrawal of the ice sheets in the Northern Hemisphere, the permafrost boundary has been receding northward and is now largely north of latitude 50°. Its preservation is aided by a protective cover of tundra vegetation, largely lichens and moss.

Types of Environmental Hazards

Hazards due to direct activity of mountain glaciers are of concern only where people or structures are located close to the ice front. As humans crowd these environments, advancing glaciers, snow and ice avalanches, and outbreaks of glacially dammed lakes will constitute growing threats.

Hazards in formerly glaciated mountains that are now largely or entirely free of ice are due to the precipitous glacially steepened slopes which encourage landslides and other mass movements.

The dangers presented by the vast ice sheets of Greenland and Antarctica are indirect. The greatest potential threat is the rise in sea level that would accompany melting of these great glaciers. Vast low-lying areas would be inundated. Melting of the glaciers would also influence climate and ocean currents. These events, however, are slow in development and do not constitute the types of hazards with which we are primarily concerned.

The deposits left by glaciers as well as the changes in drainage create other problems. The nature of the deposits determines the landscape characteristics, the surface and subsurface hydrology, and the agricultural and engineering properties of the soils. Furthermore, recognition of the topography buried under glacial deposits may spell success or failure of a dam or reservoir and may be important in assessing ground-water resources.

The periglacial regions, embracing one-fifth of the land area of the earth, are now subject to accelerating invasion, primarily in the exploration and exploitation of natural resources. Uncovering permafrost may lead to serious environmental and structural damage.

MOUNTAIN GLACIER ENVIRONMENTS

Characteristics of Mountain Glaciers

Mountain glaciers remodel preexisting stream valleys and, like streams, have tributaries (Figures 7-1 and 7-8). The position of the glacier front represents a balance between the rate of ice advance and the rate of wasting. A rapidly moving glacier may even descend into forested lowlands as in New Zealand. Some glaciers extend beyond the mountain front and spread out in fanlike forms (*piedmont glaciers*).

A number of factors may increase the rate of movement of glaciers. Excessively snowy periods may load the glacier in its upper reaches and increase downvalley pressures. Snow and debris avalanches have the same effect. Earthquakes contribute to movement by initiating avalanches and loosening frictional bonds between the glacier and its bed. Also important is the accumulation of pressure back of a stagnated glacial front. If the cause of the advance is temporary, the glacier front eventually wastes back to its former position. If the cause is a climate change, the front may remain in its new position.

Because glacial ice is brittle to a depth of 30 to 50 m (100 to 160 ft), the distension resulting from passage over irregularities in its bed, stretching around curves, or lateral spreading at the terminus, causes it to crack, forming *crevasses*.

The cross-profile of a valley glacier is convex upward because of more rapid downwasting along the margins due to heat radiation from the valley sides. Streams commonly follow the marginal depressions and may spread out as lakes. Other lakes are formed where glaciers block the mouths of tributary valleys.

Processes and Landforms

Glaciers erode by *quarrying* and *abrasion*. Quarrying is the removal of rock fragments from the leeward sides of outcrops of fractured bedrock. The fragments become frozen in the glacial ice and are carried off. Abrasion is a grinding action by rock particles frozen in the base of the moving glacier. Many outcrops are abraded smooth (see Figure 2-7).

FIGURE 7-1. A valley-glacier system. The front of the tributary glacier in valley V has melted back to the left and the main glacier has flooded laterally into the valley. Ice falls are common where glaciers overhand steep slopes, as at F . Debris along the margins of the glaciers form lateral moraines L . These join at glacier junctions to form medial moraines M . Cirques are located at C, and sharp ridges known as arêtes are at A . Eagle Glacier, Alaska. View SW from near international boundary. (*Canadian Geological Survey.*)

Erosion by a valley glacier converts the valley head into a steep-sided rocky amphitheater, or *cirque* (Figure 7-1). As the glacier creeps down its valley, it removes many of the valley side irregularities, thereby straightening the valley and converting its cross-profile to a U shape. Yosemite Valley in the Sierra Nevada of California is a classic example.

The debris carried downvalley by a valley glacier includes material eroded from the bottom and sides of the glacial trough, debris that falls, slides, or washes down from the valley sides, and debris blown by the wind onto the surface of the ice. Two distinctive forms of this debris, or *moraine*, are shown in Figure 7-1.

The debris transported by the ice is re-

leased by melting at the glacier front (Figure 7-2). Much of it collects as a ridge (*end moraine*) around the snout of the glacier. Heights exceeding 90 m (300 ft) are common. When the glacier front wastes back, debris is spread thinly in its wake. If the glacier front halts temporarily during recession, another ridge of debris accumulates (*recessional moraine*).

Debris that is melted out of the ice without size sorting or layering is a chaotic mixture of particles ranging in size from clay to boulders, known as *till* (Figure 7-3*a*). Other debris that is spread downvalley by meltwaters is sorted and stratified like any stream deposit and is known as *outwash* (Figure 7-3*b*).

Problems, Cautions, Adjustments

Anomalous Glacial Advance. Valley glaciers may experience anomalous surges when the frontal portion of a glacier becomes stagnant and blocks the actively moving ice behind. The blocked ice thickens until the accumulating stresses cause it to surge ahead. The surge generally lasts a few years and is followed by

FIGURE 7-2. **Terminal and recessional moraines (dark-colored arcs), marking successive positions of the margin of a shrinking valley glacier near Mount Iliamna, Alaska. That the glacier once extended beyond the outermost moraine shown is indicated by the abraded valley side above and to the left of the moraine. Braided streams of meltwater have breached the exposed moraines and are depositing outwash. On the glacier surface are crevasses and sinuous bands of moraine. (*Bradford Washburn.*)**

inactivity lasting decades. Other surges are due to excessive loading of the glacier by unusual accumulations of snow, by avalanches and landslides, and by reduction of friction in earthquakes.

One cannot assume that a glacier whose front has remained stationary or has receded in recent times will continue to so behave. Black Rapids Glacier, located 70 km (45 mi) south of Big Delta, Alaska, had been slowly shrinking for many years prior to 1936. Then, in 5 months, it lengthened 5 km (3 mi) at an average rate of 34 m (110 ft) per day. The anomalous advance has been attributed to an exceptional increase in snowfall during the years 1929 to 1932. If so, there was a lag of 4 to 7 years between the event which caused the spasmodic surge and its effect

FIGURE 7-3. *(a)* **Glacial till, material dumped directly from ice.** *(Eliot Blackwelder.)* *(b)* **Outwash, deposited by glacial meltwaters.** *(D. J. Easterbrook, "Principles of Geomorphology," McGraw-Hill Book Company, 1969.)*

on the ice front 16 km (10 mi) downvalley. The front of Black Rapids Glacier has subsequently receded. In the 1960s, about a dozen surges of major glaciers occurred in Alaska.

The possibility of anomalous advances is a threat to structures and roads near glacier fronts. In 1909 and 1910, the railway bridge of the Copper River and Northwestern Railway, less than $\frac{1}{2}$ mile east of the terminus of the Childs Glacier in Alaska, was threatened by a 360-m (1200-ft) advance of the glacier front (Figure 7-4). By mid-1911, the front had stabilized; it is now receding.

The problem of evaluating the possibility of anomalous glacial advances has been approached in two ways: (1) by assembling historical records on advances and retreats of glaciers, and (2) by examining the valleys below the present glaciers for evidence of recent advances.

If a valley below a glacier front displays a characteristic U shape, the glacier once extended farther downvalley. Two obvious questions then arise: "How recently did this advance occur, and is it likely to recur in the near future?"

Several criteria are employed to estimate recency of ice occupation. The valley sides for some distance downvalley from the glacier front may still be bare of soil and vegetation. Or, the valley sides above the present glacier may be barren (Figure 7-5), indicating that the glacier was recently thicker and therefore, presumably longer. The maximum age of trees on glacial deposits downvalley provide a minimum date for the advance. Radioactive dating of buried organic materials, and the relation of glacial deposits to archeological remains are also informative. Very recent advances are often indicated by tipped or overturned trees.

Periodic observations upvalley from a stagnant ice front may reveal thickening of the ice and the possibility of a surge. Long-term weather observations in the drainage basin may indicate the frequency of especially snowy periods which might stimulate surges. The monitoring of earthquakes and recording of avalanches may also help. The Earth Resources Technology Satellite (ERTS-1),[1] because of the repetitive coverage of almost every spot on earth once every 18 days, is invaluable in monitoring glacier movements and in determining snow accumulation and its possible effect on glacier surges.

Mass Movements: Ice Falls. In this section we consider only ice falls. Snow and debris avalanches are common far outside the environs of living glaciers; hence, we will delay their discussion to the section Previously Glaciated Mountains.

---

[1]The ERTS program has been renamed LANDSAT.

FIGURE 7-4. **Childs Glacier and railroad bridge, Copper River Valley, Alaska. In 1909 and 1910, an anomalous advance of this glacier threatened the bridge. (*Bradford Washburn.*)**

Ice falls may occur wherever a glacier extends over a steep slope (see Figure 7-5). The event frequently follows an anomalous advance. On August 30, 1965, the hanging front of the Allalin Glacier in Switzerland broke loose and cascaded down on the valley floor below. About 100 lives were lost. The glacier had thickened and its front had advanced 40 m (130 ft) between 1960 and 1963. The frontal portion rested on a steep slope of glacially smoothed and slippery rock, and friction was insufficient to support the glacial mass. The fall was triggered by meltwater lubrication of the bed during the summer.

Where ice falls enter bodies of water, flooding of the shoreline may occur. On July 4, 1905, Fallen Glacier, 1½ km (1 mi) long, slid out of its cirque and valley and tumbled 300 m (1000 ft) into a fiord of Yakutak Bay, Alaska. A water wave rushed up the nearby shore to a height of 35 m (110 ft).

Floods. Glacial floods may be seasonal, related to weather cycles, or random. Seasonal meltwater floods may inundate the valley floor each summer. Exceptional floods may result from longer term weather cycles involving a succession of warmer summers. Still other floods result from collapse of glacial dams. Glacial lakes, including those along ice margins, those impounded in tributary valleys, and those impounded where a glacier extends completely across another valley, may suddenly break out as the crevassed ice gives way.

At the time of this writing, the 72-km (44-mi) long Tweedsmuir Glacier in British Columbia was undergoing a spectacular advance. If it blocks the Alsek River, it will create a large lake and present a serious flood hazard. Geologic evidence indicates that a past surge of the Tweedsmuir Glacier created a lake 21 km (13 mi) long. Similar evidence suggests that about 200 years ago Lowell Glacier impounded a lake 83 km (52 mi) long, covering areas now crossed by the Alaska Highway. The outburst flood stripped the forests and denuded the valley

FIGURE 7-5. **Rapidly receding glaciers west of Bowser Lake, coastal mountains, British Columbia. The trimmed zone at TT′ and elsewhere indicates downwasting of the glaciers. The lower portions of the valleys of the tributary glaciers A, B, C, and D have been abandoned so recently that vegetation has not yet taken hold. Occupation of valley E would be hazardous. A slight advance of glacier F would block the valley converting it into a lake. Bursting of such glacial dams causes catastrophic floods. Other potential hazards are ice falls and avalanches from the hanging fronts of glaciers C and D. (*Photo by Canadian Geological Survey.*)**

sides up to 90 m (300 ft) above the present riverbed.

Much local flooding of coastal lowlands results from the breaking away of giant icebergs from glaciers. In 1910, the release of an iceberg from the Columbia Glacier near Valdez, Alaska, generated waves that inundated nearby shores to a depth of almost 7 m (more than 20 ft).

Thaw Subsidence. The outer portions of many glaciers, where wastage is rapid, may be blanketed with debris exceeding 3 m (10 ft) in thickness. If the glacier descends to a sufficiently warm level, soil may develop on this debris and support scrub or thick forests.

Buried glacial ice may also be found within the outwash deposits of glacial valleys. This implies that the glacier formerly extended farther downvalley and that, due to differential melting, ice blocks became isolated from the receding glacier front. Subsidence due to thaw would present structural hazards (Figure 7-6).

Delta Growth and Valley Sedimentation. The outwash from many valley glaciers reaches the sea to form deltas. Many of these deltas have unusual rates of growth. The front of the delta at the mouth of Hidden Glacier Valley in Seal Bay, Alaska, advanced 500 m (1600 ft) between 1899 and 1910 (Figure 7-7). Imagine the plight of a port constructed at the head of Seal Bay in 1899 only to find itself 500 m inland in 11 years.

FIGURE 7-6. **The depression in the foreground is forming by melting of buried ice in outwash below Hidden Glacier, Alaska. (*G. K. Gilbert, U.S. Geological Survey.*)**

PREVIOUSLY GLACIATED MOUNTAINS

The environmental problems of many mountain valleys are directly attributable to the former presence of glaciers.

Landforms and Processes

Glaciation of a former stream-eroded landscape changes it radically. In addition to the erosional modifications shown in Figure 7-8, there are changes due to deposition. The floors of many glaciated valleys are cluttered with debris left by receding glaciers. In many valleys, lakes are confined back of morainal ridges (Figure 7-9). Elsewhere, the valley floors have been smoothed by deposition of outwash and by the filling of former lakes.

Some valley glaciers that reached the sea deepened their troughs below sea level. As the glaciers wasted back, the sea followed the receding ice fronts inland, flooding the glacial troughs to create *fiords* (fjords). The steep fiord walls are landslide-prone, and shoreline communities are subject to landslide-generated waves.

The steep, rocky, glaciated slopes and the severe climates of high mountains make frost shattering an effective process. Rock glaciers are common in the headward reaches of many glaciated valleys where talus formation is rapid. These talus-fed masses creep slowly downvalley like rivers of stones. Movements up to 15 m (50 ft) per year have been recorded in the Alps and in Alaska.

The steep slopes of glaciated mountains are the sites of rapid large-scale mass movements. Some are spectacular in their effects (see Figure 4-13). Mass movements involving bedrock are most common in valleys glaciated during the latest glacial stages, hence still exposing large areas of bare rock. Valleys which were glaciat-

(a)

(b)

FIGURE 7-7. **Rapid growth of Hidden Glacier delta, Alaska. (a) The Harriman Expedition ship *George W. Elder* at anchor at low tide in 1899. (*Photo by G. K. Gilbert.*) (b) The front of Hidden Glacier Delta in 1910 at low tide. The place where the ship was anchored in 1899 was dry land at all stages of tide in 1910. (*Photo by G. K. Gilbert.*) (*R. S. Tarr and L. Martin, "Alaskan Glacier Studies," The National Geographic Society, 1914.*)**

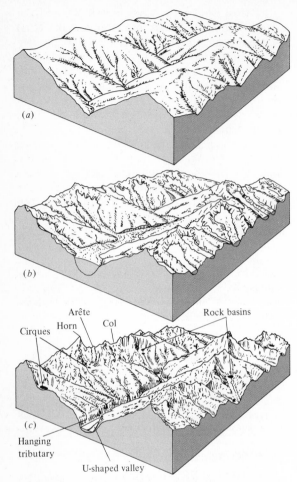

Arête
Horn   Col   Rock basins
Cirques

(c)

Hanging
tributary
                U-shaped valley

FIGURE 7-8. **Modification of mountain landscape by valley glaciers.** *(a)* **Stream-eroded landscape prior to mountain glaciation.** *(b)* **Climax of mountain glaciation. Note change in shape of valleys occupied by ice.** *(c)* **Postglacial aspect illustrating common glacial landforms.** *(After R. F. Flint and B. J. Skinner, "Physical Geology," John Wiley & Sons, Inc., 1974.)*

ed during earlier stages commonly have gentler, soil-covered slopes on which earthslides are more expectable. In winter, snow avalanches are common.

Problems, Cautions, Adjustments

Although the danger from direct falls of rock or debris is always present close to precipitous slopes, the more common hazards are snow and debris avalanches. Debris avalanches occur in any environment with steep slopes, but avalanches involving considerable snow are a cold-climate phenomenon.

High mountains may be snow-covered much of the year and almost always during the winter. As the snow thickens, the mass becomes unstable and its descent to lower levels is generally catastrophic. Tens of thousands of avalanches occur each year in Switzerland alone. Entire villages have been buried, with heavy loss of life and property. In December 1974, an avalanche in Iceland moved through a coastal town, sweeping people and dwellings into the sea.

Among the factors that account for variations in incidence of avalanches are steepness and roughness of slope; thickness, density, and internal cohesion of the snow; incidence of rain while the snow is soft and absorbent; and unseasonal thaws.

In the Alps, avalanches do not normally occur on slopes less than 22°, but exceptions occur. Most of the dangerous avalanches originate on slopes between 30 and 45°.

Snow avalanches are most common in humid mountains, where the combination of abundant snow and frequent temperature changes is common. Most occur in spring when the snow is still thick and the mean temperature is rising. Rain and meltwater add weight to the mass and reduce friction. Particularly catastrophic avalanches, however, may occur in winter when the snow pack is at its maximum and an unseasonal thaw occurs.

Avalanches tend to follow gullies. They sweep out loose debris and erode at the same time. The exposed bedrock is subsequently subject to frost shattering so that repeated fluming tends to entrench the chute. Many chutes coincide with fractures or fracture zones in the rock.

Bedrock chutes (see Figure 2-19), whether eroded by debris avalanches or by avalanches of both debris and snow, have characteristics which distinguish them from normal stream channels: (1) absence of significant rainfall-catchment areas at their heads, (2) greater width of the chute floor in comparison with stream gullies on the same slope, (3) broad and concave

FIGURE 7-9. **Fallen Leaf Lake confined in a morainal loop. Lake Tahoe and Carson Range beyond. Calfornia-Nevada.** (*Arthur D. Howard.*)

upward cross-profile, (4) relatively smooth floor, (5) chaotic accumulations of coarse, poorly sorted debris with intermixed fragments of trees and other vegetation at the ends of the chutes, and (6) extension of avalanche debris out onto the valley floor, in places blocking the drainage.

The degree of hazard from mass movements varies with location in glaciated valleys. If a talus slope stretches away from the base of the valley side, the hazards decrease with increasing distance and decreasing slope of the talus. Sites in the center of the valley may be reasonably safe. On the slopes above the valley floor, the divides between swales, the tops of hills, and areas in the lee of hills are safer than the swales themselves. Any site immediately below a steep cliff is hazardous.

Most of the problems presented by mass movements in glaciated regions are similar to those elsewhere and have been discussed in Chapter 2. They involve identifying sites of potential mass movements, preventing movements at their source, predicting occurrences, rejecting hazardous sites, and providing protection where the degree of hazard is judged acceptable.

The danger from snow avalanches can be minimized by artifically bringing the snow mass down piecemeal. This is done by gunfire from small artillery pieces or by explosives.

A number of devices employed to control or protect against avalanches are illustrated in

Figures 7-10 and 7-11. Some of these are emplaced near the edge of the upland. These consist of snow fences and other installations to prevent the snow from being swept onto the slopes below. Others are emplaced in the rupture zone where avalanches break away from the snow fields. In Norway the most common installations in this zone are sturdy steel and wood inclined fences, strongly bulwarked with footings in concrete. Protective installations along the avalanche track include avalanche sheds above roads and railroads, or tunnels where feasible. If important structures lie within an avalanche track, plow-shaped deflection walls are employed to split and divert the avalanche, much as are used to deflect falling rock (see Figure 2-10). Mounds of debris often serve the same purpose. Reforestation, where feasible, is an effective deterrent. An unusual technique is employed along the Furka-Oberalp Railway in Switzerland. The Steffenbach Bridge over an avalanche track is collapsible. During the avalanche season, the roadway is folded down and out of the way. In parts of Norway, it is common practice to evacuate hazardous sites at the height of the avalanche season.

ICE SHEET ENVIRONMENTS

Distribution and General Characteristics

About 15 million sq km (5.8 million sq mi) of the world's land land area of 150 million sq km (58 million sq mi) is under ice. The Antarctic and Greenland ice sheets make up more than 96 percent of this total, while the rest of the world's ice, the valley glaciers, piedmont glaciers, and small mountaintop ice caps, make up less than 4 percent.

The two great ice sheets locally spill through low passes in confining mountains and follow valleys down toward the sea. Some of these outlet glaciers are enormous. The Beardmore Glacier in Antarctica is 200 km (125 mi) long and 40 km (25 mi) wide. The environments of outlet glaciers are essentially the same as those of valley glaciers. The Antarctic and Greenland ice sheets are largely uninhabited. However, Greenland has many permanent towns and settlements in ice-free coastal areas.

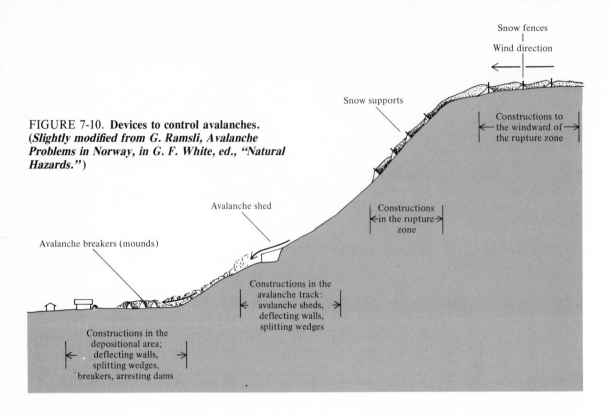

FIGURE 7-10. **Devices to control avalanches.** (*Slightly modified from G. Ramsli, Avalanche Problems in Norway, in G. F. White, ed., "Natural Hazards."*)

Snow fences

Wind direction

Snow supports

Constructions to the windward of the rupture zone

Constructions in the rupture zone

Avalanche shed

Avalanche breakers (mounds)

Constructions in the avalanche track: avalanche sheds, deflecting walls, splitting wedges

Constructions in the depositional area; deflecting walls, splitting wedges, breakers, arresting dams

(a)

FIGURE 7-11. *(a)* **Snow fences to restrain snow from accumulating on the avalanche-prone mountain side to the right of the picture. Hammerfest, Northern Norway.** (*Courtesy Gunnar Ramsli.*) *(b)* **Snow-supporting structures in the breakaway zone.** (*Courtesy Gunnar Ramsli.*)

(b)

Processes and Landforms

Ice sheets, like valley glaciers, engage in erosion, transportation, and deposition. Features such as cirques and horns (see Figure 7-8) are found only where high glacier-clad mountains rise above the ice sheet, and U-shaped valleys, at the sites of outlet glaciers.

The debris moved by ice sheets is liberated by wasting at the ice front and is spread over the ground as the glacier recedes. Outwash is spread beyond the glacier terminus as an outwash plain. The deposits of ice sheets have great lateral extent in comparison with those of valley glaciers. Because the deposits are widespread in areas not now covered by ice, we shall discuss their environmental roles in the section Previously Glaciated Lowlands.

Problems, Cautions, Adjustments

The problems presented by outlet glaciers of ice sheets are the same as those described for mountain glaciers: anomalous advances, icefalls, floods, thaw subsidence, and sedimentation. Elsewhere along the periphery of the ice, where the topography is subdued, it may be difficult to predict where advances will occur. The ice front may creep ahead at one place and recede elsewhere. The anomalous advances are related to changing weather and climate and to differences in topography. If the bedrock surface at the ice margin is uneven, with broad swales and low divides, ice lobes project ahead in the swales. The more uneven the relief, the simpler it is to determine where a possible advance might take place.

Meltwaters follow valleys leading away from the ice front. Excessive melting may therefore lead to flooding downvalley. In some places around the Greenland ice sheet, valleys drain toward the ice front. The blocked streams form lakes connected by ice-marginal channels (Figure 7-12). As we shall see, important elements of the landscape of Canada, the United States, and northern Europe are due to such relationships during the Great Ice Age.

An indirect effect of the contraction of ice sheets is the rise in sea level due to wasting of present-day glaciers. Tidal gauge records over the past century indicate a rise of sea level of about 1.5 mm per year, but with rates as high as 6 mm ($^{1}/_{4}$ in.) per year. Should sea level continue to rise, sparsely developed low-lying areas may some day have to be abandoned, and coastal cities may require protection by costly dikes.

PREVIOUSLY GLACIATED LOWLANDS

Landscape Characteristics

The landscape characteristics of areas once covered by ice sheets depend on whether erosion or deposition was the principal process. In North America, ice spread outward from Labrador and from a center west of Hudson Bay and extended south to the Ohio and Missouri rivers. The Rocky Mountains were largely buried in ice. At the same time, glaciers from Scandinavia invaded northwestern Europe and covered much of Great Britain, the north European plain, and much of European Russia.

The North American and European preglacial landscapes were significantly altered by erosion and deposition, and entirely new environments were created. Outwash formed peripheral plains beyond the ice fronts and extended down valleys such as the Mississippi as much as 1000 km (more than 600 mi).

Erosional. Erosion was dominant in the accumulation areas of the ice sheets where the ice was thickest and movement most active. In North America these regions were largely in Canada but overlapped into parts of the United States. In Europe, erosion was most vigorous in Scandinavia. The actively flowing ice removed most of the soil and some bedrock. Selective scouring by the ice produced thousands of rocky lake basins. Thus, the principal erosional areas have considerable bare rock, subdued relief, and a profusion of lakes. Much of Minnesota illustrates these characteristics.

Depositional. The debris eroded from the central areas of the ice sheets was deposited in broad peripheral zones where wasting was active. Much of it was deposited directly as till to form hummocky (morainal) topography with numerous lakes and ponds. Moraines, locally

FIGURE 7-12. **Lakes and channels along the margin of the Greenland ice sheet. (Stanford collections. Source unknown.)**

exceeding 50 m (160 ft), were formed at the terminus and during pauses in recession of the ice front; relief in intervening areas was generally less than 6 m (20 ft). The major part of the meltwater deposits was spread out beyond the ice front. Thus, areas modified by ice sheets have an inner erosional zone followed outward by morainal and outwash zones. Except locally, the glacial topography of the interior lowlands of the United States, as well as that of Europe outside Scandinavia, is depositional.

During glacial recession, ice blocks were left behind to be surrounded or buried by outwash. Some of these blocks were huge, as indicated by the size of the thaw depressions. Several in Minnesota exceed 13 km (8 mi) in length and 45 m (150 ft) in depth.

In places, till is molded into elliptical, whaleback hills, or *drumlins* (Figure 7-13). Most range from 1000 to 2000 m (3300 to 6600 ft) in length, 400 to 600 m (1300 to 2000 ft) in width, and 15 to 30 m (50 to 100 ft) in height. Drumlins usually occur in swarms aligned in the direction of ice movement. A swarm in western New York State has 10,000 individuals; another in east central Wisconsin, 5000; and one in southern New England, 3000.

Another depositional feature, the *esker* (Figure 7-14), provides a local source of sand and gravel. Eskers are curving ridges marking the courses of former subglacial streams. They range up to 50 m (160 ft) in height, to 200 m (650 ft) in width, and from less than 100 m (330 ft) to as much as 500 km (300 mi) in length.

Water-sorted and stratified debris may be deposited locally against an ice front by streams coming off the ice, or it may be deposited as delta and lake sediments in basins on the ice. On disappearance of the ice, these deposits are stranded as isolated hills known as *kames* (Figure 7-15). These range from a few meters to hundreds of meters across. Glacial alluvium is

FIGURE 7-13. **Drumlins near Madison, Wisconsin. Because of the slopes, many drumlins are left wholly or partly in forests. (Charles C. Bradley.)**

FIGURE 7-14. **This glacial-age esker, overlying ground moraine near Fort Ripley, Minnesota, consists of gravel and sand deposited in a winding tunnel at the base of an ice sheet. When the supporting ice melted away, the deposit was left as a winding ridge. (*W. S. Cooper.*)**

also deposited by meltwater streams flowing along the margins of ice in valleys. Some of these deposits form floodplainlike strips hundreds of meters wide. When the ice disappears, they are stranded as *kame terraces*. Kames and kame terraces are additional sources of sand and gravel.

Dust whipped from the treeless outwash plains and valley-floor outwash by strong winds was spread out as loess. In North America, the largest area of glacial loess extends from the Rocky Mountains eastward over the Great Plains and Central Lowland as far as western Pennsylvania (see Figure 5-19). Thicknesses of more than 30 m (100 ft) are common on the east side of the Mississippi and Missouri Valleys. Loess is scarce in eastern United States because of the sandier nature of the glacial deposits, the more rugged relief, and the moister

climate. Loess up to 60 m (200 ft) thick is reported in Alaska. Loess becomes fertile soil, but it is easily eroded.

Problems, Cautions, Adjustments

In areas of low relief, the deposits of the great ice sheets have buried the earlier topography. The glacial deposits and landforms are thus unrelated to the present milder climate or to current processes of weathering, erosion, and deposition.

Drainage. In many areas of low relief, earlier integrated drainage systems have been supplanted by new, disorganized drainage on the glacial deposits. In areas of greater relief, the ice diverted drainage without completely effacing the earlier valleys. Where the ice advanced against the flow of earlier drainage, new drainage on a grand scale was created (see Figure 7-12).

(a)

(b)

FIGURE 7-15. **Kames in eastern Montana. (*Arthur D. Howard.*) (*a*) Distant view of large kame. (*b*) Kame serving as source of sand and gravel. The kame was later overridden by ice, as indicated by the cover of till.**

The many lakes left behind by the wasting ice sheets present the same advantages and problems as lakes elsewhere. Left to themselves, they eventually disappear, the small and shallow ones first.

Swamps and meadows, the remains of former glacial lakes, are found within and adjacent to important urban areas. The Hackensack Meadows in New Jersey within the Greater New York City area is an example. The meadows are being considered for urban expansion, but there are compelling arguments for retaining at least part as ecological reserves. Bars and deltas of former glacial lakes provide sources of sand and gravel, and glacial clays provide material for brick manufacture.

The postglacial drainage of areas occupied by ice sheets may be independent of the preglacial arrangement. The extreme case is where glacial deposits completely bury the earlier topography. Usually, however, major valleys are incompletely filled and remain as linear swales in the glacial terrain. In some instances postglacial drainage has reestablished itself in these swales; in others, the postglacial drainage follows new courses. The latter behavior is illustrated by the Missouri and Ohio rivers. In preglacial times a series of independent rivers drained northward—those in the west to Hudson Bay; those in the east, to the Great Lakes–St. Lawrence drainage system. These were blocked by the ice, forming a series of lakes along the ice margin much as the drainage in parts of Greenland is blocked by the Greenland glacier today (see Figure 7-12). The lakes overflowed from one to the other along the ice front. These ice-marginal channels were eroded so deeply that, on recession of the ice, they persisted as the present Missouri and Ohio Rivers.

Engineers working in glaciated terrain on projects involving dams, reservoirs, aqueducts, and similar structures soon become aware of the problems presented by buried valleys. Leakage into the permeable sediments of these buried valleys may imperil projects. Yellowstone Canyon provides an excellent view of a filled glacial valley (Figure 7-16). Elsewhere, there may be no surface indications of buried valleys. Reservoir sites in glaciated terrain, therefore, must be investigated thoroughly. Many buried valleys, on the other hand, are

FIGURE 7-16. **Red Rock Valley, an Ice Age valley filled with glacial sediments and exposed by downcutting of Yellowstone Canyon. (*Arthur D. Howard, Geological Society of America, Special Paper no. 6, 1937.*)**

beneficial in that they are important reservoirs of ground water.

Stony Soils. In many regions, the till is very stony and presents difficulty in agriculture. The problem is not so much the presence of the boulders at the surface; these can be cleared off. The problem lies in the seasonal appearance of new crops of boulders resulting from a complicated freeze-thaw process. When the ground freezes, the expansion due to freezing of the contained water lifts all the components of the soil. On thawing, however, the ground immediately under the larger rock fragments is insulated by the rock above and remains frozen longest. As it begins eventually to thaw, the already soupy ground around it slowly invades the thawing area under the rock, providing partial support and preventing it from sinking to its former level. Each year then, the boulders stand a little higher. The stones are usually harvested annually and heaped in fences around the fields. In the midwestern United States, the boulders are collected in piles which, over the years, attain considerable size. Because of the absence of hard bedrock in this region, the boulders are broken up and used for road beds, riprap (an armor to protect the face of earth dams or other slopes), and building stones.

Problems of Loess. Loess deposits, though a benefit to the farmer because of their fertility, present hazards in construction. A dam in loess may settle as the impounded water soaks the silts and causes hydrocompaction. A road may settle because of rainwater collecting in ditches. Water from any source may create subterranean channels which enlarge and eventually cause collapse of the ground. Water filtering alongside pipes laid in loess may eventually leave them suspended in tunnellike channelways. In brief, any structure built on loess to which water has access is endangered. Loess also becomes highly mobile in severe earthquakes (see Chapter 8). When sites are properly prepared by preliminary compaction of the loess, it acquires considerable shearing strength as well as resistance to erosion.

Clay Flows. The extensive areas of clay that accumulated on the floors of glacial lakes are generally poorly drained, a condition that must be overcome before such lands can be put to use. Where dissected and exposed in slopes, the clays slump and flow easily and cause extensive sliding and ground subsidence, as discussed in the section Rapid Flowage in Chapter 2.

Ground Water. Ground water is difficult to develop in glaciated regions because of the chaotic arrangement of glacial deposits. Individual sedimentary layers are usually limited in extent, changing in short distances from a highly permeable sand or gravel to an impermeable clay or till. The variations are equally great vertically. Thus, a local aquifer may provide a bountiful supply of water, at least temporarily, while adjacent wells drilled to the same depth may end in clay or till and be dry. Whereas till is generally impermeable, the stratified sediments laid down by glacial meltwaters are highly permeable.

Outwash, including that in buried valleys, provides most of the productive wells in glacial deposits. Many wells yield 45 to 450 cu m per hour (200 to 2000 gal per minute), and some much more. Whether eskers, kames, and kame terraces contain useful quantities of water depends on their size and topographic position. If they are high topographically, they drain too rapidly.

PERIGLACIAL ENVIRONMENTS

The term "periglacial" is applied to the climatic zone peripheral to ice sheets of the past and present but includes comparable environments in high mountains at lower latitudes. The periglacial zone thus includes regions of severe frost as well as permafrost. These regions are becoming important economically in Russia, Canada, and Alaska.

Periglacial Processes

Periglacial processes, with the exception of those involving the growth and decay of layers and wedges of solid ice, are active in both the permafrost and seasonal-frost zones. The freeze-thaw cycle depends on the presence of water and on temperature fluctuations across the freezing temperature. Temperature fluctuations are determined not only by climate but by the presence of vegetation. Vegetation, particularly a dense tundra mat, insulates the ground during the warm season and retards thaw, hence the close relation of tundra and permafrost.

Frost Shattering. The temperature cycle has an important influence on the depth of frost weathering and debris produced. If, as in tropical mountains, the freeze-thaw cycle is diurnal, its effects hardly penetrate the surface and only fine debris results. In contrast, where the temperature remains below freezing for appreciable portions of the year, even water in deep fractures may freeze, enabling large blocks to be pried loose. Thus, the longer freeze-thaw cycles and greater temperature ranges in polar regions and in lofty mountains of more temperate climates result in deep frost shattering and coarse debris.

Frost Heaving. Frost heaving refers to uplift of the ground either by freezing of interstitial water or by growth of ice layers, or both. Surface uplift by freezing of interstitial water alone cannot exceed 10 percent of the thickness of the freezing layer, because this is the maximum expansion involved in conversion of water to ice. Excessive heaving, often doubling the thickness of the freezing layer, occurs where growing ice layers draw water from the surroundings. Heaving is most pronounced in silts because (1) water is retained by silts and not easily drained off as in sands and gravels, and (2) water in silts is capable of being drawn through the fine pores by capillary attraction.

Cracking of Ground. The expansion of water to ice is a well-known phenomenon, but less well known is the fact that as ice is subjected to increasingly cold temperatures, it contracts like most other solids. This explains the cracking of lake ice which allows water from below to escape and roughen up the surface. Similarly, fissures open up in frozen ground when the temperature falls well below 0°C. The fissures, in which ice wedges may form, are commonly arranged in a polygonal pattern. Fissuring may endanger shallowly emplaced structures, create drainage problems, and provide sites for accumulation of ground ice.

Icing. Icings are accumulations of ice at the surface where escaping water freezes (Figure 7-17). The water that excapes through cracks in

FIGURE 7-17. **Icing on road at Norman Wells caused by water issuing from ground between freezing active layer and permafrost. (R. J. E. Brown, "Permafrost in Canada.")**

the ice of rivers or lakes also becomes an icing. River icings are common where internal flow is impeded by freezing both from top and bottom. The water under pressure may rupture the surface ice and spread over the ice surface and parts of the surrounding floodplain. In places, the pressure within freezing rivers and lakes may force the water laterally to reappear as spring-fed icings some distance away. Other icings result where local freezing of an aquifer blocks water flow and results in sufficient pressure to rupture the surface.

As we shall see, human activities commonly interfere with ground-water drainage and cause damaging icings.

Thaw. The consequences of thaw of frozen ground depend on the materials involved, the slope, and the internal distribution of ice. In fractured rock, freeze-thaw results in frost shattering. Well-drained sands and gravels show little or no effect. On slopes of fine material, thaw water forms a slurry which creeps slowly downward. On flat ground it forms a morass. If the ice is irregularly distributed in the soil, thaw may cause unequal subsidence.

The most widespread thaw effects result from climatic change or destruction of vegetation. Irregular topography results from the melting of numerous thick ice masses in the upperpart of the permafrost zone (Figure 7-18). The melting produces caves, pits, basins, gullies, and isolated hills. Caves develop by melting of lenses of ground ice. Pits result from subsidence over thawing ice masses. Thaw basins, many occupied by lakes, result from enlargement of pits by progressive melting of the permafrost at water level. Thaw lakes grow slowly, 5 to 20 cm (2 to 8 in.) per year.

Russian scientists have reported thaw basins many square kilometers in area and 5 to 20 m (15 to 65 ft) deep on the terraces of the Lena River. Local argiculture is largely confined to these lowlands.

Where ice wedges are exposed along lake banks or other steep slopes, progressive thaw may create expanding gullies. As the ice wedges of a polygonal network are thawed out, the ground in the centers of the polygons remains as isolated hills.

FIGURE 7-18. **Buried ice lens. Fairbanks area, Alaska. (*Troy Péwé.*)**

On Banks Island in the western Canadian Arctic, thaw of ice wedges exposed in the banks of a broad, amphitheaterlike basin in stratified silts and sands has created a badland topography with a profusion of gullies and isolated hills (Figure 7-19). Some of the hills are more than 30 m (100 ft) high, with summit areas 10 m (33 ft) across. Retreat of the surrounding slopes averages 7 m (23 ft) per year.

Snowpatch Erosion. When temperatures are above freezing, meltwater wets the perimeter and base of a snow patch. If the soil is fine, it will steadily be removed by the escaping meltwaters. During cold periods, moisture freezes, and the alternation of freeze-thaw pulverizes the ground further. The progressive removal of debris deepens and enlarges the depression (Figure 7-20). The process, known as *nivation*, steepens slopes and encourages mass movements.

Solifluction. Solifluction is the most widespread mass-wasting process in periglacial regions. It is soil flow confined to the active layer that freezes and thaws seasonally. Where the active layer contains considerable fine sediment, a slurry is formed that creeps downslope often carrying large fragments of rock with it. Rates up to 13 cm (5 in.) per day have been recorded. The deposits form sheets up to 3 m (10 ft) thick

that are commonly lobate, with some lobes exceeding 30 m (100 ft) across (Figure 7-21). The debris accumulates in valley bottoms where it may be tens of meters thick. Solifluction may also take place in temperate zones,

FIGURE 7-19. **Badland topography developed in association with polygonal ground, eastern Banks Island. Thaw of ice wedges produces an irregular scarp and isolated, conical hills. (*H. M. French and P. Egginton, in T. L. Péwé and R. J. E. Brown, eds., "Permafrost," Nat. Acad. Sci., 1973.*)**

FIGURE 7-20. **Snowpatch hollows. Hayden Valley, Yellowstone National Park. (*Arthur D. Howard.*)**

without permafrost, whenever an unseasonal winter warm spell thaws the surface layer above the seasonal frost.

FIGURE 7-21. **Solifluction lobes, near Tolovana, Yukon. (*Troy Péwé.*)**

Ice Thrusting. Lake and sea ice, driven by the wind against the shore, may exert sufficient thrust to heap up an embankment (ice rampart, ice-shove ridge) well above water level, and damage structures built too close to the shore.

Formation of Patterned Ground. The two principal types of patterned ground are polygonal (see Figures 7-26 and 14-25) and striped (see Figure 2-29). Some varieties of polygonal ground result from contraction of the ground during periods of extreme cold; others may result from shrinkage due to dessication. Similar features are found outside present or past periglacial regions. Stone stripes are either polygons drawn out on slopes or streams of rock fragments below rock exposures. Polygonal ground is used with other evidence as an indication of permafrost; stone stripes indicate present or past creep.

The Permafrost Environment[2]

Definitions. The term "permafrost" (Muller, 1947) is a contraction of "permanently frozen ground," and refers to a thickness of soil or bedrock in which a temperature below freezing has existed continuously for a long time. In many areas, "long time" translates into thousands of years, as indicated by the preservation of carcasses of long-extinct wooly mammoths, rhinoceroses, and other mammals. Elsewhere, permafrost may be only a few years old and indistinguishable from seasonally frozen ground.

The upper surface of the permafrost is the *permafrost table*. The overlying *active layer*, which freezes and thaws with the seasons, varies in thickness from a fraction of a meter to more than 4 m (12 ft). Permafrost is not uniformly continuous laterally or in depth. It may contain unfrozen layers as well as large masses of solid ice. Thaw of permafrost presents serious hazards to human activities.

Climate. Permafrost is found in areas where low temperatures persist most of the year. The climate may be dry, as around the Arctic Ocean in northern Alaska, Canada, and central Siberia. Here, meager precipitation occurs during all seasons, but the winter snows are thin, easily blown away, and ineffective in protecting the ground against frost. Permafrost may also develop in cold regions with as much as 300 mm (12 in.) of precipitation, but where the mean annual temperature is even more severe, below 0°C. In both environments, there is seasonal thaw.

Vegetation. Vegetation is important in the periglacial environment. Where cold is extreme or frosts are frequent, there is no vegetation or only small plants in scattered tufts. Frost-shattered debris is common here. Where temperatures are somewhat higher, and particularly where there is a warm season, the vegetation is tundra, a community of perennial species, including lichens, mosses, grasses, and sedges (grasslike plants in tufts). The dense tundra vegetation contains much moisture, and the complex of plants forms a boggy, poorly drained mat. Much of the incident solar radiation is used up in evapotranspiration or is absorbed in the water so that the tundra mat is an excellent insulator. Because of this, tundra impedes the penetration of heat, hence favors survival of permafrost. Finally, coniferous forest covers large areas of the discontinuous permafrost beyond the tundra zone. Here, the peaty soil, which decomposes slowly, is also an excellent insulator and protects the permafrost.

Types and Distribution of Permafrost. Permafrost comes in two varieties, based on moisture content. In *dry permafrost*, the ground contains too little moisture to form an ice cement. In *icy permafrost*, ice fills the voids. Icy permafrost presents the greater hazard.

The permafrost regions of the Northern Hemisphere are shown in Figure 7-22. In the *continuous permafrost* zone, the climate is severe enough for permafrost to form even at present. Local exceptions to the continuity of the permafrost occur beneath large, deep lakes and rivers and for short distances under the sea. Beyond the zone of continuous permafrost is the zone of *discontinuous permafrost*, in which scattered unfrozen areas appear.

The zones of continuous and discontinuous permafrost occupy one-fifth of the land area of the earth, including 80 percent of Alaska, nearly half of the Soviet Union and Canada, some areas in China, and most of the small exposed areas of Greenland and Antarctica. The thickness of permafrost varies from a trace to more than 1000 m. Thick permafrost is a relic of the Pleistocene ice age.

Investigation of Permafrost Terrain

Detection of permafrost is important in many human activities. In mining, costs are increased by the necessity of blasting frozen ground; underground workings deteriorate as thaw progresses; and placer operations require special thaw procedures. In ground-water hydrology,

---

[2]This discussion of the permafrost environment relies heavily on two publications: R. J. E. Brown, "Permafrost in Canada," and T. L. Péwé and J. R. Mackay, eds., "Permafrost." (See Additional Readings.)

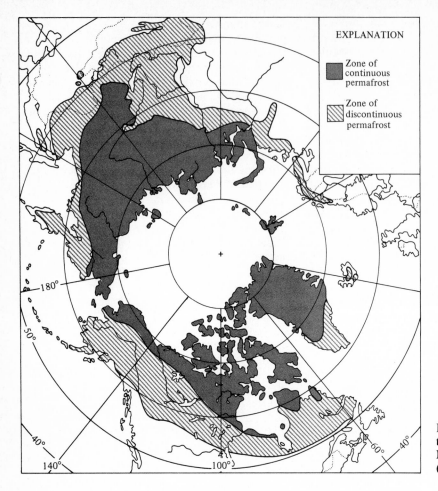

EXPLANATION

Zone of continuous permafrost

Zone of discontinuous permafrost

FIGURE 7-22. **Distribution of permafrost in the Northern Hemisphere.** (*Courtesy Troy Péwé.*)

ice is part of the water budget, and permafrost influences the location and movement of ground water. Permafrost plays an especially important role in engineering design and the stability of structures.

The detection and mapping of permafrost is difficult because of the overlying active layer and because of the rapid changes in ground conditions over short distances. Investigations in permafrost terrain generally follow traditional office and field procedures and, more rarely, geophysical techniques.

Geologic office procedures involve the examination of reports, maps, air photos, air imagery (see Chapter 14), drill records, and other available data. Maps showing different types of rock and soil may provide clues to

permafrost. For example, hazardous icy permafrost will probably be encountered in areas mapped as silt, whereas relatively problem-free dry permafrost should prevail in areas mapped as sand or gravel.

Topographic maps and air photos are particularly useful in evaluating permafrost potential. For example, in the Northern Hemisphere, shady, north-facing slopes receive less solar radiation and remain colder than south-facing slopes. Thus, permafrost tends to be thicker on north-facing slopes and may be present only on these slopes. Slopes transitional between steep hillsides and flatlands below are commonly composed of creep soils. These are favorable sites for icy permafrost. Topography and the direction of prevailing winds determine snow distribution, which then in turn affects the distribution of permafrost. Snow tends to persist in sheltered areas on lee slopes. Because it reflects back as much as 80 percent of solar radiation, it keeps the ground chilled and favors permafrost. But if very thick, it may preserve earth heat and have the opposite effect.

Drainage provides useful information. Permafrost is absent or at considerable depth below large lakes and rivers. On the other hand, we can expect permafrost along the paths of streams which show strings of ponds (Figure 7-23). The ponds indicate thaw of buried ice masses. The complete absence of drainage channels, particularly on terraces, suggests permeable, well-drained sands and gravel, hence the absence of icy permafrost. Frontal portions of terraces are better drained than portions in the rear (Figure 7-24). Hence, the permafrost table, which may be below the foot of the terrace, rises until it is close to ground level in the rear of the terrace.

Air photos reveal topographic, hydrologic, and vegetative conditions suggesting the presence of near-surface permafrost. Pertinent information includes *topographic setting*; *landforms*, including icings, thaw phenomena, solifluction lobes, and patterned ground; *drainage*,

FIGURE 7-23. **Beaded drainage, Canada. The "beads" are depressions due to thaw of ice masses. (*R. J. E. Brown, "Permafrost in Canada."*)**

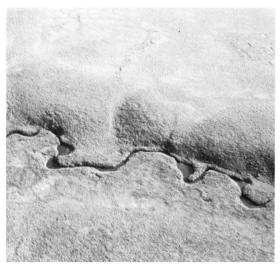

including the distribution of large rivers and lakes, the absence of surface drainage, and the presence of beaded streams; and *vegetation*. On large-scale photos it is possible to differentiate stands of shallow-rooted trees such as larch and black spruce, which grow where permafrost is close to the surface, and deep-rooted trees such as lodgepole pine, aspen, white birch, and alder, which indicate the absence of near-surface permafrost.

Field investigation is necessary to verify information inferred from office study and to provide additional data. Drilling is best concentrated at potential trouble sites as suggested in preliminary studies. Only scattered bore holes are warranted at the outer edge of a sand-gravel terrace, the crest of an esker, or in other material or rock that is subject only to dry permafrost.

Field studies are largely restricted to the summer and autumn because of rigorous winter temperatures and concealing snow cover. Because of summer thaw of the active layer, however, ground access may be difficult, and vehicles may damage the tundra and cause thaw of the permafrost.

Permafrost and Land Use

Activities that require consideration of permafrost include agriculture, lumbering, road and airstrip construction, mineral development, petroleum development, and urban siting and expansion. Where permafrost cannot be avoided, adaptation with least disturbance of the environment should be the goal. Conventional land-use procedures used in other environments are not applicable in icy permafrost because of the danger of induced thaw. Permafrost is particularly sensitive at a temperature close to 0°C, when the slightest environmental disturbance may cause melting. Conventional land-use techniques, on the other hand, are applicable in dry permafrost.

Nature, not people, causes most degradation of permafrost and soil erosion in the Arctic. The fall of a tree, by tearing up the tundra cover, leads to the development of a thaw pit which may eventually expand to a thaw lake. Tundra vegetation is destroyed along caribou trails, resulting in thaw and erosion. However,

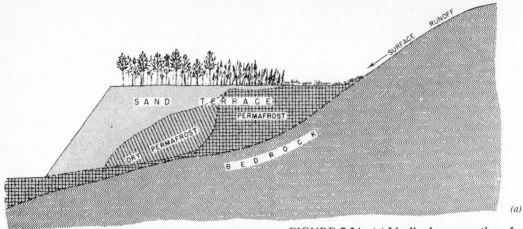

FIGURE 7-24. *(a)* **Idealized cross section of permeable terrace, showing outermost unfrozen area and inner areas of dry and icy permafrost.** *(b)* **Road on well-drained frontal portion of terrace. (*Purdue University, Engineering Experiment Station.*)**

the most important natural factor by far is lightning-induced fire. The tundra vegetation, because it has little woody matter and is moist, may burn and smolder for months. Exposure of the ground by fire or other causes results in erosion by thaw waters. In August 1968, a fire near Inuvik, Northwest Territories, Canada, destroyed the tundra cover, and in less than 1 month, streams had eroded channels 46 cm (18 in.) into the permafrost. Degradation of the permafrost continued until at least 1972, at which time the permafrost had not yet reestablished equilibrium.

Nature eventually heals most thaw damage. Even the large thaw lakes become filled in time with vegetation and sediment, permitting permafrost to accumulate once again below the surface.

Agriculture. Agriculture in polar regions encounters serious deterrents: the short growing season, poor soils, inaccessibility, and—in permafrost regions—the effects of thaw. The short growing season is partly compensated for by the greater length of the days. The small precipitation is offset by reduced evapotranspiration and the abundance of water in the seasonally thawed active layer. Fertile soils are best developed in river valleys where sediment has accumulated and along the southern fringe of the permafrost zone, where the climate is more favorable. The costs of agriculture, however, usually exceed those of importing food.

The seasonal depth of thaw influences development of the soil profile, biologic activity, and plant growth. At Norman Wells in the Mackenzie River Valley, Northwest Territories, Canada, destruction of the organic cover by cultivation lowered the permafrost table 0.3 to 1 m (1 to 3 ft). Figure 7-25 shows the amounts of lowering of the permafrost table under different surface treatments in silts near Fairbanks, Alaska.

Where the permafrost table is close to the surface, thickening of the active layer may be

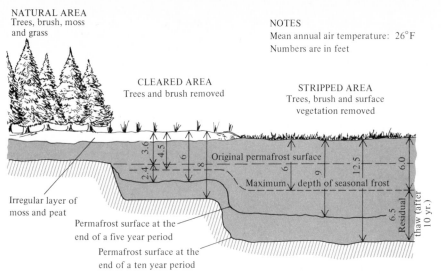

NATURAL AREA
Trees, brush, moss and grass

NOTES
Mean annual air temperature: 26°F
Numbers are in feet

CLEARED AREA
Trees and brush removed

STRIPPED AREA
Trees, brush and surface vegetation removed

Original permafrost surface

Maximum depth of seasonal frost

Irregular layer of moss and peat

Permafrost surface at the end of a five year period

Permafrost surface at the end of a ten year period

FIGURE 7-25. **Permafrost degradation under different surface treatments over a 10-year period. Near Fairbanks, Alaska.** (*After K. A. Linnell, in "Highway Engineering Handbook," section 13, McGraw-Hill Book Company, 1960.*)

necessary for agriculture. Thickening has been accomplished in the U.S.S.R. by removing the vegetation, including shade trees. This activity, however, as well as normal agricultural practices, may have unfortunate side effects. Locally, cleared ground in areas of high permafrost has become unusable after a few seasons. Thaw of ice wedges may produce ditches that make cultivation difficult, and the melting of large ice masses may make it impossible. The effect of cultivation on permafrost in the Tanana River Valley at Fairbanks, Alaska, is shown in Figure 7-26. Differential settling has resulted in a microrelief consisting of mounds 1 m (3 ft) high over considerable portions of cultivated fields, leading to their abandonment.

Where only dry permafrost exists, as on terraces of permeable material, on floodplains subjected only to occasional floods, or wherever there is good drainage of any type of soil, agriculture is possible. Thus crops are raised in restricted areas in the Mackenzie and Yukon river valleys. Even at such places, however, it may be necessary to cultivate the soil very early, especially where the length of the growing season approximates the length of the period of plant maturation. The reason for this is

FIGURE 7-26. **Polygonal ground from melting of ice wedges following clearing of fields for cultivation. Bar represents 300 m (1000 ft). Note encroachment of polygonal ground in newly cleared field in lower right. The mounds began to form within 2 to 3 years. Fairbanks, Alaska.** (*Photo by U. S. Air Force. Courtesy Troy Péwé.*)

that sustained soil temperatures below 5°C (42°F) cannot support plant growth. Early plowing is necessary to expose the soil to the warming summer temperatures. Mulches and cover crops are avoided in contrast to other climatic environments. South-facing slopes are more favorable to agriculture because of the lower permafrost table, lower moisture content, higher soil temperatures, and better drainage than north-facing slopes. The latter receive less insolation and are cold, wet, and peaty, with moss, shrubs, and scrubby trees.

The cutting or burning of trees or brush in wooded tundra has little effect as long as the tundra is preserved. Trees and brush, however, do shade the ground, intercept snow, and reduce wind velocity and should only be removed selectively and with care.

Forestry. Forests in the permafrost region are largely restricted to the discontinuous zone. Growth is slow on permafrost; in the Mackenzie River delta, trees reach maturity in about 200 years, although usable timber can be harvested in 100 years. Reforestation requires that the soil be kept warm and thawed, a major problem when rewards are delayed for 100 years. Sites where timber grows rapidly enough to be harvested in 100 years are generally those subject to periodic floods. The floods mix the forest litter with rich silts and destroy the moss. Most trees above the flood level are very small, including black spruce more than 300 years old and only 2 m (6 ft) high.

Traffic over lumbering roads or destruction of the ground cover in creating trails and fire breaks may lead to deep gullying (Figure 7-27). Firebreaks are ordinarily created by cutting through the unfrozen organic cover down to mineral soil or permafrost. To control erosion along fire lines, the runoff is controlled by low, oblique bars of earth across the fire line to divert water into side areas covered by vegetation. The exposed track is then seeded with grass, and protected by mulches. The procedure has proved successful (Figure 7-28).

Water Supply, Flooding, Sewage Disposal. Because surface waters freeze during the winter, they are not a perennial source of water. To tap the unfrozen interiors of large rivers requires

FIGURE 7-27. **Severe thermal and water erosion of a winter trail after one thaw season. (R. K. Haugen and J. Brown, in D. R. Coates, ed., "Environmental Geomorphology," 1971.)**

studies to determine whether unfrozen interiors are dependable or whether the situation changes from year to year and place to place along the river system. Where dams are built, problems may be expected due to thawing of permafrost beneath the reservoirs. Thaw at the dam may weaken foundations; elsewhere it may result in leakage. Ordinary surface springs revert to icings in winter.

Ground water is the most reliable source of water in permafrost regions. Within the discontinuous permafrost zone, water may be abundant in alluvial deposits. The Tanana River Valley has as much as 250 m (820 ft) of unconsolidated deposits near Fairbanks and 610 m (2000 ft) near Minto, 65 km (40 mi) to the west. The deposits are frozen to a depth of only 80 m (260 ft). A considerable reserve of ground water is indicated below the permafrost.

River terraces may supply ground water from below the permafrost. Permafrost is thicker and more continuous under older, higher terraces more distant from the river than under younger, lower, closer terraces and the floodplain itself.

Alluvial fans, seasonally recharged by streams, are generally underlain by discontinuous permafrost. Water quality is better than

(a)

(b)

FIGURE 7-28. **Erosion control along vehicular fire trail.** (*C. V. McVee.*) *(a)* **Erosion commencing on untreated disturbed trail in permafrost area.** *(b)* **Same trail showing growth of grasses and stabilization of soils after seeding, fertilizing, and covering with mulch.**

from the above-mentioned sources because of less organic matter, iron, and manganese.

Water may also be present in unfrozen layers within permafrost. It is impossible to locate such occurrences without drilling. Also, as elsewhere, water may be present in aquifers confined by impermeable layers.

Lakes and rivers which might freeze solid in winter are not reliable perennial sources of water. The permafrost table is low under such water bodies, however, and water may persevere throughout the year in alluvium above the permafrost. Such deposits provide the only known perennial source of fresh water in the lowlands around Beaufort Sea. At the Prudhoe Bay oil field, the water supply comes from unfrozen alluvium beneath the Sagavanirktok River.

Conditions for a perennial water supply were artificially contrived at Stein Creek, near Lisburne, Alaska. The creek is underlain by alluvium into which winter frosts formerly penetrated down to the permafrost table. To ensure a thickness of unfrozen alluvium between the seasonal frost and the permafrost table, the permafrost table was lowered by inducing an accumulation of drifting snow by fences. The thick snow cover reduced the loss of internal ground heat in winter and retarded penetration of seasonal frost. Within a few years an unfrozen zone 3 m (10 ft) thick was created in the alluvium, and a tunnel was constructed to collect the water.

Many icings occur at points of perennial discharge from aquifers; hence, knowledge of their distribution is important. Aerial reconnaissance and air photography are used to locate these icings in the spring before they melt. Ice-cored blisters, or pingos, (Figure 7-29) indicate interrupted ground-water flow. Pingos may persevere from a few years to thousands of years.

The drilling of wells through permafrost may be a frustrating experience. A water well was drilled at Fairbanks, Alaska, in April 1946. It was drilled through permafrost to a depth of 33 m (108 ft), and casing (pipe) was emplaced throughout. Because the drilled hole was wider than the casing, the hole enlarged by thaw and the water rose uncontrolled in the space around the casing, creating an enlarged pit. A long series of unsuccessful measures to complete the well ensued, made more frustrating by formation of ice plugs within the well itself. The situation was finally stabilized in 1950 by installing ground probes around the well and circulating refrigerant brine until the permafrost was restored.

FIGURE 7-29. **Olong-Erien thaw basin with a large ice-covered blister (pingo) 75 km (47 mi) west of Yakutsk, Siberia, U.S.S.R. (*Troy Péwé.*)**

Certain lessons were learned from this experience. First, casing must be tightly installed to forestall surface leakage. Secondly, it is necessary to avoid undesirable temperature interactions. For example, the water in the well, being above freezing, may thaw the permafrost outside the casing, or the permafrost outside may cause ice formation inside the casing. If the water demand is limited, the velocity and duration of flow will be small and the danger of freezing great. Protective measures may include thermal insulation of the well, introduction of heat into the well in a manner to avoid disturbing the surrounding permafrost, intentional wastage flow or continuous recycling to prevent freezing, or mechanical removal of such ice as does form.

Spring floods are common along major rivers of the Arctic. Rivers such as the Yukon and Mackenzie in North America and the Lena and Yenisei in the U.S.S.R. break up their frozen surfaces first in the warmer headwater reaches and transport the masses of floating ice until jammed up by a particularly stubborn frozen portion to the north. Great floods result from these ice jams. Flood levels exceeding 10 m (30 ft) may occur even in broad valleys. The floods deposit extensive sheets of alluvium. They cause serious bank erosion, aided by the impact of ice floes against the banks. The ice jams generally give way abruptly, creating violent surges capable of moving large blocks of rock.

Sewage disposal presents some unique problems in permafrost terrain. Sewage lines, like waterlines, must be in heated insulated housings, or utiladors (Figure 7-30). In communities utilizing utilidors, the sewage lines are included with the water, steam, power, and telephone lines. The sewage, whether treated or not, must be disposed of. Some small communities lead the raw sewage off into basins or valleys out of range of offensive odors and of possibly deleterious effects due to thaw of permafrost. Coastal communities commonly use the sea as a receptacle.

Large communities employ heated and well-insulated sewage-treatment plants. Construction is carefully designed to eliminate disturbance of the permafrost. Because septic action is so slow in cold climates, the septic tanks are themselves heated with steam to increase the digestion rate. The treated sewage, now nonpolluting, is led from the plant through insulated pipes or utilidors to an outflow site.

All these activities may affect the permafrost regime. The siting of dumps, outfalls, and treatment plants and the design of the entire sewage system must receive careful consideration.

Mining and Petroleum Activities. Prior to World War II, there were only a half dozen mining

FIGURE 7-30. **Utilidor at Inuvik bringing services to houses. Note pile foundation for house in foreground. (*R. J. E. Brown, "Permafrost in Canada."*)**

areas in the permafrost region of northern Canada. About 20 new mines have been brought into production since then, and other rich deposits have been discovered.

Permafrost presents difficulties at all stages of the mining procedure: exploration, development, and production. Any surface activity may result in thaw of permafrost, with terrain and property damage. Drilling sites and access roads become subject to thaw and erosion after abandonment. Permanent installations and mining buildings must be built in such fashion as to preserve the permafrost.

The working of surface deposits affects the terrain more than deep mining. The most common procedure for *placer mining* is the use of water under high pressure to thaw and wash down the ore-bearing gravels. Previously, the ground was thawed by (1) exposure to the sun, with each level of thawed ground removed to expose the next lower level, or (2) heating by wood fires, or (3) heating by forcing steam and hot water into the ground, a method which thaws the ground to depths of 2.5 to 3 m (8 to 10 ft) per day. Cold-water hydraulic mining has brought many low-grade deposits within economic reach. In consequence of these activities, many of the barren hydraulicked areas have reverted to badlands. In addition, *dredge mining* of valley-floor gravels has left considerable areas of rough, barren ground. On the plus side,

these now-unfrozen valley-floor gravels are good aquifers, provide good sites for highways and towns, and are excellent sources of gravel. In Siberia, thawing is accomplished by stripping the muck from the ore-bearing gravels and then flooding.

Permafrost in *strip mining* may create problems such as the need for drilling and blasting normally loose sediments, the freezing of the drill in the hole, thaw of the frozen ore prior to transportation, and secondary fragmentation following refreezing during transport. The excavation of overburden to expose the ore directly affects the environment. Not only is the pit a scar in the terrain, but it may increase in size as thawing of the frozen banks progresses. Away from the pit, all auxiliary structures including roads, railroads, tramlines, power lines, sewage systems, and buildings, because of their potential influence on the permafrost, require careful engineering.

At some mines no problems are encountered in the mining itself, but permafrost problems arise at the town sites. This was true at the Thompson, Manitoba, nickel discovery and at the Cassiar asbestos mine, 105 km (65 mi) west of Dawson, Yukon Territory. In contrast, most of the town of Yellowknife, Northwest Territories, Canada, is on glacial outwash consisting largely of sand and gravel, and few permafrost problems are encountered.

Surface and subsurface problems are encountered in *underground mining*. Thawing caused by removal of vegetation in preparing the site for a vertical shaft may create a quagmire if drainage is poor. This caused an expensive delay at the Eldorado uranium mine at Uranium City, Lake Athabaska, Canada.

A second problem in mine workings is that water may be encountered in unfrozen layers within the permafrost or in the unfrozen ground below. This water may be under hydrostatic pressure. Flooding may result until pumps or drainage tunnels are provided. Drifts which are not worked for a while may become filled with ice. The Discovery gold mine, 96 km (60 mi) north of Yellowknife in the Northwest Territories, was flooded by meltwater from thawed permafrost, and all openings choked with ice. It was reopened after siphoning water from a pond to thaw the ice, after which the mine had to be pumped dry. Sealing off aquifers is difficult because of the problem of cementing at subfreezing temperatures.

Temperatures in mines with thick permafrost may be considerably below freezing. Water that comes in contact with the metal of machines, tracks, or other equipment freezes rapidly and must be removed frequently. In summer, warm air penetrating the workings causes thaw, and the melting of ice in fractures causes walls to collapse. In other mines warm air is deliberately introduced to facilitate mining. Timbering is often necessary. The cutting of timber damages the surface cover of vegetation and may result in thaw damage to the terrain well away from the mine. Thaw becomes a dangerous factor in inclined or vertical passages (stopes) that are driven up through bedrock to the base of frozen overburden or driven along the rock-overburden interface. To prevent collapse into the workings, the ore is extracted during the winter and the space sealed off and isolated in anticipation of collapse during the thaw season.

The types of environmental damage from oil exploration and production are essentially the same as those encountered in exploration and production of solid minerals. One major difference is the need for long pipelines to carry the oil to shipping points or to markets. At present, the oil must be heated to remain fluid in its passage through the cold regions, particularly if buried in permafrost. The main fear is from oil spills from pipeline breaks caused by foundation failures over thawed permafrost.

The Alaska pipeline is routed to avoid permafrost terrain wherever possible. About half of the route is through icy permafrost. Here the pipeline is elevated on steel piles or on gravel embankments to preserve the permafrost, and—in the relatively few miles it is buried—refrigeration will keep the ground frozen. The other half of the route is in solid rock or in freely draining sand and gravel. Where it crosses rivers and floodplains, the pipe is coated in concrete and buried. The pipeline, furthermore, has been designed to sustain earthquake movements of 6 m (20 ft) horizontally and 1 m (3 ft) vertically. The entire line is monitored from a computer control center at Valdez.

Every effort has been made to minimize the pipeline's effect on wildlife. Where the pipeline is above ground, overpasses and underpasses are provided. Nesting and breeding grounds of birds and animals have been mapped, and the habits of salmon and grayling along the pipeline route have been studied. Construction was scheduled to avoid interference with their life cycles.

Construction. Careless construction in permafrost environments results in serious damage to structures and the environment. The most important cause of structural and terrain damage is unanticipated thaw of icy permafrost. Examples of the results of unanticipated thaw appear in Figures 7-31 and 7-32. Another important source of damage is differential heaving.

Two construction methods are employed in icy permafrost areas: *active* and *passive*. In the active method, the entire active zone, including the vegetation, is removed and replaced with gravel or other coarse, permeable materials. The roadway or runway is placed on this material. If temperatures rise too high, however, loss of the insulating vegetation may cause thaw.

The preferred approach is the passive method. The original vegetation is preserved

FIGURE 7-31. **Store at Dawson, Yukon Territory, Canada, in northern part of discontinuous permafrost zone. Uneven (differential) settlement is due to thawing of the underlying ice-laden permafrost. (*R. J. E. Brown, "Permafrost in Canada."*)**

and the subgrade is placed above this. The packed vegetation serves as insulation to protect the permafrost below.

Considerations in road and runway construction in permafrost are the difficulty and expense of excavation in permafrost and the possibility of interrupting normal ground-water movement and creating icings. Side ditches

FIGURE 7-32. **Tank at Norman Wells has settled because of thawing of the underlying ice-laden permafrost. The plates have buckled and oil is leaking. (*R. J. E. Brown, "Permafrost in Canada."*)**

should be narrow and deep, hence less susceptible to complete freezing, and should be as far from the road as feasible to prevent icings from spilling over the road. Culverts should be located to transfer impounded waters to the downslope side of the roadway.

The removal of material from side areas to build up roadways or runways may lead to significant side effects. An airstrip was constructed in 1962 at Sachs Harbor on Banks Island in the Canadian Arctic. Glacial sands and gravels were skimmed from a considerable area to provide material for the airstrip. This exposed frozen, interbedded sands and silts to thawing, leading to melting and differential subsidence. The area developed an irregular hummocky topography, with small interconnected linear depressions and pools of standing water.

Even under the most severe Arctic conditions water freezes only to a depth of about 4 m (12 ft). Reservoirs are therefore feasible. Depending on the foundations, dams may be masonry, rock fill, or earth. Any of these types can be built on frozen rock, but masonry dams are not built on permafrost soils. In the construction of fill dams on permafrost soils, the possible effects of the structure on the permafrost must be assessed. The dam acts as an insulating medium, causing the permafrost table to rise within it. This adds firmness to the structure and retards leakage. To maintain this high level of the permafrost table throughout the year, the dam is supercooled in winter by permitting cold air to circulate through ducts, pipes, or implanted gravel layers. The overcooling compensates for possible thaw during the summer and helps control leakage. The upstream face of the dam must be impermeable to keep water out. To prevent the reservoir water from infiltrating the dam from beneath, masonry aprons are constructed below the upstream base of the dam, extending well below the floor of the reservoir.

The water of the reservoir thaws the upper part of the permafrost to a depth of about one-fourth the depth of water. The thaw, by melting out ice accumulations in the banks, could cause leakage.

Foundations for other projects depend on the type and size of the structure, whether it is to be heated or unheated, and whether it is to be temporary or permanent. A prime design objective is to prevent thaw of the permafrost.

Small temporary buildings are commonly built on timber mudsills (the lowest timber of a building usually set in the soil) or supported on posts with concrete footings resting on beds of sand or gravel.

Permanent buildings are elevated above the ground, with air space below to prevent degradation of the permafrost. The buildings are supported by pilings (Figure 7-33) driven well down into the permafrost or by posts on pads in gravel mats up to 1.5 m (5 ft) thick (Figure 7-34). A collar is fitted on that portion of the piling within the active zone. Thus, the piling remains fixed while the collar, frozen to the active zone in winter, may move up, and later down, as the active zone expands and contracts. Ducts are embedded in the ground below large buildings, particularly if heated, to circulate winter air and compensate for any summer degradation of the permafrost. Provision is made to keep snow away from the building where it might block air circulation. Wood or metal aprons shade the base of the building on the sunny side during summer to impede thaw.

Well-drained deposits of sand and gravel, whether frozen or unfrozen, present no unusual foundation problems. If frozen, however, problems still exist in the installation of water and sewage lines.

In Alaska, laws requiring evaluation of environmental impact are supplemented by restricting summer operations in tundra regions. In Canada, similar regulations are in effect or underway.

Seasonal Frost Environments

Introduction. The depth of freezing in seasonal frost environments varies regionally and locally. In northern Maine it is 1.3 to 1.5 m (4 to 4.5 ft); in New York City it is set at 1.3 m (4 ft) by the building code; in the central part of the United States it is 1 to 1.3 m (3 to 4 ft). Local variations depend on surface deposits, vegetation cover, direction of slope exposure, and local climates.

FIGURE 7-33. **Setting piles by steaming and driving in permafrost at Inuvik, Northwest Territories, Canada. These piles extend 6 m (20 ft) into the permafrost. The gravel pad preserves the insulating mat of vegetation from destruction by the machinery. (R. J. E. Brown, "Permafrost in Canada.")**

FIGURE 7-34. **School at Inuvik built on piles. The air space under the building reduces heat flow into the ground from the building. (*R. J. E. Brown, "Permafrost in Canada."*)**

Processes. The seasonal freezing contributes to frost shattering, soil creep, soil flow, stone heaving, ground heaving, and flooding.

Because of the shorter, often diurnal duration of freeze-thaw cycles in temperate latitudes, the effects are not felt at depth. Thus frost shattering of rocks is generally superficial.

Soil creep is aided by the outward expansion of the soil during freezing and its vertical settling during thaw and by the formation and thaw of soil ice. Soil flow in fine materials may occur when an unseasonal warm spell thaws the upper part of the frozen soil, converting it to a fluid which flows slowly downslope over the still frozen and relatively impermeable ground below. If there is unseasonal thaw on a flat surface, a muddy quagmire results.

Stone heaving provides crops of stones to be cleared from many fields. Frost heaving of ground may, as in permafrost terrain, be due to freezing of either pore water alone or introduced water as well.

Flooding may result from the spring thaw of a heavy snow cover in the mountains, from ice jams farther downstream, or from heavy rains or early melting of snow while the ground is still largely frozen and relatively impermeable.

Construction. Frost-susceptible terrain presents some problems not found elsewhere in the temperate zones. The footings of structures should be below the frost depth to minimize chances of heave damage. Water and sewage pipes should be placed below the frost depth.

Unheated buildings, by masking the ground below from sunshine, may encourage freezing of the ground. Cold-storage warehouses which are maintained at temperatures well below freezing for years, have resulted in freezing depths of more than 3 m (9 ft). Because the colder temperatures are under the centers of the buildings, heave is likely to be greater here than around the periphery.

Roadways and runways must have underdrains to take care of thaw water which cannot escape downward because of still-frozen ground below. If there is inadequate drainage, the roadway may settle differentially. Attempts should be made to prevent the freezing of included water by using insulating covers wherever the ground is exposed. Sand or mulches may be employed. The exposed soil may be treated with calcium or sodium chloride to lower the freezing temperature. Or, the roadbed or runway may be constructed of permeable sand and gravel, after removing silty materials.

Dams in temperate zones are not prone to serious frost damage, but spillways, many of which are constructed of relatively thin slabs, are susceptible to frost heave. Cracking and shifting have occurred. Any material susceptible to heave should be removed from the immediate course of the proposed spillway and a thick sand and gravel blanket applied.

ADDITIONAL READINGS

Black, R. F.: Permafrost—A Review, *Bulletin of the Geological Society of America*, vol. 65, no. 9, 1954, pp. 839–856.

Brown, R. J. E.: "Permafrost in Canada," University of Toronto Press, Toronto, 1970.

Embleton, C., and C. A. M. King: "Glacial and Periglacial Geomorphology," St. Martin's Press, New York, 1968.

Flint, R. F.: "Glacial and Quaternary Geology," John Wiley & Sons, Inc., New York, 1971.

Hopkins, D. M., T. N. V. Karlstrom, et al.: "Permafrost and Ground Water in Alaska," U.S. Geological Survey, Professional Paper 264-F, 1955, pp.113–145.

Muller, S. W.: "Permafrost or Permanently Frozen Ground and Related Engineering Problems," Edwards Brothers, Inc., Ann Arbor, Mich., 1947.

Péwé, T. L., ed.: "The Periglacial Environment," McGill-Queens University Press, Montreal, 1969.

———and R. Mackay, eds., Permafrost, *North American Contribution*, Second Int. Conf., 13-18 July 1973, Yakutsk, U.S.S.R., Nat. Acad. Sci., Washington, D.C.

Ramsli, G.,: Avalanche Problems in Norway, *in* "*Natural Hazards*," G. F. White, ed., Oxford University Press, New York, 1974, pp. 175–180.

Tarr, R. S., and L. Martin: "Alaskan Glacier Studies," The National Geographic Society, Washington, D.C., 1914.

Tricart, J.: "Geomorphology of Cold Environments," St. Martin's Press, New York, 1970.

Troll, C.: Structure Soils, Solifluction, and Frost Climates of the Earth, Translation 43, *Snow Ice and Permafrost Research Establishment*, U.S. Army Corps of Engineers, 1958.

U.S. Army: "Arctic Construction," Tech. Manual TM5-560, 1952.

CHAPTER EIGHT

earthquakes
and the environment

INTRODUCTION

The Shaking Lands

It was 9:40 A.M. on All Saints Day, November 1, 1755, and most of the residents of Lisbon, Portugal, were at church. Suddenly, the earth began to shake, at first mildly, then with increasing vigor. After a brief pause, the shaking resumed with violence, and within 6 minutes of the start of the earthquake most of the city lay in ruins. Thousands died in the rubble and in fires that broke out in many places. Thousands of others sought refuge along the harbor edge, from which the sea began mysteriously to withdraw. Then, at 10:00 A.M., the sea rose in a great enveloping wave (tsunami) that reached heights of 6 to 15 m (20 to 50 ft) and moved inland as much as 1 km (0.6 mi). Nearly all the refugees were drowned as the wave carried ships and all else before it.

A second shock of great violence struck the city just as the tsunami arrived. Most of the few remaining buildings were destroyed, and numerous rockslides were triggered on adjacent mountain slopes. The area was cloaked with dust and smoke by the time a third severe shock arrived at noon. Fires, fanned by high winds, continued unchecked for 6 days. It is hardly surprising that 70,000 people out of a total population of 235,000 perished in Lisbon and its immediate environs. The earthquake was felt over an area of 40 million sq km (more than 15 million sq mi), and it set bodies of water in motion (a form of seiche) throughout much of western and northern Europe. Effects of the seismic sea wave (tsunami) were felt along the coasts of North Africa, Spain, the British Isles, Holland, and the West Indies.

Two hundred years earlier, on February 2, 1556, Shensi Province, China, was wracked by an earthquake that claimed 830,000 lives. Much of the property damage and loss of life resulted from violent shaking of the soft sediments underlying the broad, densely populated plain of the Huang Ho (Yellow River) and from flooding as the river broke through its dikes. Even greater losses, however, resulted from the mobilization of the widespread loess deposits. Thousands of lives were snuffed out in the collapse of cliffs honeycombed with cavelike dwellings (see Figure 5-18), and the silty materials rushed down into neighboring valleys, burying entire villages.

India's greatest earthquake, the Assam earthquake of June 12, 1897, caused heavy damage over more than 400,000 sq km (150,000 sq mi). The intensity of ground shaking and the devastation of structures were extreme and have since served as standards of severity on the earthquake intensity scale described later. The shock was also notable for the nature of the ground movements. Pebbles and small cobbles bounced on the ground "like peas on a drumhead," boulders rose to the surface, posts were projected directly upward out of their holes, vegetation and soils were stripped from hillslopes, and the soft sediments of the flatlands were violently shaken.

The Mississippi Valley was struck by earthquakes on December 16, 1811, and January 23 and February 7, 1812. Known as the New Madrid earthquakes, the shocks—which originated in southeast Missouri—were felt over an area of 2.6 million sq km (1 million sq mi). The ground rose and fell like swells on the sea. As the waves passed by, they tilted trees and deeply fissured the ground. Landslides swept down from bluffs and hillsides. Considerable areas were uplifted, and still larger areas sank and became covered with water which either emerged through fissures or small craters or accumulated behind obstructions to surface drainage. On the Mississippi, great waves overwhelmed many boats and washed others high up on the shore; the return currents carried thousands of trees back into the river. High banks caved into the river, and sandbars and islands disappeared in whole or part. Mobilization of subsurface materials was widespread, and tremendous quantities of sand and other debris were squeezed out upon previously fertile fields within an area of 6000 sq km (2300 sq mi). All this occurred in an area thought to be earthquake-free.

The only great seismic event to strike a large city in the United States was the San Francisco earthquake of 1906. Buildings shed huge volumes of debris or collapsed completely, streets and railway tracks were warped into strange shapes, and slides roared down hillsides. At least two dozen fires broke out, and efforts to stem them were thwarted by impassable streets and rupture of water mains. In 3 days much of San Francisco was destroyed, with 25,000 buildings burned over an area of 12.2 sq km (4.7 sq mi). The earthquake and fire claimed 500 lives within the city, and 300,000 people were left homeless. Damage amounted to $500 million, or about $2.5 billion in 1970 dollars.

It remained for the Kwantō (Tokyo), Japan, earthquake of September 1, 1923, to set the record for earthquake-related fire losses. In addition to the 128,000 houses destroyed by the shock, 447,000 others were consumed by flames. Of the 143,000 people who lost their lives, nearly 40,000 were burned to death in a relatively open area of Tokyo where they had congregated supposedly for safety only to be overwhelmed by one of the fiery whirlwinds which often originate in large conflagrations.

The hazards presented by earthquake-triggered landslides were frightfully illustrated by the Ancash, Peru, earthquake on May 31, 1970, the greatest seismic disaster of modern times in the Western Hemisphere. The quake triggered an avalanche of mud, rock, and ice from points high on the north peak of Mount Huascaran. The avalanche moved 3660 m (12,000 ft) downward over a horizontal distance of 11 km (7 mi) at an average speed of 320 km (200 mi) per hour, burying the towns of Yungay and Ranrahirca. This single avalanche accounted for one-third of the 67,000 fatalities.

The greatest differential shifts in land level ever recorded in an earthquake took place during the earthquakes of September 3 and 10, 1899, at Yakutat Bay, Alaska. The ground was uplifted and warped over an area of 1200 sq km (460 sq mi). Local uplifts of 9 to 14 m (30 to 47 ft) reflected displacements along nearby submarine faults. The earthquakes created waves up to 6 m (20 ft) high. The front of the nearby Hubbard Glacier was shattered, and enormous icebergs cluttered the sea. The Hubbard and other glaciers soon began to advance because of avalanche additions of snow and ice in their source areas.

The Prince William Sound earthquake of March 27, 1964, also known as the Great Alaska Earthquake, was felt over an area of 1.3 million sq km (500,000 sq mi), and the zone of heavy damage extended over 130,000 sq km (50,000 sq mi). The severe ground shaking, which lasted nearly 4 minutes, triggered many slides and avalanches. Deformation of the surface was widespread (Figure 8-1). Maximum uplift of 12 m (38 ft) was recorded on Montague Island at the outer margin of the Sound, and to the northwest an area of more than 100,000 sq km (39,000 sq mi) subsided. Seismic sea waves generated by movements of the seafloor caused damage along the Pacific coastline as far as Crescent City, California.

Sea waves of seismic origin have been especially destructive along populous parts of the Japanese, Hawaiian, Chilean, and Peruvian coasts. On March 2, 1933, one of the greatest earthquakes in recorded history occurred off Sanriku, on the east coast of Japan, 120 km (75 mi) north of Sendai on the island of Honshu. Thanks to its location far offshore, this shock was not directly disastrous, but it was followed

FIGURE 8-1. **Crustal deformation accompanying the Prince William Sound, Alaska, earthquake of March 27, 1964. The flat, light area is former sea bottom elevated during the earthquake. The light strip at the foot of the mountains across the water is a similar strip of raised sea floor.** (*Photo by George Plafker, U.S. Geological Survey.*)

by seismic sea waves that surged 29 m (94 ft) above sea level and drowned about 3000 people.

Earthquake Tolls

Earthquakes throughout the world during the past 4000 years may well have claimed 13 million human lives, about 2 1/2 million of them during the past 4 centuries and nearly 1 million since 1875. The rising trend in loss of life per century reflects both growth and clustering of world populations. The totals are highest for those seismic regions that are densely populated (Table 8-1).

A number of factors influence the death toll in a major shock. The time of day and season of the year determine the presence or absence of large numbers of people in dangerous areas, and climate and topography determine the extent of mudflows, landslides, and associated phenomena. In addition, building practices, fire-fighting capabilities, and the adequacy of warnings against seismic sea waves are significant. Owing to the infrequent and sporadic occurrence of earthquakes with high catastrophic potential, the trend in death tolls has been very irregular (Table 8-2).

The United States has suffered only about 1700 deaths since the beginning of the nineteenth century (Table 8-3). Nearly half of this total came from the San Francisco earthquake

Table 8-1. Lives lost from earthquakes in major seismic regions of the world, 1670–1972

India and Pakistan	420,000
Iran—Caspian Sea	350,000
China	290,000
Western Europe and Africa	270,000
Japan	250,000
Western South America	120,000
Asia Minor—Balkans	80,000
Central America	15,000
Caribbean islands—Venezuela	3,000
Mexico and Guatemala	2,300
Philippine Islands	2,000
United States	1,700
Total	1,804,000

of 1906. The 590 earthquake-related deaths from 1925 to 1975 represent only 3 percent of the total death toll from floods, hurricanes, tornadoes, and earthquakes in the United States for that period. The corresponding worldwide figure is 12 percent. Viewed over periods of decades, the small death toll from earthquakes in the United States has been equal to that from snow avalanches; 15 percent of that from hurricanes, and 10 percent or less of the toll from floods, tornadoes, and lightning.

The most obvious economic losses from earthquakes are those due to damage and destruction of property from strong shaking, fire, inundation, and seismic sea waves, as well as losses from the interruption and subsequent repair of lines of communication, supply, and travel. To these should be added the costs of emergency responses, legal actions, reduced property values, increased insurance rates, interference with normal human activities and land uses, and deterioration of personal health and morale, the costs of which are difficult to appraise.

Strong ground-shaking can reduce entire cities to rubble if buildings are weak. On February 29, 1960, hundreds of old unreinforced masonry structures and many younger but poorly built reinforced concrete structures in the resort city of Agadir, Morocco, were destroyed by a moderate earthquake. About 14,000 people were killed out of a population of 33,000 (Table 8-2). On August 31, 1968, a major shock in Iran shattered the adobe and mud-wall buildings of several villages and towns, killing 1200 people in Dasht-e-Bayaz (population 1700) and 1400 in Kahkt (population 4400). At Qir on April 10, 1972, nearly one-quarter of the population perished under the debris of collapsed buildings. The death toll from collapse of structures during earthquake shaking in Iran during the decade 1962–1972 exceeded 30,000.

During the past century more than a million homes in Japan have been shaken apart by earthquakes, consumed by earthquake-related fires, or destroyed by seismic sea waves. Even in Niigata, where principles of earthquake-resistant design had been used for many modern buildings, several high-rise structures were depressed, tilted, or toppled in 1964 when strong ground-shaking liquefied parts of their natural foundations. A nearby residential section of the city was destroyed by fires that burned unchecked for 2 weeks after it was inundated by seismic sea waves that spread burning oil spilled from ruptured refinery facilities.

Property losses can be enormous in metropolitan areas, and they tend to rise dramatically with urban growth. Fortunately, however, the ratio of loss of life to property losses tends to decrease as structural design improves. Physical damage from earthquakes in the United States has amounted to $2 billion since the year 1800 (Table 8-3), and to $1.4 billion out of a $20-billion total for floods, hurricanes, tornadoes, and earthquakes since 1925. The $500-million level has been reached or exceeded in only three earthquakes, San Francisco, California, in 1906, Prince William Sound, Alaska, in 1964, and San Fernando, California, in 1971, which thus accounts for much of the long-term total of $2 billion. Property losses from seismic events are low in relation to the costs of all natural disasters. The more than half billion dollar loss from the San Fernando earthquake of 1971 is relatively small when compared with the more than $3.1 billion in damage from Hurricane Agnes in June 1972. It is nonetheless sobering to note that earthquakes are becoming increasingly expensive with time, reflecting both changes in dollar value and growth in number and value of works susceptible to damage. For California alone, the State Division of Mines and Geology has predicted $21 billion in combined property damage and dollar equivalent of loss of life from earthquakes during the period 1970–2000, assuming that present loss-reduction practices are continued unchanged. This is about 38 percent of the estimated loss from all geologic hazards in the state during the same 30-year period.

NATURE OF EARTHQUAKES

Earthquakes are successions of waves that spread out from sources below and at the earth's surface. The passage of these waves is often visible, but far more often, the vibrations can be detected only by sensitive instruments.

Seismology and Seismicity

The science of earthquakes, *seismology*, began to emerge as a true science about 1880, with the development of the first effective instruments (seismographs) for recording earthquake waves. Imagine that a heavy weight of hundreds of kilograms is suspended by a long wire from the ceiling of a room. The weight, which almost touches the floor, has a marking pen attached to its underside, and the pen rests lightly on a piece of paper pinned to the floor. We will assume that the building is well constructed and rests on solid rock so that it will respond to earthquakes as a coherent unit. If the ground, and the building resting on it, are now shaken by a series of rapid vibrations, everything will oscillate except the weight at the end of the wire. Its inertia prevents it from whipping back and forth with each rapid vibration. The weight thus remains fixed in space while everything else moves, including the paper pinned to the floor. The movement of the paper under the stationary pen records the ground movements. In order that the record of each succeeding vibration be individually recorded and not superimposed on the earlier markings, the recording paper (or film) of a seismograph is mounted on a slowly revolving drum (Figure 8-2).

The record of the ground vibrations (seismogram) appears as a series of compressed wave forms (Figure 8-3). Fortunately, the waves transmitted in earthquakes are of different types and travel at different velocities. This enables the seismologist to determine the magnitude and location of an earthquake.

The first waves recorded on a seismogram are those that travel through the earth from the earthquake source. These *preliminary waves* are of two types: the first, the *primary*, or *P waves*, are compressional or push waves similar to sound waves. Each rock particle is pushed a minute distance against the one in front, which in turn pushes against the one ahead, and so on. Each particle springs back elastically in a vibration that continues until the energy is dissipated.

The second waves to arrive are the *secondary*, or *S, waves*. These are shear waves, such as are formed in a rope that is fastened at one

Table 8-2. Lives lost from some damaging earthquakes in various parts of the world, based on estimates

Year	Locality	Magnitude[1]	Intensity[2] (M.M.)	Lives Lost
856	Corinth, Greece	—	—	45,000
1038	Shensi, China	—	—	23,000
1057	Chihli, China	—	—	25,000
1170	Sicily	—	—	15,000
1268	Cilicia, Asia Minor	—	—	60,000
1290	Chihli, China	—	—	100,000
1293	Kamakura, Japan	—	X	30,000
1456	Naples, Italy	—	—	60,000
1531	Lisbon, Portugal	—	—	30,000
1556	Shensi, China	—	XII	830,000
1622	East Kansu, China	—	—	12,000
1667	Shemakha, Caucasia	—	XI	80,000
1693	Catania, Sicily	—	—	60,000
1693	Naples, Italy	—	—	93,000
1715	Algiers	—	—	20,000
1727	Tabriz, Iran	—	—	70,000
1731	Peking, China	—	—	100,000
1737	Bengal, India	—	XII	300,000
1755	Kashan, Iran	—	—	40,000
1755	Lisbon, Portugal	8.7	XII	70,000[3]
1783	Calabria, Italy	—	—	50,000
1797	Quito, Ecuador	—	XI	40,000
1819	Cutch, India	—	XII	2,000
1822	Aleppo, Asia Minor	—	—	22,000
1828	Echigo, Japan	—	XI	30,000
1847	Zenkojo, Japan	—	—	34,000
1868	Arica, Chile-Peru	8.5±	XI	25,000[3]
1868	Ibarra, Ecuador	—	XI	70,000
1875	Venezuela-Colombia	—	—	16,000
1891	Mino-Owari, Japan	8.4	XII	7,300
1896	Sanriku, Japan	7.6	XI	27,100[3]
1897	Assam, India	8.7	XII	1,500
1898	Japan	—	—	22,000
1905	Kangra, India	8.6	XII	19,000
1906	Kagi, Taiwan	—	—	1,300
1906	Valparaiso, Chile	—	—	1,500
1906	San Francisco, California	8.3	XI	800[4]
1908	Messina, Sicily	7.5	XI	160,000
1915	Avezzano, Italy	7.5	XI	30,000
1920	Northeast Kansu, China	8.6	XII	180,000[5]
1923	Kwantō (Tokyo), Japan	8.3	XII	143,000[4]

[1]Magnitude—Magnitude according to the Richter scale, in which each whole number represents 10 times the maximum trace amplitude of the preceding whole number (see text).

[2]M.M.—Modified Mercalli Intensity Scale with maximum intensity at XII (see Table 8-4).

Year	Locality	Magnitude	Intensity (M.M.)	Lives Lost
1927	Tango, Japan	8.0	XI	3,000
1927	Tsinghai, China	8.3	XII	100,000[5]
1930	Apennine Mts., Italy	—	—	1,500
1932	Kansu, China	7.6	X	70,000[5]
1933	Sanriku Coast, Japan	8.9	XI	3,000[3]
1934	Bihar, India	8.4	X	11,000
1935	Quetta, Pakistan	7.6	XI	60,000
1939	Chillan, Chile	8.3	XI	30,000
1939	Erzincan, Turkey	8.0	XI	40,000
1946	Ancash, Peru	7.4	XI	1,400
1948	Fukui, Japan	—	—	5,000
1950	Assam, India	8.7	XII	1,600
1953	Northwest Turkey	—	—	1,200
1954	Northern Algeria	—	—	1,600
1956	Kabul, Afghanistan	—	—	2,000
1957	Northern Iran	—	—	2,500
1957	Western Iran	—	—	1,400
1957	Outer Mongolia	—	—	1,200
1960	Agadir, Morocco	5.9	VII	14,000
1960	Southern Chile	8.5	XII	6,000[3]
1962	Northwest Iran	7.3	X	12,000
1963	Skopje, Yugoslavia	—	—	1,000
1964	Prince William Sound, Alaska	8.6	XI	131
1968	Dasht-e-Bayaz, Iran	7.2	IX	11,000±
1970	Ancash, Peru	7.8	XI	67,000[5]
1972	Qir, Iran	7.0	VIII	17,000
1972	Managua, Nicaragua	6.2	IX	12,000
1973	Hokkaido, Japan	7.7	?	?
1973	Verza Cruz, Mexico	7.2	VII–VIII	539
1974	Peru	7.6	VII	78
1975	China	7.4	?	Probably many
1975	Turkey	6.8	?	?
1976	Guatemala	7.5	XI?	23,000
1976	Italy	6.9?	?	1,000
1976	Usbek, U.S.S.R.	7.3	?	?
1976	Burma	7.0	?	?
1976	Yunnan, China	7.6	?	?
1976	Hopei, China (Tangshan)	7.8	?	650,000
1976	Philippines	7.8?	?	8,000?
1976	Turkey	7.9?	?	5,000?

[3]Includes many deaths from tsunami action.

[4]Includes losses from fires in metropolitan areas.

[5]Mainly deaths from burial by landslides or avalanches.

Table 8-3. Principal losses from United States earthquakes since 1800, based mainly on estimates published by

Year	Locality	Intensity, Modified Mercalli Scale	Lives Lost	Estimated Property Damage (thousands of contemporary dollars)
1811⎫ 1812⎭	New Madrid, Mo. (3 main shocks)	XI	20±	200
1812	San Juan Capistrano, Calif.	X	40	100
1857	Fort Tejon, Calif.	XI	1	200
1865	Santa Cruz Mountains, Calif.	IX	—	500
1868	South coast of Hawaii	X	20[1]	100
1868	Hayward, Calif.	X	30	350
1872	Owens Valley, Calif.	XI	27	250
1886	Charleston, S. Car. (2 main shocks)	X	60	23,000
1892	Vacaville, Calif.	IX	—	225
1898	Mare Island, Calif.	VII	—	1,400
1899	Yakutat Bay, Alaska	XI	—	—
1899	San Jacinto, Calif.	IX	6	150
1906	San Francisco, Calif.	XI	800±	600,000[2]
1915	Imperial Valley, Calif.	VIII	6	900
1918	Puerto Rico	—	116[3]	4,000[3]
1918	San Jacinto, Calif.	IX	—	200
1925	Manhattan, Mont.	IX	—	300
1925	Santa Barbara, Calif.	IX	13	13,000
1926	Santa Barbara, Calif.	VII	1	2,500
1932	Humboldt County, Calif.	VIII	1	1,000
1933	Long Beach, Calif.	IX	119	55,000
1934	Kosmo, Utah	VIII	2	100
1935	Helena, Mont. (2 main shocks)	VIII	4	4,000
1940	Imperial Valley, Calif.	X	9	6,000

[1]Chiefly drownings from tsunami action.

[2]Includes $100 million earthquake damage and $400 million fire damage in the City of San Francisco. The remaining damage was in the environs.

[3]Chiefly effects of tsunami generated by earthquake in Mona Passage.

end while the other end is whipped up and down or sideways. A series of undulations moves along the rope. These waves travel at only about half the speed of the primary waves.

The preliminary waves spread out spherically from the subterranean *focus* of the earthquake (Figure 8-4). Those traveling vertically upward soon reach the surface at the point called the *epicenter*. From here, *long* waves spread out like the ripples in a pond. They travel around the surface of the earth, arriving at the seismograph station after the *P* and *S* waves. They commonly display the largest amplitudes on a seismogram.

Because the primary waves travel almost twice as fast as the secondary, the greater the distance of a seismograph station from the earthquake source, the greater the time interval between the arrivals of these two sets of waves. With the aid of tables or specially prepared diagrams, it is relatively simple to determine the distance to the earthquake epicenter. If an arc with radius equal to this distance is drawn on a map, the epicenter will lie somewhere on this arc. If three stations follow this procedure, there will be only one place where all three arcs intersect. This will locate the epicenter (Figure 8-5).

Year	Locality	Intensity, Modified Mercalli Scale	Lives Lost	Estimated Property Damage (thousands of contemporary dollars)
1941	Santa Barbara, Calif.	IX	—	100
1941	Torrance-Gardena, Calif.	VIII	—	1,000
1944	Massena, New York	IX	—	2,000
1946	Aleutian Islands, Alaska	—	179[4]	27,000[5]
1949	Olympia, Wash.	X	8	25,000
1949	Terminal Island, Calif.	V	—	9,000[6]
1951	Terminal Island, Calif.	V	—	3,000[6]
1951	Kona, Hawaii	IX	2	1,000
1952	Kern County, Calif.	XI	12	60,000
1952	Bakersfield, Calif.	VIII	2	10,000
1954	Eureka-Arcata, Calif.	VII	1	2,100
1954	Wilkes-Barre, Penna.	VII	—	1,000
1955	Terminal Island, Calif.	IV	—	3,000[6]
1955	Oakland-Walnut Creek, Calif.	VI	1	1,000
1957	Andreanof Islands, Alaska	VIII	—	4,000[5]
1957	San Francisco, Calif.	VII	—	1,000
1958	Southeastern Alaska	XI	5	1,000
1959	Hebgen Lake, Mont.	X	28	17,000
1961	Terminal Island, Calif.	IV	—	4,500[6]
1964	Prince William Sound, Alaska	XI	131[7]	500,000[8]
1965	Puget Sound, Wash.	VIII	7	12,500
1966	Dulce, New Mex.	VII	—	200
1969	Santa Rosa, Calif.	VIII	—	7,500
1971	San Fernando, Calif.	XI	64	550,000
	Totals (approximate)		1700	2,000,000

[4]Includes 173 deaths in Hawaii and 1 in California from associated tsunami action.
[5]Mainly tsunami damage in Hawaii.
[6]Damage to oil wells.
[7]Almost wholly from tsunami action.
[8]Mainly landslide and tsunami damage.
[9]"Contemporary dollars" refers to dollar values at the time of the earthquake.

Earthquake activity, or *seismicity*, is by no means uniform throughout the world. Certain regions have a high incidence of earthquakes; others are relatively "quiet." Furthermore, in some regions, earthquake shocks originate at much greater depths than in others. Broad geographic contrasts in world seismicity and earthquake characteristics are so consistent that they must represent global geologic control, as discussed later.

Causes of Earthquakes

Most earthquakes, including nearly all destructive ones, are due to faulting, a phase of rock deformation or tectonic activity. Before considering tectonic earthquakes, a few minor causes of earthquakes should be noted.

Minor. Minor causes of earthquakes include landsliding, volcanic eruptions, and human activities. They are generally felt only locally, and

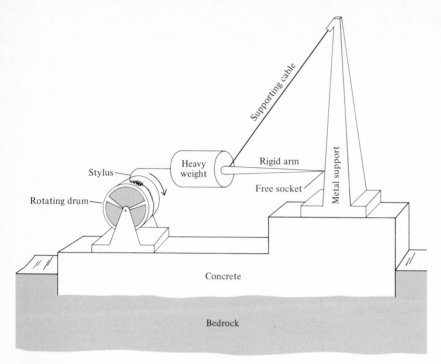

FIGURE 8-2. **Principle of horizontal-motion seismograph. If the ground moves back and forth at right angles to the page, the heavy mass remains stationary because of its inertia, while the rest of the apparatus, cemented to bedrock, moves with the ground. Since the stylus is fixed to the stationary mass, it will leave a record on the vibrating and rotating drum. The stylus may be replaced by a mirror which reflects a beam of light onto sensitized paper on the drum. If the ground were to move parallel to the paper, the rigid arm would cause the heavy mass to move also. Hence, a second seismograph, oriented at right angles to this one, is needed to ensure recording of waves from all directions of the compass. To record vertical ground movements, the weight would be suspended from a vertical spring.**

the shaking itself does not cause significant damage or loss of life.

In Chapter 2, we discussed the Vaiont landslide in northern Italy and noted that the tremors were recorded as far away as Brussels, Belgium. There is no record, however, of damage due to this ground-shaking. Compared to the energy released in tectonic earthquakes,

FIGURE 8-3. **Idealized seismogram. *P*—primary waves, *S*—secondary waves, *L*—long (surface) waves. The 11-minute interval between arrival of the *P* and *S* waves indicates a distance to the epicenter of 10,000 km (6200 mi).**

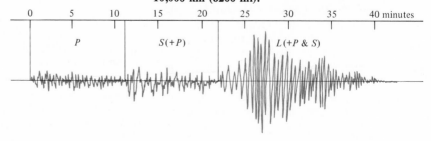

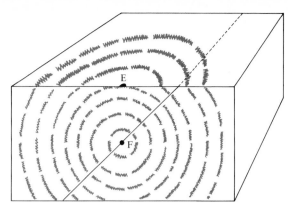

FIGURE 8-4. **Focus and epicenter.** A rupture has started along the inclined fault at F, the *focus*. The shock waves spread out spherically as preliminary waves (*P* and *S*). On reaching the surface at E, the *epicenter*, the waves spread radially outward as long surface waves (*L*). Although the surface waves travel slower than the preliminary waves, they almost always have greater amplitudes.

even such large-scale landslides are low-energy events.

Minor volcanic tremors result from small displacements caused by movement of molten rock in depth. Large-scale explosions produce more severe shaking, but again the direct effects are local.

People have created earthquakes by large-scale blasting and by underground nuclear explosions. Such explosive events can crack walls and shatter windows to distances of tens of kilometers. These effects, too, are small compared with those of major tectonic earthquakes.

People may also trigger some tectonic earthquakes. We noted in Chapter 3 the stimulation of earthquakes in the Denver, Colorado, area by injection of waste fluids into deep wells. The creation of reservoirs may have comparable effects. The enormous weight of the water, plus the increase in fluid pressure in the rocks below, may trigger faulting. Earthquakes attended the filling of Lake Mead behind Hoover Dam, the major shock (M5.0) occurring when the reservoir was 80 percent filled. The region had not previously been regarded as seismic. Of the dozen or so recorded instances of reservoir-associated earthquakes, only one, that of the Koyna Reservoir, in India, is known to have been destructive. Maximum water level was reached in 1965. Two years later, severe earthquakes occurred with magnitudes up to 6.4. The region had previously been regarded as one of the least seismic in the world. The earthquake destruction included 177 dead, 2300 injured, and extensive property damage. On the other hand, the filling of other large reservoirs in similar geologic environments has not been accompanied by earthquakes. Examples are the reservoirs behind Grand Coulee Dam in the State of Washington and the Glen Canyon Dam in northernmost Arizona.

Major: Tectonics. The major cause of earthquakes is tectonic. The most common earthquake mechanism is shown diagrammatically in Figure 8-6. A subsurface block of crustal material is divided into two units (L and R) by a vertical fault (F), an earlier surface of rupture now temporarily locked by friction. Two vertical rock layers (shaded) illustrate the deformation. Forces (arrows) are slowly applied and tend to move unit L horizontally toward the observer, relative to unit R. With continued application of these forces, the two blocks slowly respond with *elastic strain* along and near the fault. This strain is expressed by warping of the block and increasing curvature of the rock layers. It is not yet accompanied by movement along the fault, owing to frictional

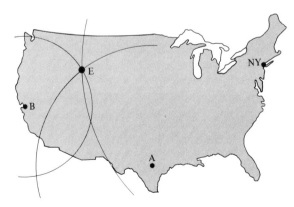

FIGURE 8-5. **Location of earthquake epicenter.** The radii of the three arcs represent the distances of the epicenter (E) from seismograph stations at Berkeley, Calfor"nia (B), Austin, Texas (A), and New York City (NY).

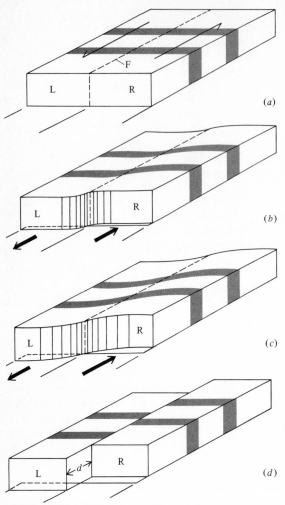

These are the essential elements of the *elastic rebound* theory. The theory was based in part upon observed horizontal offsets along the San Andreas Fault during the 1906 San Francisco earthquake (Figure 8-7) and in part on comparisons of pre-and postearthquake surveys in the region crossed by the fault.

The Nature of Faults and Fault Movements. *Kinds and Expression.* In their simplest form, faults are planar or curved surfaces, the opposite sides of which have moved differentially. They may be meters to kilometers in length and range from vertical to horizontal. Most large faults, however, are complex assemblages of parallel, subparallel, or interlacing fractures of different ages clustered in a narrow belt known as a *fault zone*. Most major fault zones (Figure 8-8) are tens to hundreds of kilometers long and up to a few kilometers wide and commonly include numerous blocks and fragments of thoroughly broken-up rock (*fault breccia*) and finely ground rock (*fault gouge*).

FIGURE 8-6. **The elastic rebound theory of earthquakes. Horizontal stretching of the crust in opposing directions builds up strain until a displacement occurs. The snapping of the rocks, like the plucking of a string, sends out shock waves in all directions.**

FIGURE 8-7. **Fence offset during 1906 San Francisco earthquake. Note that the fault trace, or fracture, is inconspicuous here. Marin County, California. (*G. K. Gilbert*)**

resistance. Finally, however, the *strain energy* becomes great enough to overcome this cohesion, causing rupture and displacement along the fault diagram *d*). The units are offset horizontally along this break by a distance d as each snaps elastically to a new position. The sudden snapping and release of energy generates earthquake waves that radiate outward from the fault.

Golden Gate

San Francisco

SA

Crystal Springs
Dam

UCS

Causeway

LCS

FIGURE 8-8. **The San Andreas fault zone on the San Francisco Peninsula, California. The Crystal Springs Dam, which is approximately parallel to the fault zone, was unaffected by the 1906 earthquake. The causeway was offset 2 m (6 ft). The width of the fault zone in the foreground is 900 m (3000 ft) from the road on the right to the broken line on the left. Large real estate developments cover the northern end of the fault zone. SA—San Andreas Lake. UCS—Upper Crystal Springs Lake. LCS—Lower Crystal Springs Lake. View northwest. (Courtesy Teledyne Geotronics.)**

Most large faults have distinctive topographic expression. This can result directly from recent displacements of the ground and indirectly from selective erosion of the crushed materials along the fault or from more rapid erosion of the materials on one side of the fault. The most obvious expressions are *fault scarps*, either low (Figure 8-9) or high (Figure 8-10). Other evidence consists of offset stream courses (Figure 8-11), straight and aligned valleys, and elongate and aligned sag ponds (Figure 8-12). These features are displayed diagrammatically in Figure 8-13.

Faults can strongly influence the circulation of ground water, some acting as passages and others as subsurface dams. Thus, many fault traces are marked by ponds, lines of springs, and concentrations of water-loving vegetation.

Fault Terminology. Several types of faults are distinguishable in terms of their attitudes and the movement along them (Figure 8-14). Faults vary from horizontal to vertical, but most are inclined at intermediate angles. The attitude of a fault can be expressed in terms of *strike*, which is the direction of the fault trace across country, and *dip*, which is the angle between the fault surface and a horizontal plane (d in Figure

8-14a). In an inclined fault, the overhanging base of the upper block is called the *hanging wall*, and the upward-facing surface of the lower block, the *footwall*.

A *normal fault* is one in which the hanging wall has moved downward *relative to* the footwall; a *reverse fault* is one in which the hanging wall has moved upward *relative to* the footwall. A reverse fault with an angle of inclination of 45° or less is ordinarily called a *thrust fault*. Vertical faults have neither footwall nor hanging wall.

Strike-slip faults are characterized by dominantly horizontal slippage (Figure 8-14d,e). If an observer stands on one side of such a fault, and the opposite side appears to have moved to the right (Figure 8-14d), the fault is a *right-*

FIGURE 8-9. 1872 earthquake fault scarp, Owens Valley, near Independence, California. The photograph was taken in 1925. Note the extent of weathering, erosion, and plant growth in 50 years. Compare Figure 8-28 (*Photo by Eliot Blackwelder.*)

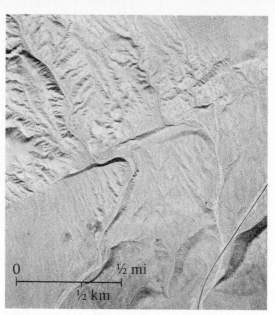

FIGURE 8-11. Offset stream courses along Garlock fault, southeastern California. The ground across the fault has moved to the left, making this a left-lateral strike-slip fault (see Figure 8-14e). (*Photo by U.S. Dept. Agriculture.*)

FIGURE 8-10. Sierra Nevada fault scarp west of Lone Pine, California. This great wall, 2400 m (8000 ft) high, was formed by hundreds of small uplifts. The older, higher parts of the range, having been "above ground" the longest, are the most eroded. (*Photo by Eliot Blackwelder.*)

FIGURE 8-12. **Sag pond along San Andreas Fault at Mustang Ridge, 32 km (20 mi) east of King City, California. Sag ponds result from differential settlement of the ground during shaking or local removal of material by ground water moving through the crush zone. (*Photo by Benjamin M. Page.*)**

lateral fault. Movement to the left (Figure 8-14*e*), as noted in this manner, identifies a strike-slip fault as *left-lateral*.

Most faults show a combination of horizontal and vertical movements. Where the components are of approximately the same order, the term *oblique-slip fault* is applicable (Figure 8-14*f*).

The distance by which a once-continuous rock unit or other identifiable feature has been offset along a fault is called *separation*, which can be referred to as horizontal, vertical, or oblique, according to the chosen plane of reference.

FIGURE 8-13. **Landforms developed along active strike-slip faults. Shutter ridges are slices of rock between parallel faults which, by differential fault movement, form a barrier across a valley (see Figure 8-11). (*After R. L. Wesson, E. J. Helley, K. R. Lajoie, and C. M. Wentworth, in R. D. Borcherdt, ed., "Studies For Seismic Zonation of the San Francisco Bay Region," U.S. Geol. Survey, 1975.*)**

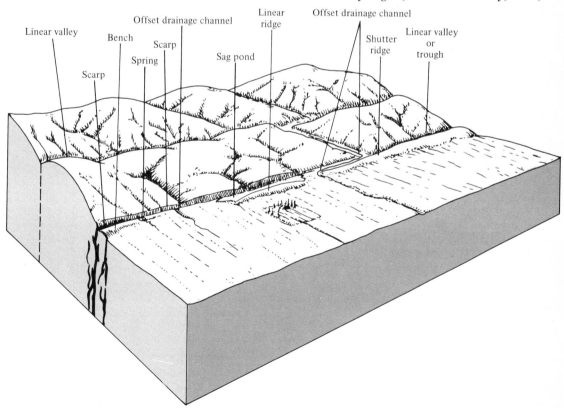

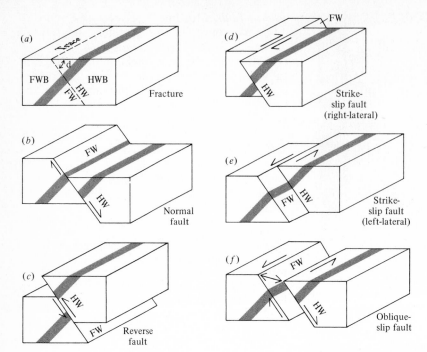

FIGURE 8-14. **Types of faults. d (diagram *a*)— angle of dip of fault; HW—hanging wall; FW— footwall; HWB—hanging wall block; FWB—foot- wall block.**

DISTRIBUTION OF MAJOR EARTHQUAKES

Relation to Plate Tectonics

The distribution and nature of seismically ac- tive zones are explained on a grand scale by the *theory of plate tectonics*, introduced in Chapter 1. The major earthquake zones occur where the gigantic plates of the earth's lithosphere slide past or under each other (Figure 8-15). Certain plate boundaries are characterized not only by frequent and large earthquakes but also by considerable volcanic activity.

In addition to the active zones at plate boundaries, there are a few enigmatic areas of seismic activity in the interiors of plates. Else- where, however, the earth's crust is relatively quiet. Even quiet regions may experience rare earthquakes, but these are usually light or mod- erate.

The theory of plate tectonics holds that the lithosphere, which is the outer, relatively brittle part of the earth, is divided areally into six to

eight large segments plus a number of smaller ones. The plates move slowly relative to one another but nevertheless fit snugly together. The boundaries between adjacent plates are of three kinds: (1) divergent, (2) transform, and (3) convergent.

Divergent plate boundaries are typified by oceanic "ridges" such as the Mid-Atlantic Ridge. The cleft axis of such a ridge results from spreading of the ocean floor. Basaltic lava rises in the fracture zone from time to time and congeals. Repeated fracturing followed by in- vasion of basalt adds increments to the oceanic crust.

Transform boundaries are characterized by horizontal strike-slip between two adjacent plates. Although most transform boundaries are in the oceans, the San Andreas Fault system of California is a transform boundary partly on land. It forms the juncture between the Pacific and North American plates.

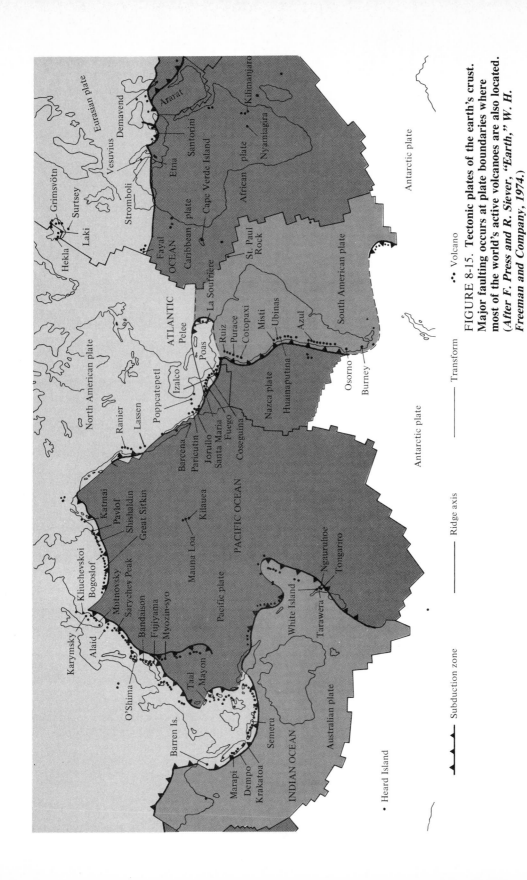

FIGURE 8-15. **Tectonic plates of the earth's crust. Major faulting occurs at plate boundaries where most of the world's active volcanoes are also located.** *(After F. Press and R. Siever, "Earth," W. H. Freeman and Company, 1974.)*

•• Volcano

◣▲◣ Subduction zone

———— Ridge axis

———— Transform

Convergent plate boundaries involve *subduction*, whereby one plate moves beneath another. The margin of the descending plate moves slowly downward along an inclined path, its leading edge being continuously consumed in depth. As a rule, there is an oceanic trench where the plate bends downward, and volcanic activity appears at the surface above that part of the slab which has descended deeply enough to induce melting.

More than 95 percent of earthquake epicenters are concentrated along the plate boundaries shown in Figure 8-15. Inasmuch as divergent zones offer ready outlets for heat and magma from the depths, and subduction commonly induces melting, both divergent and convergent plate boundaries are also favorite sites of volcanic activity.

The type of plate boundary determines the nature of the associated faults. Normal faults are prevalent along divergent plate boundaries, producing small to fairly large earthquakes but not *great* earthquakes. Strike-slip faulting occurs at transform plate boundaries, producing small to great earthquakes (e.g., San Francisco, 1906). Reverse faulting dominates over normal faulting along convergent plate boundaries, producing small to great earthquakes. It is suspected that the greatest earthquakes (in terms of energy) have originated from reverse faults in zones of convergence (e.g., Alaska, 1964).

Although plate boundaries are important loci of earthquakes, the boundaries are rarely sharp, and the affiliated fault movements and earthquakes occur in ill-defined belts of appreciable width. Furthermore, because the descending plate of some convergent zones extends far under the upper plate, it may cause distension and normal faulting of the upper plate over widths of hundreds of miles. Although major earthquakes are concentrated at plate boundaries, every state in the United States has experienced earthquakes strong enough to be felt, and some have experienced destructive events.

Distribution in Time and Space

Available records do not span enough time to provide adequate knowledge of the distribution of earthquakes in time. However, the following generalizations may be made: (1) Some regions experience earthquakes frequently, others rarely. (2) In any one area, the interval between earthquakes varies considerably, even in highly seismic localities. And (3) In any specific area, or on any particular part of a fault, there is an average recurrence interval between earthquakes of a certain size. In active areas that have been monitored instrumentally, there are a great many more small earthquakes (mostly not felt by people) than medium or large earthquakes. The recurrence interval of large earthquakes is much greater than that of small earthquakes.

In accord with the third generalization above, it has been possible to calculate average recurrence intervals for small earthquakes in active areas where sufficient numbers of events have occurred to give a statistically useful sample. It may be possible to say, for example, that a region will experience an earthquake of intensity VII (see Table 8-4) on an average of every 10 years. However, earthquakes of intensity XI or XII are rare events, and historical records are insufficient for estimates of recurrence intervals.

A map has been prepared for the United States which divides the country into zones of relative seismic risk (Figure 8-16), largely on the basis of historical records. However, truly great earthquakes may occur in a given region only at intervals of 100, 500, or 1000 years; hence, there is no assurance that a presently mapped low-risk area may not prove to be high-risk at some future date. The point is that, in many parts of the world, the records are not long enough to give the final verdict on "safe" versus "unsafe" regions. Hence, we must use geology and geophysics as further bases for assessment.

MAGNITUDE AND INTENSITY OF EARTHQUAKES

The *magnitude* of an earthquake is based on the amplitude of the earthquake waves recorded in a seismogram, with the qualifications cited below. The *intensity*, on the other hand, is a measure of the direct effects of an earthquake

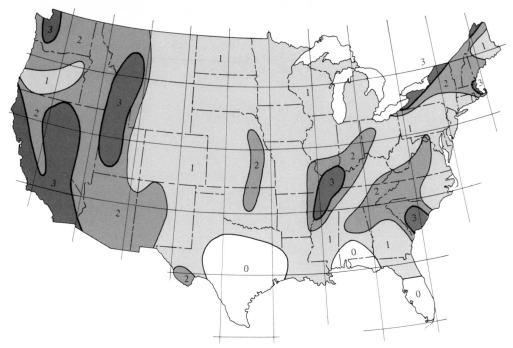

FIGURE 8-16. Earthquake risk map of the United States as of 1969 based on intensity data collected by U.S. Coast and Geodetic Survey. Zone 0—No damage. Corresponds to earthquake intensities of I to IV. Zone 1—Minor damage (intensities V and VI). Zone 2—Moderate damage (intensity VII). Zone 3—Major damage (intensities of VIII and higher). (*After S. T. Algermissen, Courtesy National Oceanographic and Atmospheric Administration Environmental Research Laboratories.*)

on people and structures. Magnitude and intensity are rated on different scales.

Magnitude is expressed on the Richter scale by an Arabic numeral preceded by "M" as, for example, M8.5. The maximum trace amplitude (see Figure 8-3) is first corrected to certain standard conditions and expressed in microns. The corrections are necessary because of differences in seismographs and differences in distance from the same earthquake. To avoid large numbers, the magnitude is expressed as the logarithm of the amplitude. For example, if an earthquake provides a corrected trace amplitude of 10 microns, its Richter magnitude would be 1 (10^1). If the amplitude were 100 microns, the Richter number would be 2 (10^2). If the amplitude were 1000 microns, the Richter number would be 3 (10^3), and so on. Scientists interested in computing the energy released by an earthquake at its focus use a simple equation in which the Richter magnitude is one factor. The energy release increases astonishingly for earthquakes of slightly higher magnitudes. The

8.3 San Francisco earthquake of 1906 released the energy equivalent of about 17 million tons of TNT; the 8.5 Alaskan earthquake of 1964 released the equivalent of 32 million tons; and the 8.7 Assam, India, quake of 1950 released the equivalent of 65 million tons. An 8.9 earthquake, such as the 1933 Sanriku earthquake off the coast of Japan, releases the energy of about 125 million tons of TNT. The Hiroshima atomic bomb was the equivalent of 20,000 tons of TNT.

The greatest measured or computed values of M are in the neighborhood of 8.9, which may correspond to the maximum amount of elastic strain energy that can accumulate in the lithosphere before faulting occurs[1]. "Large" earthquakes have magnitudes of 7 or more, and "great" earthquakes have magnitudes of 8 or more.

Earthquake intensity, in contrast to magnitude, is determined from the reactions of people and structures at specific localities. It is a function of local ground conditions, proximity to the source of the earthquake, and the energy produced in the earthquake. It is not as objective as magnitude, but it is sometimes more important in human affairs. The most commonly used scale is the Modified Mercalli (M.M.) Scale, which has 12 values denoted by roman numerals (Table 8-4). The amount of destruction, injuries, and fatalities depends on types of buildings, the materials on which they stand, the density of population, and time of day, among other factors.

The intensity scale is often applied over a broad area immediately after an earthquake, and the intensity values are plotted on a map. People describe their sensations, and interviews or reports may be solicited to complete the map. Damage, if any, can be readily observed and categorized. *Isoseismal lines* are drawn, delineating the boundaries of zones in which people experienced similar intensities (Figure 8-17). An isoseismal map with concentric zones of diminishing intensity suggests the location of the epicenter and, if the isoseismal zones are elongated, the location and trend of the fault. However, the zones of maximum intensity may be at some distance from the fault trace, depending on the physical properties of the surface materials (Figure 8-18).

Some medium to large earthquakes are preceded by smaller earthquakes, or *foreshocks*. Similarly, they may be followed by *aftershocks*. Once in a while, foreshocks and aftershocks are almost as destructive as the main shock or even more destructive if their foci are closer to a vulnerable area. Bakersfield, California, escaped severe damage in the main

Kern County earthquake of 1952 (M7.7), but it was heavily damaged by an aftershock (M5.8) 1 month later. Great earthquakes are generally followed by scores or hundreds of aftershocks which tend to become weaker and less frequent with the passage of time. In some cases the activity is detectable decades later.

ACTIVE AND INACTIVE FAULTS

Many faults, regardless of the depth at which initiated, reach the surface. Such faulting can disrupt any highway, building, or dam which crosses it, and the effects of the shaking may be felt far afield.

Most faults are inactive or "dead" and may be trusted not to misbehave, whereas others are "active." It is often critical to determine the category, and this may not be an easy task.

Definitions

Faults capable of slippage during a span of time that is on a human scale are said to be *active*. The term does not imply that all active faults are slipping continuously today, although some faults do "creep" quasicontinuously. Typically, fault slippage is episodic and interrupts prolonged intervals of nonslippage. For different faults, the intervals of nonslippage range from less than a decade to several centuries, or possibly to more than a millenium. Moreover, even for a particular fault, the intervals of nonslippage vary erratically in duration.

To determine whether a fault is active, it is customary to delve into historic records or, if historic evidence is lacking, to seek geologic evidence that slippage has occurred recently. If either kind of evidence is found, the fault may be designated active. However, sometimes it is possible to show that a fault has remained quiescent for hundreds of thousands or millions of years, in which case the fault may be considered inactive or dead.

The U.S. Nuclear Regulatory Commission, in its concern over the siting of nuclear power plants, regards as active any fault that has moved "at or near the ground surface at least

---

[1]A newly proposed magnitude scale, which includes the length of the fault involved in the displacement, raises the greatest possible magnitude to 9.5 and modifies some earlier magnitude assignments.

Table 8-4. Modified Mercalli Earthquake Intensity Scale

I. Imperceptible to all but a few people in special situations.

II. Perceptible by a few people at rest. Movement of suspended objects.

III. Felt by some, noticeably indoors. Like passing truck. Not generally recognized as earthquake.

IV. Felt by many. Some awakened at night. Dishes and windows rattle. Creaking noises.

V. Felt by nearly everyone. Many awakened at night. Breakage. Cracked plaster. Objects overturned. Poles, trees may be disturbed.

VI. Felt by everyone. Falling plaster. Damaged chimneys.

VII. Everyone runs outdoors. Considerable damage in poorly designed structures. Noticed in automobiles.

VIII. Collapse of many poorly designed structures. Falling chimneys, columns, monuments, walls. Furniture thrown about. Ejection of sand and mud.

IX. Considerable damage even in specially designed structures. Partial collapse. Buildings may be shifted off foundations. Ground cracked. Pipes broken.

X. Most masonry and frame structures, and some well-built wooden structures destroyed. Ground cracked. Rails warped. Considerable landsliding.

XI. Few masonry structures remain. Ground fissured. Pipelines cut. Landsliding. Rails bent badly.

XII. Total damage. Surface waves observed. Objects thrown in air.

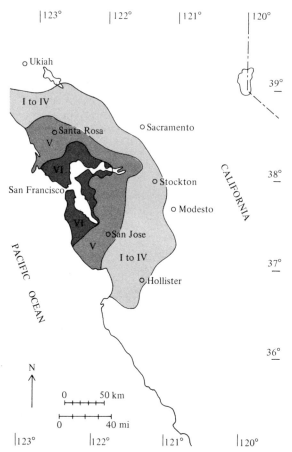

once in the past 35,000 years" or "more than once in the past 500,000 years." The Commission is concerned here with faults capable of displacing the ground surface, not just producing an earthquake. The degree of conservatism in designating faults inactive or active largely depends upon what is at stake.

Criteria for Recognition and Dating of Recent Faults

In the absence of historic records of fault activity, reliance is placed on topographic and geologic evidence.

Topographic Criteria. Much of the world's landscape is less than 2 million years old, and well preserved landforms may be very recent. The landforms include floodplains, alluvial fans, alluvial plains, hillsides, coastal features, glacial landforms, volcanic features, and many others. If such landforms are offset by faults, the faults are generally regarded as active.

The main pitfall in using geomorphic (land-

FIGURE 8-17. **Isoseismal map of earthquake of March 22, 1957, in the San Francisco Bay region.** (*Courtesy California Division of Mines and Geology.*)

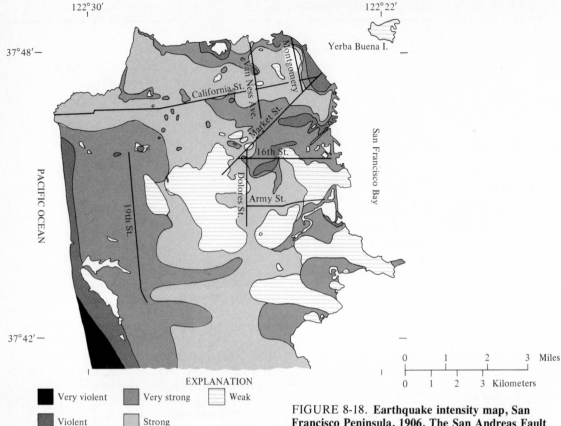

Map labels: 122°30′, 122°22′, 37°48′, 37°42′, Yerba Buena I., Montgomery, Van Ness Ave., California St., Market St., 16th St., Dolores St., Army St., 19th St., PACIFIC OCEAN, San Francisco Bay

EXPLANATION

- ■ Very violent
- ■ Violent
- ■ Very strong
- ■ Strong
- ▨ Weak

0   1   2   3   Miles

0   1   2   3   Kilometers

FIGURE 8-18. **Earthquake intensity map, San Francisco Peninsula, 1906. The San Andreas Fault lies offshore to the west. Areas of highest intensity correlate with areas of weak sediments such as bay mud, alluvium, and artificial fill. (*After H. O. Wood, in R. D. Borcherdt, ed., "Studies for Seismic Zonation of the San Francisco Bay Region," U.S. Geological Survey, 1975.*)**

form) criteria is the possibility of mistaking indirect effects of faults for direct offsets. Even a dead fault is likely to influence erosion. Stream valleys may be eroded along ancient as well as recent fault zones, and if an ancient fault separates resistant from weak rock, the latter may be eroded more rapidly, leaving an erosional scarp between the two.

The most familiar topographic feature directly produced by faulting is the *fault scarp* (Figure 8-19). These range from vertical cliffs to inconspicuous gentle slopes. Because low fault scarps are rapidly obliterated, their presence indicates recent faulting and the possibility of recurrent movement. Some fault scarps are enormous features, the result of long sequences of fault movements. The base of some of these scarps, the most recently uplifted portion, may

appear astonishingly fresh (Figure 8-20). These sites are fault hazard zones.

For land-use planning, it is crucial to distinguish fault scarps from *faultline scarps* caused by differential erosion along long-inactive faults. If the higher terrain on one side of a fault consists of resistant rock, the scarp may be due simply to more-rapid erosion of the weak rocks alongside and may not signify recent fault activity. If, on the other hand, the high side consists of soft, poorly consolidated material, the uplift has occurred so recently that there has been insufficient time for removal of this soft material.

Streams channels are offset by strike-slip faulting (see Figure 8-11). Streams which enter the fault zone, however, may show contradictory deflections. The reason is that a fault is often more influential as a guide for streams than as a deflector. Hence if two streams enter a fault zone, one may turn to the right and one may turn to the left, depending on the erodability of the crushed material occupying the zone. It is only if several streams *cross* the zone, and their channels are offset in a consistent way, that the deflection is evidence of active faulting.

Figures 8-21 to 8-23 illustrate evidence of recent faulting based on the youthfulness of the landforms offset or crossed by fault traces. Obviously, the weaker the materials offset by a fault, the more rapidly will the scarp be obliterated. In Taiwan, fault traces are visible across sand dunes, indicating very recent activity.

The foregoing examples illustrate the diverse landforms that can be used to judge the activity of faults. If, on the other hand, the fault is inactive at ground level, it will be surmounted or mantled by these landforms, which will show no disruption.

Stratigraphic Criteria. Young geologic deposits and soils are useful in assessment of fault activity. If a fault has displaced young geologic material, clearly the displacement is younger than the material and, lacking evidence to the contrary, the fault may continue to slip. In contrast, an inactive fault may be covered by deposits which show no displacement.

Unconsolidated deposits of alluvium exposed at the surface are particularly useful for the recognition of recent displacements because they are widespread and geologically young (e.g., of Quaternary age). Alluvium is

FIGURE 8-19. **Fresh scarp of normal fault on southeast side of Fairview Peak, Nevada. Relative positions of topsoil, ground surface, and snow cover indicate 4.5 m (15 ft) of vertical separation, and steeply inclined grooves on the scarp show direction of slip during earthquake of December 1954. (*Photo by Hugo Benioff.*)**

FIGURE 8-20. **The white line at the base of the Sonoma Range, Nevada, is the fresh scarp up to 4.5 m (15 ft) high, formed during the earthquake of 1915. (*Photo by Eliot Blackwelder, 1947.*)**

present in floodplains, in stream terraces, in alluvial fans, and as a mantle on erosional landforms such as pediments and marine terraces. Unbroken beds of alluvium are commonly interpreted to mean inactivity of underlying faults and relative immunity from danger of surface faulting. Figure 8-24 illustrates a thorough procedure for such fault investigation.

Unfortunately, alluvium seldom provides material for radiometric age determinations, although occasional bits of wood or bone or

peaty material useful for carbon-14 dating may be found. Remains of extinct Pleistocene vertebrates are valuable for their implications of antiquity. Comparative ages of alluvial remnants in a particular region may be judged

FIGURE 8-22. **Small fault scarp crossing glacial moraine at mouth of Independence Creek. East base of Sierra Nevada near Independence, California. The recency of the faulting is indicated not only by the relative late date of the moraine it crosses but by the freshness of the scarp in spite of the loose materials it cuts. (*Photo by Eliot Blackwelder, 1925.*)**

FIGURE 8-21. **Fresh offset of an alluvial fan. Note the low scarp crossing the fan left of center. West side Sangre de Cristo Range, Colorado. (*U.S. Dept. Agriculture photo.*)**

FIGURE 8-23. **Fault of 1972 earthquake cutting volcanic cone, Owens Valley, California. (*David B. Slemmons.*)**

qualitatively by the degree of weathering and the extent of soil formation.

Glacial deposits displaying distinctive landforms (drumlins, eskers, and kames) are relatively young, generally 100,000 years or less. In the absence of distinctive landforms, other criteria must be relied on to date glacial deposits. These include degree of weathering and erosion, radiometric dating, and correlation with other deposits. Thus, the specialist can frequently determine the approximate age of a glacial deposit. If the deposit is cut by a fault, the faulting took place after the deposit was laid

FIGURE 8-24. **Exploratory trenches at nuclear power plant site on coastal terrace adjacent to Diablo Canyon (at left), near San Luis Obispo, California. The trenches, extending through Quaternary terrace deposits into the underlying bedrock, were excavated to obtain geologic data for appraising the probability of surface faulting at the site. (*Photograph courtesy Pacific Gas and Electric Company.*)**

down. If the deposit is covered by a later deposit through which the fault does not pass, the fault is younger than the lower deposit and older than the upper deposit.

Volcanic lava and ash have been used to date fault movements or lack of movement. Ash is generally widespread and may constitute a visible datum which, if offset by a fault or if lying unbroken across a fault, can be used to date the fault. Fortunately, many ancient lava flows and ash layers can be dated by radiometric methods, and more recent ones, by historical records.

Soil is particularly helpful in some fault studies. We are using the term "soil" here in an agricultural rather than engineering sense. All other factors being equal in any one area, *time* determines the thickness and maturity of a soil. A fault which is covered by a very thick, undisturbed, mature soil probably has not offset the ground surface for at least a hundred years, possibly several thousand years. This situation applies to large areas of the Appalachian Piedmont, where thick residual soils lie undisturbed over ancient faults beneath.

Displaced structures (roads, pipelines, etc.) present clear evidence of active faulting.

EARTHQUAKE HAZARDS

Earthquake hazards include rupture of the ground, shaking, fire, mass movements, alteration of surface and ground-water regimes, seismic sea waves, seiches, glacial surges, and miscellaneous hazards due to disruption of vital services, panic, and psychological shock. The impact of these hazards varies with the magnitude of the disaster and with time and place. If the M8.3 San Francisco earthquake of 1906 were to be repeated today with the presently expanded population, the proliferation of high-rise buildings, and the considerable amount of construction on steep hillsides and on weak bay sediments and land fills, the toll in life and property would be far greater. Earthquakes that occur during working and school hours when the populace is crowded into certain buildings can result in especially heavy human tolls.

Ground Rupture

Displacements of the ground may be horizontal, vertical, or oblique. According to the elastic rebound theory, faulting originates or recurs when the deforming strains exceed the elastic limit of the rocks. In some places, however, the accumulating strain is being relieved by quasi-continuous slippage, the movement taking place in small increments but measurable over periods of time (Figure 8-25). The tiny slippages do not create discernible earthquakes. This barely detectable motion is known as *tectonic creep*. Creep is now recognized along several major faults in California, including the Calaveras, Hayward, and San Andreas Faults. Unfortunately, there has been, and still is, much construction going on in creep zones. The breaching of the Baldwin Hills Reservoir in Los Angeles, California, has been attributed to weakening of the dam by tectonic creep (Figure 8-26).

The recognition of creep along the Hayward Fault is especially significant in view of the major earthquakes that have taken place along this fault in 1836 and 1868 and the fact that the fault passes through many highly developed areas. Although creep itself can hardly cause bodily injury, the weakening and failure of structures could do so. The Hayward Fault crosses the campus of the University of California at Berkeley. A culvert beneath the football stadium is being sheared at the rate of 0.27 cm (0.11 in.) per year. There is also continuous fracturing of a water tunnel leading to the city of Berkeley and, farther south, warping of railroad tracks and shearing of the floor and walls of a warehouse. The average rate of deformation of the tracks over a 55-year period has been 0.35 cm (0.14 in.) per year.

Similar evidence for tectonic creep has been recorded along the San Andreas Fault, 11 km (7 mi) southwest of Hollister, 72 km (45 mi) south of San Francisco Bay. Here, creep is splitting the concrete walls of a winery. Measurements since 1956 indicate an average rate of creep of 1.25 cm (0.5 in.) per year.

It is not certain whether creep indicates relief of a major part of the strain that has accumulated, thereby decreasing the chance of

FIGURE 8-25. **Creep displacement of curbing along the Calaveras fault at Hollister, California. (*Photo by Benjamin M. Page.*)**

a strong earthquake, or an accumulation of excess strain in spite of slight relief. Perhaps both situations occur. In any event, sites of creep should be identified and monitored for abnormal movements. Identification of creep immediately stamps a fault as active and a factor to be considered in land-use planning.

Whether creep is present or not, accumulating strain may build up until a sudden, relatively large displacement takes place. During the 1906 San Francisco earthquake, horizontal ground displacement reached 6.4 m (about 21 ft). Maximum vertical displacement was less than 1 m (3 ft). Far greater vertical displacements have been recorded. We noted that during the Yakutat Bay, Alaska, earthquake of 1899, there were vertical offsets of up to 14 m (47 ft). Small wonder that the release of accumulated strain

FIGURE 8-26. **The empty Baldwin Hills Reservoir in Los Angeles, California, after breaching of the embankment on December 14, 1963. This failure resulted from fault creep in ground beneath the reservoir, and the surge of rapidly escaping waters created havoc in a thickly settled area of more than 13 sq km (4 sq mi) along and beyond the upper edge of this view. (*Photograph by Teledyne Geotronics, Inc.*)**

energy during offsets of this magnitude cause major earthquakes.

Although ground rupture as a cause of damage to buildings can be avoided by prohibiting construction on known faults, there is no way for linear structures such as highways, railroads, aqueducts, canals, and oil and gas pipelines to avoid them (Figure 8-27). Pipes can be constructed to provide a rather large measure of play in anticipation of ground displacement. This has been done for the Alaskan pipeline where it crosses known faults.

Once in a while, ground offsets supply a humorous note to otherwise serious events. In 1954, a 3-m (9-ft) displacement at the east base of the Stillwater Range in Nevada provided the interesting situation shown in Figure 8-28.

Fault movements and earthquakes are commonly associated with regional crustal deformation. The deformation may precede the rupture or follow it. The U.S. Geological Survey

FIGURE 8-28. **The creation of this fault scarp at the east base of the Stillwater Range, Nevada, in 1954, required use of a ladder to reach the elevated outhouse.** (*Photo by Perry Byerly, courtesy Chester Longwell.*)

FIGURE 8-27. **Deformation of railroad track during the 1906 San Francisco earthquake. South Pacific Coast Railroad near Wright's Station in the Santa Cruz Mountains southwest of Los Gatos.** (*Photograph by G. A. Waring, from Collection of R. D. Graves, San Francisco.*)

reported in February 1976 that a large area straddling the San Andreas Fault about 65 km (40 mi) north of Los Angeles is bulging upward (Figure 8-29). The swelling is believed to have begun about 1960 and has expanded eastward to include an area of 12,000 sq km (4500 sq mi). It is possible that the swelling may eventually lead to renewed faulting and earthquakes, as was true for the San Fernando, California, earthquake of 1971 and for the Niigata, Japan, earthquake of 1964. Other recorded uplifts, however, have occurred without subsequent earthquakes.

In other instances, crustal deformation was the result of faulting, as was true in the Alaska earthquake of 1964 (see Figure 8-1). About 260,000 sq km (100,000 sq mi) of seafloor and coastal plain were deformed. Local uplifts were as much as 2 m (6 ft), but on a regional scale, the warping ran as high as 16 m (50 ft).

In addition to ground displacements and warping during earthquakes, there may also be widespread fissuring. Open fissures up to 1 m or more wide constitute direct hazards. Fissuring is generally most prevalent close to the fault trace.

Ground-shaking

By far the most serious *direct* earthquake hazard is the collapse of buildings due to shaking. Buildings on firm bedrock suffer far less than those on saturated sediments. But because damage may result even on firm rock, susceptible structures should not be located too near a fault trace.

The segment of the San Andreas Fault responsible for the destruction of San Francisco in 1906 is well offshore, outside the Golden Gate. Yet, in spite of the distance from the fault trace, the intensity of destruction varied enormously within short distances. Destruction was consistently greatest on land recovered from the bay or on fill over old swamps or valley bottoms. Lurching and uneven settling of these materials caused streets to warp and crack, and buildings to sway, tilt, or collapse. The

earthquake-intensity map was a veritable patchwork quilt (see Figure 8-18).

Type of construction also influences the scale of damage and destruction. We noted that in the decade 1962–1972 in Iran alone, 30,000 people perished in the collapse of buildings never designed to withstand earthquakes. Weakest of all the structures were the adobe and mud-walled buildings, but masonry and poorly built concrete structures were also heavily damaged or destroyed. In the San Francisco earthquake of 1906, 10 percent of the total destruction was attributed directly to the shaking, with the remainder to the ensuing fire. Damage was due to poor design and construction. Walls were not rigidly secured to foundations, roofs were not adequately secured to walls, many buildings were of brick or masonry

FIGURE 8-29. Regional deformation of the crust in area crossed by San Andreas Fault north of Los Angeles, California. (*U.S. Geological Survey.*)

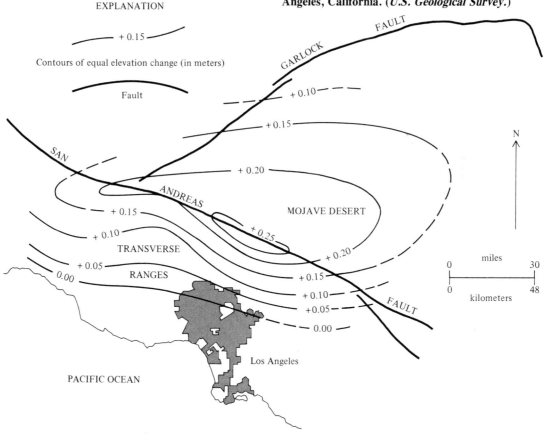

EXPLANATION

— + 0.15 —

Contours of equal elevation change (in meters)

Fault

unsupported by steel frames, others were of wood frame construction but poorly constructed. Few of these were designed to withstand the wrenching that took place during the earthquake. Some buildings were partially supported by a steel framework. During the earthquake, the frames remained as skeletons as all else fell away. The San Francisco City Hall dome was supported by a steel framework: the dome remained, while the walls and other parts of the building collapsed (Figure 8-30).

Buildings must be so designed and constructed that they move as a unit with the ground, not as a loose agglomeration of parts. This requires secure anchoring and tight bonding of foundations, frame, outer and inner walls, floors, and roof. Adequate bracing of all members is essential. Tall buildings in particular should have a certain degree of flexibility.

Wood-frame buildings, if properly constructed, can withstand strong shaking. Buildings of masonry, brick, or concrete blocks are not as resistant but are satisfactory if the mortar is good and they are reinforced with steel. Buildings with large glass windows are a hazard because of the shattering of glass under wrenching. Experimentation is in progress on the best ways to mount glass to enable it to withstand an earthquake. In some new buildings in Japan, windows are mounted in rubber.

In hilly regions, many buildings rest partly on bedrock and partly on poorly compacted fill excavated from the hillside. The fill area may settle during an earthquake, breaking the building in two.

California has a Uniform Building Code with earthquake provisions, prepared by the Structural Engineers Association of California. Community codes that adhere to and enforce these provisions ensure a relative degree of safety. The code, however, sets minimum standards and does not ensure against all possible earthquakes. For example, some of the extensively damaged buildings in the Alaskan earthquake of 1964 had been built according to the California code. The code was based on experience with the short-duration earthquakes of California. The Alaskan earthquake, however, lasted 3 to 4 minutes, with unforeseen results. Thus, duration as well as intensity are factors to be reckoned with.

Tall buildings of five or more stories are subject to another danger. They may start swaying in resonance with the earthquake vibrations until they shake themselves apart. On some occasions, tall buildings have been heavily damaged, while low, poorly constructed buildings remained undamaged. Buildings, towers, and bridges must be so designed that their vibration frequencies are unlike those of earthquakes. In contrast to buildings, tunnels and other underground structures are little affected by shaking because they move as a unit with the confining ground. They are, of course, subject to rupture where they cross faults.

Earthquake insurance is not offered by many insurance companies because of the inability to establish predictable loss tables. Insurance can be obtained, but with a deductible clause that restricts the coverage.

Fire

In many earthquakes the major devastation is from fire. Ninety percent of the destruction in San Francisco in 1906 was attributed to fire. The devastation was due to the preponderance of combustible building materials, damage to much of the fire-fighting equipment, and rupture

FIGURE 8-30. **Ruins of San Francisco City Hall, 1906.** (*From G. K. Gilbert et al., "The San Francisco Earthquake and Fire of April 18, 1906," U.S. Geological Survey, 1907.*)

of water mains. Most earthquake-induced fires start because of rupturing of power lines; damage to wood, gas, or electric stoves; damage to any other gas- or electrically powered equipment in use, and—as in the case of Tokyo in 1923—chemical reactions in industrial plants.

The great fires during the Kwantō (Tokyo) earthquake of 1923 destroyed 71 percent of the houses in Tokyo and practically all the houses in the city of Yokohama. Unlike the San Francisco fires, the Tokyo fires exacted a heavy human toll.

In earthquake regions, there must be greater emphasis on noncombustible materials and on special construction techniques so that water mains will remain unbroken under predictable displacements. San Francisco now has valves throughout its water system to protect each neighborhood's water supply. Each neighborhood, too, has its own reserve supply underground.

Mass Movements

Landslides and other types of mass movements commonly accompany earthquakes. Most landslides occur within 40 km (25 mi) of the epicenter of large earthquakes, but they may occur to distances of as much as 160 km (100 mi) in great earthquakes such as the Alaskan earthquake of 1964.

In Chapter 2 we described a number of earthquake-induced mass movements, including those at Lituya Bay, Alaska, in 1958, Madison Canyon, Montana, in 1959, and Alaska in 1964. In this chapter, we noted that landslides accompanied the Lisbon earthquake of 1755, the Shensi, China, earthquake of 1556, and the Ancash, Peru, earthquake of 1970. Some of these landslides moved at such high velocities that escape was impossible: the landslide that devastated Ancash, Peru, averaged 320 km (200 mi) per hour for the 11 km (7 mi) it traveled. The landslide that buried the town of Khait in the Tadzhik Republic, U.S.S.R., in 1949 moved at a speed of over 360 km (225 mi) per hour. Today a monument marks the site of the town 30 m (100 ft) below, where 12,000 people perished.

Other earthquake-generated mass movements are due to liquefaction of weak, saturated sediments (see Chapter 2). The residential destruction at Turnagain Heights during the Alaskan earthquake of 1964 was due to lateral flowage of saturated soils. In 1692, two-thirds of the town of Port Royal, Jamaica, West Indies, slid into the sea. Large areas of the town had been built on loose saturated sediments and landfill. These were shaken loose, carrying buildings and people into the sea. More than 20,000 lives were lost. Much of the loss of life during the Shensi, China, earthquake of 1556 was caused by mobilization of the dry loess deposits of the region. Here, the shaking converted the silty deposits into sediment-air mixtures which flowed down into the valleys like fluids, burying everything in their paths.

Other serious earthquake hazards are presented by lakes impounded behind landslide dams. A number of examples were given in Chapter 2.

Many earthquake-generated ground movements involve vertical subsidence, as in the New Madrid earthquakes of 1811 and 1812. At that time, vast areas of the Mississippi Valley subsided, creating two very large lakes and many smaller ones. Sag ponds occur along many fault zones (see Figure 8-12). Sag ponds are evidence of active faults and their environs should be avoided for most types of land use. Yet in California, small sag ponds have been filled in as sites of real estate developments.

The problems of earthquake-generated mass movements, the methods of investigation and treatment, and the problems of land use are similar to those presented by mass movements in general, as discussed in Chapter 2.

Drainage Modifications

The creation of large lakes by earthquake subsidence of river floodplains, deltas, tidal marshes, and exposed lake floors is a serious problem. In spite of the agricultural value of Mississippi River bottom land, no attempt has been made since the New Madrid earthquakes of 1811 and 1812 to reclaim the large areas that were submerged under earthquake-created St. Francis and Reelfoot Lakes.

In addition to disturbance of natural drainage, earthquakes may endanger artificial lakes and reservoirs through destruction of dams. In the San Fernando, California, earthquake of

FIGURE 8-31 **Van Norman Dam after partial failure during the San Fernando earthquake of 1971 in southern California. A portion of the embankment slumped into the reservoir leaving a clearance of only 1 m (3 ft). (***Los Angeles Times* **photograph, courtesy California Division of Mines and Geology.***)**

1971, the Van Norman Dam (Figure 8-31) almost gave way. Had the quake lasted a little longer, tens of thousands of people would have been endangered.

Surprisingly, the Alaskan earthquake of 1964 ruptured only two small, earth-fill dams south of Anchorage. Both lakes drained through ruptured dams but without loss of life. Even more surprising was the fact that only 1 of the 45 glacially dammed lakes showed any signs of earthquake-triggered drainage. Since these small isolated lakes break out periodically anyway, emptying in a few hours, earthquake release of the waters would have presented no new hazards.

The orientation of a dam relative to a nearby strike-slip fault seems to be an important factor in the ultimate safety of even rigid dams. A dam consisting of large keyed concrete blocks, 45 m (146 ft) high and 200 m (680 ft)

long, just east of the San Andreas Fault zone south of San Francisco, but paralleling the fault, was unaffected by the relative shift of more than 2 m (about 6 ft) in this area. The dam was subsequently completed to a height of 52 m (170 ft).

Ground-Water Modifications

The ground-water regime may be altered by fault displacements or by shaking. An example of the effect of a fault offset on both surface and ground-water flow occurs in Santa Clara County, California, at the south end of San

Francisco Bay. Here, the large alluvial fan of Alameda Creek has been offset by a fault 2 km (1 1/4 mi) west of the mountain front. The fresh fault scarp faces the mountains and ranges in height up to 8 m (25 ft). It has blocked much of the surface drainage toward San Francisco Bay to create a succession of ponds the largest of which, Stevens Pond, is about 3/4 km (almost 1/2 mi) long. The fault is also an effective ground-water dam. Impervious materials were raised against the permeable aquifers, blocking them almost completely. Thus, the area between the scarp and the mountains continues to receive abundant ground water, while the area to the west receives very little. The water table immediately east of the fault at the time the survey was made ranged in depth from 7 to 11 m (22 to 33 ft). Immediately west of the fault, water levels were as much as 20 m (65 ft) below ground level.

In the Hawaiian Islands, impermeable fault gouges and dikes impound enormous quantities of ground water in permeable basalt layers. These constitute prolific reservoirs.

The Alaskan earthquake of 1964 demonstrated that earthquakes may affect ground water thousands of kilometers away. Effects were recorded in more than 700 wells in North America, Europe, Africa, Asia, and Australia. A majority of wells showed a rise in water level, and—in a considerable number—the new levels were maintained. The disturbances caused local increases in transmissivity of aquifers in southeastern United States.

Tsunamis

On April 1, 1946, tsunamis from an earthquake near Unimak, Alaska, reached the Hawaiian Islands in 4 1/2 hours. They created waves 11 m (36 ft) high. The death toll of 159 and the property damage of $25 million were responsible for creation in 1948 of a warning system in Honolulu by the U.S. Coast and Geodetic Survey. When instruments in Honolulu record a strong earthquake anywhere, contact is made with other stations in western United States and Alaska to determine the location and amplitude. The possibility of a tsunami is evaluated and warnings of estimated arrival times are issued all around the Pacific basin.

The warning system is effective only at some distance from the earthquake, and it cannot supply precise data on the size of the tsunami to be expected. The danger is greatest where earthquakes originate close to low, densely inhabited coastal areas such as parts of Japan and Chile.

As with most warning systems, a certain degree of ineffectiveness results from apathy caused by occasional false alarms. Thus, in spite of adequate warning, 61 people died and 282 were injured in Hawaii as a result of tsunamis from the great Chilean earthquake of May 22, 1960. That the system can be a success, however, was demonstrated on November 4, 1962, when a submarine earthquake off the Kamchatka Peninsula generated a series of tsunamis whose time of arrival in Honolulu was accurately predicted. Not a single life was lost.

Seiches

Seiche, a pendulumlike oscillation of water mentioned in Chapter 4, may be produced by earthquakes if the ground is tilted or if the period of the long, vertical surface waves in the ground is in resonance with the period of free oscillation of the water body. The Lisbon earthquake of 1755 set waters in motion throughout northern Europe. In Loch Lomond, Scotland, the oscillation period was 10 minutes and the amplitude 0.75 m (2.5 ft).

During the Alaskan earthquake of 1964, large areas were tilted from their former positions. As a result, waters of many lakes were displaced toward the lower ends of their basins and, before coming to rest, oscillated back and forth. Kenai Lake, one of those affected, is in a glacial valley on the Kenai Peninsula 95 km (60 mi) south of Anchorage. The lake is 37 km (23 mi) long and averages 2 km (1.3 mi) wide. The basin is a flat-floored trench with a maximum water depth of 172 m (570 ft). The lake drains west by way of Kenai River. The basin was tilted up on the west 1 m (3 ft), reversing the direction of flow of Kenai River for a distance of 140 m (450 ft). A seiche was set up in the lake with a period of 36 minutes. Maximum rise or run-up was 9 m (30 ft), and maximum invasion inland was 110 m (365 ft). Trees up to 0.5 m (1.5 ft) in diameter were knocked down, turf was stripped from bedrock, and loose sediment widely scattered. No real property damage or loss of life resulted, but had the earthquake occurred 6 months later when the lake level stood 3 m (10 ft) higher, there would probably have been property damage and lives lost. Outside Alaska, the highest recorded seiche was 0.6 m (1.8 ft) in a reservoir in Michigan. The next highest was a seiche of 0.45 m (1.45 ft) on Lake Ouachita, Arkansas. Most interesting were the seiches produced in channels along the Gulf Coast in the United States. The thick sediments amplified the seismic waves to produce the local effects. In a channel near Freeport, Texas, an amplitude of 9 cm (3.5 in.) was recorded.

Glacial Hazards

Anomalous surges of glaciers may follow earthquakes due to reduction of friction by shaking and increase in weight by landslides and snow avalanches. *Outbursts of large glacial lakes*, creating great floods, may result from the weakening of ice barriers during earthquakes. Finally, earthquakes have shattered the ends of glaciers protruding into the sea and have generated large numbers of *icebergs*. These may become hazards to navigation.

Miscellaneous Hazards

One of the most crucial of the miscellaneous hazards is the *loss of vital services* such as water supply and utilities. Ninety percent of the destruction in San Francisco was due to fire which raged uncontrolled primarily through lack of water. *Interruption of power, fuel, and communication lines* may leave people stranded in high buildings because of inoperable elevators, may cause hardship and medical problems in dark and unheated homes, and may lead to dangerous delays in organizing relief and reconstruction programs through loss of communications.

Panic and lack of understanding of earthquake hazards have added heavily to earthquake tolls. The public in earthquake country must be educated to the hazards and the appropriate measures to be taken in emergencies.

There are also *social* and *psychological* impacts. As a result of the 1755 earthquake, many citizens of Lisbon began to question the existence of a divine being who could visit such suffering on devoted followers.

EARTHQUAKE PREDICTION AND WARNING

Predictions are forecasts of the time, place, magnitude, and visible effects of an earthquake. Under a federal plan, the U.S. Geological Survey would be responsible for predictions. *Warnings* of impending earthquakes would be the responsibility of local officials to whom the predictions would be transmitted. They would include recommendations or orders for defensive action, such as evacuation of buildings and lowering of reservoir levels.

Status of Investigations

The prospects for prediction of earthquakes are promising. It is hoped that predictions of major earthquakes can be made years in advance. Such a long interval could be used to strengthen existing structures, to rigidly supervise new construction, and to speed up revision of building codes. If short-term prediction becomes possible, the shorter time could be used for evacuation of weak structures, for shutting down nuclear power plants, for shutting off oil and gas pipelines, for readying plans for emergency water and power distribution, and for setting up disaster-relief programs. The individual might review safety procedures such as shutting off pilot lights and gas lines, pulling master fuses or electrical switches, and reviewing personal safety procedures for protection against falling debris.

The Chinese claim several successful predictions that involved evacuations and probable savings of life, but they also admit to failures, as in the great Tangshan earthquake. Some members of the U.S. Geological Survey had reason to anticipate the earthquake that struck near Hollister, California, on November 28, 1974, but at the present state of the art, did not feel justified in making a prediction. Nor do they expect predictions to become a reality in the next decade or so. It is reported that about

10 earthquakes in different countries have been predicted to date, but the number of failures is unknown.

Prediction Techniques

Earthquake prediction is based on the assumption that if parts of the earth's crust are subject to increased strain due to squeezing or stretching, some of the physical characteristics of the rocks will be changed. The precursory changes so far identified include tilting, horizontal and vertical shifting, changes in rock densities due to compression or tension, changes in electrical resistivity as pore spaces are evacuated or filled with water, changes in rock magnetism as magnetic minerals are dissolved out of permeable fault zones, changes in the level, temperature, and turbidity of well waters, increases in radon[2] content of well waters as permeability increases in the fault zone and more radon is introduced, and—in response to some of the above physical changes—variations in velocity of the primary waves relative to the secondary waves. The California Division of Mines and Geology has been monitoring acoustic emissions along major faults as a possible additional symptom of impending earthquakes.

Precursory tilting is determined by tiltmeters (see Chapter 9). Shifts in position, as well as tilting, may be determined by repeated surveys across faults. Anomalous uplifts were recorded in the vicinity of Niigata, Japan, for 10 years prior to the great earthquake of 1964. Changes in density are recorded by gravity meters capable of detecting minute variations in gravity. Changes in electrical resistivity are determined by feeding current into the ground and recording changes in voltage between stations across the fault. Changes in rock magnetism are recorded by magnetometers of remarkable sensitivity. Both magnetic and gravity measurements may help identify hidden faults by revealing abrupt differences in these properties across a sharp contact. Changes in level, temperature, and turbidity of well waters are

amenable to direct observation or simple instrumentation. Increase in the radon content of well waters, reported in China and the Soviet Union as a sensitive indicator of seismic activity, is easily determined with radiation counters. Of all the precursory seismic indicators, that involving changes in the velocity of the primary waves relative to the secondary is the most exciting. It seems to offer an explanation for all observed facts. This is the *dilatancy theory*, which we will consider briefly.

Russian scientists as long ago as 1960 observed that the velocity of primary or compressional waves diminished appreciably during the episode of premonitory tremors but recovered just prior to the main shock. The slowdown has been attributed to the opening of small cracks in the strained rocks, which retarded the compressional waves. The subsequent sharp increase in velocity of the primary waves just prior to the main shock is attributed by some to entry of water into the cracks. Because water is essentially incompressible and is an effective transmitter of compressional waves, the elasticity of the rock is regained. As noted earlier, this theory seems to account for all the changes in rock properties enumerated earlier. Even the increase in radon in well waters is explained by the increasing water flow which carries the radon continuously into the wells from the surrounding rocks after the opening of the small cracks. Seismologists of the U.S. Geological Survey have recently pointed out, however, that no drop in the velocity of the primary waves preceded a M5.2 earthquake in the Mojave Desert in June 1975. They suggest that the dilatancy theory may not apply all the time. However, study of seismograms recorded before the San Fernando, California, earthquake of 1971 now indicates that dilatancy did precede that earthquake.

In investigations of large active faults, arrays of sophisticated instruments are spread over areas of 100 to 1000 sq km (40 to 400 sq mi). The arrays consist of 20 or 30 seismographs buried at various depths and arranged along two lines at right angles to each other. The arrays also include tiltmeters, strain gauges, magnetometers, gravity meters, electrical resistivity devices, and instruments to record the data. There is one large array in the San Francisco

---

[2]Radon is an inert radioactive gas produced by disintegration of radium. The half-life is less than 4 days.

Bay region, another 320 km (200 mi) to the south in the Parkfield area, and a giant array of more than 600 instruments in the Yellowstone Park region. Japan has a large array northeast of Tokyo, and there are experimental networks in Scotland, Australia, and India.

Human Responses to Warnings

Some seismologists are uncertain whether the occurrences and magnitudes of earthquakes can ever be predicted with accuracy. In this event, warnings could become ineffective. A few false alarms could lead to public disregard of subsequent ones. Many people in Hilo, Hawaii, had responded to tsunami warnings in 1958 and 1959 which turned out to be false alarms. When a similar warning was issued in 1960, many people disregarded it and 61 were drowned.

The responsibility of fixing a date and magnitude is a heavy one. Should cities be evacuated? If the quake does not occur on the date predicted, for how long should the populace be kept away? Days? Weeks? Months? If the earthquake occurs earlier than predicted, what happens to the credibility of the warning system?

DEFUSING EARTHQUAKES

Inasmuch as earthquakes represent sudden relief of strain which—in the case of great earthquakes—may have been accumulating for a century or more, the problem is how to relieve this strain gradually or in a series of nondestructive earthquakes of magnitude 4 or less.

The possibility of defusing earthquakes was suggested when the U.S. Corps of Engineers inadvertently triggered a number of small earthquakes in the Denver area by injecting waste waters from the Rocky Mountain Arsenal into old wells more than 3 km (2 mi) deep. Presumably the injected water increased pore-water pressures and triggered movement along hidden faults. Note that we say the injection "triggered" the earthquakes. The natural forces involved in the long-term accumulation of strain are enormous, whereas the triggering forces are relatively small.

The nuclear blasts at the Nevada Proving Ground also triggered small earthquakes. One nuclear explosion caused thousands of aftershocks in a northeast-trending zone for a distance of 13 km (8 mi). Focal depths ranged down to 7 km (4.5 mi).

Planned and controlled triggering of earthquakes by injection of water or by setting off explosions seems within the realm of physical possibility. It has been proposed, for example, that deep holes be drilled at intervals along active faults. One segment at a time would be induced to slip by injecting water under high pressure while "pinning" the fault elsewhere by pumping out water.

There are, however, difficulties in attempting to artifically relieve accumulated stress. If the fault surface is irregular rather than plane, fluid pressure may not be sufficient to cause small displacements. In such cases, it is possible that the accumulating strain can only be relieved naturally by large fault movements and major earthquakes.

A second difficulty arises in selecting unstable segments of faults hundreds of kilometers in length and along which are aligned numerous centers of population. Many faults do not show creep at the present time, and, even if creep is present, there is no consensus as to whether creep segments are the most or least dangerous.

A third difficulty is that in many regions, such as California, there are assemblages of parallel faults. The artificial relief of stress along one fault does not ensure that the next movement may not occur along a companion fault.

Finally, the triggering of earthquakes may raise legal and fiscal problems. There is no assurance that an earthquake that is triggered will be a small one or that it will be restricted to a small area. A flood of damage suits might ensue.

EARTHQUAKE HAZARD MAPS AND SEISMIC SAFETY PLANS

The U.S. Coast and Geodetic Survey has plotted the distribution of all large and many smaller earthquakes in the coterminous United

States and has prepared an earthquake hazard map in which are outlined areas of equal expectable earthquake damage (see Figure 8-16). It is recognized that the assumption that future quakes will be distributed as in the past may not be valid. Furthermore, the risk evaluation in some areas is based on only one or two damaging earthquakes.

For detailed planning and insurance-risk evaluation, more detailed seismic maps are necessary. Naturally, they will be most detailed and accurate where earthquakes have been common and where there has been sufficient population so that local variations in intensity of past earthquakes are well recorded. It is not surprising, therefore, that California offers the best potential (see, for example, Figure 8-18).

Earthquake intensity maps are only part of the data needed to plan against earthquake hazards. Hazards other than those from ground-shaking include fire, isolation due to disruption of transportation routes and utility lines, landslides, and flooding. Mention should be made of the "Seismic Safety Plan" prepared by the Santa Clara County (California) Planning Department as one element of a general plan. Preparation of a seismic safety plan is required in all general plans in California (see Chapter 16).

Santa Clara County extends south of San Francisco Bay. The lowlands bordering the bay, and protected by dikes, are considered susceptible to flooding by tsunamis, seiches, or landslide splash waves. Other areas, below dams, are considered hazardous by reason of possible dam failures. The mountainous areas are regarded as more or less uniformly susceptible to landslides. And, in view of the intricacies of the network of transportation and utility lines, it is believed that a major earthquake could divide the county into 15 isolated areas. A series of maps has been prepared to illustrate, singly or in combination, such elements as (1) slopes susceptible to landslides (30° or more), (2) lowland areas subject to soil liquefaction, (3) known faults and epicenters, (4) areas subject to salt-water invasion, (5) areas subject to freshwater inundation, (6) the network of major water conduits, (7) the distribution of sanitation facilities, (8) the network of gas and electric lines, (9) the network of roads and railroads, (10) the distribution of hospitals, fire stations, and other public structures, (11) a map of isolatable areas, and (12) a map of zones of relative seismic safety.

Recommendations for disaster planning in the Santa Clara County Seismic Safety Plan are based on the data presented in these maps, on the distribution and presumed degree of safety of all structures in the county, and on desirable improvements in the building code and in zoning laws.

ADDITIONAL READINGS

Bolt, B. A., W. L. Horn, G. A. MacDonald, and R. F. Scott: "Geologic Hazards," Springer-Verlag New York Inc., New York, 1975.

Iacopi, R.: "Earthquake Country," Lane Books, Menlo Park, Calif., 1964.

Leet, L. D., and F. Leet: "Earthquake: Discoveries in Seismology," Dell Publishing Co., Inc., New York, 1964.

Lomnitz, C.: "Global Tectonics and Earthquake Risk," Elsevier Scientific Publishing Company, Amsterdam, 1974.

Nichols, D. R., and J. M. Buchanan-Banks: "Seismic Hazards and Land-Use Planning," U. S. Geological Survey, Circular 690, 1974.

Oakeshott, G. B.: "Volcanoes and Earthquakes: Geologic Violence," McGraw-Hill Book Company, New York, 1975.

Press, F.: Earthquake Prediction, Scientific American, vol. 232, no. 5, pp. 14–23, May, 1975.

Richter, C. F.: "Elementary Seismology," W. H. Freeman and Company, San Francisco, 1958.

Wallace, R. E.: "Goals, Strategy, and Tasks of the Earthquake Hazard Reduction Program," U.S. Geological Survey, Circular 701, 1974.

Yanev, P.: "Peace of Mind in Earthquake Country," Chronicle Books, San Francisco, 1974.

CHAPTER NINE

volcanic environments

INTRODUCTION

The United States is fortunate that its mainland volcanoes are restricted geographically and are more or less inactive at present. Volcanoes, however, are among the most capricious and unpredictable of all natural phenomena. Some may come to life after centuries of dormancy. Lassen Peak in northern California (see Figure 2-24) erupted violently in 1914 and 1915.

North of Lassen Peak is a line of lofty volcanoes forming some of the crowning summits of the Cascade Range from California to Canada (Figure 9-1). Mount Shasta in northern California, Mount Hood in Oregon, and Mount St. Helens (Figure 9-2), Mount Rainier, and Mount Baker in Washington are among the larger of these. The explorer John C. Fremont reported that Mount Baker and Mount St. Helens erupted ash in 1843. Mount Baker also erupted in 1854, 1858, and 1870.

The Aleutian Islands of Alaska consist of a string of volcanoes all active in the sense that they could erupt at any time. Mount Pavlov erupted in 1912, 1950, and 1973. Ash from the 1912 eruption reached Seattle.

In Italy, the burial of Pompeii and Herculaneum took place in A.D. 79 during an eruption of Mount Somma, the predecessor of Mount

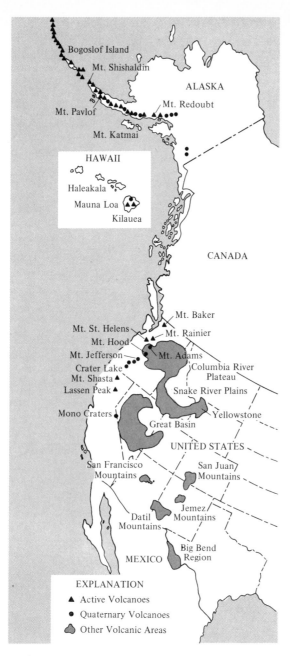

FIGURE 9-1. Active and extinct(?) volcanoes of the United States. The volcanoes and volcanic areas labeled Quaternary are ancient and presumably extinct. (*After National Science Board.*)

Vesuvius. Explosions tore off the top of Mount Somma, forming an enlarged crater (*caldera*) from whose floor modern Vesuvius rose (Figure 9-3). A violent eruption of Vesuvius in 1631 followed five centuries of quiescence. There have been other eruptions since.

Mount Pelée on the island of Martinique in the West Indies erupted in 1902 after 50 years of dormancy. Clouds of incandescent ash (*glowing avalanches*) with temperatures up to 650°C (1200°F) swept down the slopes at velocities between 80 and 180 km (50 and 110 mi) per hour, destroying all buildings and killing all but two of the 28,000 inhabitants of St. Pierre (Figure 9-4). Such glowing avalanches are common. Figure 9-5 shows one on the flank of the volcano Mayon in the Philippines.

Volcanic eruptions may take place where no volcano previously existed. This does not mean that a volcano is likely to grow in the middle of Kansas or New Jersey. Most volcanoes are restricted to relatively narrow belts. Even within these belts, however, it is impossible to predict where new eruptions will take place. The birth of the volcano Parícutin on February 20, 1943, in a corn field near the village of Parícutin 320 km (200 mi) west of

FIGURE 9-2. **Mount St. Helens in the state of Washington. The volcano is more than 3000 m (9700 ft) high. Some lava flows on the lower slopes are defined by lava levees. Mount St. Helens has been the most active volcano in the contiguous 48 states during the last few thousand years. The probability is "considered high" that it will erupt again before the end of the century. (*U.S. Geological Survey.*)**

FIGURE 9-3. **Vesuvius within the breached caldera of Mount Somma. (*U.S. Navy.*)**

FIGURE 9-4. **Ruins of St. Pierre. Mount Pelée and its spine in the background, March 1903. On May 8, 1902, a black cloud of incandescent dust and superheated steam rushed down the slope of Mount Pelée and in 2 minutes killed all but two of the inhabitants of St. Pierre. The toothlike spine rose from the crater to a height of 310 m (1020 ft) in about a year, when it began to rapidly disintegrate. Premonitory symptoms of the May 8, 1902, eruption occurred for a month previously and consisted of the appearance of steam vents, of sulfurous odors, of light ash falls, and minor earthquakes. (*Photo by A. Lacroix, from "La Montagne Pelée et ses Eruptions," by A. Lacroix, 1904. Courtesy Masson et Cie, Paris, 1904.*)**

Mexico City was unexpected (Figure 9-6). In one day the cone reached a height of 35 m (120 ft); in 1 week it was 170 m (550 ft) high; within 10 weeks it was 340 m (1100 ft) high; and in 8 months it reached a height of 450 m (1500 ft) above the original fields. One of its lava flows buried the village of Parícutin; another covered all but the church towers of San Juan Parangaricutiro (Figure 9-7).

Volcanic activity can alter global weather patterns through the solid and gaseous materials pumped into the atmosphere. The ash from the 1883 eruption of Krakatoa in the East Indies drifted around the earth, providing red sunsets for several years and lowering the earth's mean annual temperature a few degrees. One theory to account for past glacial climates relies on episodes of sustained worldwide volcanic activity.

Not all volcanic activity is bad. Volcanoes are safety valves. The consequences might be worse if the pressures built up locally were not periodically relieved by piecemeal eruptions. Furthermore, volcanic vapors contribute vital gases to the atmosphere, volcanic eruptions provide fertile soil, volcanic activity provides mineral deposits and sources of geothermal energy, lavas are ground-water reservoirs, and volcanoes add beauty to many otherwise drab landscapes and provide numerous recreational opportunities.

FIGURE 9-5. **Glowing avalanche, Mount Mayon, Philippine Islands, April 27, 1968.** (*Photo by Rufino Fernandez, Courtesy W. G. Melson and Fr. R. C. Salinas.*)

It is generally not possible to decide whether a volcano is extinct or dormant. Where doubt exists, it is better considered dormant. Some volcanologists restrict the term "active" to volcanoes that were actually observed in eruption. Others include those that still display thermal activity, relatively fresh lava flows, or other volcanic features.

Because of the unlikelihood of ever controlling the location or timing of major volcanic eruptions, environmental planning must focus on three main tasks:

1. Methods to predict which volcanoes are likely to erupt in the near future
2. Methods to forecast the time of eruption
3. Methods to minimize the impact on life and property

DISTRIBUTION AND ORIGIN OF VOLCANOES

Distribution

The principal active volcanoes of the earth are located at the boundaries of the great crustal plates (see Figure 8-15). The formation of new lithosphere between separating plates is a volcanic process and explains part of the pattern of volcano distribution. Much of the volcanic activity in the zones of separation occurs beneath the sea, but in places, as in Iceland, it is above sea level. Where an oceanic plate dips under a continental plate, the descent into the heated depths also results in volcanism. This is true of the volcanoes in the great "ring of fire" encircling the Pacific Ocean. A third pattern is provided by volcanoes of the Hawaiian type which rise from the interiors of plates. The sites of Hawaiian volcanism seem to have migrated slowly across the Pacific plate over millions of years. Actually it may be the plate that is

FIGURE 9-6. **The volcano, Parícutin, Mexico, in 1943.** (*Walter M. Parsons.*)

moving across a "hot spot," possibly a columnar plume of hot material rising from the depths.

Origin

Volcanoes erupt when molten rock called *magma* rises to the surface. The magma rises because it is lighter than the surrounding rock. It may squeeze its way upward along fractures, dissolve the rocks in its path, or eat its way upward by prying loose blocks of the roof rock and digesting these by solution. As the magma approaches the surface, the pressure due to the weight of overlying rock diminishes. It is this confining pressure that keeps the contained gases under control. As the pressure diminishes, the dissolved gases, chiefly steam but including carbon dioxide, nitrogen, hydrogen, and other gases, leave the solution and form expanding bubbles in the melt. The relative ease with which these gases can escape from the magma strongly influences the type of eruption, the eruption products, and the characteristics of the volcanic cone produced. The ease of escape of gases is in turn influenced by the fluidity of the magma.

TYPES OF ERUPTIONS

Causes of Variations

Magmas differ in fluidity, depending on their composition. If a magma flows easily (*low viscosity*), the gases that come out of solution have little difficulty rising through the melt and escaping into the air. The turbulence caused by the rapidly escaping gases helps to prevent crusting of the lava, thereby reducing the possibility of a dangerous pressure buildup. If, however, the magma is sticky and porridgelike (*high viscosity*), escape of the gases is hindered. If such magma crusts over, the gas pressure may build up to the critical point and an explosion may take place.

Even low-viscosity magmas may experience violent eruptions, but not of the magnitude of those mentioned above. Gases may be hindered in their escape for a variety of reasons. If the magma is exceptionally charged with gas, the gas bubbles may not be able to escape rapidly enough to prevent a critical pressure buildup. Release of gas pressure may also be hindered by development of a crust between pulses of activity or by restriction of gas movement in a narrow vent.

Thus, the type of eruption depends on whether gases escape easily or are confined. At one extreme are eruptions in which there is a steady escape of gases, and molten rock flows quietly away as fluid lava. At the other extreme, explosive release of trapped gases wholly disintegrates the magma, and blebs and clots of the

FIGURE 9-7. **Blocky lava flow from the base of Parícutin which overwhelmed San Juan Parangaricutiro in June 1944. (*Tad Nichols.*)**

still molten or partially solidified magma, plus fragments of the volcano itself, are blown out of the vent.

We will consider for each type of eruption the characteristics of the eruption, the products, the landforms that result, the hazards involved, and the countermeasures that may be taken.

Explosive Eruptions

Because explosive eruptions generally require viscous lavas, they are largely restricted to the continents or continental margins and are rare within the ocean basins, where the dominant lava is highly fluid dark *basalt*. The most common lava involved in violent eruptions congeals to form the intermediate-colored rock *andesite*. Another less common viscous lava forms the light-colored rock *rhyolite*.

Products. The products emitted in purely explosive eruptions are gases and solids. By far

the most abundant gas is steam. Other gases are carbon dioxide, nitrogen, hydrogen, sulfur, chlorine, and fluorine. Some of the gaseous compounds are poisonous to plant and animal life.

A considerable proportion of erupted steam is acquired from ground water by magmas in their ascent to the surface. In eruptions during which the surface water of lakes or the sea can leak downward through fractures into magma chambers or conduits, the explosiveness of eruptions may be enhanced far beyond that expectable from the nature of the magma. As for carbon dioxide, its emission from within the earth maintains, over the long term, a vital atmospheric balance without which abundant plant life could not exist. Unless replenished, the carbon dioxide in the atmosphere would be depleted by plants and by burial of carbon-containing materials in accumulating sediments.

The volcanic *solids* (*pyroclastic debris*), consist of solidified clots of lava from the vent as well as solid particles torn from the conduit and the volcano. The solid fragments range in size from dust to blocks as large as an automobile. The dust and slightly larger ash particles solidify as the rock *tuff*; the larger fragments, as *volcanic breccia*.

Dust and ash emitted in an eruption may be carried great distances by the wind and are generally cool on coming to rest. Glowing avalanches, however, hug the ground, and the components may still be hot on settling. They reach velocities of 40 to 120 km (25 to 75 mi) per hour on steep slopes, while retaining an internal temperature of 750°C (1400°F) or more. Because of the high temperature, the particles may weld together after settling.

Cinder Cones. Much of the solids, particularly the larger fragments ejected in eruptions, is heaped up around the vent as a volcanic cone. Cones composed primarily of solids are known as *cinder cones* (Figure 9-8).

Explosive activity is never sustained long enough to build lofty volcanoes. Parícutin, an exceptionally large cinder cone, is only 450 m (1500 ft) high. The cone was 340 m (1100 ft) high when the first lava broke through its flank. Most cinder cones are considerably smaller.

FIGURE 9-8. **Cinder cone. Near Little Lake, Inyo County, California. (*Eliot Blackwelder.*)**

Calderas and Explosion Pits. *Calderas* are craterlike pits that are distinguished by their great size. A normal crater is approximately the diameter of the pipe or conduit that feeds the volcano and rarely exceeds 300 m (1000 ft) in diameter. A catastrophic explosion may blow the top of a volcano off, as happened to Mount Somma in A.D. 79 and Krakatoa in 1883, producing a caldera whose diameter may be measured in kilometers. Other calderas result from collapse, as at Crater Lake, Oregon.

Not uncommonly, an explosive eruption creates an *explosion pit* with a low rim but with little or no emission of volcanic materials (Figure 9-9). Explosion pits may reach hundreds of meters in diameter. Inferno Crater, on the North Island of New Zealand, is one of a group of pits formed during explosions in 1886.

Hazards and Countermeasures. Because explosive hazards are greater in eruptions in which outflow of lava and explosive activity are both important, we will reserve discussion of the hazards and countermeasures for the section on composite eruptions. However, the following comments should dispel any notion that explosive eruptions responsible for cinder cones are of minor concern.

The volcano Parícutin not only buried the agricultural land immediately below, but ash covered an area of 52 sq km (20 sq mi) to a depth of 0.3 m (1 ft) or more. The late-stage lavas inundated 26 sq km (10 sq mi).

Auckland, New Zealand's largest city, embraces an area of 1300 sq km (500 sq mi). Within this area are 50 to 75 small volcanoes (Figures 9-10 and 9-11) formed within the last 60,000 years, some within historic time. These include cinder cones that serve as city parks, small explosion craters surrounded by rims of ash and now occupied by pleasant lakes, small, low volcanoes, including Rangitoto, which guards the main harbor and rises from shallow water offshore, and numerous small ash and lava fields. In general, each 13 sq km (5 sq mi) within the city's environs contains an old eruptive center, and each eruption devastated an area of perhaps 26 sq km (10 sq mi). The record indicates that a new eruptive center can be expected to develop about once every 1000 years. The most recent eruption of note was the one which

FIGURE 9-9. **The Laacher See, a typical explosion pit in the Eifel volcanic district west of the Rhine between Bonn and Coblenz. The crater was formed within the last 11,000 years.** (*Landesbildstelle Rheinland-Pfalz, Photo LU 12679S.*)

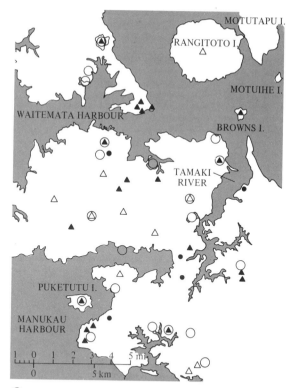

○ Dominantly explosive: tuff cones & craters

● Mainly explosive: tuff cones & minor scoria cones

▲ Mainly effusive: scoria cones & minor flows

△ Dominantly effusive: scoria cones & lava flows

formed the lava cone of Rangitoto in the harbor mouth 750 years ago. Renewed activity would thus seem likely anytime within the next few centuries—an unpleasant prospect.

Quiet Eruptions

Products. All eruptions, quiet or explosive, give off essentially the same gases mentioned earlier, but quiet eruptions do not produce glowing avalanches. As for lava, basaltic lava may flow like a river, maintain flow on gentle slopes, and extend far from the vent (Figure 9-12).

The Hawaiian lava streams flow in channels 3 to 15 m (10 to 50 ft) wide and 3 to 5 m (10 to 15 ft) deep at speeds exceeding 40 km (25 mi) per hour on steep slopes. Flows on Vesuvius have been clocked at 80 km (50 mi) per hour. This fluid lava, referred to by the Hawaiian name

FIGURE 9-10. **The Auckland volcanoes classified according to dominant types of activity.** (*E. J. Searle, "City of Volcanoes, A Geology of Auckland," Halstead Press, 1964.*)

FIGURE 9-11. Volcanoes of the Auckland district,
New Zealand. View northeast. Beyond Ourangi
Creek in the foreground is the explosion crater of
Waitomokia (Gabriels Hill). Beyond Puketutu Island
is the circular crater forming Mangere Lagoon.
Rangitoto in the far distance has erupted in historic
times. (*Whites Aviation Ltd., from E. J. Searle,
"City of Volcanoes, A Geology of Auckland,"
Halstead Press, 1964.*)

FIGURE 9-12. Mauna Loa, Hawaii, the world's largest shield volcano. Smaller craters lead up to the huge caldera at the summit. Note the dark riverlike lava flows. Similar flows in other parts of the world have reached velocities up to 100 km (60 mi) per hour and lengths exceeding 100 km (60 mi). (*Photo by U.S. Air Force.*)

pahoehoe (pah-hoy-hoy), solidifies with a smooth or slightly crumpled crust (Figure 9-13). The thickness of individual flows is generally less than 8 m (25 ft), but many are much thicker. Lavas of this sort cease motion as cooling hardens their exposed portions, but they may burst from the congealed front to form new flows. Such outflows may drain the interior of the flow, leaving elongate, sinuous lava tunnels.

More slowly moving lava crusts over during flow, and the crust is continuously broken up during movement. This rubbly lava is known by the Hawaiian name *aa* (ah-ah). The flow front (Figure 9-14) forms steep faces 10 m (30 ft) or more high. As it advances, the blocky front is overridden tractor fashion.

Extremely viscous lava, of rhyolitic composition, flows with difficulty, forming domal mounds and short, steep flows (Figure 9-15). Many rhyolitic lavas congeal to a volcanic glass called *obsidian*. The obsidian may have a light-gray upper portion, a lava froth called *pumice*.

Shield Volcanoes. Because fluid basalt is able to travel long distances on gentle slopes, the volcanic cone built up over the vent is shieldlike in profile. These *shield volcanoes* make up some of the largest volcanoes known. Explosive activity is incidental to quiet outflow and is confined to lava fountains and to sites where surface waters gain access to vents and conduits. The hazard to human life during eruptions of fluid basalt is seldom severe, because there is ample time for evacuation.

The island of Hawaii lies at the southeastern end of a long northwest-trending volcanic archipelago. The island was formed by the coalescence of five great shield volcanoes, of which the largest is Mauna Loa (see Figure 9-12) and the most active is Kilauea. Mauna Loa rises nearly 5 km (3 mi) above sea level from a base 5 km (3 mi) below sea level, but its broad flanks have gentle slopes of only a few

FIGURE 9-13. **Pahoehoe lava and steam vent, Kilauea volcano, Hawaii. The flow was erupted in 1969.** (*Donald A. Swanson, U.S. Geological Survey.*)

degrees. It and its neighbors are domal plateaus rather than mountains in the alpine sense. Steep-walled summit calderas more than 3.5 km (2 mi) in diameter and more than 100 m (300 ft) deep crown both Mauna Loa and its sister volcano Kilauea.

During major eruptions, the active vents display huge jets of incandescent lava called *lava fountains*. These rise hundreds of feet into the air. The lava falls back to earth and flows quietly away downhill, often reaching the sea. Late in the history of individual shield volcanoes in Hawaii, a change in the composition of the lava toward a more viscous type is accompanied by more explosive eruptions, and clusters of cinder cones, domes, and viscous flows appear on the broad summits of the gently sloping shields.

Cones of Viscous Lava. Not all quiet eruptions involve basaltic lava. The viscous lava rhyolite, although not as common as basalt, rises to form steep-sided cones (Figure 9-16). In some instances, the material is so viscous that it is squeezed upward like toothpaste to form spires (Figure 9-4) or domelike masses (Figure 9-17).

Volcanic Plains and Plateaus. At various times in the past, basaltic lavas from localized vents or from fissures have spread over large areas,

FIGURE 9-14. **Aa lava, Hawaii. Flow of 1889.** (*Photographer unknown.*)

FIGURE 9-15. **The light-colored flows in the foreground are of rhyolite lava. Note the rough blocky surface and the steep fronts. Mount Shasta in background. (*C. W. Chesterman, California Division of Mines and Geology.*)**

FIGURE 9-16. **East Twin Butte (left) and West Twin Butte (right), 58 km (36 mi) west of Idaho Falls, Idaho, on the Snake River Plain. The cones consist largely of rhyolite and are surrounded by basalt flows. (*Courtesy Carl N. Savage, Idaho Bureau of Mines and Geology.*)**

completely burying the earlier topography. The Columbia–Snake River Lava Plateau of Washington, Oregon, and Idaho covers 520,000 sq km (200,000 sq mi). Scores of flows are piled one on top of the other, locally reaching a thickness of more than 1200 m (4000 ft) and in one place more than 3000 m (10,000 ft). Individual flows average 30 m (100 ft) in thickness and some cover as much as 26,000 sq km (10,000 sq mi). India has a similar but larger basaltic plateau. Activity in the Snake River Plateau at Craters of the Moon may have continued into historic time.

Eruptions of basaltic lava in Iceland are similar in many respects to those in Hawaii. The lava issues from a fissure system that crosses the island from south to north. The fissure system is a northern extension of the great submarine oceanic rift that follows the crest of the Mid-Atlantic ridge, a major zone of plate separation (see Figure 8-15). Eruptions are more explosive than in Hawaii, particularly so in the Westman Island chain off the south coast, where the main volcanic axis passes out to sea and volcanoes erupt near or below sea level. Eruptive fissures in Iceland are marked by chains of cinder cones and by small shield volcanoes, but the main volcanic edifice is a broad plateau made of a thick stack of lava flows.

FIGURE 9-17. **Sugarloaf Mountain, a dome of porridgelike lava in the Cosa Range, Inyo County, California. (*Photo by C. W. Chesterman, courtesy California Division of Mines and Geology.*)**

The most voluminous historic eruption of basalt in Iceland occurred in 1783. The eruption began with explosion of solids to form cinder cones, some exceeding 100 m (300 ft) in height. This was followed by outpourings of lava which, for several months, flowed down valleys that led to the coastal plain, where the lava spread out in great lobate fans. Ash fell all over Iceland and 520 to 650 sq km (200 to 250 sq mi) were inundated by lava. The ash and volcanic gases severely damaged crops and the forage on which the livestock depended, resulting in famine.

Rhyolite, because of its lesser abundance and high viscosity, does not form large plateaus. The largest example known is in Yellowstone National Park, but a considerable part of the rhyolite is actually tuff.

Subsidence Calderas. The outpouring of lava from a magma chamber below a volcano may result in loss of support and collapse of the upper part of the volcano. Crater Lake in Oregon occupies a subsidence caldera more than 9 km (6 mi) across (see Figure 9-21). The upper 1200 m (4000 ft) of the top of the original volcano, Mount Mazama, collapsed into the depths.

Hazards and Countermeasures. Quiet eruptions of lava present little hazard to life because time is available for evacuation. Even swiftly flowing pahoehoe, because it is restricted to valley bottoms, is easily avoided. Destruction of property is another matter. We have noted

the burial of the villages of Parícutin and San Juan Parangaricutiro by lava from the new volcano Parícutin. Hilo, Hawaii, has been threatened by lava several times in recent history and has suffered some damage. The fracture or rift zones are the loci of dangerous eruptions, and the slight depressions running down the slope are the expectable paths of lava flows. Lava flows may also dam valleys, creating lakes and diverting drainage.

A number of methods to protect areas threatened by lava have been employed. In 1937 and 1942, the U.S. Army attempted to divert lava from the city of Hilo, Hawaii, by bombing threatening flows. The attempts were partly successful. Other attempts at diverting lava flows rely on the construction of barriers. The best protection for the city of Hilo is believed to be a permanent diversion barrier. The effectiveness of such barriers is indicated by the fact that lava entering villages does not knock down masonry walls. A wall of loose stones 1 m (3 ft) high in Hawaii held back a lava flow in 1920 until it rose high enough to spill over. Again, in 1950, a 1.3-m- (4-ft-) high wall of loose stones temporarily stopped a lava flow.

Examples of the effectiveness of masonry walls against advancing lava are available from Japan and Italy. In Japan, lava entered a masonry building through windows and doors, nearly filled the interior, and passed out the far side. In

Italy, on the south slope of Vesuvius, a wall surrounding a cemetery withstood almost complete burial of the cemetery itself. It has been suggested that walls 3 m (10 ft) high would provide protection against most lava flows, but local conditions may dictate the need for greater height. Walls should be located in favorable topographic positions and set diagonal to the slope to direct the flows where desired. Strategically excavated trenches to help in the diversion of lava have been employed in Japan.

Where lava rises in a crater, breaching of the crater wall to direct lava toward uninhabited ground may be possible.

A massive attempt to halt advance of lava by cooling with water was attempted during the Icelandic eruption of Kirkjufell in 1973. The eruption took place on the island of Heimaey off the south coast of Iceland, the site of Iceland's chief fishing port (Figure 9-18). The eruptions began on January 23, 1973, preceded by the opening of a fissure 1.6 km (1 mi) long. It

forced evacuation of 5300 residents to Iceland. A cinder cone 90 m (300 ft) high was built within the first 2 days. By early February, the fall of cinders had decreased, but highly heated lava, 1030 to 1055°C (1890 to 1930°F), continued to flow. Although more than 800 buildings were lost or damaged by lava and cinders, another 400 were saved. To halt the flow of lava, workers began spraying the advancing flows with water. Within 2 days the rate of advance diminished, and after 2 weeks the lava had stopped steaming. A pump ship was brought in during early March to aid the effort by delivering great quantities of cold sea water. By late April, 47 pumps were in operation and almost a million gallons of water per hour were being sprayed on advancing streams of lava. These

FIGURE 9-18. **Eruption of new volcano, Kirkjufell, in Iceland, 1973. Note the steaming lava extending into the sea at left and providing the harbor with additional protection against north Atlantic storms. On the slope, however, the lava is encroaching into the town and entombing buildings. (*Copyright, Sólarfilma, Iceland.*)**

efforts prevented the lava from completely blocking the port and causing further damage to homes and other structures.

The aim of dousing with water is to cool and thus increase the viscosity of the lava at the flow front so that the chilled front acts as a dam to divert the still fluid lava behind.

The inhabitants on their return also had to contend with harmful volcanic gases, including carbon monoxide and carbon dioxide. These gases, being heavier than air, collected in basements and in low places outdoors during calm weather. As a partial counteraction, a trench was dug along the eastern edge of town to intercept and lead the gases out of town.

On the plus side, the eruption of Kirkjufell added 2.5 sq km (1 sq mi) to the island, increasing its area by 20 percent.

Composite Eruptions

Characteristics. Most of the present eruptive activity on the lands alternates between explosive and quiet. Ordinarily, the cycle starts with an explosive eruption in which large quantities of solid particles are ejected. Once the initial gas pressure is released, lava is extruded. There may then be a long period of quiescence and cooling of the magma in the vent. Accumulating gas pressure may once again lead to an explosion, starting the cycle again. The sequence is most common where lavas of intermediate viscosity, primarily andesite, are involved.

Because the bulk of the solids and the largest fragments fall close to the vent, the slope at the vent is steepest, as much as 40°. The slope diminishes outward as the solids diminish in size and quantity. Lava spilling out of the crater or erupting on the flanks of the volcano contributes to the gentle lower slopes.

Stratovolcanoes. Volcanoes built by composite eruptions and composed of interbedded explosion products and lava are known as *stratovolcanoes*, or *composite cones*. They include all the great Cascade volcanoes, such as Mount Shasta, Mount Hood, and Mount Rainier, all the lofty volcanoes of Central and South America, including Cotopaxi in Ecuador, famous Mount Fuji in Japan, and Mount Mayon in the Philippines. The stratovolcanoes form volcanic chains where the edges of oceanic plates dip under continental margins.

Some of the volcanoes, like those of the Andes and Central America, stand near the edge of the continent; others form island chains offshore. In the western Pacific, the populous island countries of Indonesia, the Philippines, and Japan are dotted with active volcanoes of this type, which also occur at intervals along other major island chains, including the Kurils, Ryukyus, Marianas, Solomons, and New Hebrides. A continuation of the volcanic belt of the western Pacific extends up the Kamchatka Peninsula of Siberia and then along the Aleutian Islands to southern Alaska. The Cascade Range includes a line of 15 active and dormant stratovolcanoes stretching from southern British Columbia across Washington and Oregon to northern California. The volcanoes of the Lesser Antilles in the Caribbean and those of southern Italy and the Aegean Sea in the Mediterranean are of similar origin and exhibit similar behavior.

Composite eruptions from many closely spaced volcanoes have built whole mountain ranges. Examples are the San Juan Mountains in Colorado and the Absaroka Range in Wyoming (Figure 9-19). Presumably, the centers of eruption kept shifting during the accumulation of these thick, extensive piles of volcanic debris.

Hazards and Countermeasures. Stratovolcanoes, because of the scale of their eruptions and the secondary phenomena that result, present great hazards to people and their works. The hazards include base surges; asphyxiation; burial by explosion debris, mudflows, and lava; landslides; "tidal waves"; drainage changes; topographic changes; and glacial-meltwater floods.

Base Surges. Base surges are ground-hugging blasts of hot, relatively ash-free air. A base surge has been described as a "ring-shaped basal cloud which sweeps outward as a density flow from the base of the vertical explosion column." They may extend outward several kilometers and attain velocities of 20 to 30 m per second (45 to 70 mi per hour). They are

FIGURE 9-19. **A view in the Absaroka Range, Wyoming. The pile of rocks includes ancient lava flows, layers of cinders, and mudflows.** (*U.S. Geological Survey.*)

commonly associated with water-stimulated explosions and may involve only water and gases and leave no deposits. It is reported that a party of Hawaiian warriors was caught in such a hot air blast during an eruption of Kilauea in 1790. The bodies, including some standing groups, were scorched but not covered with ash.

Asphyxiating Gases. Some eruptions involve poisonous gases. We have noted that among the gases emitted by volcanoes are carbon monoxide and carbon dioxide. After the Kirkjufell eruption of 1973, several returning townspeople were overcome by these gases and one died.

Not all volcanic clouds contain lethal gases, as indicated by observers who have been enveloped in relatively cool volcanic clouds of gas and ash from initial eruptions for as much as 15 minutes without harm.

Explosion Debris. Coarse debris, consisting of large fragments, is a direct hazard only close to the vent. Fine debris, however, which we will refer to collectively as *ash*, may either drift great distances on the upper winds or flash down the flanks of a volcano as ash flows.

The wind-drifted ash may have disastrous short-term effects. Where the fall is heavy, roofs of buildings may collapse and vegetation may be destroyed (Figure 9-20). In the eruption of Katmai in Alaska in 1912, 21 cu km (5 cu mi) of ash were ejected. Thousands of square kilometers were deeply buried. At Kodiak, 160 km (100 mi) away, the weight of ash on rooftops caused collapse of houses. Approximately 16,000 Romans died at Pompeii in A.D. 79, presumably from suffocation in the ash-laden air. The ash eruptions of Irazu in Costa Rica in 1964 caused $150 million in damage.

Burial is only one hazard from volcanic ash. Surface-water supplies may become contaminated and drainage channels clogged. The respiratory and digestive systems of animals may be affected by breathing the ash-laden air or eating ash-laden vegetation. Eighty thousand people died of famine when the 1815 eruption of Tambora in Indonesia caused destruction of crops.

The lethal combination of high speed and high temperature make glowing avalanches one of the most feared volcanic phenomena. Like mud and lava flows, they tend to follow valleys.

FIGURE 9-20. **Homes in various stages of burial in ash. The water jets were used to chill and halt the advancing lava. Eruption of Kirkjufell, Iceland, 1973. (*Copyright, Sólarfilma, Iceland.*)**

Thus, human habitation should avoid such likely paths. If this is impracticable, the density of habitation should be kept to a minimum, and evacuation should be considered at the first indication of volcanic unrest. Yet, Mount Somma had given the Pompeiians 5 years of warning through numerous micro-earthquakes prior to the eruption of A.D. 79. Mount Pelée had been expelling ash several weeks before its catastrophic eruption, but few people took flight and the government took no steps to ensure their safety.

Since the disaster at St. Pierre, the initiation of explosive activity at volcanoes that normally erupt viscous lava has been accepted as a warning of possible glowing avalanches, particularly if large domes begin to rise in the craters. In spite of this, disasters have continued. An eruption at Mount Lamington in Papua in 1951 devastated 190 sq km (75 sq mi) of jungle with scattered villages and killed 3000 people in 12 hours. There were no records of previous volcanism at Lamington, and radiocarbon dates suggest that the last previous ash flow took place 12,500 years ago. The only warnings were landslides within the crater, swarms of local earthquakes, and minor emission of ash during the 5 days preceding the main eruption. At the climax, an ash-laden cloud of mushroom shape rose 16 km (10 mi) within minutes, and simultaneously, a frothing emulsion of gas and ash overflowed the crater, mainly through a notch on one side, to become ash flows that rushed down the sides of the mountain to the plains below. For 6 weeks thereafter, eruptions spread ash over a large area and a dome grew to a height of 450 m (1500 ft) above the vent as lesser ash flows swept down the flanks of the mountain. The basal avalanches followed stream valleys and fed volcanic mudflows that moved even farther downstream.

We mentioned welded ash-flow deposits that are emplaced so hot and in such thick sheets that the particles at lower levels fuse to form a compact rock. An eruption of partly *welded tuff* took place in 1912 near the foot of Mount Katmai in Alaska's Valley of Ten Thousand Smokes. The eruptions filled a previously glaciated valley 19 km (12 mi) long to a depth of 30 m (100 ft) with a deposit of pumice and ash 1.6 to 8 km (1 to 5 mi) wide.

On the North Island of New Zealand, 26,000 sq km (10,000 sq mi) of the central plateau are underlain by welded and unwelded tuffs. The total sequence is hundreds of meters thick. Individual welded tuffs range up to 100 m (300 ft) thick and extend laterally for scores of kilometers. The plateau is scarred by several large calderas with associated domes. The last major eruptions of ash from two of the calderas took place about A.D. 125 and A.D. 1000. Each eruption spread pumice and ash over 13,000 to 26,000 sq km (5000 to 10,000 sq mi) to depths up to 1.6 m (5 ft).

There are examples in mainland United States. The Bishop tuff, named after the town of Bishop on the east side of the Sierra Nevada, was erupted 750,000 years ago and covered 1300 sq km (500 sq mi) to an average depth of about 75 m (250 ft). Airborne ash from the eruption spread over much of the Southwest and as far away as the Great Plains.

The Yellowstone Plateau in Yellowstone National Park is in large part welded tuff. On three occasions within the past 2 million years, a vast caldera 40 to 80 km (25 to 50 mi) in diameter has been filled like a glowing cauldron with hundreds of meters of ash. Explosions spread the ash far to the east over the Great Plains. The most recent ash fall took place half a million years ago.

No one knows how to respond to the threat of major ash flows. Volcanologists remain alert for indications of explosive eruptions and are aware that failure to detect premonitory symptoms in time to issue warnings could presage catastrophy.

Volcanic Mudflows. The generation of volcanic mudflows is not always linked directly to an eruption. Nevertheless, many follow closely on the heels of eruptions. Flowage is initiated when highly unstable slopes of freshly erupted ash are drenched by torrential rains, or when eruption of lava or hot ash causes rapid melting of snowfields, or when the wall of a lake-filled crater or caldera is breached during an eruption. Earthquakes that accompany eruptions also help jar loose materials on steep slopes. Some mudflows originate when ash flows enter streams.

Mudflow speeds of 40 km (25 mi) per hour are common, and the distance traveled may be 80 km (50 mi) or more. The three mudflows that buried Herculaneum in A.D. 79 totaled 24 m (80 ft) in thickness. Those from the volcano Kelut on Java in 1919 destroyed nearly 130 sq km (50 sq mi) of farmland and killed 5000 people. The Kelut event illustrates the destructive effects of mudflows that form during eruptions in crater lakes. About 30 million cu m (1 billion cu ft) of water mixed with ash rushed down the valleys, engulfing entire villages.

Ice Floods. Ice floods are violent, ice-laden torrents caused by volcanic activity in craters beneath ice fields. The term "glacier burst" is also used but includes the violent outbursts of glacially dammed lakes. The ice floods of 1934 in the Skeidará River Valley of Iceland originated in the crater Sviagigur. As meltwater accumulated under a blanket of ice, and pressures mounted, the water was forced over the ice-covered lip of the crater and down under the Skeidará Glacier. The surging masses of subglacial water lifted the entire glacier, broke up the ice, and caused the glacier to undulate like swells on the sea. The floods carried ice blocks up to 30 m (100 ft) high out onto the broad plains between the glacier and the sea and created waves that ranged hundreds of kilometers along the coast.

Inasmuch as many of the world's lofty volcanoes, including those of the Cascade Range, are capped by snow and ice, the danger of volcanic ice floods is always present. The only sure precaution against these violent floods, as with glowing avalanches and mudflows, is to avoid the valleys that provide potential paths, as well as the lowlands immediately beyond.

Explosion Pits and Caldera Formation. The creation of explosion pits with little or no volcanic debris presents local problems in volcanic terrain. Because of their small size, however, generally less than 400 m (1300 ft) across, they do not present the same degree of hazard as calderas.

In the creation of explosive calderas, part

of the volcano is blown off. Subsidence calderas, on the other hand, result from loss of support due to evacuation of material from the magma chamber. Most calderas appear to be due to an initial explosive phase followed by subsidence. The great submarine caldera formed in the eruption of Krakatoa in 1883 is of this origin. Some scientists believe that the subsidence of a caldera on the volcanic island of Santorin in the Aegean Sea more than 3000 years ago completely submerged a large city and that this event, plus the havoc created by a huge tsunami which raked the coastal cities of Crete and Asia Minor, contributed to the decline of the Minoan civilization.

The most famous caldera in the United States is occupied by Crater Lake in Oregon (Figure 9-21). The lake lies within the stump of Mount Mazama, which rose as a majestic stratovolcano to an elevation exceeding 3600 m (12,000 ft), or 1200 m (4000 ft) above the present caldera rim. About 7000 years ago, voluminous eruptions spread ash over thousands of square kilometers; the pumice on the slopes of Mount Mazama itself was more than 15 m (50 ft) thick. Glowing avalanches raced down previously glaciated valleys on the mountain's flanks and moved across adjacent plateaus or down river canyons for more than 40 km (25 mi). The rapid ejection of so much material from the magma

chamber caused collapse of the peak to form a caldera more than 9 km (6 mi) across and nearly 1.6 km (1 mi) deep. The lake itself is 600 m (2000 ft) deep. The small Wizard Island volcano later grew within this caldera, the basin of Crater Lake. The similarity of ancient Mount Mazama to many modern peaks of the Cascade Range raises the specter of similar events.

Earthquakes and Landslides. In spite of the enormity of some volcanic eruptions, perceptible shaking of the ground is restricted to within a few tens of kilometers of the source. As with all earthquakes, landslides are common, particularly on the volcano but also in any nearby loose materials. The loose deposits forming the steep upper slopes of many volcanoes are in a precarious state of stability. They rest at the maximum angle of repose of dumped material, the slope sometimes exceeding 40 °. Even without eruptions, these slopes may fail, creating avalanches. Shaking during eruptions facilitates the process and, together with heavy rain, may create mudflows.

When explosive volcanic activity or collapse creates new steep slopes, landslides generally follow until a more stable slope is achieved. The adjustments may continue for

FIGURE 9-21. **Crater Lake, Oregon. Wizard Island is slightly above and left of center.** (***U.S. Air Force.***)

weeks or longer, as after the moderately violent eruption of Taal Volcano in the Philippines in late September 1965.

Water Waves. Even during the relatively minor eruption of Taal Volcano, the explosive shocks generated water waves in Lake Taal that extended 4.7 m (15 ft) above lake level and swept inland 80 m (250 ft). The waves capsized some of the rescue boats taking people from the island, accounting for many fatalities. These waves, however, were but ripples compared to those generated in the great eruption of Krakatoa in 1883. A giant wave 40 m (130 ft) high rushed inland over densely populated low-lying coasts and claimed 36,000 lives.

Modifications of Drainage and Topography. Volcanic alterations of drainage and topography are not in themselves of widespread environmental concern. People accommodate to most of these changes. The adjustment may involve abandonment of affected areas. Deposits burying former fields may have to lie fallow for years until new soil is created. To drain a valley blocked by a lava flow and submerged under lake waters may prove impracticable. Where such changes affect an urban area, however, action may be justified. The inhabitants of Vestmannaejar, the fishing port in Iceland partly buried in the eruption of Kirkjufell volcano in 1973, felt justified in digging their city out of the cinders. If the harbor on which their livelihood depended had been completely filled in by the flow, another decision might have been considered.

Hydrothermal Phenomena

Hydrothermal phenomena, hot springs, geysers, fumaroles, and certain mud volcanoes are the dying phases of volcanic activity. A *hot spring* is a quiet spring at the top of a roomy vent in which heated waters from below can easily rise to the surface, cool, and descend again. A *geyser* is a violent eruption of superheated water confined to a narrow tortuous vent. A *fumarole* is a vent from which only vapors emerge. *Mud volcanoes* are small cones formed at vents where mud-charged waters reach the surface.

HUMAN ADJUSTMENT TO VOLCANIC ENVIRONMENTS

The Lure

People are drawn to volcanoes for a variety of reasons. Volcanoes provide some of the earth's most spectacular scenery. Because many attain great heights and are well watered, volcanoes attract climbers, hikers, skiers, fishermen, and nature lovers in general.

Many volcanic regions are attractive for agriculture. Volcanic regions such as Java in the East Indies, are renowned for their fertile soils periodically replenished by eruptions. On the other hand, Borneo, without volcanoes, has poorer soils. As a result, Java, in spite of its volcanic hazards, has a population density of 460 people per square km (1200 per sq mi), whereas Borneo, without volcanic hazards, has only 2 per sq km (5 per sq mi). Many volcanic hazards are limited to the slopes of the volcanoes. If the slopes are avoided, the dangers are reduced considerably.

Volcanic regions are receiving attention as sources of geothermal energy (see Chapter 12). Industrial installations will have to be established at the sources of energy and housing provided for workers. The need for energy being what it is, settlement of volcanic areas will probably increase.

Adjustment

The inability to determine whether an apparently extinct volcano is only dormant complicates the problem of human adjustment. It seems unwise to gamble on the proposition that a volcano is extinct. At the time of the eruption of Mount Somma in A.D. 79 the flanks of the volcano were cultivated to the very summit. Perhaps the crestal areas of all volcanoes should be reserved for recreational and observational activities.

The United States has, in a sense, engaged in volcanic zoning. Most of the known active volcanoes, except for those in the Aleutian Islands, have been set aside as national parks or national monuments. The Yellowstone thermal area and the Valley of Ten Thousand Smokes in Alaska have also been withdrawn from general habitation. But other volcanoes may only be dormant instead of extinct. Perhaps these too should be set aside as national reserves.

Sites on the slopes of volcanoes, such as lookouts, inns, and observatories, should be situated on high ground to reduce risk from lava, ash flows, and mudflows, which follow valleys. Observatories and accommodations on crater rims should avoid low places over which rising lava in the crater might spill. To avoid collapse of buildings under the weight of accumulating ash, roofs should be steeply pitched as in areas of heavy snow.

Where crater lakes exist and the danger of volcanic explosion threatens, breaching of the crater rim to lower the lake level and diminish the threat of mudflows might prove feasible. In the 1919 eruption of Mount Kelut in Java, the crater lake supplied much of the water for the mudflows that killed more than 5000 people. In 1926, pipes and tunnels were installed to siphon the lake waters and keep the lake level relatively low. The effectiveness was demonstrated in 1951 when an eruption caused only seven deaths.

PREDICTION

The problem of prediction of volcanic eruptions is twofold. First is the problem of predicting eruptions from known active volcanoes; second is the problem of determining the probability of renewed eruption of a volcano long dormant (extinct?).

We have noted that the type of volcano will indicate the kind of eruption to be expected. There is no reason to expect Krakatoa-like explosions from the shield volcanoes of Hawaii or vast far-reaching outpouring of fluid lava from stratovolcanoes like Fujiyama. Particularly with stratovolcanoes, however, considerable variation in eruptive type is possible. Appreciation of the magnitude of the possible variations may require careful study of the anatomy of the volcano and of the types of eruptions that have occurred at similar volcanoes.

We will consider first the measures used to predict eruptions from known active volcanoes. These include study of historical records and monitoring of the volcanoes.

Known Active Volcanoes

Historical Records. Historical records may be helpful guides to possible future volcanic activity, although it may be inadvisable to rely too heavily on human memory, even if recorded in the literature. Accounts of two observers of the same event may differ significantly.

Some active volcanoes erupt more or less continuously or at short intervals, such as Kilauea, in Hawaii; Stromboli, on an island between Italy and Sicily; and Vulcano, south of Stromboli. The nature of these frequent eruptions is well known, the hazards are obvious, and the precautionary procedures are well organized.

Some investigators believe that historical records indicate a cyclical repetition of volcanic eruptions. The idea is that after release of energy in one eruption, a number of years is required for regeneration of the energy. However, others believe that the magmatic system is too complex and variable to result in uniform periods between eruptions.

Monitoring. Volcanoes frequently give warning of impending eruptions. Possible signals include premonitory eruptions, earthquakes, tilting of the ground, appearance of new hot springs and fumaroles or changes in temperature and composition of existing ones, changes in ground temperature, and changes in rock behavior at depth. Most volcanic eruptions are preceded by earthquakes and ground-tilting.

Premonitory eruptions may consist of emission or increased emission of volcanic clouds, appearance of hydrothermal phenomena, appearance of lava in craters or fissures, and actual emission of lava or explosion products. Thus, a month before the eruption of Mount Pelée on May 8, 1902, fumaroles appeared and were followed within the next few weeks by light falls of ash, sulfurous odors, and earthquake shocks.

Mauna Loa, on the island of Hawaii, erupted on July 5 and 6, 1975. Although a small eruption by Mauna Loa standards, involving only 30 million cu m (40 million cu yd) of lava, historical records suggest that it is probably just a precursor to a much larger eruption. The U.S. Geological Survey predicts a major eruption before July 1978 and recommends measures to protect the city of Hilo in the path of the expected lava flows.

In many eruptions, there are few if any preliminary visual or audible manifestations. Thus, there is need for delicate instruments to record the ordinarily imperceptible warning signals.

A volcano watch has been established by the United States Geological Survey on 16 volcanoes in Washington, California, Hawaii, and Alaska and with the aid of cooperating scientists, in Iceland, Guatemala, El Salvador, and Nicaragua (Figure 9-22). Cotopaxi, in Ecuador, was added to this list in 1976 at the request of the government of Ecuador. The 5897-m (19,347-ft) volcano began emitting steam in November 1975 with melting of ice and snow. Devastating mudflows, such as occurred in the last major eruption in 1877, are feared.

The primary instrument at the monitored volcanoes is the *earthquake counter*. The heart of the earthquake counter is a geophone (G, Figure 9-23). It consists of an enclosed, spring-supported weight which, because of its inertia, remains fixed in space as the container vibrates with the ground. The rapid alternations in dis-

tance between the geophone weight and the container walls are translated into electric impulses.

The importance of the earthquake counter stems from the fact that swarms of earthquakes precede many eruptions. The activity may result from movement of magma in depth. Earthquake counters on the volcano Fuego in Guatemala reported a high incidence of earthquakes as much as 6 days before the eruption of February 27, 1973. However, earthquakes do not in themselves imply imminent eruption, but rather that the volcano is alive and bears watching.

Supplementing the earthquake counters at some of the monitored volcanoes are one or more *tiltmeters*. These are sophisticated bubble-type levels. The position of the bubble is electronically determined to record changes in slope of as little as one in a million. This is equivalent to a slope of 1 cm in 10 km (1 in. in 16 mi). The instrument easily detects the slight swelling that precedes many eruptions caused by upward movement of magma under pressure. Provision is made in the instrument to filter out extraneous recordings due to local

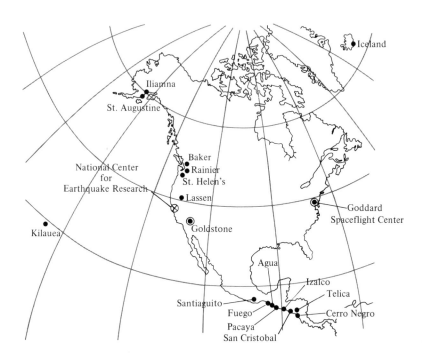

FIGURE 9-22. **Volcanoes monitored by U.S. Geological Survey and collaborating foreign institutions. All sites have earthquake counters, but only Mount Lassen, Kilauea, Fuego, and Pacaya have tiltmeters. (*After U.S. Geological Survey.*)**

FIGURE 9-23. **Earthquake counter station at St. Augustine, Alaska. The geophone (G) is in the container in the small pit in the foreground 25 m (80 ft) from the steel box (B) containing the electronic earthquake counter, batteries, and radio transmitter. The plexiglass dome (D) protects the antenna.** (*U.S. Geological Survey.*)

those at Parícutin in 1943 began 2 weeks before the eruption; those at Mount Pelée in 1902 began 5 weeks in advance; those at Krakatoa in 1883 preceded the eruption by 14 weeks; and those at Vesuvius in A.D. 79 started 16 years earlier. Not only is the time uncertain, but the earthquakes provide no information on the intensity of the eruption to be expected.

The above uncertainties raise interesting questions. Should populated areas be evacuated at the first warnings from a monitoring system? If an eruption fails to materialize within a period of days, weeks, or months, should the populace be permitted to return? Suppose an evacuation is ordered and only a trivial eruption ensues. It would take only a few unfulfilled alarms to render the public apathetic.

Other warning signals that may precede eruptions are the appearance of new hot springs and fumaroles or changes in chemistry or temperature of those already present. Ground temperatures may also rise. Such measurements may now be made from orbiting satellites which record infrared heat radiation.

Other types of investigations are aimed at determining whether impending eruptions are preceded by changes in the magnetic and electrical properties of the rocks of the volcano. In addition, seismic surveys are used in which artifically created earthquakes monitor changing conditions below the surface.

events such as volume changes in the soil caused by wetting or temperature changes. A typical tiltmeter station is shown in Figure 9-24.

The earthquake counters and tiltmeters are automatic and report their activity by radio via the Earth Resources Technology Satellite (ERTS-1) directly to Goddard Spaceflight Center in Maryland, or indirectly via Goldstone in California (Figure 9-22). The data are processed at Goddard and relayed by teletype within 90 minutes to the Center for Earthquake Research in Menlo Park, California.

The chances of predicting future eruptions at instrumented volcanoes are good. However, the time between the incidence of earthquakes and an eruption may be unpredictable. Thus, the first recorded earthquakes preceded the 1973 eruption of Fuego in Guatemala by 6 days;

Long-Inactive Volcanoes

There is no way of knowing whether long-inactive volcanoes are extinct or simply dormant, and there is no easy way of predicting their possible behavior. What then can be done about these problem volcanoes? Should all such volcanic sites be set aside for national preserves?

In an attempt to arrive at an assessment of the possible hazards if one of the Cascade volcanoes were to erupt, the U.S. Geological Survey has been investigating the evidence of ancient activity as revealed in the rock record. Crandell and Mullineaux (see Additional Readings) have described the hazards of Mount Rainier, and Crandell and Waldron (see Additional Readings) have summarized the potential hazards for the Cascade Range as a whole.

FIGURE 9-24. **Tiltmeter station on slope of Pacaya Volcano, Guatemala. The tiltmeter and its electronic accessories are in the steel-encased pit (P). The battery and transmitter are in the steel box (B) 6 m (20 ft) away and connected to the pit by cables in a flexible steel conduit. The antenna (A) transmits data to Goddard or Goldstone space stations via the Earth Resources Technology Satellite (ERTS, now LANDSAT).** (*U.S. Geological Survey.*)

The procedure followed by the Survey geologists consisted of identifying the deposits of past eruptions, dating them, mapping their extent, and characterizing the type and frequency of the volcanic activity. The assumption is that future eruptions may follow past patterns.

It is believed that ash eruptions and mudflows are the greatest hazards. The degree of hazard from ash eruptions depends on amount of material blown out, rate at which emitted, strength and direction of the winds during the eruption, and distance to populated centers.

Figure 9-25 shows the area covered by an ashfall from Crater Lake 7000 years ago. The shaded area shows where 15 cm (6 in.) or more of ash accumulated. Six-inch ash-depth lines have also been drawn around the other

volcanoes. Figure 9-26 shows an exposed section of overlapping beds of volcanic ash from Mount Rainier, Mount Mazama (Crater Lake), and Mount St. Helens. Recent studies indicate that St. Helens has been more active in the last few thousand years than any other volcano in the contiguous 48 states.

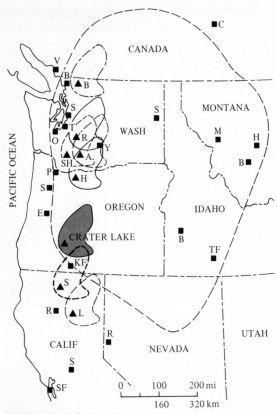

height of eruption, particle size and density, wind direction and velocity at different levels, and rainfall.

To investigate the probable influence of winds on the distribution of Katmai ash, R.E. Wilcox prepared a number of wind roses or diagrams in which arrows indicate wind directions and their lengths represent percentages of time. They were based on data recorded by the U.S. Weather Bureau at Kodiak, Alaska, and represent a summation for the years 1948–1953. The upper level wind patterns and the actual ashfall pattern of the 1912 eruption were in agreement. Wilcox believes that such wind roses will at least indicate the probable distribution of ash from volcanic eruptions.

FIGURE 9-25. Ash distribution from eruption at Crater Lake 7000 years ago and sites of other Cascade eruptions. The outer line shows the recognizable limit of the Crater Lake ashfall. The shaded area is the area covered by 15 cm (6 in.) or more of ash. The 15-cm ash areas for other Cascade volcanoes are also shown. The triangles indicate the volcanoes. From north to south these are B—Baker, R—Rainier, SH—St. Helens, A—Adams, H—Hood, S—Shasta, and L—Lassen. The squares represent important cities. (*After D. R. Crandell and H. H. Waldron, Volcanic Hazards in the Cascade Range, in "Geologic Hazards and Public Problems."*)

FIGURE 9-26. **Volcanic ash deposits on the southeast flank of Mount Rainier. A—ash layers (light) from Mount St. Helens, 80 km (50 mi) to the southwest. B—pumice (light) and scoria (dark) from Mount Rainier. C—ash from Mount Mazama (Crater Lake, Oregon) 440 km (275 mi) to the south. (*D. R. Mullineaux, U.S. Geological Survey.*)**

Figure 9-27 shows the distribution of ash from the great eruption of Mount Katmai and nearby vents in Alaska in June 1912. Depths of ash varied from more than 16 m (50 ft) in the vicinity of Mount Katmai to 25 cm (10 in.) at Kodiak, 160 km (100 mi) to the southeast. Even at that distance, the ash damaged buildings and crops. The distribution of ash is influenced by

The weight of ash on structures can be considerable. A 2.5-cm (1-in.) layer of ash weighs 50 kg (kilograms) per sq m (10 lb per sq ft). A roof 140 sq m (1500 sq ft) in area would thus be supporting an additional load of 7000 kg (15,400 lb). A 15-cm (6-in.) layer would add a total of 42,000 kg (92,400 lb). Soaking by rain increases the load. Steeply pitched roofs are advisable where ash fall hazards exist.

Dense ash falls also contaminate surface-water supplies with sediment and temporarily increase their acidity.

Mudflows are high-potential hazards in the Cascades. The danger is especially great because most of the volcanoes have a permanent snow cover to contribute large volumes of meltwater. Mudflows generally move faster than floods of water and can extend scores of kilometers (tens of miles) downvalley. Crandell and Mullineaux have identified 55 large mudflows that took place on Mount Rainier in the last 10,000 years, as well as a dozen ash eruptions, several hot avalanches, and at least one

period of lava emission. The last eruption of any consequence was 2000 years ago. Based on the above statistics, the principal future hazard will be mudflows. Although many will be restricted to valley floors close to the volcano, others may equal the magnitude of two ancient mudflows which extended to great distances (Figure 9-28). The Osceola Mudflow, which originated 5000 years ago, involved 1.9 billion cu m (2.5 billion cu yd or half a cu mi) of debris. The mudflow rushed down the valley of White River for 80 km (50 mi) at inferred speeds of 32 km (20 mi) per hour. The flow is more than 150 m (500 ft) thick in White River canyon. On entering the Puget Sound Lowland, it spread widely, accounting for much of its total 310 sq km (125 sq mi) coverage, and deposited debris to depths of 23 m (75 ft). About 30,000 people

FIGURE 9-27. **Ash distribution from eruption of Mount Katmai, Alaska, June, 1912.** (*After R. E. Wilcox, Some Effects of Recent Ash Falls with Special Reference to Alaska.*)

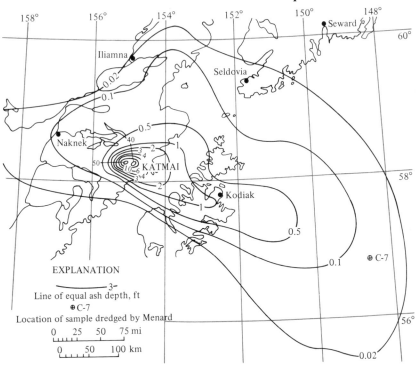

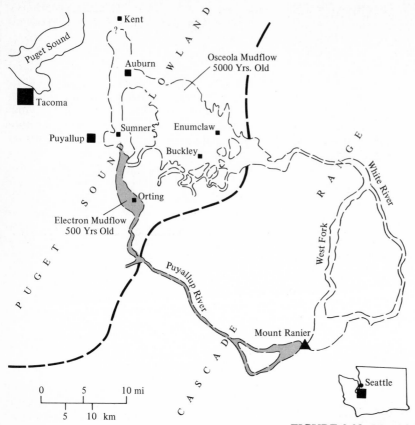

FIGURE 9-28. **Map of Mount Rainier and vicinity, showing extent of the Osceola Mudflow in the White River valley (dashed boundary) and the Electron Mudflow (pattern) in the Puyallup River valley.** (*After D. R. Crandell and H. H. Waldron, Volcanic Hazards in the Cascade Range, in "Geologic Hazards and Public Problems."*)

are located on top of this flow, including the inhabitants of Enumclaw and other towns. An outcrop of this ancient bouldery mudflow is shown in Figure 9-29.

The Electron Mudflow, only 500 years old, involved 150 million cu m (200 million cu yd). It extended 32 km (20 mi) down Puyallup Valley and another 24 km (15 mi) out onto the Puget Lowland. Between 2000 and 3000 people live on this flow today. In brief, comparable mudflows in a new eruption would imperil between 32,000 and 33,000 people.

Direct burial under mudflows is only part of the hazard. Two of the valleys heading on Mount Rainier have dams in them. Mud Mountain Dam in the White River Valley was built for flood control and is generally kept empty. It has a capacity of 130 million cu m (170 million cu yd). Alder Dam, in Nisqually River Valley, maintains a reservoir for hydroelectric genera-

tion. Its capacity is about 275 million cu m (375 million cu yd). A mudflow of the dimensions of the Osceola Mudflow would completely fill either of the reservoir basins without seriously curtailing the extent of the mudflow below the dams. On entering the Alder Reservoir, the flow would drive the water ahead of it, and a catastrophic flood would precede the advancing mudflow.

Crandell and Waldron have estimated the following average spacing of volcanic events at Mount Rainier: eruptions of lava near the summit, 10,000 years; major rubble avalanches on

FIGURE 9-29. **Bouldery Osceola Mudflow exposed in a terrace scarp. The mudflow extends to a depth of about 60 m (200 ft) in this part of the valley. The largest boulder in the exposure is 2 m (6 ft) in diameter. (*D. R. Crandell, U.S. Geological Survey.*)**

the flanks of the volcano, 5000 years; pumice and ashfalls on the surrounding uplands, 2500 years; major mudflows reaching the lowlands, 1000 years; and minor mudflows in the heads of valleys around the volcano, 1 to 10 years. Clearly, a close watch for unstable internal and external conditions must be maintained.

ADDITIONAL READINGS

Bolt, B. A., W. L. Horn, G. A. MacDonald, and R. F. Scott: "Geologic Hazards," Springer-Verlag New York, Inc., New York, 1975.

Bullard, F. M.: "Volcanoes in History, in Theory, and in Eruption," University of Texas Press, Austin, 1962.

Cotton, C. A.: "Volcanoes as Landscape Forms, New Zealand," Whitcombe and Tombs Ltd., 1944.

Crandell, D. R.: Map showing potential hazards from future eruptions of Mount Rainier, Washington, U.S. Geological Survey, Miscellaneous Geologic Investigations Map, I-836, 1973.

——— and D. R. Mullineaux: "Volcanic Hazards at Mount Rainier," U.S. Geological Survey, Bulletin 1238, 1967.

——— and H. H. Waldron: Volcanic Hazards in the Cascade Range, *in* "Geologic Hazards and Public Problems," Conference Proceedings, R. A. Olson and M. M. Wallace, eds., Santa Rosa, Calif., Office of Emergency Preparedness, pp. 5–28, 1969.

MacDonald, G. A.: Barriers to Protect Hilo from Lava Flows, *Pacific Science*, vol. 12, 1958, pp. 258-277.

———: "Volcanoes," Prentice-Hall, Inc., Englewood Cliffs, N.J., 1972.

Oakeshott, G. B.: "Volcanoes and Earthquakes: Geologic Violence," McGraw-Hill Book Company, New York, 1975.

Rittman, A.: "Volcanoes and Their Activity" (translated by E. A. Vincent), John Wiley & Sons, Inc., New York, 1962.

Searle, E. J.: "City of Volcanoes, A Geology of Auckland," Halstead Press, Sidney, 1964.

Stearns, H. J.: "Geology of the State of Hawaii," Pacific Books, Publishers, Palo Alto, Calif., 1966.

Wilcox, R. E.: "Some Effects of Recent Ash Falls with Special Reference to Alaska," U.S. Geological Survey, Bulletin 1028-N, 1959, pp. 409-476.

Williams, H.: Volcanoes, *Scientific American*, vol. 185, no. 5, 1951, pp. 45-53. (*Readings in the Earth Sciences*, W. H. Freeman and Company, San Francisco, Reprint 822, 1969, pp. 163-172.)

special problems
and methods

CHAPTER TEN

geochemistry and environmental impact

INTRODUCTION

Environmental Geochemistry

In 1953, some of the villagers in the vicinity of Minamata on the island of Kyushu, Japan, developed strange symptons, including tottering, jerky gaits, loss of manual dexterity, impairment of speech, and commonly, deafness and blindness. Twenty-five percent of those afflicted died. All victims had eaten fish contaminated by mercury. The mercury was traced to a factory which a year earlier had begun using mercury compounds in processing procedures. The industrial wastes included methyl mercury, an organic compound far more toxic than the element itself. The wastes were emptied into the sea where they contaminated the food fish.

The insidiousness of methyl mercury is its ability to penetrate and erode brain tissue, particularly the areas that control sight, hearing, and equilibrium. Once the factory at Minamata ceased its discharges into the bay, mercury levels dropped, fishing returned to normal, and further outbreaks of the disease ceased.

In the United States, both tuna and swordfish were found to be contaminated with mercury. The Food and Drug Administration established one-half part per million (ppm) as the maximum safe limit for mercury in fish, and the Department of Agriculture has banned the use of mercury in seeds. As a result of government pressures, mercury pollution of water at 50 surveyed industrial plants has been decreased 86 percent.

Mercury is only one of the substances which, in excess, can impair health. We shall discuss others, but the effects of many will not be known for a long time.

In some areas, health-affecting geochemical abnormalities occur naturally. Place names such as Badwater, Alakali Flat, and Alum Creek, indicate such abnormalities. Early settlers often found local water to be toxic or local soils to be useless for agriculture.

Some substances in excessive amounts may impair the environment without directly creating health problems. At Copperhill in southeastern Tennessee, smelter fumes killed off vegetation in an area of 130 sq km (50 sq mi). Loss of ground cover led to intensive erosion so that the area now resembles the badlands of the arid west in degree of dissection and desolation.

The consequences of altering the chemistry of the earth's surface and subsurface waters may be even more severe. The dumping of mining, industrial, and domestic wastes, and the indiscriminate use of chemical fertilizers, results in deterioration of water quality, unsightly chemical precipitates, and the growth of algae. Water may be rendered unfit for domestic use and may be unable to sustain aquatic life. Changes in the chemistry of ground water may affect engineering stability of structures by altering the chemistry of clay minerals and affecting the load-bearing capacity of the soil.

Nature has been a long time in establishing global balances in the chemistry of the atmosphere, lithosphere, and hydrosphere. There are, however, natural local imbalances, and people have introduced others. In this chapter, we explore the impact of the chemistry of the environment on land use, and the effects of human activities on the chemical environment (*environmental geochemistry*). We shall start by clarifying the meaning of pollution and contamination.

Pollution and Contamination

Pollution is the fouling of the environment by natural phenomena or human activities. Pollu-

tion need not involve constituents hazardous to health. They may merely be offensive to sight, taste, or smell. Pollution connotes a certain concentration of impurity relative to the intended uses of land, water, or air. Consequently there are conflicting interpretations of the degree of fouling that constitutes pollution.

Contamination is pollution involving constituents that are hazardous to health because of their nature or quantity. Water may be considered polluted if it is contaminated or if it contains substances that are offensive to the senses or detrimental to its usefulness. An example of the latter is the presence of iron, which stains laundry and leaves a scale in hot-water heaters.

Pollution can be controlled by scientific and engineering methods, although complete elimination is economically infeasible. The degree of pollution abatement that will be attained depends on what we are willing to accept and what we are willing to work and pay for. Thus, an individual or small community may accommodate to drinking and household water that tastes poor or is somewhat mineralized, rather than undertake the expense of corrective measures.

NORMAL COMPOSITIONS OF EARTH MATERIALS

To recognize an anomalously high or low concentration of any element in rock, soil, water, or vegetation we need a basis for comparison. The global average compositions of these materials can serve this purpose.

The earth is composed of about 100 chemical elements. Table 10-1 lists some of the common elements and their relative concentrations in the atmosphere, hydrosphere, and lithosphere. The fourth domain, the biosphere, consisting of all living things and the earth zones they inhabit, will be considered later.

The Atmosphere

The atmosphere consists chiefly of nitrogen and oxygen. It includes small but important amounts of water vapor and carbon dioxide and traces of other chemicals and dust. The free

Table 10-1. Average compositions of earth materials. (Based on data from M. Fleischer, J. D. Hem, K. B. Krauskopf, and H. T. Shacklette et al. See Additional Readings

	Earth's Crust	Atmosphere	Igneous Rocks			Sedimentary Rocks			Soils	Waters		
			Granites	Basalts	Ultramafics¹	Shales	Sandstones	Limestones		Rain and Snow	Rivers	Oceans
Major Elements												
	Weight Percent									Parts per million (ppm)		
Aluminum	8.2	Nil	7.7	8.8	0.45	8.0	3.2	0.9	4.5	?	?	0.01
Calcium	4.1	Nil	1.6	6.7	0.7	2.5	2.2	27.2	0.88	?	15.	400
Iron	5.6	Nil	2.7	8.6	9.8	4.7	1.9	0.8	5.6	?	0.67	0.01
Magnesium	2.3	Nil	0.16	4.5	25.9	1.34	0.8	4.5	0.47	0.1	4.1	0.135
Potassium	2.1	Nil	3.3	0.83	0.03	2.3	1.3	0.2	1.2	0.05	2.3	380.
Silicon	28.2	Nil	32.3	24.0	19.0	23.8	35.9	0.003		?	?	3.
Sodium	2.4	Nil	2.8	1.9	0.57	0.66	0.4	0.04	0.4	0.5	6.3	10,500.
Titanium	0.57	Nil	0.23	0.9	0.03	0.45	0.2	0.04	0.25	?	?	0.001
Hydrogen	0.14	Variable	Nil	Nil	Nil	3.4	Nil	Nil				
Oxygen	46.4	20.1 (O_2)	48.7	43.5	42.5	52.8	52.7	54.9				
Selected Minor Elements												
	Weight Percent		Parts per Million (ppm) Except Where Indicated as Parts per Billion (ppb)									
Arsenic	1.8	Nil	1.5	2.	0.5	6.6	1.	0.9	?	?	?	0.003
Boron	10.	Nil	15.	5.	1.	100.	90.	16.	26.	?	0.013	0.0006 ppb
Cadmium	0.2	Nil	0.2	0.2	0.05	0.3	0.02	0.05	.05 to .5	?	1 to > 10 ppb	0.11 ppb
Carbon	200.	0.03 (CO_2)	300.	100.	100.	1000.	14,000	114,000.	?	28.	11.	5.5
Chromium	100.	Nil	4.	200.	2000.	100.	120.	7.1	37.	?	0.1 to 10 ppb	0.05 ppb
Copper	55.	Nil	10.	100.	20.	57.	15.	4.	18.	?	10 ppb	0.003 ppb
Fluorine	625.	Nil	850.	400.	100.	500.	220.	112.	?	?	1.3	1.3
Lead	125.	Nil	20.	5.	0.1	20.	14.	16.	16.	?	<1	0.03 ppb
Manganese	950	Nil	400.	1500.	1500.	850.	392.	842.	340.	?	1 to 10 ppb	0.002
Mercury	0.08	Nil	0.08	0.08	0.01	0.4	0.06	0.05	0.08	?	?	0.03 ppb
Molybdenum	1.5	Nil	2.	1.	0.2	2.	0.5	0.8	2.	?	0.09 ppb	0.01
Nitrogen	20.	78.1 (N_2)	20.	20.	6.	60.	?	?	?	0.3	<10	0.5
Selenium	0.05	Nil	0.05	0.05	0.05	0.6	0.5	0.3	?	?	0.23	0.4 ppb
Sulfur	260.	Nil	270.	250.	100.	220.	945.	4550.	?	1.	3.7	885.
Uranium	2.7	Nil	4.8	0.6	0.003	3.2	1.	2.2	?	?	?	0.003
Zinc	70.	Nil	40.	100.	30.	80.	16.	16.	44.	?	10.	0.01

¹Ultramafic rocks have an even higher concentration of dark iron-bearing minerals than basalt.

oxygen is primarily a byproduct of the metabolism of plants. The carbon dioxide is a byproduct of the metabolism of animals and of natural and industrial combustion.

Many pollutants are released by natural processes, but at low concentration. Any excess is removed naturally from the atmosphere. The chemical balance in the global atmosphere is maintained by balanced generation and withdrawal of components.

The Hydrosphere

All water contains dissolved solids and gases. Rainwater contains oxygen and carbon dioxide, atmospheric dust, and salts dissolved from the dust. It also contains traces of nitrogen oxides produced by lightning discharges. These are important sources of nitrogen for plants. In industralized areas, rainwater is likely to contain dissolved sulfur compounds and nitrogen oxides. These may be so concentrated that the rain is acid and will corrode limestone in buildings and statuary, irritate the eyes, and cause respiratory problems. In areas of heavy traffic, the dust load is large and the air contains lead compounds, hydrocarbons from automotive exhausts, and asbestos from brake linings.

Fresh water in the ground and in lakes and streams contains substances dissolved from the atmosphere and from the rocks and soils with which it is in contact. Therefore, surface water and ground water contain far more extraneous components than rainwater.

Part of the water that is precipitated on the lands evaporates and part returns to the sea, flowing either on or below the surface. The dissolved materials carried into the oceans or into undrained lakes such as the Dead Sea in the Middle East and Great Salt Lake in Utah become concentrated as the water evaporates and may precipitate and mingle with accumulating sediments. The floors of saline lakes and shallow seas thus become sites of accumulation of salts.

Sediments on the ocean floor are saturated with brine. Ancient marine sediments now preserved on land may still be saturated. Such brines are often encountered in oil wells and geothermal stream wells.

The compositions of rain, river, and ocean waters are sufficiently distinctive to permit their differentiation. Average compositions are given in Table 10-1.

The Lithosphere

Average compositions of several common igneous and sedimentary rocks are included in Table 10-1. Metamorphic rocks are not included because they consist of altered igneous and sedimentary rocks.

All known natural chemical elements are present in rocks, but the chemical compounds or minerals in which they occur differ among the major rock groups. Most of the minerals of igneous rocks, for example, are considerably altered during chemical weathering so that the sedimentary rocks which form from this debris differ in composition from the original igneous rock. The minerals of sedimentary rocks may in turn be altered to new minerals when deeply buried and exposed to high temperature, high pressure, and chemical fluids. Some chemical elements may be lost and others gained during these changes.

The loss of specific elements in weathering and soil formation is generally not complete. Thus, a soil derived from a rock with copper-bearing minerals is likely to contain more copper than other soils. Such soils are toxic to vegetation. An average soil composition with typical trace element concentrates is shown in Table 10-1. Soil analysis is useful in prospecting because many soils reflect the composition of the rock below, including the presence of useful elements. Erosion, soil creep, and ground-water flow may spread the traces of the element over an area many times larger than the area of concentration in the parent rock. This makes it easier to detect than the smaller, though more concentrated, occurrence in the parent rock.

Sedimentary rocks can be divided into three geochemical classes. First are rocks such as sandstone, which consist predominantly of mineral fragments originally eroded from other rocks and transported to the site of deposition without major chemical change. Second are the rocks such as shale, which consist predominantly of clays and/or the hydrous compounds of silicon, aluminum, iron, or manganese formed by chemical weathering or interreactions with water. Third, represented by rocks such as limestone, are those whose components precipitate out of water in response to chemical reactions or biologic activity. Average analyses of these three rock types appear in Table 10-1. Shales are especially rich in heavy metals and other environmentally important elements.

Earth Materials as Sources of Contaminants

We are herein concerned mainly with earth materials as sources of contaminants. The compositions in Table 10-1 are average compositions. However, actual compositions at any place or time may deviate widely from these averages. Ore deposits are examples of exceptional concentrations.

Mercury provides an example of the variability that may be found in the lithosphere. The average mercury content of ultramafic rocks (see Table 10-1) is 0.01 ppm and that of sedimentary rocks, 0.05 to 0.4 ppm. Yet, altered rocks formed where the originally molten ultramafic rock invaded sedimentary rocks in California contain 5 to 45 percent mercury by weight. Soils in the vicinity of these rocks contain 10 times the average amount of mercury of soils in general, and if the soils are very close to those rocks, they may contain as much mercury as the rock mined.

Any earth material that contains significantly more than the average amount of an element or compound is a potential source of contamination of soil, water, flora, or fauna. Whether or not it actually does function as a contaminating source depends upon whether or not the anomalously abundant element or mineral is in an *available form*: some of the compounds are so unreactive that they are harmless even when abundant. Similarly, crops grown on soils of normal chemical composition may suffer nutrient deficiency if an essential element is present in a form that cannot be assimilated by plants. Iron-deficiency disorders are common in soils developed on many limestones because the iron is in a nonavailable form.

Availability varies with the geochemical environment. Metallic sulfides, for example, are stable in rocks protected from air or from oxygenated ground water. When exposed to oxygen, as by excavation of a road or quarry, the sulfides react with the oxygen, transforming them into a much more mobile, available state. Thus, a rock of normal composition may become a source of pollutants through human activity. Cases are known in which grazing land has been made toxic for cattle simply by opening a clay quarry. Certain heavy metals present as normal constitutents of the clay, but in a form unavailable to ground water, were converted to available compounds by exposure to oxygen and were then dispersed by surface waters.

Living Organisms and Biogeochemical Cycles

Organisms have the ability to capture and store solar energy and to establish and maintain a number of very important biogeochemical cycles involving both major and minor elements.

Biogeochemical cycles involve oxidation and reduction (addition and extraction of oxygen). For example, the primary foods of plants—carbon dioxide, nitrates, and sulfates—are reduced to carbon, nitrogen, and sulfur and converted to organic matter after they are taken up by the plants. On the other hand, animals obtain energy by oxidation of organic matter, reversing the process by forming carbon dioxide, nitrates, and sulfates. Living cells are able to maintain oxidizing and reducing conditions simultaneously in separate compartments within their own structures and use the energy provided by the interaction.

Interdependence of Atmosphere, Hydrosphere, Lithosphere, and Biosphere

The linkages joining the various domains are the *transport processes*, including precipitation, mass movements, running water, wind, and ice. The means by which material or energy is made available to these transport processes are *release mechanisms*, including weathering and erosion and chemical and biochemical reactions. The mechanisms which remove materials from transport are *retention, or fixation, mechanisms*. These may be physical (deposition), chemical (precipitation), or biochemical (organic deposition).

CHEMICAL IMPACTS ON THE ENVIRONMENT

Chemical impacts on the environment are either natural or induced by humans. For example, sources of trace metals are of either natural origin (chemical weathering of rock, volcanic and hydrothermal activity, natural combustion, etc.) or are manufactured or concentrated by humans (burning fossil fuels, mining operations, industrial uses of trace metals, creating waste dumps, etc.).

Natural Processes

Weathering may release potential contaminants from otherwise passive rocks. An example is oxidation of sulfide minerals, with the generation of sulfuric acid. Decaying vegetation provides the noxious-smelling gas hydrogen sulfide, which may also be converted to sulfuric acid. Volcanic activity commonly releases noxious and lethal gases such as chlorine, fluorine, sulfurous compounds, carbon monoxide, and carbon dioxide. Erosion also liberates potential contaminants from rocks and distributes them widely. Without the natural supply of many of these elements, humanity probably could not have persevered on earth.

The rates of release of potential contaminants by natural processes such as weathering and erosion are very slow from the human standpoint. Volcanic activity, on the other hand, is rapid and may cause immediate pollution.

Human Activities

Normal cycling of materials in nature is such that there is no significant loss. Each living thing borrows its substance briefly, returns much of it immediately to nature, and is itself recycled on death. However, in the process of refining, fashioning, and converting raw materials into a usable form, humans discard a large fraction. Some is reused for other purposes, but inevitably there are waste products. Some waste products are dumped in spoil banks; some are burned, with end products exhausted to the atmosphere; some are buried; and an increasing amount is discharged into the natural water systems and ultimately into the ocean.

In many urbanized regions, and in areas where there has been considerable excavation, strip mining, and road building, the exposed and loose material weathers at an accelerated rate. The exposure of fresh rock to the atmosphere in highly mineralized mining regions accelerates the mobilization of trace metals into natural aquatic systems, providing direct linkage between human activities and natural processes.

Toxic Materials

General Discussion. Certain elements are essential to life; they are *vital nutrients* (Table 10-2). At some concentration, however, each nutrient element can become toxic or even lethal (last two columns, Table 10-2). A substance is said to be *toxic* if it inhibits the growth or metabolism of an organism when present above a certain concentration. All elements are toxic at high concentrations, and some are poisons even at low concentrations. Copper, for example, is highly toxic at relatively low concentrations and is widely used in solutions to kill algae.

Toxicity may be due to an extremely low concentration of a highly toxic substance or to an unusually high concentration of slightly toxic or even normally required materials. The symptomatic response to toxic materials is acute for high-dosage and chronic for low-dosage, long-term exposure.

Cumulative toxins, substances which are retained rather than readily excreted, are especially dangerous and difficult to deal with. Selenium and cadmium are examples. In addition, many substances such as DDT and other synthetic organic compounds, are incredibly stable and harmful in the natural environment. For example, accumulated DDT interferes with the calcium metabolism of brown pelicans and results in fragile, easily damaged egg shells and, hence, a lower birth rate.

Synergistic interactions of toxic substances, that is, interactions which achieve an effect that each substance individually is incapable of, pose problems when dealing with complex natural environments into which synthetic materi-

Table 10-2 Quantities of elements in the human diet. (Data from H. J. M. Bowen, "Trace Elements in Biochemistry," Academic Press, 1966)

Element	Symbol	Man Mean body weight (70 kg) Weight of dry diet (750 g/day) Intake Level Below Which Deficiencies Occur	Normal Intake	Intake Level Above Which Toxicity Occurs	Lethal Intake
		(All Quantities in Milligrams per Day)			
Arsenic	As		0.1–0.3	5–50	100–300
Boron	B		10–20	4000	
Cadmium	Cd		0.5	3	
Chlorine	Cl	70	2400–4000		
Chromium	Cr		0.05	200	3000
Cobalt	Co		0.0002	500	
Copper	Cu		2–5	250–500	
Fluorine	F		0.5	20	2000
Iodine	I	0.015	0.2	1000	
Iron	Fe		12–15		
Lead	Pb		0.3–0.4		10,000
Manganese	Mn		3–9		
Mercury	Hg		0.005–0.02		150–300
Molybdenum	Mo		0.5		
Selenium	Se	0.015	0.03–0.075	3.0	
Silver	Ag		0.06–0.08	60	1300
Sodium	Na	45	1600–2700		

als have been introduced. As an illustration, 8 ppm of dissolved zinc is toxic to trout when zinc is the only heavy metal present, while 0.2 ppm of dissolved copper is toxic to trout when copper acts alone. However, even when the concentration of zinc is only 1 ppm and that of copper 0.03 ppm, the toxicity is much higher when the metals are present together. Each metal increases the toxicity of the other.

The *form of the offending material* is especially important in toxicity. For example, compounds of mercury or lead with hydrocarbons such as tetraethyl lead, used in some gasolines, are considerably more toxic than inorganic mercury or lead compounds such as calomel, a medicinal compound of mercury and chlorine. In contrast, organic copper compounds are less toxic than the inorganic compounds. Comparisons of toxic levels and required nutrient levels for some heavy metals are given in Table 10-2.

A few examples of elements that may become toxic are discussed in the following sections.

Molybdenum. The metal molybdenum is poisonous in large doses. However, like many other metals, trace amounts are essential in animal nutrition and, by inference, in human nutrition. The metal also assists soil microorganisms in fixing nitrogen from the air. Molybdenum is added to fertilizers in many parts of the world to correct for molybdenum deficiency in the soil and water. Molybdenum deficiency detracts from the quality of forage for range animals and the quality of crops such as cantaloupes and cirtus fruit.

Plants do not seem to suffer from concentrations of molybdenum up to 100 ppm. Many plants, particularly legumes such as alfalfa and clover, accumulate high concentrations of the metal when grown in high molybdenum soils or when irrigated with high molybdenum waters. This selective accumulation is used in geochemical exploration for ores. For example, one investigator found plant concentrations of 10 to 300 ppm molybdenum where the soils contained only 1 to 15 ppm. He found further, that the concentration in plants varied directly with concentration in the soil. The areas of highest concentration presumably indicated rocks with high concentrations of molybdenum below the soil.

In animal feed that contains ordinarily adequate amounts of copper, an excess of molybdenum interferes with copper metabolism, and animals may show symptoms of copper deficiency. Livestock that consume forage in which the molybdenum content is greater than 10 to 20 ppm develop molybdenosis, a disease characterized by loss of appetite, diarrhea, loss of sexual interest, abnormalities in the joints, and sometimes death. It can be corrected by supplemental copper.

Figure 10-1 shows the approximate boundaries of regions in the western United States in which the natural molybdenum concentrations are above the average for the earth's crust, and areas in which occurrences of molybdenosis have been reported. Most reported instances of molybdenosis are either within a high-natural-molybdenum area or downstream from a high-molybdenum area. Those occurrences which are farther afield are all characterized by highly organic, wet soils which are capable of entrapping metals by binding them to organic soil particles. The wet soils concentrate the metals

FIGURE 10-1. **Regions of high-molybdenum rocks and of occurrences of molybdenosis in the northwestern United States. The outlined areas are regions with molybdenum-rich rocks. The small, dark areas are those in which molybdenosis has been reported. (*Based on data from R. U. King; J. Kubota and W. H. Allaway; and J. D. Ridge.*)**

brought in by water which itself may have only slightly higher than normal content of molybdenum.

The possibility of molybdenosis must be considered in land-use planning downstream from rocks or soil anomalously high in molybdenum. It must be considered also in using land which is high in organic matter and mucky or swampy.

Fluorine. The element fluorine is essential in small amounts but can be detrimental or even toxic in excessive quantities. Calcium phosphate is the major compound making up the bones and teeth of most animals. When fluorine is present in moderate amounts (up to 1 ppm) in the water supply, bones and teeth are strengthened by the inclusion of fluoride (fluorine-containing compounds) in the calcium phosphate. At concentrations much in excess of 1 ppm, however, the inclusion of excessive fluoride in the tooth material results in a discoloration and in a weakening of the tooth enamel.

Fluorides in surface or ground waters come from natural fluorine-bearing minerals, volcanic emanations, and industrial processes such as phosphate fertilizer production, manufacture of aluminum, brick manufacture, and smelting of fluorine-bearing ores. Where high-temperature processes are involved, the fluorine is released as a compound of fluorine with hydrogen—hydrogen fluoride or hydrofluoric acid.

Hydrogen fluoride is extremely toxic and can have long-term effects because it is taken out of the air, concentrated in plants, and later ingested by animals. Concentrations as low as 1 part per billion (ppb) are damaging to some plant species. A few parts per million in plants indicate that the atmosphere is locally contaminated. Livestock ingesting high-fluorine vegetation develop fluorosis. Fluorosis causes cow's teeth to mottle and wear faster than normal. Their bones become deformed and brittle and break easily. The poisoning of cattle by the emission of fluorine in the smoke from brickworks is well documented.

A graphic example of the effects of natural fluorine concentration was provided by the eruption of the volcano Hekla in Iceland on May 5, 1970. Within the first 2 hours of activity, ash fell over one-fifth of the island. The coarser ash that fell near Hekla contained 100 ppm of fluorine; finer particles of ash, which spread over important grazing areas 200 km (120 mi) or more to the north and northwest, contained up to 2000 ppm fluorine. About 100,000 sheep and an unspecified number of cattle were poisoned from eating grass on which the ash had settled.

Fluorine in surface and ground water usually comes from solution of natural fluoride-bearing minerals or from hydrothermal activity. When a water source containing more than 1 ppm fluorine is to be used for drinking purposes, the concentration of fluorine should be reduced. This is usually accomplished by exposing the water to solid alumina or bone ash.

Mercury. The complex processes controlling the behavior of trace metals in natural aquatic systems can best be understood by examining the behavior of a specific metal, mercury.

An aquatic environment consists of water, the air above it, the sediment beneath it, and the living organisms in all three. Both the air and water invariably contain a variety of particles of such small size that normal turbulence keeps them suspended for a long time. The bottom sediment is also composed of particles, but the sediment is saturated with *interstitial*, or *pore*, *water*.

Materials, including mercury, enter and leave an aquatic system continuously, carried by moving water or air or by mobile animals. For example, a fish contaminated with mercury in one part of the system may swim into another part and die there. Bird droppings might contaminate the system locally. A schematic illustration of such a system, its inputs and outputs and its internal transformations and redistribution processes, is presented as Figure 10-2.

Mercury can exist as solid, liquid, or vapor. Under normal conditions it is a liquid which evaporates and adds mercury vapor to the gaseous environment around it. The solubility of mercury is sufficient to exceed some water quality standards. In combination with other elements, mercury forms a wide variety of solid, liquid, and gaseous compounds. Most of these are also slightly soluble in water. The state (solid, liquid, gaseous, or dissolved) and the chemical forms in which mercury is present

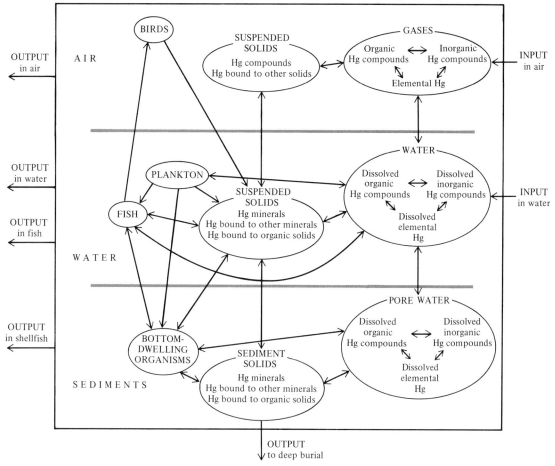

in air

A I R

OUTPUT
in water

OUTPUT
in fish

W A T E R

OUTPUT
in shellfish

S E D I M E N T S

BIRDS

SUSPENDED
SOLIDS
Hg compounds
Hg bound to other solids

GASES
Organic ⟷ Inorganic
Hg compounds   Hg compounds
Elemental Hg

INPUT
in air

WATER
Dissolved
organic ⟷ Dissolved
inorganic
Hg compounds   Hg compounds
Dissolved
elemental
Hg

INPUT
in water

PLANKTON

FISH

SUSPENDED
SOLIDS
Hg minerals
Hg bound to other minerals
Hg bound to organic solids

BOTTOM-
DWELLING
ORGANISMS

SEDIMENT
SOLIDS
Hg minerals
Hg bound to other minerals
Hg bound to organic solids

PORE WATER
Dissolved
organic ⟷ Dissolved
inorganic
Hg compounds   Hg compounds
Dissolved
elemental
Hg

OUTPUT
to deep burial

FIGURE 10-2. Conceptual model for transport and transformations of mercury in an aquatic system. This type of model assumes a dynamic system at steady state. Plankton-aquatic organisms, largely minute, drift or swim weakly. Mercury in water or gases is represented in the right-hand column. Mercury in solid particles is in the central column. Finally, mercury in organisms is represented in the left-hand part of the diagram.

in the system are important. Some states and chemical forms are more readily available to or assimilated by organisms than others, and once assimilated, some are more toxic than others.

Discharge of mercury-containing industrial waste waters is a common cause of anomalously high concentrations of mercury in aquatic systems. Mixing waste water with natural water changes the chemical environment of the mercury. Natural processes, including biological, redistribute the mercury in different chemical forms. Some of the chemical forms, especially elemental mercury and many organic mercury compounds, are volatile and evaporate into the air above the water. All dissolved forms are taken up (or bound to) suspended solids to some degree. The solids include microscopic

plants and animals (plankton). In many systems, the mercury concentration in suspended solids is at least 25 times that of the associated water.

Consumption of plankton by fish, and these fish by still larger fish and animals, transfers mercury up the food chain. Because larger species eat more and live longer than the smaller ones, they accumulate higher concentrations of mercury. Fish of commercial size often contain concentrations of mercury thousands of times greater than that of the water and the suspended solids within it. Commercial fishing removes mercury from the system and transfers it to the human system. Settling of suspended solids and dead organisms transfers mercury from the water to the bottom sediment. Because chemical and physical conditions in the sediment are different from those in the water, chemical redistribution occurs again, and the extent of absorption by solids is likely to change. Solid mercury compounds or minerals may precipitate. Bottom-living animals absorb some of the mercury, especially those that feed by filtering solids from water. Bottom dwellers often contain thousands of times the mercury concentration of the sediment they inhabit. Most commonly, these processes lead to a net transfer of mercury to the bottom sediment where continued sedimentation buries it, hopefully out of reach of chemical communication with biological systems.

All the processes just described are reversible. Mercury bound to suspended solids in the water of a freshwater stream may be released to the dissolved state when the stream discharges into the ocean because of the large change in salinity. Mercury in buried sediment may be remobilized by dredging or natural erosion. Changing conditions can convert a repository to a source. For this reason, most of the arrows representing the various chemical and physical processes in Figure 10-2 are double-ended.

Cadmium. Some years ago, in the Jintsu Valley in Japan, more than 100 people became afflicted with a disease referred to as Itai-Itai. The disease, caused by cadmium poisoning, resulted in brittle bones and extreme pain. The source of the cadmium was rice grown in irrigation waters drawn from drainage that had passed through a zinc-mining area.

Cadmium (Cd) is a rare element; except for shale, its content in the atmosphere, rocks, soils, and waters is 0.2 ppm or less (Table 10-1). Other than a few minor deposits of cadmium sulfide and cadmium carbonate, there are no predominantly cadmium mineral deposits known. However, cadmium is a minor con-

stituent in sulfide ores of other heavy metals such as zinc and lead. It is also concentrated in organic-rich shales. The total natural flow of cadmium to the oceans is 500 tons per year. About 11,000 tons are used annually, primarily in electroplating, paint pigments, plastics, and batteries.

Cadmium is more volatile than most heavy metals, with a boiling point of 790°C (1454°F). Because of this, significant quantities of cadmium are released to the atmosphere during processing of zinc and lead ores, predominately as a gas. The gas is rapidly oxidized and settles as fine particles over the surrounding areas. This is a major source of cadmium to the terrestrial environment. Other important sources include the cadmium in phosphate fertilizers and in the sewage sludge used in landfills.

The major processes by which cadmium is moved into aquatic ecosystems include the natural leaching of soil and rocks and the disposal of domestic and industrial waste waters. The concentration of dissolved cadmium in streams and ground water ranges from 1 to more than 10 ppb. However, the concentration in bottom sediments is 1000 to 10,000 times greater than in the water. Sediments in industrial or mining areas may contain 10 to 20 ppm or more.

Cadmium has no known biological function and is thought to act as a poison primarily by displacing the essential metal zinc in protein and enzyme metabolism. Cadmium poisoning has been suggested as contributing to hypertension, kidney disorders, emphysema, anemia, and the disease, Itai-itai. Acute, single-dose cadmium poisoning is unlikely, but cadmium accumulates slowly over the course of a lifetime, and there is no known mechanism for ridding the body of it.

The principal source of cadmium in humans is tobacco and food. The average diet contains about 0.5 milligrams (mg) of cadmium per day, of which about 5 percent is absorbed through the intestinal wall. A cigarette contains only about 0.001 milligram cadmium, but cadmium is absorbed much more readily in lung tissue than in the intestines. Once cadmium is in the blood stream, it is transported to the kidneys and liver, where about two-thirds of the total body cadmium is retained.

Selenium. Selenium is an element that is required in the diet of animals (see Table 10-2) at a minimum level of about 0.03 mg per day and is beneficial to levels of about 0.075 mg per day. Above 3 mg per day it becomes toxic to most animals, including humans. Selenium deficiency is a more common cause of poor crops than selenium toxicity, and selenium deficiency is more likely in the diet of animals, including humans, than selenium toxicity.

Selenium is found in many metallic sulfides. Its concentration in black shales, coal, and petroleum is 10 to 20 times that of its average crustal abundance (0.05 ppm).

Seleniferous black shales are the parent materials of the widespread seleniferous soils of the western plains in the United States. Unfortunately for cattle, horses, and sheep, there are certain forage plants, collectively known as *locoweed*, which accumulate and store selenium in their tissues. The concentration may run as high as 15,000 ppm. When ingested in these concentrations, selenium affects the motor and sensory capabilities and may result in death. Fortunately, cultivated crops and native grasses do not store enough selenium to threaten animal life.

The annual release of selenium from the combustion of coal and oil in the United States is 3.6 million kg (almost 8 million lb). This is 6 times the 1964 production of selenium in North America and 4 times the world production for the same year. These figures might lead one to suspect that industrialized areas are highly contaminated with selenium. The fact is, however, that 65 percent of the forage crops in the industrial eastern part of the United States contain insufficient selenium for the growth of healthy animals. The reason for this is that the selenium compounds formed during combustion are relatively insoluble. In this case, a potentially toxic material introduced into the environment as a waste product is essentially unavailable due to its chemical form.

Physical Effects of Chemical Changes

Clay minerals contain sodium, potassium, calcium, and magnesium in variable ratios. Unlike most minerals, their chemical compositions change fairly rapidly in response to changes in composition of associated water. In contact with water rich in sodium salts, the ratio of sodium to other constituents will be high. In water that is rich in calcium, the ratio of calcium to other constituents will be high. The physical properties of clay-rich soils reflect these changes.

Sodium-rich clay soils are "heavy," difficult to work, and have poor permeability; they absorb and transmit water very slowly if at all. These are agricultural disadvantages. Conversely, the cohesiveness which makes for poor workability in soils is an advantage in clay-rich formations which must support heavy natural or structural loads.

Calcium-rich clay soils are more likely to have a crumbly texture, are more workable, and have better permeability. They are better able to absorb water and drain off excess water. However, because they have less cohesiveness than their sodium-rich counterparts, they may be more subject to landslides and may fail more easily under natural or artificial loads.

Loss of cohesiveness in the clay which underlies much of Anchorage, Alaska, was the key factor responsible for the severe damage caused by the earthquake of March 27, 1964. The clay in most places is high in sodium because it is of marine origin and the formation water is still saline. However, fresh ground water has locally leached away the readily soluble sodium salts. In these places, the loss of cohesiveness was responsible for the large earth slides triggered by the earthquake. The clay served as a lubricant and sliding surface upon which overlying materials moved. Similar clays of marine or saline-lake origin underlying buildings may lose cohesiveness and load-bearing ability if irrigation water, reinjected waste water, or ground-water recharge is allowed to alter the exchangeable elements of the clays.

ADDITIONAL READINGS

American Chemical Society: "Cleaning Our Environment, The Chemical Basis for Action," Washington, D.C. 1969.

Fleischer, M., ed.: "Data of Geochemistry," 6th ed., U.S. Geological Survey, Professional Paper 440 (in preparation; portions published separately).

Garrels, R. M. F. T. MacKenzie, and C. A. Hunt: "Chemical Cycles and the Global Environment—Assessing Human Influences," William Kaufmann, Inc., Los Altos, Calif., 1975.

Hem, J. D.: "Study and Interpretation of Chemical Characteristics of Natural Water," U.S. Geological Survey, Water Supply Paper 1473, 1970.

Krauskopf, K. B.: "Introduction to Geochemistry," McGraw-Hill Book Company, New York, 1967.

McKee, W. D., ed.: "Environmental Problems in Medicine," Charles C Thomas, Springfield, Ill., 1974.

Odum, E. D.: "Fundamentals of Ecology," 3d ed., W. B. Saunders Company, Philadelphia, 1971.

Schacklette, H. T., J. C. Hamilton, J. G. Boerngen, and J. M. Bowles: "Elemental Compositions of Surficial Materials in the Conterminous United States," U.S. Geological Survey, Professional Paper 574-D, 1971.

management of solid mineral resources and solid wastes

INTRODUCTION

Degradation of the environment accompanies exploitation of mineral resources and the disposition of solid urban wastes. A compromise between these activities and environmental concerns is essential, considering the increasing importance of these activities.

SOLID MINERAL RESOURCES

The magnitude of our resource problem is illustrated in Table 11-1. Column 2 shows the resources consumed in the United States in the first four months of 1974. Column 3 shows total past consumption of resources of the nation since 1776, and column 4 shows the future resource needs during the lifetimes of persons now living. Even assuming no change in population or per capita consumption, the nation will develop and use more resources in the next three generations than in all its previous history.

Thus, three major challenges face the nation—the need to locate and develop vast amounts of mineral and energy resources; the need to protect the environment from the consequences of such production; and

Table 11-1. Past, present, and predicted resource use. (U.S. Geological Survey)

(1) Resource	(2) Amount Consumed in the U.S. between January 1, 1974, and May 4, 1974	(3) Total Past Consumption in the U.S. since 1776	(4) Future Consumption in U.S. during Lifetimes of Persons Now Living
Iron ore (tons)	48 million	6 billion	6 billion
Aluminum ore (tons)	5.25 million	290 million	698 million
Copper ore (tons)	638 thousand	72 million	86 million
Sand and gravel (tons)	313.5 million	30 billion	42 billion
Energy (equivalent barrels of oil)	4.5 billion	400 billion	585 billion
Water (gallons)	44.3 trillion	4.7 quadrillion	4.9 quadrillion

FIGURE 11-1. **Rock structure displayed by mining of coal along trend of inclined beds near Tremont, Pennsylvania. (*J. S. Shelton.*)**

the need to conserve.

Figures 11-1 and 11-2 show environmental damage from strip mining of coal in two geologically different areas. They illustrate the environmental problems that arise in satisfying resource needs and emphasize the need to resolve these problems.

The United States does not produce important quantities of chromium, manganese, tin, asbestos, nickel, and platinum. Its degree of self-sufficiency has shrunk in regard to aluminum, potash, fluorspar, iron, zinc, copper, lead, uranium, natural gas, and petroleum. Figure 11-3 illustrates our dependency on foreign sources for a number of important minerals as of 1972. Some of the import percentages have risen since then, particularly petroleum.

Foreign mineral resources are becoming increasingly difficult and expensive to obtain. The success of the oil-exporting nations in increasing the price of oil fourfold has prompted other mineral-exporting nations to follow suit. Considerable price increases have already resulted, as in the case of bauxite, an aluminum ore. Meanwhile, mineral-importing nations are planning steps to compete more effectively for their mineral needs, and a growing number of countries are nationalizing their mineral resources.

FIGURE 11-2. **Giant furrows formed by power shovel in strip mining in western Kentucky. The thick bed of coal exposed on the bottom of the trench is mined before the next furrow is formed.** (*Division of Forestries, Fisheries, and Wildlife Development, Tennessee Valley Authority.*)

Our dependence on vital resources requires intensified exploration and exploitation within our own political boundaries. This brings us face to face with the environmental problems of resource development and solid-wastes disposal. Solid wastes accumulate at every stage of mining, manufacturing, and use. Such problems are closely related to problems of disposal of municipal, agricultural, and other industrial wastes. Solid wastes create unsightly dumps, pollute streams, and represent wasted resources.

Special Features of Mineral Resources

Agricultural, forest, and water resources are *renewable resources*. When managed properly, the field and forest will produce crops on a continuing basis, and water supplies are renewed constantly by circulation through the hydrologic cycle. In contrast, most mineral resources are *nonrenewable*. When they are

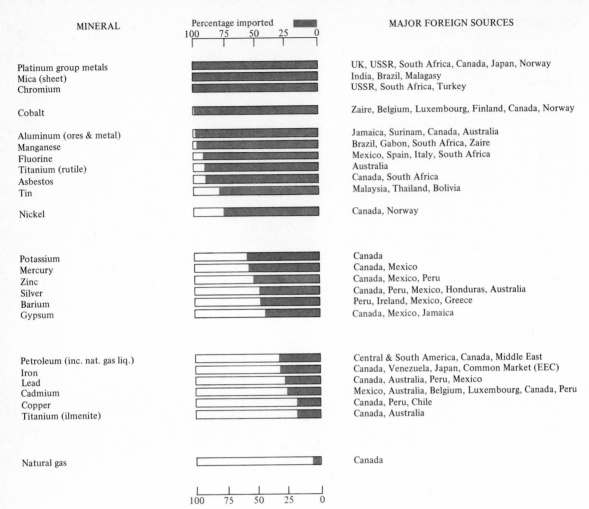

MINERAL | Percentage imported | MAJOR FOREIGN SOURCES
100 75 50 25 0

Platinum group metals — UK, USSR, South Africa, Canada, Japan, Norway
Mica (sheet) — India, Brazil, Malagasy
Chromium — USSR, South Africa, Turkey

Cobalt — Zaire, Belgium, Luxembourg, Finland, Canada, Norway

Aluminum (ores & metal) — Jamaica, Surinam, Canada, Australia
Manganese — Brazil, Gabon, South Africa, Zaire
Fluorine — Mexico, Spain, Italy, South Africa
Titanium (rutile) — Australia
Asbestos — Canada, South Africa
Tin — Malaysia, Thailand, Bolivia

Nickel — Canada, Norway

Potassium — Canada
Mercury — Canada, Mexico
Zinc — Canada, Mexico, Peru
Silver — Canada, Peru, Mexico, Honduras, Australia
Barium — Peru, Ireland, Mexico, Greece
Gypsum — Canada, Mexico, Jamaica

Petroleum (inc. nat. gas liq.) — Central & South America, Canada, Middle East
Iron — Canada, Venezuela, Japan, Common Market (EEC)
Lead — Canada, Australia, Peru, Mexico
Cadmium — Mexico, Australia, Belgium, Luxembourg, Canada, Peru
Copper — Canada, Peru, Chile
Titanium (ilmenite) — Canada, Australia

Natural gas — Canada

100 75 50 25 0

FIGURE 11-3. **Imports that supplied significant percentages of total United States demand in 1972.** (*Report of Secretary of the Interior for 1973.*)

removed from the ground, the deposits are depleted. Of course, when metals are used nonconsumptively, it may be possible to recycle them. About one-third of the iron that is used is recycled by the scrap metal industry. On the other hand, titanium or zinc that is used in paint, a consumptive use, cannot be recycled. Unfortunately, a large proportion of our solid mineral resources are used consumptively or end up in waste dumps.

There are two classes of solid mineral resources. The first has value that is critically dependent upon its location. For example, sand, gravel, and cement rock are of low unit value, and transportation accounts for one-half or more of the costs. Several cities will soon find themselves without available sand and gravel within 120 km (75 mi), a situation that will spread as cities expand. The costs of construction will be greatly increased. The second class of mineral resources has sufficient value to justify local concentration and long distance shipment. Copper, iron, lead, zinc, gold, and silver are members of this group. Actually, most mineral resources are intermediate types, with the ratio of bulk to value determining the economic feasibility of shipping to distant markets.

Political problems of mineral resource development are becoming increasingly critical. In the past, the finder and developer of a deposit had little interference from the government in whose jurisdiction he operated. Now, nations with resources needed by other countries are nationalizing industries, demanding majority ownership, or combining to manipulate prices.

Importance of the Mining Industry

The environment has suffered in many mineral producing and processing areas, evoking strong criticism. However, suggestions to seriously curtail or eliminate mining and mineral processing are unrealistic. Except for agricultural, fishery, and forest products, virtually everything used in the modern world is extracted from the earth. This includes oil and gas, coal, metals, and the many products made from chemicals refined from natural mineral deposits. Our homes and factories could not be built without the sand, gravel, cement rock, building stone, copper, iron, and other products taken from the ground.

A large population requires high-yield agriculture, which is not possible without mineral fertilizers to replenish the soil. A mining industry is necessary to supply adequate power, electricity, machinery, tools, vehicles, hard roads, bridges, fireproof buildings, and railroads—in short, the fuels and materials of modern living.

To curtail our mining industry would be unwise for another reason. Foreign nations, on whom we would have to depend, would not only object to exploiting their depletable mineral reserves for our benefit but demand prices that would be difficult to bear. The area of choice is how much we are willing to pay to make our mining industry less destructive to the natural environment.

Unique Status of the United States

The United States is by far the world's largest producer and consumer of minerals and fuels (Table 11-2). It is also the largest importer. This dependence on foreign resources will increase because we are depleting our mineral resources faster than the rest of the world. Any economies we may accomplish by reduction of per capita requirements or by reclamation of wastes will hardly counterbalance the demands of expected population growth.

The public has displayed little inclination to adopt austerity. It is also doubtful that those groups in our society which have been deprived of material affluence in the past will take kindly to any preachment of self-denial. In short, the standard of living in this country is high and will almost certainly continue so, and no projection for the future can contemplate a shrunken mining industry.

To compound the problems of mineral and energy resources, the desires of underdeveloped nations to achieve greater affluence is changing the world mineral market from one in which oversupplies have been common to one

Table 11-2. Value of minerals and fuels produced and consumed in the United States and in the rest of the world during 1969. Production and consumption are in billions of dollars; per capita consumption is in dollars. (F. C. Kruger, "Our Mineral Heritage—Overindulgence or Self-Denial," Mining Congress Journal, Washington, D.C., 1971)

	Production		Consumption		Per Capita Consumption	
	U.S.A.	Rest of the World	U.S.A.	Rest of the World	U.S.A.	Rest of the World
Fuels	16.6	52.8	18.0	51.4	90	15
Metals	3.4	20.6	6.3	17.7	32	5
Nonmetals	5.4	37.3	6.1	36.6	31	10
Total	25.4	110.7	30.4	105.7	153	30

in which shortages are becoming conspicuous. We shall have to compete more vigorously for imports. Substitutions, recycling, a decrease in planned obsolescence of manufactured products, and more conservative use of products and energy will, it is hoped, ameliorate the basic trend of expanding mineral demand, but they will not reverse it.

Restricted Distribution of Mineral Deposits

Occurrences of economic mineral deposits in the accessible portions of the earth's crust are rare. Most economic deposits are the rare products of relatively obscure concentration processes. They may be due to unusual sedimentary conditions, such as the concentration of placer gold at selected sites along streams or the deposit of minerals in favored places by hot waters from depth. Even oil and gas, derived from fairly common organic-rich source materials, have accumulated only where special geologic "traps" are present.

There are vast quantities of useful elements thinly scattered within the earth's crust. Lacking the necessary concentration by natural processes, however, they are unavailable for human use. Table 11-3 shows the crustal abundance of some important elements and the concentration factors necessary before they

Table 11-3. Crustal abundance of some economically important elements and the concentration factors needed for profitable mining. The present writers believe that concentration factors of 4000 for lead and 8000 for gold would be more realistic. (F. Press and R. Siever, 1974, "Earth," W. H. Freeman and Company. Copyright 1974)

Element	Crustal Abundance (% by weight)	Concentration Factor
Aluminum	8.00	3–4
Iron	5.8	5–10
Copper	0.0058	80–100
Nickel	0.0072	150
Zinc	0.0082	300
Uranium	0.00016	1,200
Lead	0.00010	2,000
Gold	0.0000002	4,000
Mercury	0.000002	100,000

become available for development under present technology. In all cases, the economic concentrations are only tiny fractions of the total contents of the crust, and only very small fractions of these are within minable depths. Considering the limited occurrences of economic deposits of minerals, if we forgo taking them from available sites, we must pay for the choice in terms of lessened resource availability and higher prices.

Environmental Impacts of Mining

Pollution and defacement of the environment result from all aspects of mining activity: exploration, mining, processing, and miscellaneous activities.

Mineral Exploration. Initial exploration requires investigation of the surface geology, often followed by ground and air geophysical surveys and by geochemical sampling of stream beds and soil. The laying out of survey grids requires cutting of lines of sight through forested country and a certain amount of digging in favorable spots. More conspicuous damage results when favorable findings call for trenching with bulldozers or for blasting, drilling, or underground work. Such activities require roads, campsites, excavations, and preparation of drilling sites. Much of the ground displaced during exploration can be replaced, and natural conditions are ordinarily restored in a few years. The costs of reasonable rehabilitation of environments disturbed during exploration will not add materially to the costs of the minerals produced.

There remains the question of the extent to which exploration should be banned in national parks, wilderness areas, and national forests. To curtail prospecting will reduce mineral supplies and raise costs. This must be weighed against the need and popularity of unspoiled primitive areas.

Mine Production. Important disruptions of the environment occur in the *development and production* stages when heavy machinery and large amounts of materials are moved. Many people are employed, many buildings are needed, and if the operations are not near towns, townsites may have to be established. These requirements call for roads and, in many cases, railroads.

If the mine is to be an open pit, preparation for production will involve the stripping of a large volume of overburden which must be deposited nearby. If the mine is underground, shafts or access workings are required. These excavations also call for dumping large volumes of waste rock on the surface. In addition, many mining operations produce substantial quantities of waste rock from ore treatment.

Mineral Processing. Disposal of the solid wastes obtained during *mineral processing* presents unique problems. Mineral processing may consist of simple washing to remove clay or sand, or it may involve crushing, grinding, and separation of valueless rock from the desirable materials. The valueless material may be as little as 10 percent of the mine output or as much as 99.9 percent. Thus, the volume of tailings (wastes) from ore treatment is often large, and this must be dumped or piped into local valleys or into artifically created ponds. If the tailings are fine-grained, they may be a source of dust when dry. If a retaining dam breaks, tailings may wash into nearby streams or lakes.

The floodplain of the Coeur d'Alene River in northern Idaho is covered with mine tailings to a depth of 1 m or more. A dam was constructed on the lower reaches of the river to prevent the tailings from being discharged into scenic Coeur d'Alene Lake. After a mine is abandoned, the tailings dam may become breached. This has happened at an abandoned copper mine adjacent to Lake Chelan, Washington. Tailings are now washing into the lake from the old dump.

Dumps from coal mines sometimes become ignited by spontaneous combustion. Although no flames are produced, polluting smoke is emitted. The fires are extinguished only with great difficulty.

Operators of many underground mines fill their workings with waste rock and with mill tailings. This lessens the possibility of ground subsidence. However, even if the underground workings were completely filled, all the waste material would not be used unless the salable product comprised a substantial part of the

crude ore. This is because breaking, crushing, and grinding expands the volume of solid rock by 30 to 40 percent.

Underground solution of certain ores may eliminate many environmental concerns. The procedure would apply only to soluble ores and in formations where the ground water would not be adversely affected. It is estimated that 30 to 50 percent of the uranium deposits in sandstone formations of the American West would lend themselves to this process. The ore would be dissolved by solutions introduced in some drill holes and recovered in others. Most ores, however, are not readily soluble.

Failure of coal waste embankments is especially serious because of their large number and size. It took a major tragedy to bring this problem into the national limelight. In February 1972, flooding caused by failure of a coal waste dam at Buffalo Creek, West Virginia, claimed 125 lives. Since that time, waste embankments are strictly regulated. Mine inspections now include waste embankment evaluations.

Other Impacts. Besides the dumps, pits, and unsightly structures, environmental degradation includes noise and shaking from blasting; noise, vibrations, and dust from heavy trucks; and traffic congestion.

Serious environmental problems result from the pumping of many mines and the drainage from abandoned mines. Mine waters are frequently acid and iron-bearing. The beds and banks of many streams become coated with sticky brown, iron-bearing precipitates. In dry seasons when stream discharge is low, some of the rivers of West Virginia become highly acid with drainage from abandoned coal mines, and ground waters become unusable without treatment. Government agencies have spent considerable money and time in sealing off old mines from surface and subsurface waters and in water treatment.

Mine waters and rainfall percolating through dumps may carry small amounts of metals such as copper or zinc into surface waters. Copper, in particular, is lethal to fish, but if copper-bearing waters can be collected before draining into streams, the copper can be precipitated at a profit. In a few cases, chemicals from ore-treatment plants accidentally get into the streams. If cyanide, a very poisonous substance, were to escape from some gold and silver processing plants, it could do great harm. Luckily, cyanide oxidizes rather quickly to harmless substances.

Special Impacts. Three special mining techniques, hydraulicking, dredging, and strip mining, have unique environmental impacts. Unique problems also result from the extraction of sand and gravel in urban areas and from abandoned excavations.

Hydraulicking. Hydraulicking is the washing down of banks of sand, gravel, and other loose sediments with large jets of water. The material is run through riffled sluices to recover heavy minerals such as gold, and sand, gravel, and clay are discharged downstream. The sediment may bury valley floors to 1 m or more for many kilometers downstream. Hydraulicking has been outlawed in settled areas, but still persists in remote places.

Dredging. Dredging may be undertaken in lakes or the ocean, along beaches and floodplains, or on terraces where old beaches or stream deposits are preserved. Ponds may have to be created by excavating below the water table. The pond is moved forward with the dredge as it excavates ahead and deposits the tailings behind.

In most floodplain and terrace dredging (Figure 11-4), the dredge works back and forth along the edges of the ponds, dumping the fine wastes in the water and the coarse material in large windrows extending above the surface. The windrows are rough, and the coarse, washed gravel and cobbles are infertile and do not retain water for plant use.

In open water or along beaches, suction dredging is common. A large pipe sucks up large volumes of water and sediment from the bottom. The wastes may be dumped overboard, stored in the dredge hull for transportation elsewhere, or discharged via pontoons and overland pipes to a waste pond.

Strip Mining. Shallow *strip mining* on level or fairly level ground also leaves the surface covered with large windrows of waste material (see

FIGURE 11-4. **Air view of gold dredges in Yuba County, California, 1937. The dredges are indicated by the letter D. (*Photo by U.S. Dept. Agriculture.*)**

Figure 11-2). Large excavating machines work in or beside a long trench, digging overburden from one side of the trench and depositing it on the other. The valuable material, usually coal, is taken from the floor of the trench, and the trench floor is covered by a later windrow. The abandoned terrain is thus a series of parallel ridges of mixed rock and earth.

Contour stripping (Figure 11-5), is practiced in hilly or mountainous terrain. This is almost exclusively a coal mining activity in such areas as the Appalachian Plateau where the beds lie flat. The stripping machinery follows the coal

FIGURE 11-5. **Contour strip mining, Mingo County, West Virginia. The stereo pair may be viewed with a pocket stereoscope to obtain a three-dimensional view. Scale 1:20,700. Note the aprons of waste below the working bench.** (*University of Illinois Committee on Aerial Photography from U.S. Dept. Agriculture photos.*)

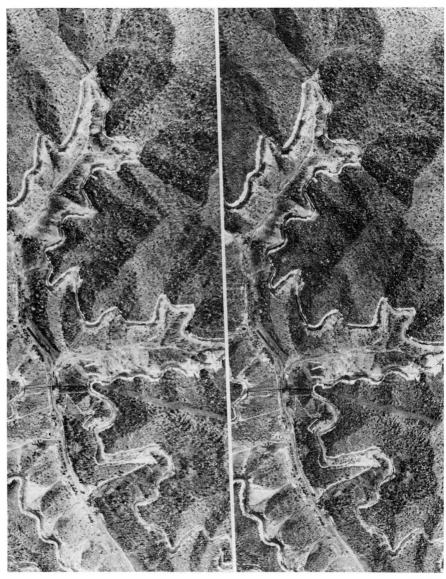

seam along its outcrop, casting the barren overburden dowhhill. The contour strips may extend for kilometers.

Figure 11-1 shows a different type of surface mining. The deep scars result from the mining of coal from dipping formations. Each seam is worked downward from the surface, instead of horizontally.

Other products extracted by surface mining are ores of iron, aluminum, chromium, manganese, tin, gold, and platinum, and bulk products such as sand and gravel, clay, phosphates, and stone. Figure 11-6 shows a deep open-pit copper mine at Bingham Canyon, Utah.

Extraction of iron ore at Hibbing, Minnesota, has left a gulch 160 m (530 ft) deep and more than 5.5 km (3.5 mi) long. Waste has been piled in mounds exceeding 100 m (330 ft) in height.

In Appalachia, contour strip mining has resulted in 32,000 km (20,000 mi) of "high walls" on the flanks of steep slopes. Figure 11-7 shows the acreage affected by surface mining in the United States as of 1965.

For some years, strip miners on flat ground have been practicing some reclamation. This has consisted of partial leveling of the wind-rows, planting legumes to stimulate soil formation, and planting trees. A few operators have converted the trenches or pits into fishing ponds with surrounding park land for public use. Unfortunately, considerable time is required for the development of new soil, especially in arid regions.

In recent years, the public outcry against landscape defacement has confronted the mining industry with the need for more elaborate rehabilitation practices. Legislation has been enacted at federal and state levels to curb the more objectionable aspects of strip mining and to require some degree of reclamation.

In March 1976, the Interior Department made public a final environmental impact statement proposing regulations for the leasing, exploration, mining, and reclamation of federal coal lands. The regulations state that "no new leases will be issued, nor will mining plans for existing leases be approved, unless reclamation of the affected lands to prescribed standards is attainable and assured."

FIGURE 11-6. **Open-pit copper mine, Bingham Canyon, Utah. For scale, note the long ore trains on the benches. (*Courtesy Kennecott Copper Corporation.*)**

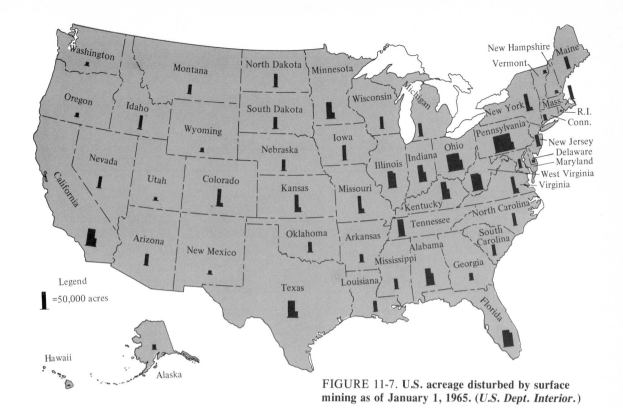

FIGURE 11-7. **U.S. acreage disturbed by surface mining as of January 1, 1965.** (*U.S. Dept. Interior.*)

Mining Gravel from Urban Stream Beds. It is reported that the mining of sand and gravel has become the largest mining business in the United States. Much of the sand and gravel is obtained from convenient stream channels. Sources close to urban areas increase profit margins, because even 32 km (20 mi) of transport doubles the cost of aggregate.

Table 11-4 shows the effects of urban activities on water and sediment discharge in ephemeral (intermittent) streams. The bottom line shows that the mining of stream-bed sand and gravel causes minimal changes in water and sediment discharge.

Abandoned Excavations. Abandoned excavations include prospect pits, shafts, adits (horizontal entries), and water wells. Unguarded excavations of which there are probably 2000 or more in California alone, are hazards to the reckless, the unwary, or the uninitiated. The popularity of "all-terrain vehicles" has increased the physical danger because of the

suddenness with which such excavations are encountered.

In 1949, 3-year-old Kathy Fiscus dropped into a 36-m (120-ft) deep well in San Marino, California, and died there in spite of heroic efforts to rescue her. The entire nation was made aware of the danger of unguarded excavations. Certain provisions of California state law which require property owners to cover, fill, or fence hazardous excavations have come to be known as the "Kathy Fiscus Law." Pits and shafts must now be covered or protected by fences and posted with warning signs. Adits must be sealed or secured with strong doors.

Excavations on privately owned land are the responsibility of present owners, but many abandoned excavations are scattered throughout the vast areas of public and state lands. There are federal and state laws that apply to these hazardous excavations, but they apply generally to excavations abandoned after enactment of the laws.

Table 11-4. Effects of urban activities on water and sediment discharge in ephemeral streams. (W. B. B. Bull and K. M. Scott, "Impact of Mining Gravel from Urban Stream Beds in the Southwestern United States," Geology, Geological Society America, 1974)

Urban Activity	Effect on Stream Discharge					Other Effects
	Water		Sediment			
	Flood Peaks	Base Flow	Suspended Load	Bed Load		
Exposure and disturbance of soil during construction	Increase	Decrease	Increase	Increase		Decreased vegetation and aesthetic values
Construction of impervious surfaces (streets, parking lots, buildings)	Increase	Decrease	Decrease or increase	Decrease or increase		Increased rate of conveyance of water to stream channels; sediment load decreases in areas of impervious surfaces, but increase in flood peaks also increases potential sediment discharge downstream, where flow passes through reaches where abundant sediment is available
Planting of gardens, lawns, and parks	Decrease	Increase	Decrease	Decrease		Increased aesthetic values
Construction of unlined storm drains	Increase	Decrease	Decrease and increase	Decrease or increase		Decreased recharge to ground water; increased rate of conveyance of water to streams; effect on sediment is variable
Lowering of ground-water table	Decrease	Decrease	Minimal change	Minimal change		Decreased vegetation along streams; increased recharge to ground water
Mining of stream-bed sand and gravel	Minimal change	Minimal change	Minimal change	Decrease		Lowering of stream bed; decreased extent of flooding; increased bank erosion; decreased recharge to ground water

Remedial Procedures in Mining. An acceptable degree of *rehabilitation of mined-over areas* is possible, but at considerable cost. The surplus waste, left on the surface, can be smoothed and landscaped to blend with the surroundings. Because of the danger of soil erosion, rapid replanting is necessary. This requires covering the site with soil.

Almost complete elimination of environmental degradation could probably be achieved, but at even greater cost. Active mining sites could be screened, landscaped, and otherwise beautified, noise and dust could be largely eliminated, the use of large explosive charges could be held to a minimum in populous areas, and aqueous discharges could be almost completely purified. The topography and vegetation could be restored after completion of the operation. However, the costs may be hard to justify, especially since most mines are in remote places. A reasonable compromise might require the conversion of old pits to ponds or lakes for public use, the smoothing of old strip mining areas, the planting of crops, timber, or other vegetation, and treatment of waste waters. Hydraulicking is already outlawed, and onshore dredging is unlikely to be important in the future. When dredging is practiced, leveling and resoiling are possible. In fact, with a growing shortage of rock aggregate near cities, the dredged cobbles could be crushed and sold at a profit, and the smaller pebbles used directly in construction and in concrete.

Where contour strip mining is practiced, rehabilitation presents special problems. In the past, most overburden was dumped downhill onto timbered slopes. It would be difficult and expensive to recover this waste material. However, methods have been devised that retain most of the waste from current strip mining on the working bench by backfilling as the excavating machine moves forward. Most of the existing scars can be covered or screened with vegetation.

Environmental Impacts of Mineral Processing and Power Production

Reduction Plants. Many metals, such as copper, lead, and zinc, must be extracted from the ores at *reduction plants*. These metals are commonly in the form of sulfides. After crushing and separation of the ore minerals from the barren minerals, the ore minerals are roasted to convert the sulfides to more easily treated oxides. The roasting releases a considerable amount of sulfur dioxide, a noxious gas that is destructive of vegetation and unhealthy for animals.

The devastation to vegetation from uncontrolled emissions of sulfur dioxide from smelters can still be seen in many places, even though the damage was done decades ago. The quantity of sulfur dioxide produced in zinc and copper smelting is so great that it is collected and made into sulfuric acid where markets are available. Elsewhere, it is dispersed into the atmosphere via tall stacks. This practice is now forbidden, and smelters must collect the sulfur dioxide in practically all areas of the United States. This is a costly process, and operators complain that economically adequate technology is not yet available. They contend that reasonable results can be obtained with tall stacks by monitoring atmospheric conditions and banking their fires when atmospheric inversions prevent the upward dissipation of stack effluents. Sulfur dioxide does not remain in the atmosphere; it forms dilute acid which is removed by rain.

The treatment of iron ore produces less noxious gases because iron ore is low in sulfur. It does, however, create dust, heat, and soot, and emit carbon dioxide.

In some instances, people and animals have suffered lead poisoning in the vicinity of lead smelters because of lapses in the effectiveness of dust- and fume-collecting devices.

Power Plants. Coal-burning *power plants* and plants that produce coke are in much the same category as smelters. Two-thirds of the coal mined east of the Mississippi River is rated as high-sulfur, unsuitable for use under present standards. Electric utilities have been forced to convert to low-sulfur fuel oil or to scramble for low-sulfur coal from Wyoming and Montana, and many Eastern coal mines are facing loss of their markets. Figure 3-3 shows how soil acidity caused by fumes from beehive coke ovens using high-sulfur coal has killed the vegetation and caused gullying on a Pennsylvania hillside.

The sulfur in coal occurs in iron sulfides and in organic compounds. Most of the iron sulfides can be extracted in advance of burning, but organic sulfur cannot be eliminated in this fashion. The power plants that burn high-sulfur coal have to extract the sulfur dioxide from their stacks. The most common method has been to dust quicklime into the stack gases, producing calcium sulfate, but the technique is not yet reliable. In a few years, power plants will be faced with requirements to control nitrogen oxide emissions, a much more difficult problem.

Aluminum and cement. *Aluminum reduction plants* extract aluminum electrolytically from molten alumina, an aluminum oxide. A fluorine-bearing mineral, cryolite, or a synthetic equivalent, is used to promote fusion. Some decomposition occurs during electrolysis, and emanations of fluorine are given off. Fluorine in excessive quantities is harmful to people and animals living in the vicinity, and it may poison milk if cattle graze nearby. As a preventive, careful collection and treatment of furnace emissions are necessary.

Cement plants create substantial quantities of dust. The dust is not toxic, but it can be a nuisance. Various means of collection are employed, but because of the expense, many cement plants have depended on tall stacks to disperse the dust.

Slag and Ash Disposal. Smelters also have solid wastes which they must discard. These are typically *slag*, the glassy rocklike material formed from the waste substances in the furnaces. The volume of slag is likely to be at least equivalent to the volume of marketable products.

Slag may be used as a soil conditioner, as raw material for rock wool, or as building blocks, but most smelters are too remote from settled areas to market these products economically. Even in settled areas, the market may not be able to absorb the quantities of slag produced. Thus, most smelters have simply dumped their slag, or the unmarketable portions, locally. Slag dumps can be aesthetically objectionable, and even though many slags

contain plant nutrients, these food materials are not available for many years until weathering produces soil.

Coal-burning power plants produce large volumes of *ash*. Because these plants are largely in urbanized areas close to markets, a number of uses and by-products have developed. Of special concern is fly ash, the fine materials carried out of the furnace in the draft. Current research is appraising the ground-water-pollution potential of such materials when used as surface fill.

The wastes from nuclear reactors are discussed in Chapter 12.

Remedial Procedures. The environmental impacts of mineral processing and power production are now viewed critically by an environmentally sensitive public. As in mining, the public must decide what degree of control it is willing to pay for in higher prices. Poisonous effluents, such as lead and fluorine, can and should be removed from emissions; 95, even 98 percent of the sulfur and dust can be eliminated, but at a very high price. A practical solution might be elimination of 85 to 90 percent of the pollution. Reduction to these levels can probably be accomplished without raising treatment costs excessively and would maintain a healthy environment.

General Conclusions

Impending Conflicts. The Alaska pipeline dispute provided an example of what may happen when environmental concerns and basic needs conflict. The energy crisis might not have been quite as severe had Alaskan oil been on its way a few years ago. On the other hand, there is no way of determining the potential environmental damage that may have been averted.

Controversies also impend over the mining and utilization of Western coal, the allocation of water rights attendant thereon, the regulation of smelter and power plant effluents, the withdrawal of public lands from mineral exploitation, and the development of an oil shale industry.

The sand, gravel, and crushed stone industry faces considerable future difficulty. The products are widely used, particularly in urban areas, and almost universally are from local sources. They sell for low prices, and transportation costs account for about one-half of what the consumer pays. Local sources of these common building materials are disappearing under urban growth, and producers are finding it impossible to develop new sources from convenient sites. Many sites have been condemned, and it is probable that over the next 10 years, prices in constant dollars will double. Few communities have been farsighted enough to protect their sources of common building materials by special zoning.

International Ramifications. The United States, with 6 percent of the world population, consumes one-third of the world mineral production, including fuels (see Table 11-2). This one-third had a value of $36 billion in 1972, of which $32 billion, or 30 percent of the world total, is produced domestically. To import an increasing part of domestic requirements from the reserves of foreign nations will raise mineral prices, apart from attempts by exporters to charge what the traffic will bear. General price increases resulting from our disproportionate demands will cause resentment among other importing nations. Also, excessive imports by the United States will convince the exporting nations that we are exploiting their resources to avoid environmental problems of our own.

Decreased self-sufficiency will have an adverse impact on our national security and balance of payments. The present energy crisis, attributable in part to our dependency on Middle East oil and the vulnerability of this source of supply, provides an example of what could occur with other mineral commodities, including iron, copper, aluminum, zinc, nickel, tungsten, tin, asbestos, mercury, fluorspar, and potash. The recent quadrupling of the price of Jamaican bauxite may be symptomatic of future events.

The Broader Problems and Long-Term Perspectives. Despite the great importance of the domestic mineral industry, including fuels, it accounts for only 3 percent of the total national product. One might argue, therefore, that doubling or tripling the cost of minerals in order to enjoy an almost completely clean environment

would be no great handicap. A justification for maintaining a less clean but tolerable environment at a much smaller cost is based on the argument that the relative cost of minerals in constant dollars will probably double in the next few decades anyway as a result of depletion. Also, the adverse environmental impacts of the mineral industry constitute only a small fraction of the total environmental impacts from all sources.

Because of domestic and worldwide depletion, many mineral commodities which we view as essential to our economy and national security will gradually become too expensive, and substitutes will have to be found. Substitution will continue until we become primarily dependent on a group of elements that are so abundant that they can never be depleted, even though we may have to go to the ocean to obtain many of them. Luckily for the future of civilization, there are enough such elements to support a progressive, reasonably comfortable, cultured society. The elements include oxygen, hydrogen, nitrogen, carbon, sulfur, chlorine, calcium, iron, aluminum, magnesium, manganese, nickel, sodium, phosphorus, potassium, silicon, cobalt, copper, titanium, bromine, and iodine. As we converge toward dependence on these elements, there will be a return toward national self-sufficiency. On the other hand, extracting many of these elements will be costly in terms of human effort and power consumption. Thus, to move in this direction is logical only when the world in general is in a similar resource situation and is similarly compelled. To do this unilaterally would destroy our competitive status because of the high cost of both the minerals and the products manufactured from them.

The power aspect of future technology deserves special emphasis. We can continue to enlarge our mineral reserves by mining lower grade materials and deep-sea nodules, and by higher rates of mineral extraction from ores, but the energy requirements will increase significantly. Our descendants will presumably be able to obtain iron, aluminum, and potassium from common rocks and extract many more elements from sea water than is now practicable, but again the energy required will be vastly increased. Thus, *the key to the future is a continued and growing abundance of energy at a digestible cost.*

SOLID WASTES AND SANITARY LANDFILL

We have seen that the most serious environmental problems of the mineral resources industries relate to waste disposal. Industrial, municipal, and agricultural solid wastes present many of the same problems. Furthermore, they represent wasted resources. If they could be reclaimed, either by more efficient use, by recycling before disposal, or by mining or leaching of disposal sites, the demands on the mineral resources industries would be lessened.

Magnitude of the Solid-Wastes Problem

Table 11-5 shows the main categories of *solid wastes*. The relative quantities of the different kinds vary from area to area. In 1967, *municipal* solid wastes in the San Francisco Bay area amounted to 2.6 kg (5.7 lb) per capita per day, *industrial* solid wastes to 0.7 kg (1.6 lb) per capita per day, and *agricultural* solid wastes to 2.8 kg (6.1 lb) per capita per day. Each individual daily generated 6.1 kg (13.4 lb) of solid wastes. This added up to a staggering annual total of 12.5 million metric tons (13.8 million short tons) for the population of the San Francisco region.

Methods of Solid-Wastes Disposal

Practices available for disposal of solid wastes include landfills, recycling, composting, incineration, and marine disposal. Ninety percent of the solid wastes collected in the United States are disposed of in some type of landfill.

The most common type of landfill is the open dump. The wastes are deposited with little regard for pollution control or aesthetics. They may be left untouched or they may be burned. The open dump is a source of pollution, health hazards, and environmental degradation. It is undesirable and should be avoided. *Sanitary landfill* is a preferred method of solid-wastes disposal.

Sanitary Landfill

Definition. Sanitary landfill is a method of disposing of refuse on land by confining it to the smallest practical area and covering it with a layer of earth each day. The compacting and earth moving are done by bulldozers or related heavy equipment. Placement of solid wastes may be preceded by shredding or compaction and bailing. Alternatively, it may be preceded by incineration, which reduces solid-waste volumes substantially and extends the life of the landfill.

Advantages. Figure 11-8 shows a well-managed sanitary landfill. Homes were intentionally built overlooking the landfill which was to become a park and golf course after filling was completed.

Table 11-5. The main categories of solid wastes. (Modified from J. Cornelius and L. A. Burch. "Solid Wastes and Water Quality: A Study of Solid Wastes Disposal and their Effect on Water Quality in the San Francisco Bay–Delta Area," California Dept. Public Health, 1968)

I. Municipal
  1. Residential
    a. Household garbage and rubbish
    b. Lawn clippings and prunings
  2. Commercial
    a. Refuse from stores, markets, offices, etc.
    b. Refuse from schools, hospitals, and other institutions
  3. Special
    a. Street refuse (sweepings, leaves, tree trimmings)
    b. Demolition and construction wastes
    c. Sewage treatment residue (sludge and screenings)
    d. Dead animals (dogs, cats, etc.)
II. Industrial
  1. Mining and mineral-processing wastes
  2. Manufacturing-process wastes
  3. Cannery wastes
  4. Petroleum and chemical sludges
  5. Miscellaneous food processing wastes (meat packing, beverage, etc.)
III. Agricultural
  1. Livestock manure
  2. Fruit and nut crop wastes
  3. Field and row crop wastes

FIGURE 11-8. A well-managed sanitary landfill that was completed as a golf course. (*T. J. Sorg and H. L. Hickman, Jr., "Sanitary Landfill Facts."*)

One of the obvious advantages is that a marginal site can be upgraded and converted to a beneficial use if planned properly.

Management Methods. Sanitary landfilling involves spreading, compacting, and covering the fill. Two general methods have evolved, the area method and the trench method (Figure 11-9 and 11-10).

In an *area sanitary landfill*, the solid wastes are placed on the land, a bulldozer spreads and compacts it, the wastes are covered with a layer of earth, and finally the earth cover is compacted. The area method is used wherever suitable depressions exist. Normally the earth-cover material is hauled in or obtained from adjacent areas.

In a *trench sanitary landfill*, a trench is cut in the ground and the solid wastes are placed in it. The solid wastes are spread in thin layers, compacted, and covered with the earth excavated from the trench. The trench method is best suited where the water table is not near the ground surface.

Pollution Potential. Sanitary landfills can be causes of solid pollution, liquid pollution, gas pollution, biological pollution, and visual pollution.

Solid pollution results largely from materials blowing out of the site. This may be controlled by wind fences, as in Figure 11-9. A landfill located in a drainage way may be completely washed out during an intense storm and the debris deposited in nearby surface waters.

Liquid pollution may result when water (generally from rainfall) filters through the solid-waste material, dissolves noxious substances, and carries these into ground water or surface water. Such water with materials dissolved from the wastes is known as *leachate*.

Gas pollution results from the generation of gases such as methane (marsh gas), which constitutes a fire hazard, and carbon dioxide, which moves downward to cause minor ground-water pollution.

Biological pollution from disease-carrying rodents and insects is minimal in properly managed sanitary landfills. The technique of spreading, compacting, and covering eliminates holes as refuges for rodents and eliminates water-logged depressions which serve as breeding grounds for insects.

Visual pollution stems from open dumps and improperly managed sanitary landfills,

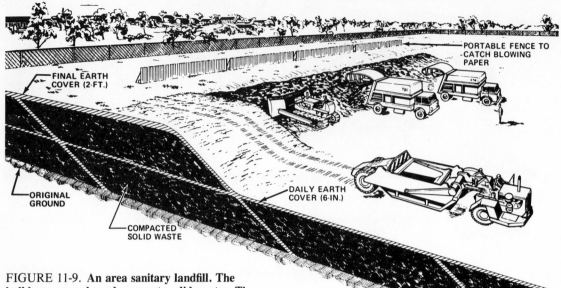

FIGURE 11-9. **An area sanitary landfill. The bulldozer spreads and compacts solid wastes. The scraper spreads the cover material at the end of the day. The portable fence catches blowing debris.** (*T. J. Sorg and H. L. Hickman, Jr., "Sanitary Landfill Facts."*)

which can be blights on the landscape. However, a well-managed sanitary landfill may be quite acceptable, especially in anticipation of an aesthetically pleasing end product such as a park or garden.

Sanitary-Landfill Site Selection and Environmental Problems

Sanitary-landfill sites must be selected with great care if environmental problems are to be minimized. Some constraints are operational; others are ecological; and still others are topographic, geologic, or hydrologic.

Operational Considerations. The first consideration is the availability of a sufficiently large area to accept the desired volume of wastes over the projected life of the landfill. Next, zoning regulations must permit the project, and the completed landfill must conform with city or regional plans. Finally, the landfill must be accessible to trucks during all weather conditions, and the haul distance must be within economic limits.

Ecological Considerations. Many landfills have been used to convert marshland and other so-called marginal lands into developable real estate. Now, many marginal lands are recognized as valuable botanical and faunal preserves. Therefore, the ecological consequences of all landfill operations must be appraised.

Topographic, Geologic, and Hydrologic Considerations. Topography determines the operational method to be used. Gullies and drainage ways are susceptible to erosion and flooding, and the configuration of the ground determines the length of the path of leachate flow before reaching ground water or surface water.

The availability and type of cover material is important. A sandy-silty material that is workable but fairly impervious to rainfall when compacted is most desirable. Table 11-6 shows the suitability of different soil types as cover materials.

Sanitary landfills should be located above the water table. In humid environments where a leachate is expectable, the subsurface hydrology must be appraised. If it appears that the leachate may reach a stream or aquifer, the effects of pollution on the water quality of the stream or aquifer must be evaluated. The volume of leachate may be reduced by diverting

DAILY EARTH COVER (6-IN.)

EARTH COVER OBTAINED
BY EXCAVATION
IN TRENCH

ORIGINAL
GROUND

COMPACTED
SOLID WASTE

FIGURE 11-10. A trench sanitary landfill. The bulldozer spreads and compacts the wastes. At the end of the day, a dragline removes soil from the future trench for use as the cover material. (*T. J. Sorg and H. L. Hickman, Jr., "Sanitary Landfill Facts."*)

surface runoff from the landfill by use of impermeable cover materials and careful grading. Establishing a cover of transpiring vegetation also reduces the volume of leachate.

The most desirable sites are those safe from erosion, underlain by non-water-bearing rocks, and distant from bodies of surface water.

A Water-Quality Classification of Solid-Wastes Disposal Sites

The California Department of Water Resources classifies solid-wastes disposal sites on the basis of potential for impairing the quality of surface and ground water. The classification is based on the fact that the physical characteristics of a disposal site control the type of wastes that can be safely disposed of without resulting in water pollution. The system considers the geology, hydrology, and topography of each site in relation to pollution potential.

Table 11-6. Suitability of different soil types as cover materials for sanitary landfill. The ratings for rodent control are based in part on the ease with which rodents can burrow into the soil and construct nests. The ratings for inhibiting filtration, minimizing or allowing gas venting, and growing vegetation depend in part on the permeability, which is the capacity of the soil to transmit water and gases. The appearance of a completed landfill is less dependent on the soil type than on the completion techniques and plantings. (Modified from D. R. Brunner and D. J. Keller, "Sanitary Landfill Design and Operation," U.S. Environmental Protection Agency, 1971)

Function	Sandy Gravels	Silty Gravels	Gravelly Sands	Silty Sands Clayey Sands	Inorganic Silts	Inorganic Clays
Rodent control	G	F–G	G	P	P	P
Inhibit moisture filtration	P	F–G	P	G–E	G–E	E
Minimize gas venting	P	F–G	P	G–E	G–E	E
Appearance	E	E	E	E	E	E
Grow vegetation	P	G	P–F	E	G–E	F–G
Allow gas venting	E	P	G	P	P	P

E = Excellent; G = Good; F = Fair; P = Poor.

Three general classes of disposal sites have been established:

Class 1 Disposal Sites are located on non-water-bearing rocks or on rocks containing isolated bodies of unusable ground water. Sites are located more than 150 m (500 ft) from adjacent surface water, and facilities are provided to divert streams around the site. Except for radioactive materials, there is no limitation on the type of material, liquid or solid, that may be dumped.

Class 2 Disposal Sites may be underlain by usable, confined, or free ground water where the lowest elevation of the disposal site is at least 2 ft above the highest anticipated ground-water level. The exact distance separating the bottom of the fill and the highest ground-water level is determined by drilling and evaluated on a case-by-case basis.

Adjacent surface water must be excluded from the site as in the Class 1 Disposal Site and discharge to surface water is prohibited.

The materials that can be disposed of at these sites include decomposable organic wastes as well as other solid wastes. Specific examples include garbage, dead animals, human fecal matter, plant wastes from agriculture, lumber and wood, abandoned vehicles, miscellaneous metals, and paint sludge from which the major portion of liquids has been removed.

Class 3 Disposal Sites afford little or no protection to underlying or adjacent waters. Debris control must be practiced. Only non-water-soluble, nondecomposable inert solids may be disposed of at the site. Examples are natural earth, rock, sand and gravel, paving fragments, concrete, glass, and inert demolition and construction materials.

Any material that can be disposed of in a Class 2 site can also be disposed of in a Class 1 site. Similarly, anything that can be disposed of in a Class 3 site is acceptable in either a Class 1 or Class 2 site.

URBAN ORE

We have discussed the worsening resources situation and the environmental problems resulting from mining, mineral processing, and waste disposal. For such reasons, F. F. Davis (see Additional Readings) suggests that society must readjust its thinking about solid wastes and treat them as the resource they represent, *urban ore*. The metals and minerals in the wastes should be viewed as renewable resources and returned to the economic cycle.

Mineral-recovery techniques can concentrate such materials as ferrous metals, nonferrous metals, and glass. Some combustibles may be converted to oil or used directly as a fuel. To accomplish these objectives, an integrated recovery system should be developed, incorporating the best features of many processes. Tax incentives or subsidies might be granted during the start-up period to encourage the development of expensive recovery systems.

Besides being a mineral conservation measure, development and use of a comprehensive urban ore beneficiation system would have significant environmental benefits. Some major landfill operations could be eliminated along with associated problems such as long-term subsidence, generation of leachate which may reach ground water, generation of explosive or noxious fumes, and underground fires.

The above suggestions probably cannot be implemented under current economic constraints. However, if present trends and shortages of resources continue, they can become economically feasible.

ADDITIONAL READINGS

Brunner, D. R., and Keller, D. J.: "Sanitary Landfill Design and Operation," U.S. Environmental Protection Agency, Report SW-65 TS, 1971.

Cornelius, J., and L. A. Burch: "Solid Wastes and Water Quality: A Study of Solid Wastes Disposal and Their Effect on Water Quality in the San Francisco Bay–Delta Area," California Department of Public Health, Bureau of Vector Control, 1968.

Davis, F. F.: "Urban Ore—A New Resource Opportunity," *California Geology*, California Division of Mines and Geology, vol. 25, no. 5, May 1972. pp. 99-112.

Frasche, D.: Mineral Resources, National Academy of Science, Publ. 1000-C, Washington, D.C., 1962.

Given, I., et al.: "Mining Engineers Handbook," American Institute of Mining, Metallurgical and Petroleum Engineers, New York, 1973.

Just, E.: "Minerals for a Teeming World," Mineral Information Service, California Division of Mines and Geology, September, 1964.

National Commission on Materials Policy: "Mineral Needs and the Environment, Today and Tomorrow," U.S. Government Printing Office, Washington, D.C., 1973.

Park, C. F.: "Affluence in Jeopardy," Freeman, Cooper, and Company, San Francisco, 1968.

Pfleider, E., et al.: "Surface mining," American Institute of Mining, Metallurgical, and Petroleum Engineers, New York, 1968.

Sorg, T. J., and H. L. Hickman, Jr.: "Sanitary Landfill Facts," Bureau of Solid Waste Management, U.S. Department of Health, Education, and Welfare, Rep. SW-4ts, 1968.

Williams, R. E.: "Waste Disposal in the Mining, Milling and Metallurgical Industries," Miller Freeman Publications, San Francisco, 1975.

CHAPTER TWELVE

special environmental problems of the energy industries

INTRODUCTION

Costs of gasoline, heating fuels, fuels for power generation, and fuels for the myriad products of the petrochemical industry are steadily rising. Ironically there is no *global* shortage of energy *at this time*, even of petroleum. Petroleum is currently in short supply in the United States because of diminishing domestic reserves, international relations, wasteful usage, federal and state practices that discourage exploration, and delays occasioned by environmental concerns.

It may be decades before some of the alternate sources of energy can contribute significantly to national power supplies. These alternate sources include solar energy, the wind, the tides, ocean currents, and geothermal power. Development of alternate power sources should be complemented by conservation. Unless, however, we are to lower living standards significantly, conservation by itself cannot solve the problem. Energy is indispensable in converting inert resources into basic needs: food, clothing, and shelter. Where manual labor is the principal source of energy, living standards are low. There is a close correlation between national per capita energy consumed and national per capita income.

It is unlikely that the citizens of the more developed nations will consent to a significant lowering of their living standards. To give up the use of time-saving home appliances such as washers and dryers, refrigerators, and smaller kitchen devices, would mean a return to drudgery for the homemaker. The abandonment of labor-saving devices in industry would mean a large-scale return to manual labor, with increased costs for everything now produced cheaply by machines.

Restrictions in the use of energy that would lead to a serious lowering of living standards may not be necessary. Provided that population can be controlled, energy supplies from coal, oil, gas, oil shale, tar sands, uranium and other radioactive minerals, geothermal energy, solar radiation, the energy of the winds, the tides, and ocean currents, and the thermal energy and hydrogen in ocean waters could eventually fulfill all needs. The use of some of these resources, however, could have considerable environmental impact.

As a prelude to discussion of energy resources, let us consider a few elementary aspects of energy and power, and how energy relates to human welfare.

ENERGY AND PEOPLE

Energy is the ability to do work. The rate at which we do work is *power*. Two people carrying equal burdens up a hill expend the same amount of energy even though one does the job in a minute and the other in an hour. The faster-moving person, however, is using body energy at a more rapid rate and may feel "drained" of energy at the summit. This person has been operating at a higher power level.

Mechanical power is commonly measured in *horsepower*, the approximate rate at which a horse can work. On an average, a draft horse can lift a weight of 550 lb a distance of 1 ft in 1 second. A man is rated at 1/20th of a horsepower. Another power unit is the *watt*, a measure of electrical power. One horsepower is equivalent to 746 watts; hence a man rates 37 watts. However, such simple expressions of energy equivalents in the conversion of one form of energy to another (mechanical, electrical, thermal) are unrealistic in that they neglect heat losses. Thus, the efficiency in conversion of electrical to mechanical energy exceeds 90 per-

cent, whereas the efficiency in conversion of mechanical to electrical energy is generally less than 50 percent.

To illustrate the advantage of electrical power over manpower, consider that a medium-size color television set requires 250 watts of power. In terms of manual labor, it would appear that the energy production of seven men would be required to run the television set in the same way that a bicycle rider, if pedaling fast enough, can keep the bicycle's headlamp burning. Actually, because conversion of mechanical to electrical energy in this case is only 50 percent efficient, it would require 14 men to perform the menial task of keeping the television set in operation. It is legitimate to ask, of course, whether all our energy-consuming devices are essential. Let us examine first what the use of energy has accomplished, and why many are willing to accept development of energy industries in spite of some environmental degradation.

One factor limiting population growth until modern times has been the limited energy supply, mainly human and animal labor. When fossil fuels began to be employed with work-producing machines, people were freed from the bondage of subsistence agriculture. It was the combination of abundant energy, new machines, skilled labor, improved transportation, and adequate capital that accounted for the industrial and social changes of the last century.

In recent centuries, world population has increased 2 percent per year, while the global consumption of energy has increased 4 percent annually. The increased availability of energy has resulted in a steady increase in the standard of living throughout much of the world. The increase has not been uniform, however. With 6 percent of the world's population, the United States consumes 33 percent of the world's energy production.

The correspondence between per capita income and energy usage in 1968 is shown in Figure 12-1 for 16 nations. Energy is expressed in *Btu*'s, or *British thermal units*, the amount of heat required to raise the temperature of 1 lb of water 1°F. The Btu values may be roughly converted to equivalent barrels of oil by dividing by 5.8 million. Thus, the *annual per capita*

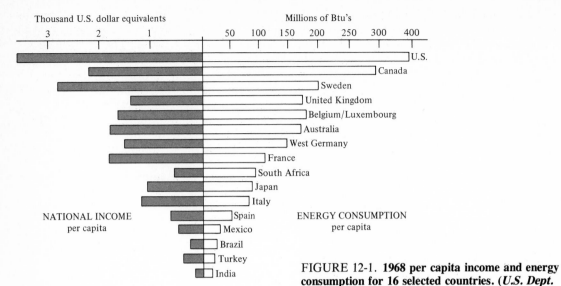

Thousand U.S. dollar equivalents   Millions of Btu's

NATIONAL INCOME per capita   ENERGY CONSUMPTION per capita

FIGURE 12-1. **1968 per capita income and energy consumption for 16 selected countries. (*U.S. Dept. Interior, "United States Energy: A Summary Review," 1972.*)**

consumption of 400 million Btu's in the United States is equivalent to about 70 barrels of oil. Much of this energy, however, is supplied by coal, gas, and hydroelectric power. The diagram reveals a strong correlation between energy consumption and income. Canada's greater expenditure of energy relative to income is due to its great distances, its climate, and its many heavy industries, including mining. Sweden, in contrast, uses less energy because it is small and compact, depends more on light industry and services, and emphasizes conservation.

The correlation between energy consumption and income has been upset by rapid increases in the price of crude oil made by the OPEC (Oil Producing and Exporting Countries) in the 1970s. These countries possess large incomes but are presently low energy consumers.

The shift from agrarian to industrial economies in developing nations requires increases in energy. During the current decade the estimated rate of increase in the United States averages about 4 percent, while that of the rest of the free world is estimated at 6 percent. Clearly, the developing countries will compete strongly for available supplies of energy in the future.

A brief review of world and United States trends in population, gross national product, and production and consumption of energy

resources will provide fuller appreciation of the problems created by rising energy demands.

WORLD TRENDS

Population growth rates of many of the underdeveloped nations exceed that of the United States and will accelerate to the year 2000. Projections indicate a world population in the year 2000 of 6.4 billion compared to 4 billion in 1976. This means an average increase of 100 million people per year.

Projections of world gross national product indicate an even more spectacular rate of increase, from under $4 trillion U.S. in 1971 to $11.5 trillion in the year 2000. Unlike the projections for population growth, the rate of increase of gross national product will be greater in the more developed nations. Total world energy requirements in the year 2000 may be more than 3 times those of 1970.

TRENDS IN THE UNITED STATES

Increase in gross national product has far outpaced population growth in the United States.

This has paralleled a rapid increase in services and in the quality of life. The value of the physical structure materials[1] produced is about the same as the value of the energy produced. Together, these add up to only about half of the gross national product. The remainder is devoted to services and to nonstructure materials such as agricultural and fishery foods.

Our primary energy sources consist of petroleum, natural gas, and coal, with smaller contributions from running water and atomic fuels. Petroleum and natural gas, bolstered by imports, supplied more than 75 percent of domestic energy requirements in 1972.

Table 12-1 presents an energy balance for the United States for the years 1970–1985. During this period, energy consumption is expected to grow 4 percent per year. The largest growth is projected for the electric utility sector. The imports needed to make up deficiencies are shown in the bottom two rows.

Energy consumption exceeded energy production from domestic sources in the United States for the first time in the late 1940s. In 1970, domestic production satisfied only 88 percent of total energy consumption. The gap has been steadily widening and is now a matter of national concern.

---

[1]Physical structure materials are raw materials that provide the structure of the things we make and use. They include minerals; forest products; agricultural nonfoods such as cotton and tobacco; fishery nonfoods, such as whale oil and seal fur; and wildlife products such as fur and ivory.

Table 12-1. U.S. energy balance, 1970–1985. (National Petroleum Council, "U.S. Energy Outlook: An Interim Report," November, 1971)[1]

Energy Source	1970	Projected		
		1975	1980	1985
Projected Domestic Supply				
Oil (millions of barrels per day)				
Conventional	11.3	11.1	11.8	11.1
Synthetic (from shale)	—	—	—	0.1
Total	11.3	11.1	11.8	11.2
Gas (trillions of cubic feet per year)				
Conventional	21.82	19.80	17.47	14.50
Synthetic	—	0.37	0.55	0.91
Total	21.82	20.17	18.02	15.41
Coal (millions of short tons per year)				
For domestic use	519	651	799	933
For export	71	92	111	138
Total	590	743	910	1071
Other (billions of kilowatt-hours per year)				
Hydroelectric	249	271	296	316
Nuclear	23	326	926	2067
Geothermal	0.7	12	34	51
Total	272.7	609	1256	2434
Imports Required to Balance				
Oil (millions of barrels per day)	3.4	7.3	10.7	14.8
Gas (trillions of cubic feet per year)	0.92	1.55	3.75	6.08

[1]The projections should not be taken too literally. For example, the projections for 1975 for oil, gas, and nuclear energy proved to be too high (Energy Perspective 2, U.S. Department of the Interior, June 1976.)

Petroleum is the principal source of energy in the United States. According to Table 12-1, domestic supply will remain reasonably constant at 11 million barrels per day to 1985. On the other hand, imports of crude oil will grow from 3.4 million barrels per day in 1970 to 14.8 million barrels per day by 1985.

Domestic production of natural gas is expected to decline from 22 trillion cu ft per year in 1970 to less than 15 trillion cu ft per year in 1985. Imports needed to balance needs will have to increase from about 1 trillion cu ft per year in 1970 to 6 trillion in 1985. Thus, imported natural gas will amount to 40 percent of domestic production in 1985, another cause for concern.

The domestic use of coal is expected to increase from 519 million short tons in 1970 to 933 million short tons in 1985. Total production will probably increase from 590 million tons in 1970 to over a billion tons by 1985, an increase of more than 80 percent.

Finally, Table 12-1 includes other energy sources. Production is in billions of kilowatt-hours per year. Note that the greatest projected expansion is in nuclear power. It is not certain that either the nuclear or geothermal objectives will be met. By December 1976, 64 nuclear plants were in operation and several times that number were in various stages of planning or construction. There is considerable delay in the program, however, due to environmental concerns, high construction costs, high fuel costs, high costs of repairs, and frequent shutdowns.

Figure 12-2 summarizes the uses of energy resources in the United States as of 1970. The flow of energy is from the sources on the left through its conversion into work and waste heat toward the right. The overall efficiency of the system in 1970 was 51 percent. The efficiency of electrical generation and transmission was 31 percent; that of direct fuel use in transportation, 25 percent; and that of fuel use in other applications, such as space heating, 50 to 75 percent. Reduction of energy loss due to waste is one of the prime goals in conservation.

The nation now faces a number of energy-related problems. These include not only the problem of diminishing domestic supplies of natural gas and petroleum but also decisions relating to the acceptability of alternate energy sources; the siting, financing, and construction of new power plants and related facilities; the problem of increasing dependence on imported oil and gas, with its adverse effects on the balance of payments and on national security; and the problems of environmental degradation. A cause of increasing concern are estimates that the annual United States per capita energy consumption will more than triple by the year 2000.

References to shortages of nonrenewable energy resources such as oil, gas, coal, and uranium must be properly qualified. It is essential to distinguish between a physical resource and a reserve and to recognize that their status may change with time as discussed below.

PHYSICAL RESOURCES AND RESERVES

In the mineral industries, *resources* are the inferred quantities of useful materials. *Reserves* are those resources whose extent is known and which are *currently extractable at a profit.* Confusion between these terms often leads to apparent contradictions regarding supplies of minerals.

The following example will illustrate the difference. Practically every year during the last 50 years, the statement has been made that the United States' domestic reserves of liquid petroleum were sufficient to last at most 10 to 15 years at then current rates of consumption. During this same period, production of petroleum rose by more than 75 percent. Clearly, the dire predictions were based on reserves as identified at that time. But each year, the quantity of petroleum found by exploration drilling was essentially equal to that consumed during the year. In other words, portions of the undiscovered resources were continuously added to the identified reserves in sufficient quantity to offset the reserves consumed during the year. At present, however, demand is outstripping discovery, and liquid petroleum reserves are truly diminishing. This situation by no means applies to all minerals. The reserves of some often increase so rapidly that the market becomes glutted.

A commodity is considered in short supply when it is not available at a reasonable price to

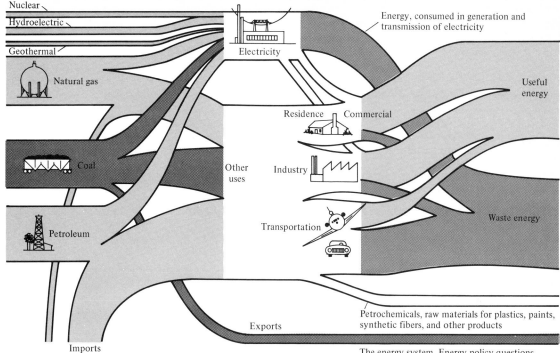

Nuclear

Hydroelectric

Geothermal

Natural gas

Coal

Petroleum

Electricity

Other uses

Residence Commercial

Industry

Transportation

Energy, consumed in generation and transmission of electricity

Useful energy

Waste energy

Petrochemicals, raw materials for plastics, paints, synthetic fibers, and other products

Exports

Imports

U.S. ENERGY: WHAT WE HAVE AND WHERE IT GOES

The energy system. Energy policy questions eventually come back to this chart. On the left are the energy supplies—predominantly coal, natural gas, and petroleum; in the center the ways we consume it; and on the right, perhaps most revealing of all, how much of it is wasted.

FIGURE 12-2. The flow of energy in the United States in 1970 in quadrillion Btu. The widths of the bands are proportional to the quantities of energy involved. The domestic supplies of natural gas and petroleum are bolstered by imports, whereas part of our abundant coal production is exported. Note at the far right that the amount of energy wasted under present technology is nearly equal to the amount doing useful work. (*E. Cook, The Flow of Energy in an Industrial Society, Scientific American, vol. 225, no. 3, pp. 135–142, 144, September 1971.*)

perform a needed function. The "reasonableness" of the price is usually set by the price of the next most desirable substitute. Society tends to select substitutes with as much regard for price and availability as for a specific property.

Natural gas provides a good example of the above. Its portability and cleanliness as a fuel led to a growing demand. Normally, this would have meant an increasing price and an increasing exploration drive. However, federal price

regulation eliminated the incentives. Gas producers, faced with fixed prices, turned to more profitable enterprises, including oil production.

The cost of exploring for oil and gas reserves is high and the risks are great. The success ratio for finding a *commercial* oil well is 1 in 40. The expense of unsuccessful exploration is shared by investors, and ultimately by the public through price increases. Actually, the risks in exploration are so great that in view of public needs, government encouragement may be desirable. A single, deep, dry hole on land may cost millions of dollars. Offshore drilling may be even more expensive.

Table 12-2 presents data on the United States' reserves and resources of energy materials. The reserves are listed at 1971 prices and the identified and hypothetical resources are estimated. The probable primary demand for

Table 12-2. United States reserves and resources of selected energy mineral commodities. (Modified from National Commission on Materials Policy, "Material Needs and the Environment Today and Tomorrow," 1973)

Commodity	Units	Probable Cumulative Primary Mineral Demand 1971–2000	Reserves at 1971 Prices[1]	Resources[2]	
				Identified[3]	Hypothetical[4]
Coal	Billion short tons	21	Adequate	Huge	Huge
Peat	Million short tons	43	Adequate	Huge	KDI
Petroleum	Billion barrels	276	38	Large	Large
Natural gas	Trillion cu ft	1098	279	Moderate	Large
Uranium	Thousand short tons	1240	130	Large	Large
Thorium	Thousand short tons	21	2	Very large	KDI

[1]As estimated by U.S. Bureau of Mines, 1973.

[2]Resource appraisal terms:

Huge—Domestic resources are greater than 10 times the minimum anticipated cumulative demand (MACD) between the years 1971 and 2000.

Very large—Domestic resources are 2 to 10 times the MACD.

Large—Domestic resources are 75 percent of to twice the MACD.

Moderate—Domestic resources are 35 to 75 percent of the MACD.

Small—Domestic resources are 10 to 35 percent of the MACD.

Insignificant—Domestic resources are less than 10 percent of the MACD.

KDI (known data insufficient)—Resources not estimated because of insufficient geological knowledge of surface or subsurface area.

[3]Identified resources include reserves and materials other than reserves which are essentially well known as to location, extent, and grade and which may be exploitable in the future under more favorable economic conditions or with improvements in technology.

[4]Hypothetical resources are undiscovered, but geologically predictable, deposits of materials similar to identified resources.

coal from 1971 to 2000 is estimated at 21 billion short tons. The reserves appear adequate. On the other hand, the probable demand for petroleum for the years 1971 to 2000 is estimated at 276 billion barrels. The United States' reserves at 1971 prices are 38 billion barrels. The identified and hypothetical resources are indicated as large. Of course, these resources could not become reserves except at considerably higher prices.

The probable demand for natural gas from 1971 to 2000 is estimated in Table 12-2 at 1098 trillion cu ft. Only 279 trillion cu ft are shown as reserves. Identified resources are moderate, and hypothetical resources, large. As for uranium, the probable cumulative primary demand for the years 1971 to 2000 is estimated at 1240 thousand short tons. The reserves at 1971 prices are only 130 thousand short tons, although identified resources and hypothetical resources are large. Finally, the demand for peat from

1971 to 2000 is expected to be 43 million short tons. Reserves are considered adequate, and identified and hypothetical resources are huge.

ENERGY RESOURCES AND ENVIRONMENTAL IMPACTS

There is controversy regarding the development of some energy resources because of possible environmental degradation. A few cases will illustrate.

In 1962, Consolidated Edison Company of New York announced plans for a 2000-megawatt[2] hydroelectric facility at Storm King

---

[2]A megawatt is 1 million watts. A generating facility of 1000 megawatts can satisfy the domestic and industrial electrical needs of a population of 1 million.

Mountain in the Hudson River Highlands 65 km (40 mi) north of New York City. The plan was to pump water from the Hudson River into a reservoir in the mountains at an elevation of 300 m (1000 ft). The pumping was to be done at night when energy demands are low, the water to be released for electrical generation during times of peak demand. Original plans called for an aboveground powerhouse which would have required a deep cut in the flank of Storm King Mountain, a scenic landmark along the Rhine-like gorge of the Hudson River. The aesthetic threat created strong environmental opposition.

The project was redesigned so that the powerhouse and water conduits were below-ground. New objections were raised, however, on the grounds of potential harm to recreational and commercial fisheries. The question was also raised as to whether the proposed system was the most economical. The case is still unsettled after more than a dozen years in spite of power shortages in New York City.

Opposition has led to the abandonment of proposed nuclear reactor sites. A proposed nuclear facility at Bodega Head on the California coast 80 km (50 mi) north of San Francisco was abandoned because of proximity to the San Andreas Fault. Construction of nuclear reactors at Monticello, Minnesota, on the Mississippi River 55 km (34 mi) upstream from Minneapolis; on Biscayne Bay on the southeast coast of Florida; and on Lake Michigan west of Kalamazoo were all delayed because of legitimate public concerns. One of the issues raised at the Monticello site was the discharge of low-level radioactive wastes into the Mississippi River. The conflict was resolved by enactment of stricter standards for waste discharge. The concern at Biscayne Bay was primarily over the ecological effects of direct thermal discharge into the confined waters of the bay. The conflict was resolved when a cooling pond was substituted for a direct outlet into the bay. The Lake Michigan reactor created concern because of uncertainty of the impact of thermal waters on the ecology of the lake, already upset by sewage and industrial wastes. The issue was resolved by agreement to install cooling towers, as well as devices to reduce radioactive emissions.

Environmental concerns have also delayed the development of new coal lands, and only recently has the government leased parts of its oil shale lands. Surface mining of coal, oil shale, and other commodities, can have significant environmental impact. As of 1965, 0.14 percent of the land area of the United States had been directly disturbed by surface mining. The distribution by states is shown in Figure 11-7. Figure 12-3 shows the federal lands and the amount disturbed by mining. The federal lands total 3,100,000 sq km (1,200,000 sq mi), or approximately one-third of the total land area of the 50 states. To date, 0.3 percent of our federal lands have been disturbed by mining (coal, oil, gas, stone, sand, gravel, cement rock, and metallic and nonmetallic ores). The relative size of the disturbed area is indicated diagrammatically by the small circle. One-third of the mined area has

FIGURE 12-3. **Public lands of the United States and diagrammatic representation of area disturbed by mining. (Bureau of Mines and National Commission on Materials Policy, 1973, "Material Needs and the Environment Today and Tomorrow.")**

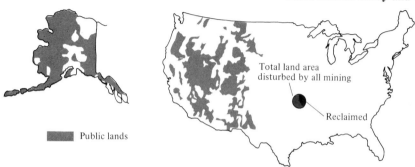

Public lands

Total land area disturbed by all mining

Reclaimed

been reclaimed, or restored by nature. Of course, the aesthetic and hydrologic impacts extend farther afield.

Another controversial facet of the resource problem concerns the large areas of public lands that have been withdrawn from mineral exploration and development. The withdrawn acreage is equivalent to all the land east of the Mississippi River exclusive of Maine, and the rate of withdrawal is increasing. At least one authority predicts a self-imposed malnutrition of minerals.

Let us now consider the environmental impacts associated with the different energy sources.

Hydroelectric

Dams. Almost all hydroelectric power is generated at dams. Power production is clean and cheap. There is no combustion or waste and no significant thermal effects. Adverse effects are inundation of scenic valleys or valuable bottom land, ecologic disturbance, disturbance of the river regimen downvalley, and possibly the triggering of earthquakes in seismic regions. A further objection is that the life span of reservoirs is limited to a few decades or centuries. The benefits of dams and reservoirs, aside from supplying a cheap source of power, are their contributions to a stable and abundant water supply, to flood control, and to recreation.

Most of the desirable dam sites in the United States have already been preempted, and this energy source can only contribute in a small way to future needs.

Tides. Hydroelectric power may be generated by the tides at coastal sites where there is an appreciable tidal range. Maximum tidal ranges occur in funnel-shaped bays and estuaries in which the incoming tide is progressively restricted. Tidal energy systems require a damlike barrier through which the incoming tide is permitted to pass but which can be closed to trap the high-tide waters in the embayment. As the tide drops outside the barrier, an increasing difference in elevation (head) develops between the waters of the embayment and the open waters outside the barrier. The trapped high waters are passed through the barrier and turn turbines connected to generators.

Unfortunately, there are few suitable embayments of this sort. Total global economic power production from tidal sources is estimated at 13,000 megawatts, equivalent to the output of about a dozen modern power plants using coal or nuclear energy. If all feasible tidal resources were used, making use of both outflow and inflow, they would add only 1 percent to potential water-power generation.

The only large modern operating plant in the world, built in 1966, is in the Rance River estuary on the Brittany coast of France (Figure 12-4). The tidal range is 9 m (30 ft). The barrier includes 24 generating units, each with a capacity of 10 megawatts.

In North America, the most promising sites are in the Bay of Fundy between the Canadian provinces of Nova Scotia and New Brunswick and in Passamaquoddy Bay between New Brunswick and the state of Maine. The maximum tidal range at two sites at the head of the Bay of Fundy is 13 m (44 ft), and at a third site, 16 m (53 ft). Various schemes for the efficient use of the tidal energy have been considered, but they have all been found uneconomic.

Russian technologists are investigating the potential of the White Sea, an embayment of the Arctic Ocean in northwesternmost Russia, where tides reach 9 m (30 ft) in places. A small experimental plant was recently placed in service in Kislaya Bay, with a tidal range of 3.5 m (11 ft).

The environmental impacts of exploitation of tidal energy are minimal but do involve the construction of a tidal barrier with locks for shipping, and this might disturb the ecology of the bay.

One of the disadvantages of tidal energy is that high-water stages for maximum production of power are generally out of phase with periods of maximum power needs. Some feasible method of energy storage is required, particularly since optimum conditions for power generation come only twice a month during the flood tides.

Ocean Currents. The Florida Current, that part of the Gulf Stream that hugs the Florida coast, is a possible source of usable energy. About

FIGURE 12-4. **Tidal power hydroelectric plant, Rance River, Brittany coast of France.** (*Courtesy, French Embassy and Phototheque EDF-Michel Brigaud.*)

2000 megawatts are judged recoverable by practical systems close to the coast, which might exploit 4 percent of the flow. This is equivalent to two modern power plants.

A proposed system to tap this energy source requires a permanently anchored ship and some sort of device to be motivated by the current. It is believed that costs would be competitive with other energy systems. If feasible, such systems could help meet energy needs in areas where swift currents pass close to shore.

To summarize hydroelectric energy resources, few suitable dam sites remain in this country, and promising sites of tidal and current-generated power are severely limited.

Oil and Gas

General Considerations. Oil and gas, together with coal, account for 95 percent of United States energy consumption and 82 percent of its electrical generation. Because oil and gas are produced by similar methods and often from the same well, they will be discussed together.

The 1975 estimates of recoverable oil and gas in the United States are much more conservative than earlier ones. In considering the statistics that follow, bear in mind that we consume 6 billion barrels of oil annually, or 30 barrels per person. Daily consumption in mid-1975 was 16.7 million barrels, of which 40 percent was imported (Figure 12-5). In March 1976, daily consumption almost reached 18 million barrels, of which imports accounted for 45 percent.

United States consumption of natural gas in 1972 totaled 22 trillion cu ft. Four percent of this was imported.

The 1975 estimates of United States oil and gas reserves, both onshore and offshore to the 200-m (650-ft) water depth, are:

1. Estimates based on actual exploratory drilling:

Crude Oil	39 billion barrels
Gas	237 trillion cu ft (energy equivalent of about 45 billion barrels oil)
NGL (natural gas liquids)	6 billion barrels

2. Inferred resources in fields not fully defined by drilling:

Crude Oil	23 billion barrels
Gas	202 trillion cu ft (equivalent to about 40 billion barrels oil)
NGL	6 billion barrels

3. Undiscovered recoverable resources:

Crude Oil	50 to 127 billion barrels
Gas	322 to 655 trillion cu ft (equivalent to 64 to 130 billion barrels oil)
NGL	11 to 22 billion barrels

Before becoming too enthusiastic about the as yet undiscovered resources, it should be kept in mind that any new discovery requires 2 to 10 years to bring into production. Furthermore, on the average, only 1 well in 60 leads to discovery of a field exceeding 1 million barrels (1 to 2 hours supply for the United States), and only 1 in 2000 leads to discovery of a field of 25 to 50 million barrels, the lower economic limit in remote areas.

The above resource estimates do not include oil that is too viscous for present recovery, oil that is tied up in oil shales or tar sands, gas that is held in tight formations or locked up in coal, and oil and gas offshore in water depths exceeding 200 m (650 ft). Hence, the estimates are too pessimistic.

It has been predicted that, at expected rates of consumption, domestic supplies of oil and gas will be gone between 1990 and 2000. We will then be entirely dependent on alternate energy resources or on foreign supplies.

The possibility of materially increasing our land-based supplies of oil and gas is not good, except in Alaska. A number of new Alaskan discoveries during 1975 have gone unnoticed in the publicity attending exploitation of the giant Prudhoe Bay field. There has been a discovery at Gwydyr Bay, 27 km (17 mi) west of Prudhoe Bay, and another off shore in the Beaufort Sea. At Prudhoe Bay itself, new deep production zones can conceivably double the proven reserves of 10 billion barrels. And Naval Petroleum Reserve Number 4 (PET 4), with a potential of 10 to 33 billion barrels of oil and 80 trillion cu ft of gas, could provide 2 million barrels of oil daily by 1985. With the 2-million-barrel-per-day output from Prudhoe Bay, almost one-half of present-day imports could be satisfied. The situation is equally promising for natural gas. The offshore possibilities in the coterminous 48 states may be even better.

Offshore Resources. Figure 12-6 shows the extent of our offshore areas locally favorable for petroleum, as well as the small areas leased for exploration and production by 1973.

The Outer Continental Shelf Lands Act of August 7, 1953, specified that lands beyond the 3-mile limit are subject to federal control, and the Secretary of the Interior was authorized to grant mineral leases. This is done through the Bureau of Land Management with technical

FIGURE 12-5. **Widening United States gap between oil consumption and production. Figures are in millions of barrels of oil per day. The deficit in 1969 was 23 1/2 percent of consumption. In March 1976, demand was slightly less than 18 million barrels per day, with imports making up 45 percent of the total.** (*Modified from a study based on latest data by U. S. News and World Report, August 25, 1975.*)

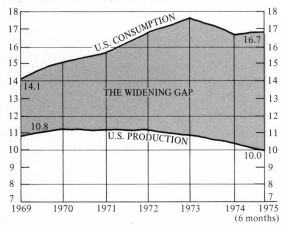

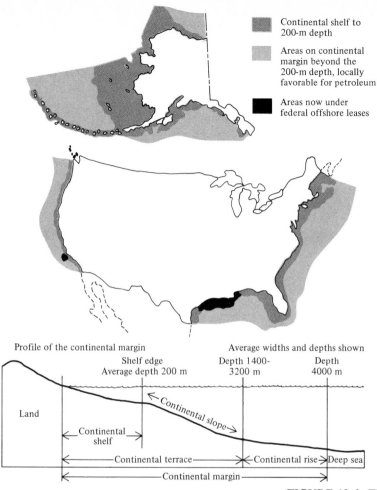

Profile of the continental margin   Average widths and depths shown

Shelf edge   Depth 1400-   Depth
Average depth 200 m   3200 m   4000 m

Land

Continental shelf ←Continental slope→

←————Continental terrace————→|←Continental rise→|Deep sea|
←————————————Continental margin————————————→

FIGURE 12-6. **The United States continental margin and areas favorable for petroleum. (*National Commission on Materials Policy, 1973.*)**

assistance from the Geological Survey. The leasing procedure requires environmental impact statements and public hearings.

The potential of the continental shelf off the Pacific coast is relatively untested. Good production has already been obtained in the Santa Barbara channel and may be the forerunner of additional discoveries. The Gulf Coast is already an important producing area. The Atlantic offshore, however, is untouched. Three areas are promising (Figure 12-7). The first is the Georges Bank Trough, an area of 36,000 sq km (14,000 sq mi). The second is the Baltimore Canyon Trough, involving 31,000 sq km (12,000 sq mi). The third is the Southeast Georgia

Embayment and the adjacent Blake Plateau Basin, involving 180,000 sq km (70,000 sq mi). In November 1975, the U.S. Geological Survey approved the drilling of two deep exploratory wells, one in the Georges Bank area and one in the Baltimore Canyon area.

Part of the opposition to offshore drilling along the Atlantic coast is based on opposition to construction of the necessary refineries. Another environmental concern is the possibility of oil spills due to blowouts or other equipment

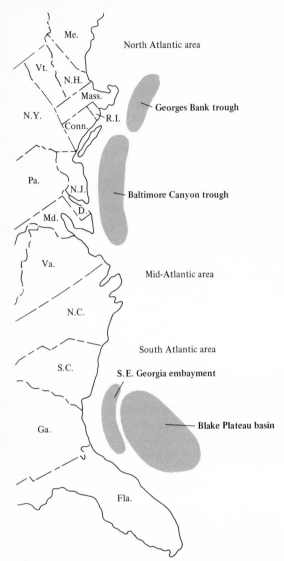

FIGURE 12-7. **Petroleum provinces of the Atlantic offshore. (*Exxon, U.S.A., 1974.*)**

20,000 barrels of oil, but the wreck of a single tanker, the Torrey Canyon, off Cornwall, England, in 1967, released 700,000 barrels.[3]

The major cause of marine oil pollution is from shipping. Offshore drilling, therefore, is far safer environmentally than an increase in shipping to bring foreign oil into American waters. The pollution from shipping results not only from damage to oil tankers, but from accidents during loading and unloading (35 percent of all oil spillage), and from thoughtless disposal of waste as as in pumping out bilges or cleaning tanks. Spillage amounts to 12 million barrels annually. Although the percentage is small, considering that between 30 and 35 million barrels are on the seas *daily* in 6800 tankers, spillage must be reduced.

Some marine pollution is due to natural oil seeps identical to those found on land. Early explorers along the California coast noted oil slicks. Worldwide marine seepage exceeds 4 million barrels a year. Additional pollution is contributed from the lands by polluted streams and air.

Evidence is growing that oil pollution does not have lasting effects except, perhaps, in confined areas. The Orielton Field Center, a research group in Great Britain, studied more than 100 cases of oil pollution from offshore wells and tanker accidents from 1960 to 1971 and found that only 2 caused damage that lasted more than a year. Both were tanker spills, the Torrey Canyon spill off the Cornwall Coast, and a spill off West Falmouth, Massachusetts.[4] Some scientists feel that the study did not delve

---

failures, as at Santa Barbara, California, in 1969. While locally significant, failure at drilling sites is a minor cause of oil spills, accounting for only 1.3 percent of the total. Through 1970, there were 14,000 offshore wells, 9000 of which were on the outer continental shelf. Only 25 blowouts had occurred, and only 3 involved serious spills. The Santa Barbara event spilled

[3]On April 22, 1977 the first blowout in the North Sea occurred. The well, in Norway's Ekofisk Field, gushed uninterruptedly for $7^{1}/_{2}$ days at the rate of 28,000 barrels per day. A lucky combination of winds and currents kept the oil slick from neighboring coasts and ecological damage appears to have been slight.

[4]At the time of this writing, two other major tanker spills had occurred. The wreck of the tanker *Urguiola* at La Coruna, Spain, in May 1976, released about 35,000 barrels of oil. In December 1976, the Liberian tanker *Argo Merchant* ran aground off Nantucket Island and spilled a large part of its 180,000 barrels of oil, threatening important fishing grounds.

deeply enough into more subtle aspects of the marine life cycle, particularly in confined estuaries. However, marine biologists who have studied the ecology of the Santa Barbara Channel found no significant changes in bottom life as compared to areas outside the confines of the spill.

Because ships are the principal source of marine oil pollution, a number of nations regulate oil tankers, not only from the standpoint of security against leakage but also of cleaning procedures. The pumping of bilges and disposal of wastes are strictly regulated. As a result, controlled tankers contribute only 1.5 percent of marine pollution. Other nations, however, do not have such regulations, and tankers and other vessels contribute seriously to pollution. Obviously, international cooperation is necessary.

There are conceivable global repercussions should pollution get out of hand. A veneer of oil over the oceans would reduce evaporation and diminish the supply of moisture for transport over the lands. By inhibiting evaporation, the oil would affect heat distribution in both air and water and would affect the hydrologic cycle. Changes in the heat budget might also affect the ocean currents, at least to the extent of modifying rates of movement. A veneer of oil could reduce wave energy and lead to reduced mixing and aeration of waters, with undesirable ecologic effects. Although these global impacts appear far-fetched, vigilance is desirable, particularly in small, largely enclosed seas, bays, and estuaries.

Coal

Primary Fuel. Coal provides the fuel for generation of much of our electrical energy. A smaller proportion is used in heating buildings and in industrial uses such as smelting and refining.

United States coal resources are enormous, estimates ranging to several trillion tons. Even if consumption in the year 2000 reaches 3 billion tons annually, 2 to 3 times present consumption, domestic supplies should last several centuries.

Coal introduces environmental problems both in mining and in burning. Most of these problems were considered in Chapter 11. Where underground mining is practiced, there is danger of surface collapse or slow subsidence (see Chapter 2). Operators must now keep up-to-date plans of underground workings. The underground ramifications of many old abandoned mines, however, are unrecorded. Many buildings have collapsed into unsuspected mine workings.

Federal and state regulations now curb the more objectionable environment-defacing practices and require some reclamation. We noted in Chapter 11 that a considerable degree of reclamation of stripped lands is possible, but at considerable cost.

We also noted in Chapter 11 that much of our coal is high-sulfur, and releases sulfur fumes on burning. Unfortunately, equipment for effective removal of these objectionable fumes is not yet completely dependable. Nevertheless, present regulations require their use, and periodic modifications or replacements by more advanced systems create a financial drain.

Gasification. Possible solutions to the problem of how to use the vast resources of coal efficiently and with minimum disturbance of the environment are gasification and liquefaction. The desirability of gasification of coal stems from ease of transportation, particularly when the goal is gasified in place, and low environment impact. Production of electric power by combustion of gas has little environmental impact. Water pollution, other than thermal discharge, is negligible, and air pollution is far less than coal- and oil-fired systems.

Gasification of coal has been practiced in Europe for many years. Recent cost estimates run 2 to 3 times the energy costs in the United States. However, gas from coal may eventually compete effectively with natural gas because of rising prices of natural gas and improvements in gasification techniques.

In gasification, air and steam are driven through a bed of coal heated to about 1030°C (1800°F), about half its normal combustion temperature. The air and steam react with the carbon to produce combustible gases, but the heat value of the gas is only one-sixth that of natural gas and hence is uneconomical to transport long distances. Other problems relate to the huge quantities of coal that must be processed and the tarry residues that result.

Liquefaction. The production of gasoline from coal is uneconomic but is resorted to under certain conditions. South Africa produces gasoline in this manner because it has limited petroleum resources. The United States is interested in liquefaction primarily as a source of low-sulfur, low-ash fuel for power plants. Research is also underway for efficient conversion of the liquids back to solids without the ash and sulfur of the original coal ("solvent-refined coal").

Liquefaction involves mixing the coal with a liquid solvent, heating the mixture, and exposing it to hydrogen under high pressure. Hydrogen sulfide and excess hydrogen are then removed, the mixture is filtered, the solvent is distilled for reuse, and the final product is recovered either as liquid or solid. The end product is a clean power plant fuel with a sulfur content of only 0.5 percent.

Coal liquefaction, like gasification, will take advantage of the enormous United States coal reserves, will provide a relatively pollution-free fuel, and will help conserve our dwindling petroleum resources for more specialized uses.

Oil Shales

Enormous quantities of petroleum are locked up in organic-rich shales in many parts of the world. Much of the world's liquid petroleum may have originated in such shales. Oil has already been produced from these shales in Scotland, China, Australia, the eastern Baltic, and South Africa.

The reserves of oil in oil shales are enormous. The Green River Formation of Colorado, Utah, and Wyoming (Figures 12-8 and 12-9) contains 2 trillion barrels in shale beds capable of providing 15 gallons of oil per ton of shale. A ton of shale is equivalent to a block less than the bulk of two 3-drawer letter-file cabinets. Of the 2 trillion barrels, 600 billion occur in beds at least 3 m (10 ft) thick and containing 25 gallons of oil per ton; about 400 billion barrels are in beds that can yield 30 gallons per ton; and some beds have a potential yield of up to 150 gallons per ton. Under present economic and techno-

logic conditions, between 80 and 90 billion barrels of oil can be recovered.

Eighty percent of the oil shale lands are federally owned. In 1971, a prototype oil shale leasing program was announced by the Department of the Interior. At the same time a draft environmental impact statement was released. Several leases have been awarded on the basis of competitive bids.

Exploitation of oil shales presents problems in mining and processing. Where the rich shale beds are close to the surface, they may be mined by open-pit methods. This results in nearly 100 percent recovery. Where the beds are far below the surface, underground mining, in which pillars are left for support, yields up to 60 percent recovery.

The environmental impacts of surface mining would be essentially the same as those involved in strip mining for coal. Unsightly excavations can be backfilled with spent shale, but the latter has 20 to 30 percent more volume than the original solid rock. There would still be excess wastes. Shale mined underground could be processed in place by passing gases and liquids through rock broken up by explosives.

The production of 100,000 barrels of oil per day would provide, in 20 years, 570 million cu m (740 million cu yd) of spent shale. If 50 percent of this, a reasonable average for both underground and surface workings, were disposed of on the surface, it would require a space 9 km (5 mi) long, 600 m (2000 ft) wide, and 60 m (200 ft) deep.

The processing of oil shale requires crushing and heating (retorting) to distill off the oil. The oil is too viscous to transmit through pipelines; hence, it must be partially refined at the site.

The availability of water is a serious concern. Three barrels of water are needed to produce one barrel of oil. A 100,000 barrel per day plant would need 12,000 to 18,000 acre·ft of water per year, and associated urban development might increase this to 13,000 to 20,000. Presumably, much water would be recycled. Air quality should not suffer inasmuch as shale oil has a low sulfur content.

There is increasing pessimism about shale oil becoming a substitute for conventional petroleum in the near future. Aside from the

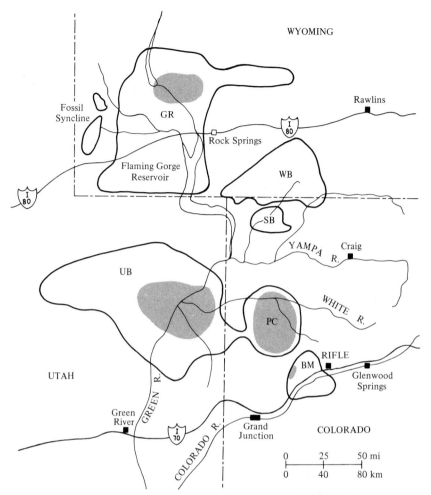

FIGURE 12-8. **Principal oil shale deposits of the United States. Gray areas contain beds more than 5 m (15 ft) thick with at least 25 gallons of oil per ton. Surrounding and outlying areas are unassayed or have smaller yields. Only a few small areas of this dominantly Federal land have been leased for private development. GR—Green River Basin; WB—Washakie Basin; SB—Sandwash Basin; UB—Unita Basin; PC—Piceance Basin; BM—Battlement Mesa area.**

environmental problems are those of economics. The capital investment for leases and production of a mere 100,000 barrels of oil per day would exceed $1 billion. Shale oil will probably not contribute significantly to our petroleum reserves for at least 10 or 20 years.

Tar Sands

Tar sands contain very heavy oil tars, generally the residual product left after volatile hydrocarbons have escaped.

The Athabasca tar sand deposit in Alberta, Canada, is the largest one known. Total resources of enclosed oil are estimated at 600 billion barrels, and the potentially recoverable oil, at 300 to 350 billion barrels. Current production is only 16 million barrels per year, but a number of new plants are in the planning stage. United States resources of tar sand oil are 29 billion barrels, but no estimate of the recover-

FIGURE 12-9. **Typical view in the oil shale province, Garfield County, Colorado. Some of the beds contain more than 25 gallons of oil per ton.** (*Arthur D. Howard*)

able amount has been made. As with oil shales, large capital outlays are required. In addition, the plants must operate near capacity continuously to be economic. A favorable investment climate is needed.

Environmental impacts are essentially the same as for oil shale.

Nuclear

General Comments. In December 1976, 64 nuclear reactors were contributing about 10 percent of the electric power generated in this country. Between 2 and 3 times that number were in planning or construction stages. It is estimated that by the year 2000, nuclear plants may be contributing as much as 60 percent of our *electricity* (not total energy).

Present nuclear plants operate by atomic fission. The fission process is based on the property of uranium 235 to split into fragments when struck by a heavy atomic particle (neutron). Before the collision, considerable energy is tied up in holding the various parts of the atom together. On splitting, this energy is released. The fission process produces additional

neutrons. These collide with other uranium 235 atoms, producing still more fission events, more energy, and more neutrons to collide with other uranium 235 atoms in a chain reaction. *A single pound of uranium 235, through such chain reactions, can yield the same amount of energy as 1500 tons of coal.*

Uranium 235 constitutes only 0.7 percent of the uranium atoms in a uranium sample. However, two other isotopes capable of providing chain reactions have been created artificially by neutron bombardment of the element thorium 232 to create uranium 233, and of uranium 238 to create plutonium 239. Thorium 232 and uranium 238 are hundreds of times more abundant in nature than uranium 235, hence offer much greater potential as fissionable fuel.

If the chain reaction were to proceed unimpeded in a mass of pure fissionable material, a catastrophic explosion would result. If, however, the fissionable material is sparsely disseminated in nonreactive materials, the latter soak up many of the neutrons released in fission and control the reaction. *In fission reactors the fuel is so diluted that an atomic explosion is impossible.* There are, however, other hazards that exist.

Reactors in Current Use. The common reactor uses uranium as fuel and water as a coolant. Fission releases heat which is used to create steam. Before the uranium is ready for a reactor, a series of preliminary activities that affect the environment take place. First, the ore is mined. The environmental impacts of exploration, development, and operation of mines were discussed in the preceding chapter. A unique hazard in uranium mining is radon gas, a by-product of radioactive disintegration which is heavier than air and remains in the mine where it is breathed by the miners. Milling of the ore also releases a small amount of radiation to the air and to the waters used in processing. Uranium 238, the bulk of the ore, is only feebly radioactive and is not fissionable. However, certain by-products, such as radium, and radon, are far more radioactive.

In extracting the 4 lb of uranium oxide from 1 ton of ore at the mill, between 850 and 900 gal of chemically toxic and radioactive liquid waste are formed. Most of this formerly went into surface drainage, but the procedure is now forbidden.

At present prices, domestic uranium reserves are limited and will probably last less than a century at projected rates of production. If prices continue to rise, resources sufficient to last for centuries will be added to the reserves. The goal of the proposed *breeder reactor* is to create fissionable fuel faster than it is consumed.

Because the fissionable uranium 235 constitutes only 0.7 percent of uranium ore, the ore must be enriched to about 3 percent uranium 235. The enriched fuel is packed into thimble-size metal pellets, each of which is capable of providing the heat energy of a ton of coal. About 200 pellets are packed into a 3-m (10-ft) rod of stainless steel. A cluster of up to 200 of these rods in a container constitutes a fuel assembly; and 100 to 600 assemblies constitute the reactor core. The core is encased in a housing of carbon steel and in turn enclosed in a dome of heavily reinforced concrete.

In the Boiling-Water Reactor, the water is passed through the fuel assemblies, converted to steam, and led directly to a turbine connected to a generator. In the Pressurized-Water-Reactor, the water is kept under pressure to prevent boiling and is used to create steam in a separate water system.

Inevitably, the water circulating through the reactor picks up some radioactive materials, such as the gas tritium, through microscopic pores in the fuel rods. The radioactive products find their way into the final coolant. The disposition of these relatively low-level radioactive waters is not a major problem. Of greater public concern is the possibility of meltdown of the reactor core due to failure of the many safeguards.

A serious problem is the disposal of the highly radioactive wastes in the core assemblies. About one-third of these need to be replaced each year as the uranium 235 is used up. The withdrawn assemblies are stored in water at the reactor site for several weeks to permit decay of the shorter lived radioactive elements. They are then shipped to reprocessing plants where they are again stored in water. They are finally broken up and dumped into nitric acid, which dissolves the radioactive elements.

Most of the uranium 235, and 99.5 percent of the small amount of plutonium 239, are extracted and purified. Remaining in the nitric acid are a variety of highly radioactive wastes. These are stored at the reprocessing plant as liquids for 5 years and as solids for another 5 years. They are then ready for disposal. *The annual waste from a 1000-megawatt plant, when reduced to solid form, would be a block only about average desk size.* However, its potential for damage is such that the radiation cannot be allowed to escape to the environment. Furthermore, the volume of fuel wastes is growing and disposal is a problem.

The present temporary procedure of storing liquid wastes in tanks is not a permanent solution because of the limited lifetimes of tanks and the possibility of leakage. Leakage has occurred at Hanford, Washington.

A second proposal is to pump liquid wastes into deep underground excavations in granite or other bedrock 1000 m or more below the surface. The wastes would be allowed to boil, the steam would be piped to the surface, condensed to water, and piped back to the chamber below. After 25 years, the wastes would be allowed to boil dry and the cavity would be sealed off from the surface.

A third proposal is to inject the liquid wastes into deep formations via drill holes. The wastes would be mixed with cement and forced under high pressure into impermeable shales. The high pressure would fracture the shales enough to provide entry. This scheme is being used for the disposal of intermediate-level wastes.

A fourth proposal is to store the wastes in solidified form in maintained underground storerooms, using air cooling to prevent melting.

A fifth proposal would dispose of the wastes in the sea floor, either in bedrock in a stable part of the ocean floor, in bedrock where rapid sedimentation would bury it further, or in places where the descending crustal plates might slowly carry the wastes into the depths.

A sixth proposal is to bury the wastes under the Greenland and Antarctic ice caps. The containers of waste, previously fused into solid glass for safe transport would be left on the ice to melt their way downward to the rock surface below. The hole would close by refreezing and by plastic flow of the ice, and the enclosing ice would absorb the waste heat. Figure 12-10 shows a tentative waste-disposal site in Antarctica. The site is 600 to 1000 km (400 to 600 mi) inland, and the ice may be 2000 to 3000 m (6600 to 9800 ft) thick. Costs would be considerably higher for ice-cap disposal than for land-based disposal. Furthermore, under present international agreements, Antarctica cannot be used as a disposal site.

The proposal which has been regarded as the most promising is disposal of wastes in salt mines. West Germany has already begun deposition of low- and intermediate-level wastes in a salt mine. The United States has engaged in intensive testing of a salt mine at Lyons, Kansas. Salt is plastic; that is, it will deform under pressure, sealing fractures and other voids. Salt formations are dry because of the absence of voids; hence, ground-water contamination should not be a problem. The thermal conductivity of salt is moderately high so that heat

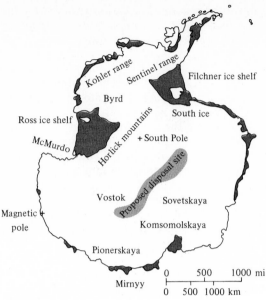

FIGURE 12-10. **Antarctic radioactive waste disposal site. (*From California Geology, August 1973, after E. J. Zeller, D. F. Saunders, and E. E. Angino, "Proposal for an International Radionuclide Depository in Antarctica," Science and Public Affairs, Bull. Atomic Scientist, vol. xxix, no. 1.*)**

would be conducted away from the waste containers. Investigation, however, is not yet complete and a final decision has not yet been made.

A proposal to propel the wastes into space encounters almost insuperable objections and need not concern us.

Thus far we have said nothing of the potential hazards to reactors from *external* causes, including earthquakes, windstorms, floods, and airplane crashes. Earthquakes involving ground motions that would justify precautionary shutting down of a reactor, even though reactors are designed to withstand such motions, are extremely rare. Because reactors are prohibited near active faults, accidents due to surface rupture are not expected. Obviously, construction standards must be higher in active earthquake-prone areas such as California than in stable areas such as New England.

All reactors in the United States are now designed to withstand the impact of tornadoes with winds up to 480 km (300 mi) per hour. Siting and construction of reactors are also rigidly constrained with respect to floods. Present construction, plus distance from airports, provides security against airplane crashes.

Fast Breeder Reactors. In contrast to present reactors, which use only 1 1/2 percent of the energy in the fuel, fast breeder reactors would use 40 to 50 percent.

The first breeder reactor was put in operation in the Soviet Union in 1972 on the eastern shore of the Caspian Sea. It produces 350 megawatts of electrical power. The United States may have a commercial breeder reactor available in the 1980s, unless present plans are jettisoned.

The fast breeder reactor produces fissionable fuel at a greater rate than it uses it. The reactor requires a supply of appropriate material that can be converted to fuel readily. Uranium 238, which constitutes 99.3 percent of natural uranium, and thorium 232, which is also common, provide abundant sources of the needed materials.

Breeder reactors will be more difficult to handle and more dangerous than conventional reactors. A very efficient coolant is required and it must be a poor, rather than an effective, moderator of neutrons. It must remain liquid at high temperatures. Liquid sodium satisfies the requirements. The sodium, which circulates at 400 to 540°C (750 to 1000°F), must be shielded from contact with water to avoid violent chemical reactions. It becomes highly radioactive, and the danger from leakage is much greater than for current reactors.

Any malfunction of the cooling system would result in rapid melting of the core because of the absence of moderators, and the possibility of an atomic explosion exists. The handling and reprocessing of breeder wastes pose additional difficult problems. It is doubtful that breeder reactors will dominate the scene before the year 2000.

Thermonuclear Fusion Reactors. Thermonuclear fusion reactors, which duplicate the energy-producing process of the sun, derive energy by the fusion of the nuclei of light elements rather than by the fission of heavy elements. For fuel, the fusion process would probably rely on the heavy isotopes of hydrogen; deuterium, with an atomic weight of 2, and tritium, with an atomic weight of 3. Deuterium is directly available in seawater in inexhaustible amounts, and tritium can be manufactured in the reactor from lithium, a metal also present in inexhaustible amounts in seawater. The seawater from which the deuterium and lithium is extracted would be returned to the oceans without known adverse effects.

The environmental impacts of fusion reactors would be small compared to fission reactors. The fusion fuels and their reaction products, except for tritium, are not radioactive. Control of emissions of tritium presents no insurmountable problems. Even if all the tritium were to escape, the biological hazard is regarded as very small. The radioactive materials created by impact of neutrons with the materials of the reactor structure would present a biological hazard less than one-thousandth of that of a comparable fission reactor. Long-lived radioactive waste would also be a small fraction of that from fission reactors.

Runaway accidents would be impossible because any malfunction would automatically quench the reaction through loss of temperature, confinement, or both. There is, however, the same possibility of a chemical explosion due to the molten lithium coming into contact with water as there would be in sodium-cooled breeder reactors. Unfortunately, the development of commercial thermonuclear fusion reactors is a long way off.

Initial capital costs of all nuclear plants are high. This restricts such plants primarily to the satisfaction of continuous power requirements, not the meeting of peak energy needs. The latter would still have to be satisfied by other types of plants.

Offshore Reactors. Offshore nuclear reactors have been proposed to eliminate some of the environmental concerns associated with land-based installations. According to one concept (Figure 12-11), the reactors would be mounted on barges in water 9 to 21 m (31 to 70 ft) deep and no more than 5 km (3 mi) from shore. They would float in a lagoon protected by rock or concrete breakwaters. The barges, with a draft

FIGURE 12-11. **Artists' concept of a proposed offshore nuclear reactor. (*Courtesy Offshore Power Systems: A Westinghouse Enterprise.*)**

of 9 m (30 ft) would be watertight to 12 m (40 ft) above water level. Thus, in the event the barges were to sink, only the watertight portions would be under water and no flooding would take place. Advantages include lack of defacement of land areas, reasonable isolation, and abundant water for cooling.

Geothermal

Geothermal energy derives from heat emanating from the earth. The average rate of thermal emission over the earth is so low that it has no practical value. The geothermal gradient (increase of temperature with depth) averages only 1°C per 55 m of depth (1°F per 100 ft) in deep wells or mines. Volcanic areas have abnormally high heat flows, however, and some can be tapped as sources of energy. The energy is in the form of steam or hot brines in hydrothermal systems, or simply as heat in dry rocks. There are less than 60 high-temperature hydrothermal areas in the world. Most are located in areas of present or recent volcanism or at the boundaries of tectonic plates. Within the past few decades geothermal energy has been used practically in New Zealand, Japan, Soviet Russia, Mexico, Iceland, the United States, and a

few other countries. As of 1974, 19 nations had plans for utilizing this form of energy.

Geothermal heat may be emitted as steam, as in fumaroles, or in liquid form, as in hot springs and geysers. In the vapor-dominated systems, the steam may be delivered directly to turbines which turn generators. The liquid-dominated operations require generation of steam from the hot water. Because geothermal temperatures are low compared to temperatures in conventional steam-powered electric plants, the power-generating efficiency is lower.

Geothermal steam has been used for power production at Larderello, Italy, since 1904. Production is presently 404 megawatts. The only producing field in the United States, also a steam plant, is at The Geysers, 120 km (75 mi) north of San Francisco, California. At the time of this writing, it was producing 520 megawatts from 80 steam wells. In a few years the output should double, supplying enough electricity to sustain a city the size of San Francisco, the equivalent of the power produced by Hoover Dam on the Colorado River. The steam fields at

Lardarello and The Geysers produce more electricity than all the hot-water fields combined.

Liquid-dominated (hot-water) operations involve a mixture of hot brine and steam. If temperatures are high enough, the liquid is flashed into steam by reduction of pressure. The oldest and most important such operation is at Wairakei, New Zealand, producing 200 megawatts. Many similar operations have been started in the last decade.

Extraction of steam and hot water may deplete a geothermal reservoir. The wells at Larderello provide satisfactory heat output for only 10 or 12 years, and new wells have to be drilled. The total world potential is estimated at 60,000 megawatts per year for a useful period of 50 years.

In addition to the steam and hot-water geothermal resources, there are large areas of relatively impermeable, hot, dry rock, usually at depths of 3300 m (11,000 ft) or more. The large areas involved, and the large number of deep wells that would be required for significant energy recovery, would mean complex and costly operations. One approach to energy recovery would be to drill to sufficient geothermal depth, fracture the rock to provide greater surface area, pump water down at high pressure, and recover hot water through a separate hole.

Problems with geothermal heat involve dissolved substances which may corrode and clog pipes and pollute surface streams. Noxious gases such as ammonia, hydrogen sulfide, and carbon dioxide are emitted in many geothermal areas. In some localities, as in the Imperial Valley, California, the waters are highly saline, and handling huge quantities of residual salts could be a problem. Reinjection into the ground is a possible solution. Wherever liquids under high pressure are withdrawn from the ground, surface subsidence may present an additional hazard. Subsidence at Wairakei has averaged 0.3 m (1 ft) per year in the center of the field and has produced local earthquakes. Subsidence is not expectable where only steam is involved because the steam is at relatively low pressure and further reduction does not lead to compaction of the rocks.

In spite of the above limitations, geothermal energy is a relatively clean source of power and should prove welcome in areas lacking other power resources or as supplements to other power sources.

In some parts of the world, natural hot water is piped to heat homes, industrial buildings, and other installations. One of the oddities that strikes visitors to Reykjavik and other large cities in Iceland is the absence of chimneys. Iceland is a volcanic area. In Reykjavik, the main pumping plants draw hot water from as much as 1.6 km (1 mi) deep, allow it to cool to 95°C (190°F), and pump it to homes and other buildings for heating. On return to the pumping station, the water is mixed with additional hot water from the depths to continue the cycle. The hot water is also used to heat greenhouses where even grapes and bananas are cultivated in this Arctic climate. The piped hot water eliminates the need for furnaces, and there are only minimum space requirements and costs for water pipes. There is no air pollution and no fire hazard. Very little of the geothermal heat is used in generation of electricity; only 3 megawatts were produced in 1974. Geothermal water has also been used for space heating in Korea, the United States, and other countries. France is planning a large development in the Paris basin.

In the United States, federal lands were made available for geothermal leasing by the Geothermal Steam Act of December 1970. Final leasing and operating regulations were put into effect on January 1, 1974. Under these regulations, the Bureau of Land Management is authorized to issue geothermal leases, and the U.S. Geological Survey is responsible for evaluating environmental impact.

Because known exploitable geothermal areas are scarce, geothermal energy will satisfy only a small part of energy needs. It will be important where available in quantity near large markets, in remote areas far from other sources of energy, or in developing countries where either huge electric power plants are not needed or where fossil fuels are scarce.

Solar Radiation

Solar energy is being used experimentally for heating and cooling buildings. In our sunny Southwestern deserts, a collector area of 56 to

75 sq m (600 to 800 sq ft) can satisfy the heating and cooling needs of a 140 sq m (1500 sq ft) home. Solar energy should be used even in areas where the sunlight is intermittent.

For air-conditioning by solar radiation, homes need three main elements. The first is a heat-collecting device. A common device consists of pipes meandering across panels of metal or wood painted a heat-absorbent black and covered by glass to conserve the heat as in a greenhouse. The second element is a heat reservoir, generally an insulated tank into which the heated fluid is led. Heat can also be stored in a bed of rocks or chemically. The final element is a system of pipes throughout the building for circulation of the fluid for heating purposes. For cooling purposes, the system would operate like a gas-fired refrigerator, the heat providing the energy for the cycle. Backup systems will be needed when neither the sun nor the stored heat can provide sufficient energy.

For sizable power plants, the collecting areas will have to be large because solar energy reaching the surface of the earth is diffuse. Consequently, the large array of collectors and auxiliary structures will be expensive. In addition, to store energy in large amounts for use when the sun is not shining will be difficult and costly.

The collectors could be photoelectric cells which convert sunlight directly into electricity. The high cost of photoelectric cells probably means that they will not provide the basis of large-scale power generation.

Another idea employs mirrors to concentrate solar energy. Figure 12-12 is one concept of such a solar electric power system. A 500-megawatt system of this type would require 2.6 sq km (1 sq mi) of reflectors and a tower 460 m (1500 ft) high. The parabolic mirrors would follow the sun in its passage across the sky. They would focus the sunlight onto the top of the tower where it would heat a fluid to 600°C (1100°F) or more. Steam would be led down the tower to an electrical generator. The Energy Research and Development Administration announced in January 1977 selection of a site near Barstow, California, for a large-scale pilot plant.

A *total energy concept* would use some of the waste heat as well as the electricity. Many communities with shopping centers and commercial clusters require from 100 to 300 megawatts of energy for heating, lighting, air-conditioning, and running machines. Fortunately the ratio of heat to electricity requirements is about the same as would be produced by a solar power plant. Perhaps an effective use of energy could be achieved by planning new communities to fit the energy production of solar power plants.

Because incident solar energy is intermittent, a storage system is essential if the plant is to serve as a primary energy source. The most promising proposal is the *electrolysis* (electrical decomposition) of water to produce hydrogen and oxygen which can be stored and later burned as desired. The burning of hydrogen is clean, and the exhaust is hot water. Both hydrogen and oxygen can be transported by pipeline and can be liquefied and stored in low-temperature containers.

The principal environmental impact of solar energy plants concerns the large collecting areas required. Thermal pollution should be smaller than with other systems. Costs will probably be several times higher than present power systems because of the large land area and number of collectors involved, the sun-tracking system, the huge energy storage and recovery systems, and the power and heat transmission lines, ducts, and cables.

If all new homes and single-story commercial buildings built between now and the year 2000 have solar heating and cooling systems built in, 4.5 percent of the country's total energy needs would be satisfied. Many states have tax incentives to encourage installation of solar devices.

Wind

During the years 1880–1930, nearly 6 million windmills were in use in the United States to pump water, operate mills, and generate small amounts of electric power. Only a few remain. But the need to draw on all energy resources may bring the windmill back. The Federal government allotted $12 million for wind-power research in fiscal 1976. Wind, however, is regarded by many as even less promising than the tides as a major contributor to satisfy energy

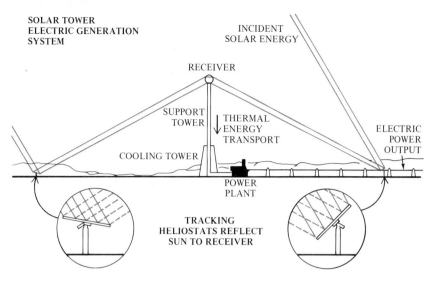

SOLAR TOWER
ELECTRIC GENERATION
SYSTEM

INCIDENT
SOLAR ENERGY

RECEIVER

SUPPORT
TOWER

THERMAL
ENERGY
TRANSPORT

COOLING TOWER

ELECTRIC
POWER
OUTPUT

POWER
PLANT

TRACKING
HELIOSTATS REFLECT
SUN TO RECEIVER

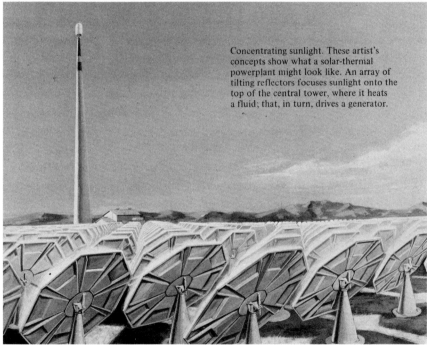

Concentrating sunlight. These artist's concepts show what a solar-thermal powerplant might look like. An array of tilting reflectors focuses sunlight onto the top of the central tower, where it heats a fluid; that, in turn, drives a generator.

FIGURE 12-12. **Concept of a solar tower electric generation system. (*Courtesy McDonnell-Douglas Corporation and National Science Foundation, "Mosaic," vol. 5, no. 2, 1974.*)**

needs. One optimistic estimate, however, suggests that the wind may contribute up to 20 percent of our energy supplies in the year 2000.

Wind is unreliable. Average wind speed at the surface is low, and considering that air is light, its effectiveness for generation of power is limited. Because wind velocities are greatest over the oceans, it has been proposed that wind-power systems be installed offshore on the continental shelf. Provided storage of wind-generated energy is achieved on a large scale, strength of wind, rather than steadiness, provides the greatest power return. The system that offers special promise proposes that the electricity produced by the wind be used to decompose water into hydrogen and oxygen, as mentioned in the discussion of solar energy plants.

Inland windmill installations would have to be 200 to 300 m (650 to 1000 ft) high to reach reasonably constant strong winds. Offshore installations would probably exceed oil-drilling rigs in height. One proposal foresees huge arrays of propellers, each propeller 18 m (60 ft) in diameter, on a frame extending as much as 140 m (450 ft) above the sea surface. Experiments are also being conducted with windmills in which propellers lie horizontally or with blades like an eggbeater.

Small systems could feed their electricity into a regional network continuously and there would be no need for storage facilities. Large systems, however, would require storage.

Although wind power cannot presently compete with other power sources, the situation might change if large-scale production of the necessary hardware reduced costs sufficiently. Wind power would have special appeal in regions where other energy resources are scarce or absent. Wind power, for example, could be a boon in New England and Hawaii, which rely on distant sources for oil supplies. Individual homes and farms are already installing wind generators, and oil companies are installing wind turbines to supply electrical power for offshore platforms. A 100-kilowatt experimental wind-turbine generating plant is being built on the shore of Lake Erie to determine the relation of cost to performance.

Use of the wind on a large scale is not without environmental problems. The estab-

lishment of thousands of gigantic towers, particularly on land, may be aesthetically objectionable. There is also the question of the effects of thousands of high wind devices on the weather, although the effects could be negligible, and the possibility of creating havoc among flocks of migrating birds and a danger to aircraft. Such considerations could reduce the number of acceptable wind power sites.

Oceanic Heat Exchangers

The most intense solar radiation reaches the earth between the Tropics of Cancer and Capricorn, where the sun is directly overhead at some time during the year. Solar radiation does not penetrate deeply into the ocean waters; hence, the warm surface waters are underlain at shallow depth by cold waters. These thermal differences provide a large potential source of energy.

In a thermal-difference power plant, the heat of the warm surface water would be used to convert a refrigerant fluid into vapor. Farther along in the system of pipes, cold water brought up from depths of hundreds of meters would chill and condense these vapors to create a low-pressure area toward which newly created vapors would move, driving turbines in their path. A thermal difference of as little as 17°C (27°F) would work at an efficiency of 1.5 percent. The electricity produced would be transmitted by cable or used to produce hydrogen which would be piped to the mainland. Realization is well off in the future.

Solid Wastes

The burning of solid wastes will contribute increasingly to energy supplies. Municipal wastes have a fuel value one-third that of coal. The burning of all the municipal solid wastes of the United States at 33 percent plant efficiency would satisfy 12 percent of our energy needs. Paris has been burning its solid wastes for years. There are now a number of solid-waste power plants in the United States. A major problem is control of the noxious gases liberated.

Farm wastes can also contribute to satisfying energy needs. This is particularly true of manure, from which gas can be extracted, leaving a residue that may prove of economic value as a fertilizer.

Methanol (wood alcohol) is regarded by many as the answer to the energy problem. Methanol may be used in properly modified power plants, factories, homes, and cars. Most methanol is made chemically from natural gas and coal, but it can be made from wood or plants grown as crops and from urban and agricultural organic wastes. The organic material is shredded and heated to cause decomposition into gases and liquids including or convertible into methanol. When burned, methanol creates little pollution. If used at 100 percent concentration, the amount of unburned fuel would be only 5 percent of that left by gasoline, and the carbon dioxide, only 10 percent. Even if used as a mixture with gasoline, it would reduce pollution significantly. In power plants, it would eliminate smoke and sulfur emissions from the smoke stacks, significantly reduce nitrogen oxides and carbon monoxide as compared to fuel oil, and release a negligible amount of unburned hydrocarbons.

Nebraska has reduced its tax on gasoline consisting of at least 10 percent alcohol made from agricultural products. Some cities are considering using garbage to make methanol for city-owned cars. A California bill proposes 40 plants to convert 28 million tons of solid wastes to methanol to be added to all gasoline by 1980.

Problems with methanol are that cars would require bigger tanks, cold engines would be harder to start, methanol absorbs moisture, causing corrosion and vapor lock, and performance is not equal to gasoline.

The price of methanol presently runs from 50¢ to $1.20 per gal, but large-scale production would probably lower costs. One authority believes that methanol from urban, farm, and timber wastes could satisfy 12 percent of American energy needs.

It is not clear how much energy is needed for the production process itself. Neither is it clear what effect continuous removal of organic wastes that were formerly plowed into the soil would have on soil fertility. There is also the problem of the acreage required for cultivation of energy crops.

Odessa, Texas, is the first American city to recycle all its daily wastes. Sewer effluents and garbage are converted to fertilizers, and extracted metals are sold. Odessa hopes soon to return a profit on these operations.

CONSERVATION

Large amounts of energy are wasted, and some energy resources are not used wisely. We drive large and heavy cars. We ride one-to-a-car and avoid public transportation. We use automobiles when walking or bicycling might do. We transport cargo by air when trucks are 5 times as efficient. Trucks, in turn, are only 20 to 50 percent as efficient as trains. Our homes are poorly insulated, and we keep them warmer than necessary in winter and cooler than necessary in summer. We commonly use wasteful electric heating when direct gas- or other fuel-burning devices are more efficient. In the smelting of ores, electric furnaces have been substituted for those burning fossil fuels. Yet the thermal efficiency of direct fuel-burning furnaces is 2 to 3 times greater than that of electric furnaces. We throw away reusable items that require materials and energy to manufacture. We use aluminum containers in spite of the fact that refining aluminum requires 5 times as much electricity as the production of the steel of "tin cans." We use synthetic fabrics which require considerable energy for their manufacture when natural products such as cotton and wool might suffice.

Research and development should aim at reduction of power needs in providing essentials. We need products that last longer. We need to reduce the weight of autos, trains, and aircraft by using lighter materials.

Many of the above comments indicate a need for intensive research and development in materials science. When the use of superconductors in generators and transmission lines becomes practicable, there will be less loss of electricity in overcoming resistance. Turbines that could operate at high temperatures with increased thermal efficiencies will require new materials to withstand these high temperatures and stresses.

Priorities in the use of energy resources need reconsideration. Oil is burned up in the generation of electricity, even though oil will be vital for years to come for propelling autos and airplanes. This is done despite the availability of more abundant energy resources for generation of electricity. Furthermore, both oil and gas are vital to the petrochemical industry.

In brief, rational selection and allotment of energy resources, coupled with research and development aimed at increasing the roles of environmentally clean resources, and abetted by sincere attempts at conservation, seems a necessary approach to the energy problem.

ADDITIONAL READINGS

American Association for the Advancement of Science: Energy Issue, *Science*, vol. 184, no. 4134, 1974.

Ellis, A.J.: Geothermal Systems and Power Development, *American Scientist*, vol. 63, September–October, 1975.

Holdren, J., and P. Herrera: "Energy," Sierra Club, San Francisco, 1971.

Hubbert, M. K.: Energy Resources, *in* "Resources and Man," W. H. Freeman and Company, San Francisco, 1969, pp. 157–242.

Kruger, P., and C. Otte: Resources, Production and Stimulation, *in* "Geothermal Energy," Stanford University Press, Stanford, 1973.

Landsberg, H. H., L. L. Fischman, and J. L. Fisher: "Resources in America's Future," The Johns Hopkins Press, Baltimore, 1963.

National Science Foundation: Energy for America's Third Century, *Mosaic*, vol. 5, no. 2, 1974.

Park, C. F.: "Earthbound: Minerals, Energy, and Man's Future," Freeman, Cooper and Company, San Francisco, 1975.

Reed, C. B.: "Fuels, Minerals and Human Survival," Ann Arbor Science Publishers, Inc., Ann Arbor, 1975.

U.S. Geological Survey: "Geologic estimates of undiscovered recoverable oil and gas resources in the United States," U.S. Geological Survey Circular 725, 1975.

U.S. Nuclear Regulatory Commission: Reactor Safety Study—An Assessment of Accident Risk in U.S. Commercial Nuclear Power Plants, Wash—1400, 1975 (The Rasmussen Report).

White, D. E.: "Geothermal energy," U.S. Geological Survey, Circular 519, 1969.

Shasta Dam and Reservoir. (*Irwin Remson.*)

CHAPTER THIRTEEN

investigation and management of hydrologic units

INTRODUCTION

Objectives

Chapter 3 describes the environmental role of surface and subsurface waters. This chapter introduces some of the procedures available for a preliminary assessment of the hydrology of an environment. We are primarily concerned with the development of water resources and with the minimization of pollution and other undesirable impacts of human occupation.

Summary of Approach

A useful initial step in the appraisal of the hydrology of a watershed is the estimation of gross water yield. This is the total flow (surface water and ground water) out of the watershed during a specified period without consideration of variations in flow during that period or in different parts of the watershed. Suppose gross water yield proves to be less than the water requirements of a proposed project. It will be necessary to recycle or import water, abandon or redesign the project, or reduce water losses, as from evapotranspiration. The two methods presented for estimating gross water yield are based on streamflow data and climatological data.

A detailed analysis of water distribution at different times and places in a watershed can be achieved if streamflow and ground-water data are available. The methods include the use of flow duration and flow frequency data. Ground-water analysis requires an understanding of geologic and hydrologic conditions and of the occurrence and movement of water in different hydrologic situations. Darcy's law aids in ground-water studies. Finally, pollutant analysis can provide estimates of possible pollution loads.

GROSS WATER YIELD

Introduction

To illustrate the meaning of *gross water yield*, let us consider the Pilarcitos Creek watershed in coastside San Mateo County, California (see Figure 18-1). Pilarcitos Creek emerges from the mountains onto coastal terraces along the north edge of the community of Half Moon Bay. The watershed is underlain by relatively impermeable bedrock. Ground-water seepage (*underflow*) around the margins of the watershed and through the narrow and relatively thin strip of stream-deposited sediment under the creek bed is small compared to surface flow. Part of the rain that falls in the watershed is returned to the atmosphere by evapotranspiration. The remainder follows surface and subsurface paths down the watershed. Sooner or later, the residues of this and other rainfalls reach Pilarcitos Creek and flow to the ocean. This flow is the gross water yield. It is the maximum amount of water that can be developed from the Pilarcitos Creek watershed during an average year, because gross water yield is based on long-term averages.

The gross water yield is determined with greatest accuracy from steamflow data. Where streamflow data are unavailable, climatic data may be used.

Gross Water Yield from Streamflow Data

For watersheds with small underflow the best way to determine gross water yield is from *records of stream flow*. If flow measurements

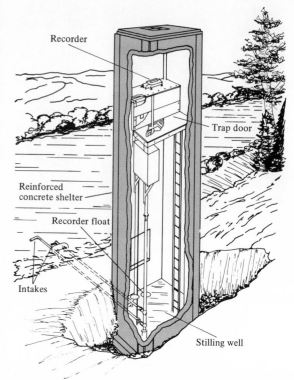

FIGURE 13-1. **Stream-gauging station. (*Modified from U.S. Geological Survey.*)**

have been made over a number of years (say, 30 years), the records may be averaged and projections made. It is assumed that the hydrologic and climatologic characteristics of the watershed remain unchanged. Serious problems in making projections are the lack of data for many watersheds and the fact that watershed characteristics may alter significantly during urbanization.

We will first consider methods for measuring and summarizing streamflow data. Reexamination of the Pilarcitos Creek situation will then show why lack of data forces the hydrologist on occasion to seek sources of data other than streamflow records for estimates of gross water yield.

Stream Gauging. Figure 13-1 shows a *stream-gauging* station. Intake pipes maintain the water level in a "stilling well" under the shelter at the same level as the water in the stream behind a low dam. A clock-driven mechanism records the time and the position of a float on the water in the stilling well. Thus, the equipment provides a record of the elevation of the water surface in the stream behind the dam.

Water-surface elevations are converted to streamflows using a rating curve. A rating curve is prepared by determining flows corresponding to different water levels. A technician measures water velocities at regular intervals across the stream by using a current meter (Figure 13-2). Stream discharge, or flow, is then determined by multiplying the velocity in each interval by the respective cross-section area. Because stream geometry can change, rating curves must be checked periodically.

Published Streamflow Data. Streamflow data are published by the U.S. Geological Survey in *Water Supply Papers.* Figure 13-3 shows the format used. It tabulates streamflow data for Bazile Creek near Niobrara, Nebraska, for the 1960 water year (October 1, 1959, through September 30, 1960). Bazile Creek is a perennial stream with discharge every day of the year.

Stream discharge in Figure 13-3 is in cubic feet per second (cfs). This is the number of cubic feet of water flowing through a stream cross section every second. One cfs amounts to almost 650,000 gal per day (gpd). Thus, 1 cfs could provide 6500 people each with about 100 gal (380 liters) of water per day.

Examples of Streamflow Data. Figure 13-4 is a *hydrograph* showing monthly mean values of daily streamflow rates in cubic feet per second of the Salem River at Woodstown, New Jersey. It was constructed from published data such as those in Figure 13-3.

The Salem River drains a humid area with mild winters. Precipitation is fairly constant throughout the year, averaging about 10 cm (4 in.) per month. The decrease in summer runoff is from increased evapotranspiration.

If daily streamflow were shown instead of monthly averages, the hydrograph would show individual peaks from runoff during and after storms. The smaller flow between storms would represent ground-water seepage to the stream and its tributaries. It is this ground-water seepage that keeps streams flowing in humid areas during lengthy dry-weather periods.

Figure 13-5 shows monthly and yearly mean discharge for the Mojave River at Barstow, California, over a 20-year period. It is apparent from the absence of discharge in many months that the Mojave River drains an arid region. In fact, the record shows entire years without streamflow.

The runoff in Figure 13-5 is related directly to precipitation. Therefore, the stream has no flow between the rare storms. Such a stream is *intermittent* (ephemeral) in contrast to *perennial streams* such as that represented in Figure 13-3.

Figure 13-6 shows monthly and yearly mean discharge for the Red River of the North at Drayton, North Dakota, over a 12-year period. The Red River of the North drains a region with severe winters. Winter precipitation falls as snow when streams are frozen. Therefore, Figure 13-6 shows an absence of streamflow during the winter. On the other hand, large flows occur during the spring, when winter snows and river ice melt. Flood forecasting requires periodic surveys of snow accumulation.

Example of an Area Lacking Streamflow Data. There is no stream-gauging station on Pilarcitos Creek (see Figure 18-1), and no systematically collected flow data are available for deter-

FIGURE 13-2. **Current meter. (*Modified from U.S. Geological Survey.*)**

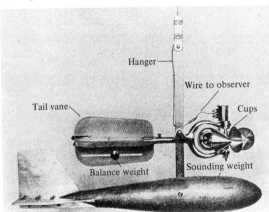

4665. Bazile Creek near Niobrara, Nebr.

Location. —Lat 42°45'00″, long 97°56'10″, in NE¼sec. 18, T. 32 N., R. 5 W., on downstream side of
left pier of bridge on State Highway 12, 2½ miles upstream from mouth and 4½ miles east of
Niobrara.

Drainage area. —440 sq mi, approximately.

Records available. —May 1952 to September 1960. Records for October 1931 to September 1932,
published in WSP 731, have been found to be unreliable and should not be used.

Gage. —Water-stage recorder for stages above 4.3 ft and wire-weight gage read once daily. Datum of
gage is 1,210.81 ft above mean sea level, datum of 1929, supplementary adjustment of 1949.
Prior to Dec. 16, 1952, wire-weight gage only and Dec. 16, 1952, to June 16, 1957, water-stage
recorder, above 4.2 ft, at present site at datum 4 ft higher. June 17, 1957, to Sept. 14, 1958,
water-stage recorder above 8.2 ft at present datum.

Average discharge. —8 years, 103 cfs (74,570 acre-ft per year).

Extremes. —Maximum discharge during year, 10,700 cfs Aug. 28 (gage height, 13.62 ft), from
rating curve extended above 5,600 cfs as explained below; minimum daily, 2.5 cfs Dec. 29.
  1952-60: Maximum discharge, 68,600 cfs June 16, 1957 (gage height, 19.96 ft, present datum,
from high point on surge), from rating curve extended above 6,500 cfs on basis of contracted-
opening measurements at gage heights 15.36 and 19.96 ft, present datum, minimum daily, that of
Dec. 29, 1959.
  Flood of June 19, 1951, reached a stage of 15.36 ft, present datum, from floodmarks
(discharge, 24,400 cfs on basis of contracted-opening measurement of peak flow).

Remarks. —Records fair except those for period of ice effect and those above 1,000 cfs, which are
poor. Discharge measurements generally made every two weeks. Minor diversions for irrigation
above station.

Revisions (water years). —WSP 1279: 1952. See also Records available.

Discharge, in cubic feet per second, water year October 1959 to September 1960.

Day	Oct.	Nov.	Dec.	Jan.	Feb.	Mar.	Apr.	May	June	July	Aug.	Sept.
1	39	62	71	17	70	32	4,670	146	174	195	38	109
2	39	53	63	22	71	31	1,380	146	120	132	36	92
3	35	56	92	25	50	31	195	114	109	132	34	56
4	36	56	84	22	76	25	795	132	92	80	34	52
5	35	42	73	21	80	26	1,460	322	89	153	36	33
6	34	30	74	22	76	28	1,680	1,890	89	92	46	34
7	49	40	85	38	80	28	984	840	86	83	46	29
8	103	50	70	55	80	31	480	488	89	74	50	139
9	48	75	70	54	87	28	244	251	223	80	54	89
10	35	90	71	54	80	29	114	83	846	71	46	52
11	42	70	57	56	52	29	132	80	585	62	44	36
12	40	60	57	59	53	28	1,170	46	398	96	35	54
13	42	55	61	61	55	30	1,480	59	315	74	34	52
14	44	50	62	59	63	29	578	59	197	59	35	68
15	44	53	59	39	64	39	345	50	216	62	40	48
16	44	52	68	36	63	42	258	251	896	80	35	65
17	45	32	87	50	59	52	209	237	322	173	42	68
18	42	52	52	46	62	61	109	322	244	567	109	62
19	42	54	56	44	50	59	132	2,060	195	279	77	77
20	44	60	58	38	41	58	96	1,170	322	65	68	74
21	49	72	65	33	48	61	89	1,550	382	48	54	65
22	54	76	60	27	50	68	74	578	223	52	42	71
23	44	74	48	32	45	68	65	472	188	59	46	114
24	41	73	38	33	49	64	77	315	139	50	450	74
25	40	69	48	37	45	65	68	577	104	48	300	68
26	45	49	25	40	44	72	62	895	80	44	181	62
27	54	45	9.0	39	39	300	80	412	132	48	146	74
28	48	57	5.4	44	38	1,000	114	308	805	56	4,830	68
29	64	70	2.5	49	34	1,100	272	265	587	80	2,200	62
30	94	83	2.9	52	---	1,850	195	216	286	56	668	59
31	71	---	18	64	---	1,650	---	195	---	44	237	---
Total	1,486	1,760	1,691.8	1,268	1,704	7,014	17,607	14,529	8,533	3,194	10,093	2,006
Mean	47.9	58.7	54.6	40.9	58.8	226	587	469	284	103	326	66.9
Ac-ft	2,950	3,490	3,360	2,520	3,380	13,910	34,920	28,820	16,920	6,340	20,020	3,980

Calendar year 1959: Max 4,790   Min 2.5   Mean 89.1   Ac-ft 64,500
Water year 1959-60: Max 4,830   Min 2.5   Mean 194   Ac-ft 140,600

Peak discharge (base, 1,000 cfs). —Apr. 1 (12 m.) 7,810 cfs (12.80 ft); Apr. 6 (4 a.m.) 2,450 cfs
(9.26 ft); Apr. 12 (11:30 p.m.) 3,800 cfs (10.60 ft); May 6 (2 a.m.) 3,580 cfs (10.40 ft); May 19
(5 a.m.) 7,750 cfs (12.78 ft); May 25 (11 p.m.) 1,340 cfs (7.88 ft); June 10 (11:30 a.m.)
1,710 cfs (8.36 ft); June 16 (11 a.m.) 1,580 cfs (8.20 ft); June 28 (3 p.m.) 1,520 cfs (8.13 ft);
July 18 (3 a.m.) 1,160 cfs (7.64 ft); Aug. 28 (5:30 p.m.) 10,700 cfs (13.62 ft).

Note. —Stage-discharge relation affected by ice Nov. 6 to Mar. 30.

FIGURE 13-3. (Opposite). Page from Water Supply Paper 1709, showing streamflow records for Bazile Creek near Niobrara, Nebraska. (The station is located using the U.S. Land Office System of Land Subdivision as described by J. W. Low, "Plane Table Mapping," New York, Harper and Brothers, 1952.) (*U.S. Geological Survey.*)

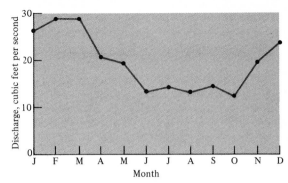

FIGURE 13-4. **Monthly mean values of daily streamflow rates for the Salem River at Woodstown, N.J., for the 9 years from October 1, 1941, through September 30, 1950.**

mination of gross water yield. Therefore, the hydrologist seeks to synthesize a streamflow record for Pilarcitos Creek by comparison with records from nearby gauged watersheds having comparable climatic and hydrologic characteristics. Unfortunately for this objective, this is a region of great climatic variability, with rainfall varying from 62.2 cm (24.5 in.) per year at the coast to 114 cm (45 in.) per year at the mountain crest only 6 to 8 km (4 to 5 mi) away. The vegetation, topography, and geology are equally variable. Furthermore, no nearby streams are

FIGURE 13-5. **Part of page from Water Supply Paper 1314, showing streamflow records for the Mojave River at Barstow, California. (*U.S. Geological Survey.*)**

273. Mojave River at Barstow, Calif.

Location. —Lat 34°54′25″, long 117°01′20″, in SW¼SE¼ sec. 31, T. 10 N., R. 1 W., 75 ft upstream from bridge on U. S. Highway 91 at Barstow.

Gage. —Water-stage recorder. Altitude of gage is 2,090 ft (from topographic map).

Average discharge. —20 years (1930-50), 37.4 cfs; median of yearly mean discharges, 1.0 cfs.

Extremes. —1930-50: Maximum discharge, 64,300 cfs Mar. 3, 1938 (gage height, 8.60 ft), by slope-area determination of peak flow; no flow for several months each year.

Remarks. —Diversions above station for irrigation of about 2,000 acres; slight regulation by Lake Arrowhead (capacity, 48,000 acre-ft, used principally for recreation).

Monthly and yearly mean discharge, in cubic feet per second.

Water year	Oct.	Nov.	Dec.	Jan.	Feb.	Mar.	Apr.	May	June	July	Aug.	Sept.	The year
1931	0	0	0	0	0	0	0	0	0	0	0	0	0
1932	0	0	0	0	*407	162	66.2	*2.34	0	0	0	0	*51.6
1933	0	0	0	0	0	0	0	0	0	0	0	0	0
1934	0	0	0	0	0	0	0	0	0	0	0	0	0
1935	0	0	0	0	0	0	19.9	0	0	0	0	0	1.64
1936	0	0	0	0	0	0	0	0	0	0	0	0	0
1937	0	0	0	0	534	770	383	66.0	0	0	0	0	143
1938	0	0	0	0	0	1,962	230	61.7	0	0	0	0	191
1939	0	0	0	0	0	.17	9.07	0	0	0	0	0	.76
1940	0	0	0	0	0	0	0	0	0	0	0	0	0
1941	0	0	0	0	222	738	547	93.5	0	0	0	0	133
1942	0	0	0	0	.5	1.0	.2	0	0	0	0	0	.14
1943	0	0	0	492	281	601	131	6.19	0	0	0	0	126
1944	0	0	0	0	86.5	318	177	19.0	0	0	0	0	49.9
1945	0	0	0	0	55.2	187	125	1.27	0	0	0	0	30.5
1946	0	0	2.98	0	0	61.0	145	0	0	0	0	0	17.4
1947	0	4.30	31.2	10.8	.42	.19	0	0	0	0	0	0	3.97
1948	0	0	0	0	0	0	0	0	0	0	0	0	0
1949	0	0	0	0	0	0	0	0	0	0	0	0	0
1950	0	0	0	0	0	0	0	0	0	0	0	0	0

*Revised

Location. —Lat 48°33′40″, long 97°10′30″, in NW¼SE¼ sec. 26, T. 159 N., R. 51 W., on
highway bridge at Drayton, at mile 208.2.

Drainage area. —34,800 sq mi, approximately (includes 3,940 sq mi in closed Devils Lake basin).

Gage. —Wire-weight gage. Datum of gage is 756.59 ft above mean sea level, datum of 1929.

Extremes. —1936-37, 1941-50. Maximum discharge, 86,500 cfs May 12, 1950 (gage height, 41.58 ft);
minimum observed, 7.7 cfs Oct. 16, 1936 (gage height, 1.75 ft).
Maximum discharge known since 1860, that of May 12, 1950. Flood of April 1897 reached a
stage of about 41 ft.

Remarks. —Some regulation by reservoirs on the tributaries.

Monthly and yearly mean discharge, in cubic feet per second, of Red River of the North at Drayton, N. Dak.

Water year	Oct.	Nov.	Dec.	Jan.	Feb.	Mar.	Apr.	May	June	July	Aug.	Sept.	The year
1936	–	–	–	–	–	–	5,768	1,826	399	118	50.1	27.4	–
1937	13.8	–	–	–	–	–	1,729	1,952	968	–	–	–	–
1941	–	–	–	–	–	–	–	–	–	–	–	–	–
1942	–	–	–	–	–	–	11,400	†7,003	–	–	–	–	–
1943	–	–	–	–	–	–	†20,640	–	–	–	–	–	–
1944	–	–	–	–	–	–	–	–	–	–	–	–	–
1945	–	–	–	–	–	–	15,870	6,780	3,504	–	–	–	–
1946	–	–	–	–	–	–	12,430	3,687	2,132	–	–	–	–
1947	–	–	–	–	–	–	–	10,360	11,140	5,267	1,762	–	–
1948	–	–	–	–	–	–	–	12,700	3,096	2,067	–	–	–
1949	–	–	–	–	–	–	11,440	2,997	*5,460	*3,425	*3,265	*1,737	–
1950	*1,125	1,283	*1,069	*798	*730	*979	31,120	58,890	15,360	8,463	3,325	2,300	*10,510

*Revised

†Not previously published; estimated on basis of adjacent record and records for stations at Oslo and at Emerson.

FIGURE 13-6. **Parts of pages from Water Supply Paper 1308, showing streamflow records for the Red River of the North at Drayton, North Dakota. (*U.S. Geological Survey.*)**

gauged. Therefore, it is not possible to use records from other stations to estimate runoff in this watershed. Fortunately, there are more weather stations than stream-gauging stations, and one can use climatological data to give estimates of gross water yield.

Gross Water Yield from Climatological Data

Climatological water-balance methods involve the computation of a water budget for the soil. Computed average monthly evapotranspiration is subtracted from the amount of water stored in the soil. Average monthly infiltrated precipitation is added to the stored water. Any water input in excess of the soil-water storage capacity is rejected by the soil to become direct stream runoff or percolates downward to become ground water.

A number of techniques are available for computing a climatological water balance. The *Thornthwaite method* requires only records of precipitation, mean daily temperature, and station latitude. While this method can lead to errors in specific applications, it provides at least a first approximation of a region's hydrology. The details of the method are presented in Thornthwaite's classic paper (see Additional Readings).

ANALYSIS OF STREAMFLOW DATA

Introduction

A knowledge of gross water yield cannot answer all hydrologic questions. For example, suppose the gross water yield is ample for the water requirements of a proposed project but the water is available only during one season or in a restricted part of the watershed. Then, detailed data will be needed, both on the distribution of the water yield through time and space and on variations in water requirements.

If streamflow records are adequate, a useful approach is to derive flow-duration and flow-frequency data.

Flow-Duration Data

Table 13-1 shows the derivation of the *flow-duration data* for the 21-year period from October 1931 through September 1952 for the Flint River at Genesee, Michigan. In this basin, the average stream discharge during a single month varied from 23 cfs (0.65 cu m per sec) to nearly 3000 cfs (84.9 cu m per sec). Column 2 shows the number of months for which the measured average stream discharge fell within a given range of values. For example, the average streamflow was between 23 and 49 cfs (0.65 and 1.39 cu m per sec) for 10 months in the 21 years. Average flows of 2000 to 2999 cfs (57 to 84.9 cu m per sec) occurred only during 4 months in the 21-year period.

The number of occurrences is accumulated from the bottom upward in column 3. Thus, in the 21-year period, there were 150 months in which the average stream flow exceeded 150 cfs (4.2 cu m per sec). Finally, column 4 lists the percent of the total time that the given discharge range was equaled or exceeded. Thus, the streamflow equaled or exceeded 23 to 49 cfs (0.65 to 1.39 cu m per sec) 100 percent of the time. If this is all the water that is required, it could have been obtained at any time during the 21-year period. However, if one requires 500 cfs (14.2 cu m per sec), one can obtain this amount of water (or more) only 23 percent of the time. Therefore, any development plan will either have to accommodate to the lower water supplies or arrange for storage facilities to provide water during the deficiency periods.

Figure 13-7 shows flow-duration curves for two other Michigan streams. For the Manistee River, the variation in runoff is only from 0.75

Table 13-1. Derivation of flow-duration data for the Flint River at Genesee, Michigan, for October 1931 through September 1952. (C. O. Wissler and E. F. Brater, "Hydrology," John Wiley & Sons, Inc, New York, 1963)

(1) Average Monthly Discharge, cfs	(2) Occurrences in 21-yr period	(3) Accumulated Monthly Occurrences	(4) Accumulated Percent of Total Occurrences
23–49	10	252	100.0
50–99	54	242	96.0
100–149	38	188	74.6
150–199	16	150	59.5
200–249	20	134	53.7
250–299	14	114	45.2
300–349	10	100	39.7
350–399	9	90	35.7
400–499	23	81	32.1
500–599	11	58	23.0
600–699	8	47	18.7
700–799	6	39	15.5
800–899	5	33	13.1
900–999	4	28	11.1
1000–1999	20	24	9.5
2000–2999	4	4	1.6
Total	252		

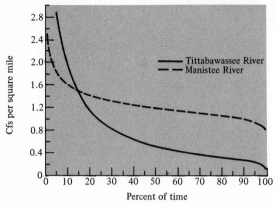

FIGURE 13-7. Flow-duration curves for the Manistee and Tittabawassee Rivers, Michigan. (*Modified from C. O. Wissler and E. F. Brater, "Hydrology," John Wiley & Sons, Inc., 1963.*)

to 2.5 cfs (0.02 to 0.07 cu m per sec) per sq mi. This stream has significant flow at all times and very few floods. The geologic conditions are such that a significant portion of the rainfall infiltrates the ground and becomes ground water. The ground water is available for discharge to streams in dry periods between storms.

The Tittabawassee River in Figure 13-7 has a streamflow-duration curve that varies from 0.1 to 2.9 cfs (0.003 to 0.08 cu m per sec) per sq mi. It has smaller dry weather flows and larger floods than the Manistee River. Such a stream is said to be more "flashy." Less rainfall from storm events enters ground-water storage. Direct storm runoff is more intense, and fair-weather ground-water discharge to the stream is smaller.

Flow-Frequency Data

Flow-frequency data indicate how often one might expect a flood or a drought of a given intensity. These data are needed to design road culverts and flood-control works. Drought frequency data are needed in planning water supplies. The example below deals with a drought situation; the same procedures may be used for floods.

Assume that one has a 100-year record of streamflow at a gauging station. Assume further that one is interested in droughts of 120-day

duration. Then, one would locate the driest 120-day period in the 100-year record and assign it the number "1" as in Table 13-2. During this period, the average flow was 500,000 gal per day (gpd) (1,900,000 liters per day). The second driest 120-day period in the 100-year record had an average flow of 700,000 gpd (2,700,000 lpd). This is assigned the number "2". When all the 120-day droughts are assigned numbers and listed in order of decreasing severity, the first 2 columns in Table 13-2 are obtained. Column 3 shows the computation of the *recurrence interval*. Thus, a 120-day drought in which the average flow is only 500,000 gpd would be expected on the average once in 100 years. A 120-day drought in which the average flow reaches 980,000 gpd would be expected to occur on the average once in 10 years.

Assume that a water supply for 10,000 people, each using 100 gpd (380 lpd), is required. Then, 1 million gpd (3.8 million lpd) must be developed from the stream. Assume further that the water supply is designed to carry the population through the 120-day, 33-year drought but not the two more severe but rarer droughts. Whereas 1 million gpd is required, Table 13-2 shows that only 800,000 gpd (3 million lpd) would be available during this 120-day, 33-year drought. Therefore, storage would have to be built into the system to provide 200,000 gpd (760,000 lpd) over a period of 120 days. On the average of once every 33 years, the stored water would be used completely to supply the community through the drought. Should one of the rarer droughts intervene, the stored water would be inadequate. Presumably, it would be cheaper to truck water in or restrict water consumption during the 50-year and 100-year droughts than to build the larger water-storage facility required to meet the rare shortages.

GROUND WATER

Introduction

The role of ground water in matters of environmental concern was discussed in Chapter 3. In this section, we consider the appraisal of the ground water hydrology of a watershed, includ-

Table 13-2. Computation of recurrence intervals for 120-day drought flows. The recurrence interval increases with the severity of the drought

Number of the drought, m	Average 120-day flow, gpd	Recurrence interval, years: $T_r = \dfrac{t}{m}$
1	500,000	$\frac{100}{1} = 100$
2	700,000	$\frac{100}{2} = 50$
3	800,000	$\frac{100}{3} = 33$
4	850,000	$\frac{100}{4} = 25$
5	890,000	$\frac{100}{5} = 20$
6	920,000	$\frac{100}{6} = 17$
7	940,000	$\frac{100}{7} = 14$
8	955,000	$\frac{100}{8} = 12$
9	968,000	$\frac{100}{9} = 11$
10	980,000	$\frac{100}{10} = 10$

T_r = recurrence interval, years.

t = total number of years of record.

m = number of the drought. Droughts are numbered consecutively, beginning with the most severe. The higher the number of the drought, the less severe it is.

ing (1) hydraulic properties of the geologic materials, (2) geologic and hydrologic characteristics of the watershed as a whole, and (3) analysis and prediction of ground-water yield.

Hydraulic Properties of Geologic Materials

At an early stage in ground-water investigation, it is necessary to determine the nature and hydraulic properties of the earth materials that make up the "underground plumbing." The important hydraulic properties describe the capacity of the rock or sediment to (1) transmit water and (2) to store and release water.

Hydraulic conductivity is a measure of the capacity of a rock or sediment to transmit water under specified conditions. Hydraulic conductivity might be 1 cm per sec (34,000 ft per day) for a gravel, 0.1 cm per sec (3400 ft per day) for a clean sand, or 0.00001 cm per sec (0.34 ft per day) for a clay.

Storativity tells how much water can be obtained by lowering ground-water levels. For example, if the storativity of a sand aquifer is 10 percent, lowering the water level 5 m (16 ft) over an area would produce a volume of water equal to a depth of 0.5 m (1.6 ft) over the area.

The environmentalist can obtain much hydrogeologic information from a geologic map of the area showing the distribution of the unconsolidated and consolidated materials. Unconsolidated deposits consist of loose sediment. If the material is a clean sand or gravel, it can transmit, store, and release water effectively because of the interconnected pore spaces. If present in adequate thicknesses, such sediments can be used for ground-water supplies. On the other hand, a clay has poor hydraulic conductivity. It can store large amounts of water but will not release it to wells. Therefore, it has poor storativity. Such material would have poor ground-water possibilities. However, it often confines ground water in an artesian, or confined, aquifer as in Figure 13-8. All gradations are possible between the two types of unconsolidated material mentioned.

A consolidated rock such as slate or granite lacks the interconnected pore spaces needed to transmit, store, and release water. Such materials ordinarily constitute poor ground-water terrain. If the rocks are extensively cracked, however, their transmission properties may be

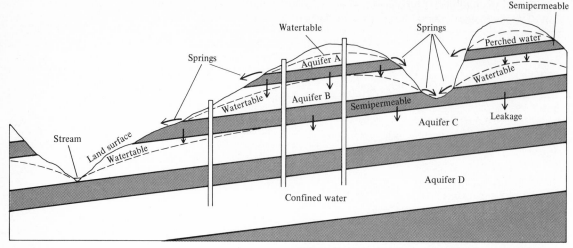

FIGURE 13-8. **Perched, unconfined, and confined ground water in a sequence of permeable sandstones (unshaded) and relatively impermeable shales (shaded).** (*Modified from S. N. Davis and R. J. M. DeWiest, "Hydrogeology," John Wiley & Sons, Inc., 1966.*)

improved greatly. Because their storage properties are rarely good, such fractured rocks yield large ground-water supplies on a continuing basis only if their fractures penetrate to a lake, a river, or some other source of stored water.

Limestones tend to dissolve and form sink holes, caves, and associated features. These solution cavities can be large and contain underground streams. The upper portions of some lavas are highly permeable due to cavities formed by escaping gas bubbles or to fracturing during flow. These permeable layers make excellent aquifers, as in the thick sequence of lava flows in the Columbia Plateau of northwestern United States. Some lava tunnels also contain streams. Underground streams, however, are relatively rare. Virtually all other ground water occurs in pores of unconsolidated sediments or in myriad cracks in consolidated rocks and moves at rates of only a few meters to a few hundred meters per year.

Geologic and Hydrologic Characteristics of the Watershed

The geologic characteristics of interest to the ground-water investigator are the kind and configuration of rock units. Are they layered and dipping as in Figure 13-8? Are they compact with negligible hydraulic conductivity and stor-

ativity? Are they disrupted by faults across which ground-water levels are markedly different? In essence, what is the nature of the plumbing in which the ground water occurs and moves?

The best plumbing, however, will stand empty if no water is available. Therefore, the hydrologic inputs as determined by the climate are the next concern. This may be illustrated by comparing ground water in a humid climate with that in a semiarid climate.

Figures 3-1 and 13-8 show watersheds in humid areas. Soil-moisture surpluses in excess of evapotranspiration percolate to and recharge the water table. The ground water moves in the general direction of the continuous arrows in Figure 3-1 and eventually discharges into the stream. Thus, the stream always receives ground-water discharge and is perennial. The water table is a subdued replica of the topography.

A typical hydrogeologic situation in a semiarid region is provided at Stanford University, 35 mi (56 km) south of San Francisco, California. The Santa Cruz Mountains west of Stanford are underlain by consolidated to poorly consolidated rocks. Stanford is in the San Francisco Bay Lowland, a down-faulted area that is filled with alluvial sediments from the adjacent mountains. The university is located on an alluvial fan. The fan consists of coarse sediment close to the Santa Cruz Mountains and progressively finer sediments outward toward the center of the valley. Similar alluvial fans (in a drier climate) are shown in Figure 5-4.

The annual precipitation of 38 cm (15 in.) on the fan takes place almost entirely in winter. Evapotranspiration uses up the infiltrated rainfall, and little or no ground-water recharge occurs from this source. Field studies show that infiltrated rainfall does not penetrate below a depth of 3.5 m (11 ft), whereas the water table is considerably deeper. In contrast with humid areas, the water table is well below the intermittent stream beds on the fans. These streams receive no ground-water inflow and flow only when supplied by direct storm runoff.

Hydrologic conditions are quite different in the adjacent mountains. Precipitation increases to over 76 cm (30 in.) per year toward the crest of the range. Temperatures and evapotranspiration are less than on the alluvial fan. Soils are thin with small water-holding capacities. Soil-moisture surpluses are generated, and recharge to ground water occurs during the winter and spring. The streams are perennial close to their heads, where they are fed by ground-water seepage throughout the year. During the summer and fall, none of this streamflow comes out of the mountains on to the fans. Rather, it is consumed by evapotranspiration and by seepage into the stream beds in the lower drier portions of the range.

The runoff from winter storms is heavy on the steep slopes of the Santa Cruz Mountains. After winter storms, streams rising in the mountains carry sufficient water to flow across the alluvial fans. At the heads of fans, the water percolates into the coarse sediments and recharges the ground water under the fans. This is the mechanism of ground-water recharge for the extensive aquifers in valley-fill sediments in the semiarid regions of the Western United States.

Analysis and Prediction

With knowledge of the geology and hydrology, it is possible to analyze a ground-water system and predict its response to proposed land-management schemes. The most important analytical relationship is *Darcy's law*. Through this law, the rate of ground-water flow q can be determined if hydraulic conductivity of the aquifer K, the ground-water gradient I, and the cross-sectional area of flow A, are known. Then, a modified form may be written

$$q = KIA$$

For example, suppose we are interested in a sand aquifer having a hydraulic conductivity of 900 gal per day per sq ft (0.04 liter per sec per sq cm), and 100 ft (30 m) thick. Suppose further that ground-water elevations in artesian wells fall 10 ft (3 m) in a horizontal distance of 1000 ft (300 m). We want to determine the flow through a 10-ft (3-m) width of aquifer.

Substituting the above values in the modified form of Darcy's law:

$$q = \overset{K}{} \qquad \overset{I}{} \qquad \overbrace{}^{A}$$
$$= 900 \times \frac{10}{1000} \times 100 \times 10$$

$$= 9000 \text{ gal per day (341 cu m per day)}$$

WATER QUALITY

Introduction

Problems of water quality were introduced in Chapter 10. A few additional topics of environmental interest merit consideration.

Coliform Bacteria

Escherichia coli (fecal coli) is a bacterium that is not harmful to humans. It spends part of its life cycle in the intestinal tract of a warm-blooded animal. It is easily detected by laboratory tests; hence it is an excellent indicator of the presence of animal wastes. The wastes, of course, may carry other organisms such as those that cause typhoid fever and other deadly diseases.

It would require tremendous amounts of water to dilute untreated human wastes sufficiently to meet U.S. Public Health Service standards for drinking water. Fortunately, the number of coliform and other organisms are reduced in waste-water treatment and by destruction in natural hydrologic systems.

In a study of the Upper East Branch of Brandywine Creek in Pennsylvania, measurements yielded high coli counts. Investigators attributed this to discharge of raw sewage directly into the stream. Coli counts can be reduced dramatically if such wastes pass first through septic tanks and then percolate through the soil before reaching the stream.

Dissolved Oxygen

Dissolved oxygen (DO) is a determinant of water quality. If water has no dissolved oxygen, only anaerobic organisms (organisms that can exist without dissolved oxygen) thrive. These organisms generate methane, ammonia, hydro-

gen sulfide, and other unpleasant substances. Water with a dissolved oxygen content of 4 ppm can support some fish, but a higher content is needed to provide sparkling water with abundant game fish.

Consider an outlet discharging raw sewage into a stream. The sewage will have a certain *biochemical oxygen demand (BOD)*. This is the amount of oxygen required to stabilize the sewage by oxidation and convert it to plant nutrients. Aerobic bacteria, which are users of dissolved oxygen, play a fundamental role in this process. As time passes and the sewage moves downstream, this biochemical use of oxygen reduces the dissolved oxygen content of the water. At the same time as this deoxygenation goes on, there is reoxygenation, the restoration of oxygen to the stream through diffusion from the atmosphere and by the photosynthesis of primitive plants, the algae. The dissolved oxygen content of a stream is the result of both processes.

Sanitary engineers have made extensive studies of the deoxygenation and reoxygenation processes. Using such information, they can specify management procedures to maintain

Table 13-3. Milligrams of different contaminants added to each liter of ground water and surface water from different sources in southeastern Pennsylvania. (Personal communication from R. J. Schoenberger, Drexel University, 1967)

Contributions to Ground Water, mg/1				
Source	Ammonia	Nitrate	B.O.D.	Phosphate (PO$_4$)
Atmosphere	0.75	2.0		0.09
Agriculture	1.60	5.9		0.80
Domestic sewage	30.00	25.0	250	8.00
Forest and grass		3.0		0.92
Lawns (estimated)	0.80	4.5		0.86

Contributions to Surface Water, mg/1				
Source	Ammonia	Nitrate	B.O.D.	Phosphate (PO$_4$)
Atmosphere	0.75	2.0		0.09
Urban runoff	0.28	3.58		0.31
Domestic sewage	30.00	25.0	250	8.00
Agriculture	2.70	9.7		1.30
Forest and grass	0.10	0.88		0.07
Lawns (estimated)	1.40	5.30		0.70

bodies of water at specified levels of dissolved oxygen. Such procedures commonly require sewage treatment facilities.

An Example of Water-Quality Prediction

Table 13-3 was used to predict the water quality of the Upper East Branch of Brandywine Creek in Pennsylvania under specified land-management procedures. It includes estimates in milligrams per liter of different contaminants contributed to ground water and surface water from different sources.

The total watershed area devoted to different uses was first determined. From the study of streamflow data and precipitation records, it was possible to determine what portion of the streamflow was direct runoff and what portion was seepage from ground water. Using this information and the table, the amount of the specific contaminants added to the known proportions of ground water and streamflow were determined.

Of the sources in Table 13-3, domestic sewage is the largest contributor of pollution per volume of water. For the study basin, BOD at the stream outlet was computed at 33.6 mg per liter. Using standard sewage treatment, 95 percent BOD compensation was planned, and

this ceased to be a problem. However, ammonia, nitrate, and phosphate would not be reduced by this treatment and would be respectively 6.65, 12.5, and 2 mg per liter. These pollutant levels would create a *eutrophication* problem. In short, these *plant nutrients* would fertilize and facilitate the growth of stream algae. A chemical precipitation step in the proposed sewage treatment procedure was required to remedy this problem.

ADDITIONAL READINGS

Davis, S. N., and R. J. M. DeWiest: "Hydrogeology," John Wiley & Sons, Inc., New York, 1966.

Johnson Division, Universal Oil Products Co.: "Ground Water and Wells," Edward E. Johnson, Inc., St. Paul, Minn., 1972.

Leopold, L. B., and W. B. Langbein: "A Primer on Water," U.S. Geological Survey, 1960.

Linsley, R. K., and J. B. Franzini: "Water-Resources Engineering," McGraw-Hill Book Co., New York, 1964.

Meinzer, O. E.: "The Occurrence of Ground Water in the United States," *U. S. Geological Survey Water Supply Paper 489, 1923.*

Thornthwaite, C. W.: An Approach Toward a Rational Classification of Climate, *Geographical Review*, vol. 38, 1948, pp. 55–94.

CHAPTER FOURTEEN

remote sensing in environmental investigations

INTRODUCTION

Much information about the environment is now collected via air photos and other aerial pictorial representations, rather than by physical contact. This is *remote sensing*. Remote sensing is particularly useful for preliminary environmental reconnaissance, for early recognition of potential hazards, and for planning detailed field studies.

Cameras are *passive remote-sensing systems* in that they record only the solar radiation that is reflected from nonluminous objects and the radiation given off by burning or other luminous objects. Ordinary black-and-white film (*panchromatic*) records about the same range of radiation as the human eye. *Infrared film* extends the range beyond the red end of the color spectrum to the *near infrared*

Active systems are those that produce, transmit, and receive their own energy. *Infrared scanners* detect *far or thermal infrared*, the radiation beyond the near infrared. *Radar* exploits the radiation beyond the far infrared. The information transmitted and received by active systems is presented as TV-like *imagery* (Figure 14-1), that is, aggregates of tiny dots. Active systems are effective day or night. Photos taken from unmanned satellites are also transmitted to earth as imagery.

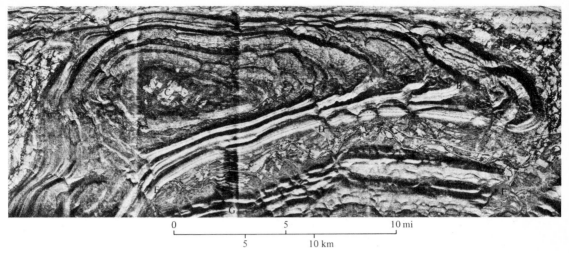

FIGURE 14-1. **Radar imagery of Tuskahoma Syncline, Oklahoma. The syncline (downfold) is indicated by the asymmetric ridges as at A and B. The gentler slopes indicate dips toward the center of the downfold at C. Note fault displacements of the strata at D and E. Interpretation is more difficult between F and G. (*Imagery by NASA.*)**

The most useful sensors in environmental geology are those that provide pictorial representations, namely cameras, thermal infrared devices, and radar. The advantages of these sensors are their ability to record data over large areas in a short time and to produce images suitable for interpretation. Ordinary black-and-white panchromatic photographs provide the major source of large-scale aerial data because they are readily available and relatively cheap.

In this chapter we shall emphasize the principles of interpretation from photos or photolike imagery.

Most panchromatic black-and-white air photos are vertical photos, that is, photos taken with the camera pointed directly downward. The most common print size is 23 by 23 cm (9 by 9 in.). The most common scale is 1:20,000 in which 1 mm on the photo is equal to 20 m (65 ft) on the ground, or 1 in. equals about 1700 ft. Each photo covers 25 sq km (9 sq mi). Overlapping photos can be viewed with a stereoscope to provide three-dimensional views. This helps in identification and mapping of topographic features, drainage lines, soils, vegetation, and geology.

Space photos and imagery provide regional views not otherwise obtainable (Figure 14-2). In addition, the repetitive imagery provided by orbital satellites provides valuable information on the nature and progress of environmental change.

A number of guides are used in interpretation of ground conditions from photos and imagery. The guides are applicable to photos of all scales and, with certain variations and restrictions, to much imagery.

INTERPRETATION GUIDES IN REMOTE SENSING

The three most important guides are photographic tone and color, drainage characteristics, and topography. Only those applications will be emphasized that provide data of interest to the environmentalist. Interpretations are greatly aided by stereoscopic examination.

Tone and Color

Tone refers to the shades of gray in black-and-white photos and imagery. Its significance in conventional black-and-white photographs dif-

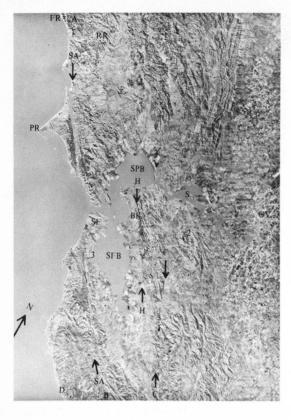

15 0 15 30 mi

15 0 15 30 km

FIGURE 14-2. **High-altitude photograph of the San Francisco Bay region. The area is 200 km (125 mi) long and 130 km (80 mi) wide. It extends from Fort Ross (FR) in the north to Davenport (D) in the south, and from San Francisco (SF) in the west almost to Sacramento in the east. Other places indicated are Point Reyes (PR), Russian River (RR), San Francisco Bay (SFB), San Pablo Bay (SPB), Suisun Bay (S), Berkeley (BK), and the Great Valley of California (GV) with its high-density agriculture. Arrows indicate faults: SA—San Andreas Fault; H—Hayward Fault, C—Calaveras Fault. Curved valleys, visible under a magnifying glass west of the number 3 indicate that the ground west of the right-lateral San Andreas Fault has moved to the north. Stratified rocks dipping gently toward the Great Valley are exposed along the mountain front at 1, and a domal structure appears at 2. (Photo Courtesy Lockheed Missiles and Space Corporation, 1956? Annotation by Arthur D. Howard.)**

fers from that in infrared photography, infrared thermal imagery, and radar imagery.

Conventional Black-and-White Photos. The tonal differences in ordinary black-and-white (panchromatic) photography are a measure of relative brightness. Brightness in turn largely depends on how effectively the object reflects sunlight.

The photointerpreter ordinarily accepts tonal variations as evidence of differences in ground materials and does not spend too much time trying to run down the exact causes of the variations in brightness. The photointerpreter recognizes, however, that in black-and-white photos:

1. Black and white soils and rocks ordinarily photograph black and white.

2. Yellow materials photograph brighter than those of other colors, with blue and red photographing dark.

3. Rough surfaces scatter the light in all directions and hence appear in medium grays.

4. Smooth surfaces may appear light or dark, depending on whether they reflect the sun's rays directly into the camera (glare) or away from the camera.

5. Moisture results in darker tones due to absorption of light.

In arid terrain, where surface materials are dry, different tones ordinarily indicate differently colored formations. The situation is more complex in humid environments.

Tonal patterns are also informative. In some photos the tone is uniform over large areas, indicating uniform ground conditions. In other photos, tonal differences are arranged in bands, suggesting the presence of unlike layered units. In still other photos, the tonal pattern is mottled or blotchy, indicating interspersed dry and moist areas.

Infrared Photos. Black-and-white infrared photos may be taken on infrared film that is sensitive to red wavelengths slightly longer than the visible red (*near infrared*).

Tones in near-infrared photos record variations in infrared radiation rather than differences in brightness. Infrared has the ability to penetrate light haze; hence, clear photographs can be obtained when conventional photographs are obscure. Infrared photos, however, do not show detail through fog, heavy dust, or clouds, and the shadows are dark and obscure detail.

Water absorbs much infrared radiation; hence, water bodies appear black and their boundaries can be mapped with great accuracy (Figure 14-3). On the other hand, the dark water tones conceal bottom topography which is revealed on ordinary black-and-white or color film. Infrared film is also useful in detecting variations in moisture content of soils; the higher the moisture content, the darker the

tone. It also has the ability to differentiate general tree types. Conifers or softwood trees, such as the pine and hemlock, reflect less infrared radiation than do broad-leaved hardwood trees such as the maple. Hence, conifers appear dark on infrared prints, while the broad-leaved trees appear light. This simple distinction is a great asset in forest inventory and management.

Color and False-Color Photos. Color photography permits even finer distinctions of ground materials than does black-and-white and has the advantage of revealing underwater topography to depths of as much as 30 m (100 ft). Furthermore, pollution in water is more clearly revealed than in black-and-white film.

False colors often emphasize surface contrasts. False color may be obtained using ordinary color film with filters to cut out particular parts of the color spectrum. A special false color film (camouflage detection film) is available commercially. This is useful in agriculture and forestry because healthy vegetation appears red, and deviations indicate diseased or dead vegetation.

Thermal, or Far-Infrared, Imagery. In the *thermal*, or *far-infrared*, wavelengths, beyond the photographic or near infrared, much of the radiation is absorbed by normal lenses; hence, special lenses or mirrors must be used. In the airborne scanning thermographic sensor, a narrow strip of terrain at right angles to the flight line is scanned by a rotating mirror. The radiation from each spot along the scan line is focused by the mirror on a detector which transmits an electric signal which is amplified and converted to visible light. The light is projected onto photographic film by a second

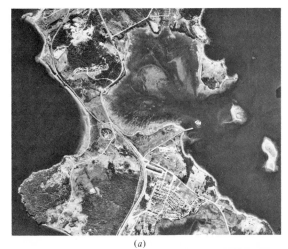

(a)

(b)

FIGURE 14-3. **Panchromatic (a) and infrared (b) photography, Mouth of St. Croix River, Maine. September 30, 1956. Note how sharp the shoreline appears in the infrared photo as compared to the panchromatic view. Note, too, in the infrared photograph the clear distinction between evergreens (dark) and deciduous trees (light) in the forested areas. (Aerial photographs courtesy U.S. Coast and Geodetic Survey.)**

scanning mirror while the film advances line by line as the aircraft moves ahead.

In thermal imagery, warmer areas are recorded in light tones and cooler in dark tones. Major causes of temperature differences are variations in geothermal heat and in moisture.

Radar Imagery. Radar, like thermal scanning devices, scans the ground below and records the results as a pictorial image. Some systems scan from horizon to horizon, but the most popular system is Side-Looking Airborne Radar (SLAR), which scans out to one horizon only. Radar has the ability to penetrate light clouds.

The tones of radar imagery are determined by the amount of transmitted energy that is reflected back to the instrument. The reflectivity depends on the angle of transmission of the impulse, the slope of the ground, surface roughness, and other factors. Radar has little potential in identification of materials because of the difficulty of determining which of the variables are dominant at any one place. The principal advantage is the high-contrast image it provides (see Figure 14-1). Regional geologic structures are vividly revealed by the tonal patterns. Building, roads, and other cultural features are highly reflective and appear light.

Disadvantages of radar are (1) scarcity of imagery, (2) high cost, (3) relatively low resolution (4) black shadows, and (5) limited opportunity for stereoscopic viewing.

Drainage

We will emphasize only those aspects of drainage that may contribute to environmental investigations. These include (1) drainage patterns, (2) drainage texture, and (3) gully characteristics.

Patterns. *Regional drainage patterns* are the designs formed by the aggregate of drainage-ways in an area. Because of their regional coverage, they have little application in environmental studies of the type herein considered.

Individual stream patterns may be more informative. Those with many right-angle turns, for example, commonly indicate intersecting fractures. An anomalously long straight stretch in an otherwise winding valley may indicate a fault. Ponds or marshes along a valley may be due to a variety of causes, including blockage by creep or landslides. And we noted in Chapter 7 that a beaded pattern in Arctic streams is indicative of permafrost.

Texture of Drainage. The texture or spacing of drainage provides useful preliminary data on the characteristics of surface materials. In some places, drainage lines are widely spaced; in others, they are densely crowded. The terms "coarse," "medium," and "fine" are commonly used to express the relative density.

Although drainage density is influenced by a variety of factors, variations within any small area indicate differences in the texture of the surface materials and in their permeability. In loose sediments, drainage density is related to grain size (Figure 14-4). Small rills can move particles of clay and silt and develop myriad small channels. On similar slopes, however, the removal of sand and gravel requires larger streams; that is, the accumulated drainage from larger watersheds, and the channels are therefore more widely spaced. In areas of hard rock, the decisive factor in drainage density is the size of fragments provided by weathering. Whereas a loose silt displays a fine texture of drainage, a hard siltstone breaking into large pieces presents a coarser texture.

Except on steep slopes, highly permeable materials such as sand and gravel soak up rainwater rapidly so that few if any surface channels develop. The texture of drainage is coarse. Clay and silt deposits, on the other hand, are relatively impermeable. Most of the rainfall cannot infiltrate and becomes surface runoff. The opportunity for channeling is therefore increased and the drainage texture is fine.

Vegetation, for reasons outlined in Chapter 3, reduces runoff and thereby discourages rill formation. In general, the texture of drainage in humid climates is coarser than in arid and is coarser on heavily vegetated than on relatively barren slopes.

Gully Characteristics. Gullies are channels or ravines too deep to be obliterated in ordinary tilling of the soil. Their characteristics provide information on soil properties that is useful in

FIGURE 14-4. Contrasts in drainage texture. The thin upper formation of coarse sands and gravels is highly permeable and little runoff takes place. The texture of drainage is therefore very coarse. The lower formation of silts and clays is impermeable and runoff is heavy. The fine sediment is easily eroded by closely spaced rills. Near Panaca, Nevada. (*Stanford photo collection.*)

first-stage environmental studies. Gullies are common in relatively impermeable materials where runoff is high; hence, they are most common in fine-textured materials with considerable clay or silt. They may occur in sandy and gravelly soils but only on steep slopes, such as terrace scarps, where runoff persists in spite of infiltration.

The informative characteristics of gullies are the cross profile, the heights of vertical faces, the long profile, and the surface pattern. Clay-rich soils are characterized by gullies in which the side slopes are convex upward (Figure 14-5*a*), reflecting the resistance of the cohesive materials to slumping. In moderately cohesive materials, the cross-profiles are U shaped (Figure 14-5*b*). The heads of U-shaped gullies, like the side walls, are vertical, and the gully floor gradients remain low to the very base of the steep heads.

In loose materials such as sand and gravel, the gully slopes represent the angle of repose, and the gully is V shaped. The long profile is short and uniformly steep. Only on short steep slopes can streams persist without disappearing entirely into the permeable materials (Figure 14-6).

Topography

Topography may provide important clues to materials and geologic structure. In flat areas, conventional vertical photographs taken at high sun angles provide little topographic information because of the absence of shadows (Figure 14-7). Conventional vertical photography taken at low sun angles, on the other hand, reveals a wealth of intricate detail (Figure 14-8).

Erosional Topography. Uniform slopes indicate materials of uniform resistance to weathering and erosion. Scattered haphazard breaks in slope are due to undercutting, landslides, or fractures.

Regularly interrupted slopes, on the other hand, display a systematic sequence of benches arranged contourlike on the slopes (see Figure

(a)

(b)

FIGURE 14-5. **Gullies in clay-rich deposits (*a*) and
in semiconsolidated sediments (*b*). (*U.S. Dept.
Agriculture.*)**

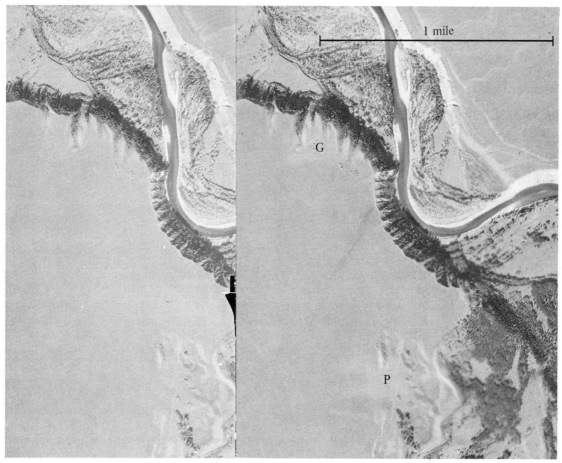

1 mile

FIGURE 14-6. **Stereoscopic view. Jackson Hole, Teton County, Wyoming. The terraces consist of loose, highly permeable sediments such as sand and gravel. The evidence consists of the absence of surface drainage, the steep, V-shaped gullies (G) at the terrace edge, and the excavations at P. Scale 1:25,000. (***Stereopair by Arthur D. Howard from* U.S. Air Force Photos.**)**

5-6). Individual benches may continue for long distances, or their remnants may be so numerous as to indicate former continuity. The benches are underlain by resistant strata; the slopes in between, by weak layers.

If the strata are dipping (inclined), and the slope of a resistant layer coincides with the dip, the slope will be planar, like the slope to the right of the tunnel entrance in Figure 2-12. If the surface slope is steeper, it will cut across the lower edges of the beds, as to the left of the tunnel entrance along the railroad and along the road above. And if the surface slope is gentler than the dipping layers, as in the upper parts of Figure 2-12, the layers will display an archlike

pattern between adjacent valleys. Where this pattern indicates a dip toward the open slope, the danger of rock slides down the layers is great. If the pattern indicates that the rock layers dip back into the slope, the danger of rock slides is less.

Dipping formations, when deeply eroded, may give rise to parallel ridges and valleys, the

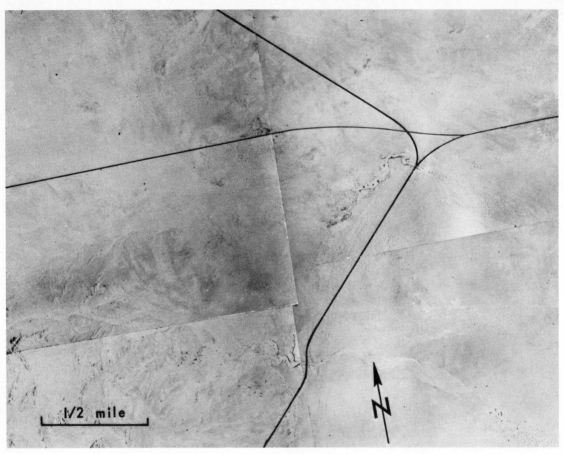

½ mile

N

FIGURE 14-7. **Vertical, high-sun-angle photography of test area 8 mi northwest of Imperial, California. The washed-out appearance results from absence of shadows. The paved road running from top to bottom was not in existence at the time of photography but has been added to assist in comparison with Figure 14-8. (*Photography by Robinson Aerial Surveys for U.S. Geological Survey, May 1953.*)**

ridges composed of resistant formations and the valleys, of weak formations. The profile and pattern of the ridges provide information on the direction and steepness of dip and enable the geologist to determine the geologic structure below ground.

Many ridges are asymmetric in cross-profile, with one side gentle and the other steep. The gentle slope generally indicates the direction of dip. On the dip side, sheetlike rockslides may take place if the slope is oversteepened, whereas rockfalls and complex landslides may result from undermining the scarp side.

Depositional Topography. Depositional landforms may be diagnostic of the ground materials and therefore of properties such as strength, permeability, erodibility, and fertility. These depositional features include landslide deposits, alluvial fans, floodplains, deltas, sand dunes, moraines, and volcanic features. Later, we shall illustrate some of these deposits with stereoscopic three-dimensional views.

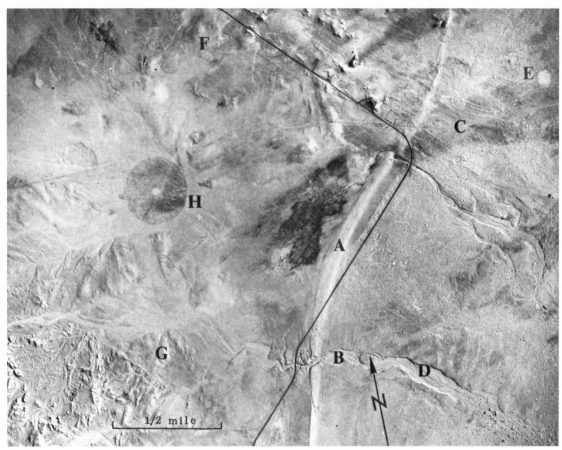

FIGURE 14-8. **Vertical, low-sun-angle photograph of area in Figure 14-7. Note the wealth of detail revealed. The highest shoreline of an ancient lake is at A. The area to the east is part of the former lake floor. Lower shorelines are visible in several places, as north and south of the channel at B. Possible bedrock trends are indicated in the vicinity of C. A pattern reminiscent of a geologic fold appears north of D, but this may be a coincidental combination of shoreline traces and flood swales. The oval pattern at E is a barren patch on the desert floor. Excellent detail is exhibited in the hilly area in the southwest. Between these hills and the paved road to the north the topography appears smooth. This effect has probably been artifically produced by grading operations within this military range. Evidence of such grading may be indicated by the furrowlike patterns at F and G. The circular feature at H is probably a target area. (*Photo by J. Mercado Aerial Surveys, Redwood City, California. Annotation by Arthur D. Howard.*)**

EXAMPLES OF REMOTE SENSING IN ENVIRONMENTAL INVESTIGATIONS

The applications of remote sensing herein presented relate to the major environmental fields discussed in Part Two of this text. We shall concern outselves primarily with the readily available black-and-white vertical air photographs and particularly with overlapping photos amenable to three-dimensional viewing.[1] Infrared photos and infrared and radar imagery will be considered where applicable.

---

[1]Inexpensive pocket stereoscopes are available for viewing the stereopairs herein presented.

Mass Wasting

Mass movements vary from obvious to obscure in air photos. Figure 14-9 is a stereopair of a landslide in western Wyoming. Once the landslide characteristics are recognized, the siting of structures almost anywhere in the area must be regarded as hazardous in view of the possibility of recurrent movement of the old slide or of new slides nearby. Even the floor of the main valley is hazardous because of the possibility of new movements blocking the valley, causing inundation upstream, and raising the threat of later catastrophic flooding.

A more obscure situation is revealed in Figure 14-10 in the coastal range south of San Francisco. The primary hazard is provided by small slides any one of which could be dangerous to a single dwelling or group of dwellings. Some of the slides outlined are only 30 m or so in width; others are more than 100 m across. Some are too small to recognize at this scale (1:20,000), and others are concealed in the brush and trees.

Large-scale infrared photography and thermal scanning imagery assist in recognition of obscure landslides because of the ability of infrared to detect moisture differences (Figure 14-11). Note how the landslide lobes are outlined in the predawn infrared image.

Temperature and vegetation patterns in thermal imagery and air photos around Bartow, Florida, are being investigated as possible clues

FIGURE 14-9. **Landslide, Teton County, Wyoming. Note peripheral cracks and slump blocks in upper reaches and the lower irregular topography resulting from turbulent flow. Some of the debris has flowed out of the landslide valley, partially blocking the main valley. (*Stereopair by Arthur D. Howard from U.S. Dept. Agriculture photos.*)**

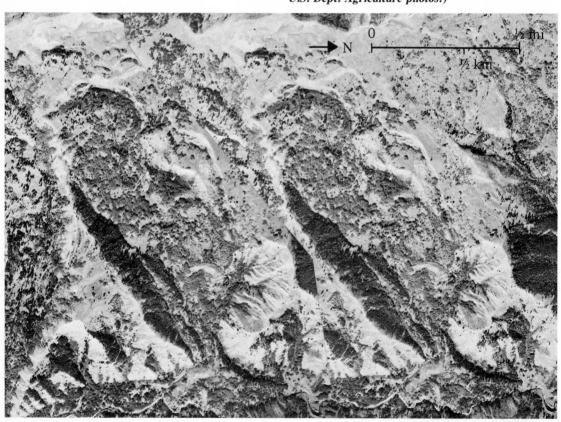

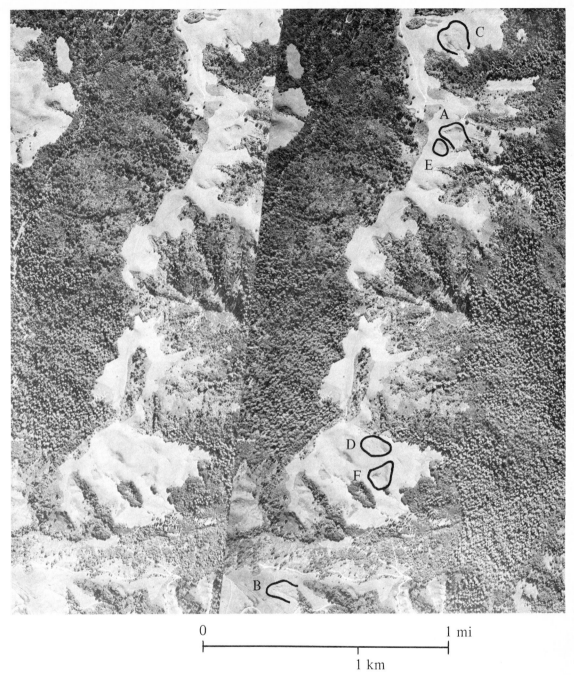

0   1 mi

1 km

FIGURE 14-10. **Small landslides, San Mateo County, California, near Stanford University. The letters may be disregarded. Scale 1:24,000.** (*Stereopair by Arthur D. Howard from photos by Fairchild Aerial Surveys.*)

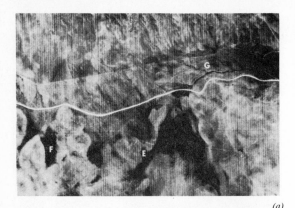

(a)

(b)

FIGURE 14-11. **Use of thermal infrared imagery in detecting landslides. (a) Presunrise infrared image. The light-toned areas extending up from the bottom of the view are lobelike landslide masses, whereas the dark intervening areas are undrained or poorly drained depressions. The dark cooler tones of the depressions are attributed to greater moisture content and a denser cover of vegetation than in the landslide areas. The road skirts the ends of the landslide lobes. (b) Aerial photograph showing the same slide areas with little indication of the slides. The San Andreas Fault follows the base of the hills. (*Prepared by NASA and U.S. Geological Survey.*)**

to the presence of limestone caverns subject to collapse.

Hydrology

Conventional vertical air photos are reasonably accurate maps from which many of the factors that determine the hydrologic budget of a watershed may be assessed. These factors include the area of the watershed; the number, length, and density of drainageways; the declivities of slopes; the kinds and distribution of vegetation; the types of soil with respect to permeability and erodibility; the extent of agriculture; and the distribution of human works.

Figure 14-12 shows a floodplain and adjacent ground in Quitman County, Mississippi. Suppose the area labeled A is being considered as a development site. In favor of the site is the sandy, permeable soil as indicated by the light tone and the absence of surface drainage lines. Note, however, the faint concentric arcuate pattern to the left of A and within the broad curve at B. These are bars (light lines) and swales (darker traces) representing stages in the growth and migration of the meanders. These features, plus the scattered oxbow lakes and traces of older cutoffs (C, D, E), indicate that the river is not only shifting position gradually by lateral erosion but also suddenly by flooding across narrow meander necks as at F. The meander neck at G is in imminent danger of being cut through in a flood. The old cutoffs, completely filled as at E and F, consist of finer sediment, commonly clay-rich with abundant organic matter, and are poorly drained. The oxbow at C, and the group at D, display various stages of filling and revegetation. This river bottom, with its bars, swales, and cutoffs, is clearly susceptible to floods, as well as to bank erosion. A much safer site for development is the terrace H, where there is no evidence of modern floodwater invasion.

Black-and-white air photos commonly reveal plumes of sediment- or pollutant-laden waters in rivers and lakes. Figure 14-13 shows dark plumes of raw sewage entering the Anacostia River in Washington, D.C., in April 1961. This case was unusual in that no color difference was detected either by visual observation or in color film. In most other instances, color and false-color photos are more revealing.

Infrared thermal imagery, taken at night when extraneous thermal influences are at a minimum, is also useful because it records differences in water temperature. Plumes of cooler water from the land entering the warmer waters of a lake or the sea provide information on new sources of fresh water in coastal areas

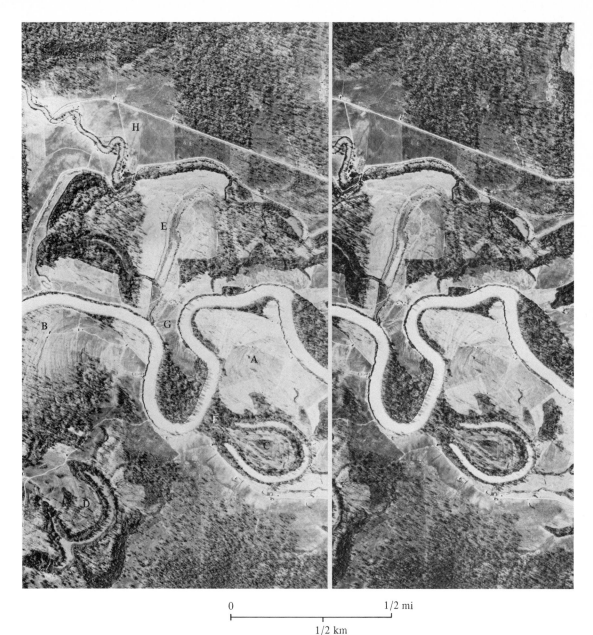

0   1/2 mi
├─────────────────────────┼──────────────┤
                    1/2 km

FIGURE 14-12. Floodplain, Quitman County,
Mississippi. A, B—parts of floodplain, showing
curved bars deposited during lateral shifting of the
stream. C— oxbow lake, a cutoff meander. D, E—
older cutoffs. F, G—meander necks, the constricted
portions of meanders. H—terrace. (*Stereogram
prepared by University of Illinois Committee on
Aerial Photography. Annotation by Arthur D.
Howard.*)

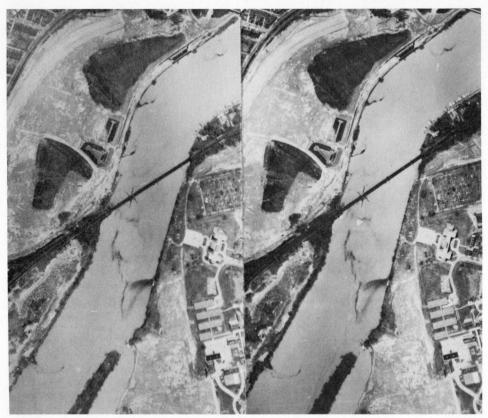

FIGURE 14-13. Raw sewage entering Anacostia
River in Washington, D.C., April 1961. (*Air
Photographic Inc., Wheaton, Md. Courtesy Carl
Strandberg, Itek Corp.*)

FIGURE 14-14. Thermal infrared image of
freshwater discharge into the ocean. Balayan Bay,
Luzon Island, Philippines. The cool, hence dark,
surface and ground waters contrast with the
warmer, light-toned ocean and estuarine waters.
(*U.S. Air Force, 1967. Interpreted by U.S. Geological
Survey.*)

(Figure 14-14). Such thermal imagery has been used in mapping submarine freshwater outflows along stretches of the coast of Long Island. In summer, the escaping ground water appears dark because it is cooler than the ocean water; the situation is reversed in winter.

Thermal imagery has also revealed changes in stream temperatures due to human activities, has delineated fault traces, and has led to detection of springs, particularly warm springs. The detection of springs should be advantageous in cold regions as clues to sources of ground water and to sites of winter icings.

The value of satellite imagery in regional hydrologic investigations is illustrated in Figure 14-15. The left view shows the Illinois, Missouri, and Mississippi Rivers at normal water levels on October 2, 1972. The right view shows the areas submerged under still-rising floodwaters on March 31, 1973. Visual records of such great floods are useful in planning for future floodplain development, in preliminary assessments of damage, and in evaluating flood-control systems.

Hydrologic investigations in many parts of the world must also consider the impact of precipitation in the form of snow. Satellite imagery is especially useful in this respect. The recurrent overviews provide information on the growth of the snowpack and, through previous ground experience, the probable thickness. The start and the rate of melting can be inferred

FIGURE 14-15. Satellite views of the St. Louis, Missouri, area before (left) and during (right) the Mississippi River flood of 1973. The flood crest at St. Louis was 12 m (38 ft), the highest since 1903. The flood view will serve as a permanent record of areas subject to inundation in the once-in-a-lifetime Mississippi Valley flood and will assist in floodplain planning. (*U.S. Geological Survey/NASA. Earth Resources Satellite Number 1007-18143.*)

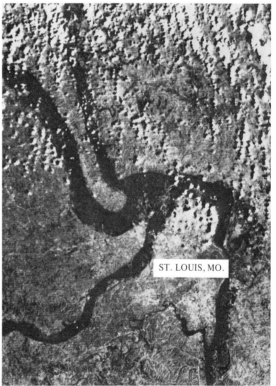

from the rate of shrinkage of the snow cover, and this information may be useful in assessing possible flood dangers.

Coastal Environments

We have noted the value of infrared photography in precise shoreline mapping (see Figure 14-3). Photographs and imagery are also useful in (1) providing general and three-dimensional views of coastal topography, (2) detecting coastal currents, (3) providing visual data on waves and wave erosion, (4) providing data on bottom topography in shallow water, (5) indicating sites of sediment accumulation, (6) revealing effluents from the land, and (7) mapping of estuarine and other wetlands.

Figure 14-16 demonstrates the value of air photos in revealing the magnitude of hazards along barrier islands. Prior to early 1933 there was no inlet between A, the southern limit of the barrier island (Fenwick Island) on which Ocean City, Maryland, lies, and C, the north end of the barrier island known as Assateague Island. Formerly, Fenwick and Assateague Islands were joined in an unbroken barrier beach 65 km (40 mi) long. The inlet was opened in a hurricane in August 1933 and was stabilized in 1935 when jetties were built out on both sides of the inlet. A northern jetty at A, extending out of the view to the right, plus the many groins along the coast to the north, stopped much of the southward littoral drift, robbing Assateague Island of the sand with which it was formerly able to repair storm damage. Within 20 years of losing its natural supply of sand, Assateague Island beach had receded 450 m (1500 ft). By 1961, it had been beaten back past the west end of the southern jetty at C, leaving open water at the inner end of the jetty. On March 5, 1962, a great storm created the havoc shown in the figure.

Figure 14-17 is a stereopair of a low plains coast, the former bed of a large glacial lake. Because of the very low relief, stereoscopic examination adds little to the view. The straight, light-toned features are low, relatively dry beach ridges, whereas the mottled areas are largely waterlogged.

Air photos also reveal bottom topography. With stereoscopic mapping devices, underwa-

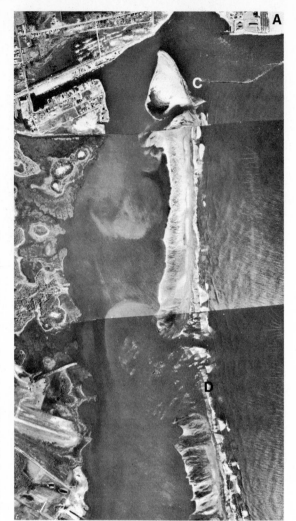

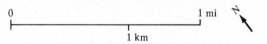

0   1 mi

1 km

FIGURE 14-16. **The hazardous environment provided by barrier islands is illustrated in this photo mosaic. Prior to early March 1962, the barrier island (Assateague Island) was an unbroken strip of sand 350 to 400 m (1150 to 1300 ft) in width. A hurricane on March 5 created the havoc shown in this view, taken on March 22. New inlets were opened, many overspill sluice channels were created, and much of the surface was so lowered by erosion as to become shoals. Ocean City, Maryland. (*U.S. Coast and Geodetic Survey. Modified from F. P. Shepard and H. R. Wanless, "Our Changing Coastlines," McGraw-Hill Book Company, 1971.*)**

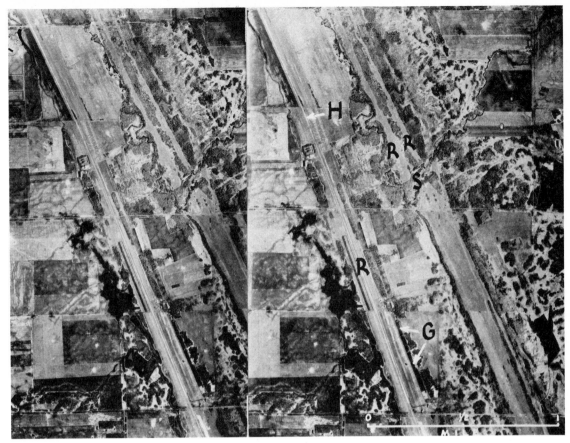

FIGURE 14-17. **Part of the floor of an ancient glacial lake, southern Manitoba, Canada. The beach ridges (R) show up in relatively light tones because they are higher and better drained than many of the intervening strips. The gullies at G and near H are too small in this view to reveal details, but their density suggests that the materials of the ridges are silt or fine sand rather than coarse sand or gravel. The darkly mottled intervening areas are much moister. The stream (S) has cut through the eastern ridges. Note that well-drained beach ridges form natural sites for highways (H). (*Photos by Royal Canadian Air Force. Stereopair prepared by Cornell University Center for Aerial Photographic Studies, Landform Series, vol. 4-ONR, 1951.*)**

ter topography in clear water can be surveyed with almost the accuracy achieved by similar methods on land. Color photographs provide the most accurate results. Under ideal circumstances, they can reveal the bottom to depths of over 30 m (100 ft).

Color is also useful in delineating salt-water invasions along low-lying coasts. This is possible because of the distinctive colors of plant communities which thrive in different salinities and under different tidal conditions. Color is also effective in mapping water currents, water pollution, and kelp beds.

The New Jersey Wetlands Act of 1970 requires mapping and inventory of all wetlands along the coast to ensure proper management. Preliminary investigations were carried on at two test sites using remote sensing (see Ander-

son and Wobber in Additional Readings). Natural color and color infrared photography were obtained at a scale of 1:12,000. Maps were prepared showing the upper wetlands limit, the line of biological mean high water to establish state riparian lands, and major plant associations of 2 ha (5 acres) or more. Vegetation types proved to be sensitive indicators, recording small changes in moisture and salinity. Thus, some grasses flourish in environments flooded only during the bimonthly "spring" tides, others in somewhat more frequent floodings, and still others in daily intertidal environments. The 21 final maps for the two sites, at a scale of 1:24,000, were completed in 105 days.

Arid Environments

As in humid environments, photos and imagery assist in the mapping of watersheds and drainageways, in determining slope declivities, in mapping vegetation, in determining flood limits, and in mapping fault traces. Specialized uses include the location of sites favorable for wells, the locations of springs and seeps, the identification and mapping of water-wasteful phreatophytes, the determination of sites of fluvial deposition, the location of areas of ground subsidence, and the assessment of the role of the wind in erosion, movement, and deposition of sand.

Many large alluvial fans bordering desert mountains are capable of providing copious supplies of ground water. Figure 14-18 shows a large alluvial fan in Death Valley. Note the fringes of vegetation starting well down the slopes; the beginning of the fringes indicates the seep line where ground water migrating through coarse, permeable beds in the upper part of the fan is blocked by finer, relatively impermeable deposits farther down. The possibilities for tapping ground water are best in the more permeable beds above this line. Air photos are helpful in identifying other landforms that are potential ground water sources and in eliminating others that are not.

On *panchromatic film* most soil tones in an arid environment are light. The presence of moisture, however, results in darker tones. In nighttime *thermal infrared* imagery, water and moist areas appear light. This is because it is heat radiation that is mapped and the temperature of the water is higher than the nighttime temperatures of the dry soil and rock. Figure 14-19 is a predawn thermal infrared image of a site where alluvial fans meet the

shoreline of a former lake. The light strips, splotches, and rectangles represent moist ground where ground-water discharge and seepage occurs or where fields have recently been irrigated.

Figure 14-20 is a stereoscopic view of a long-inactive and largely effaced dune area. The level topography encouraged farmers to clear off the vegetation for agriculture. This exposed the roots of the ancient dunes. Examination of views such as this would have led to recognition of the ancient dune pattern and the poor sandy soil, prior to cultivation.

Rain Forest Environments

Rain forest environments are as diversified as those elsewhere. Even in relatively flat areas, the forests do not completely conceal the drainage lines or hide the topography. They tend to form a more or less uniform cover whose surface irregularities simulate the topography below. Drainage lines, even in flat areas, may be easily traced by characteristics of the vegetation which thrives in flood conditions (Figure 14-21).

Figure 14-22 is an area of much greater relief. The flat-lying formation in the lower half of the view is limestone as indicated by the many sinks, the advanced karst topography at K_1, the knobby border of the upland with outlying knobby remnants at K_2 and K_3, and the disappearing and reappearing river R1. The river disappears underground at R1, reappears on the floor of the larger sink S1, disappears underground again to reappear on the floor of sink, S2, and passes underground to reappear in sink S3, from which position it follows a continuous valley. Collapse of the surface over hidden caverns constitutes a hazard, even though habitations are far between as in the clearing C.

The rocks in the upper half of the view are normal ridges and valleys without undrained depressions. The streams flow toward the limestone upland where they disappear underground. The rocks in the upper part of the view are clearly not limestone and do not present danger of collapse.

Infrared photography is useful in the rain forest because of its ability to penetrate haze. Although superior to panchromatic in this re-

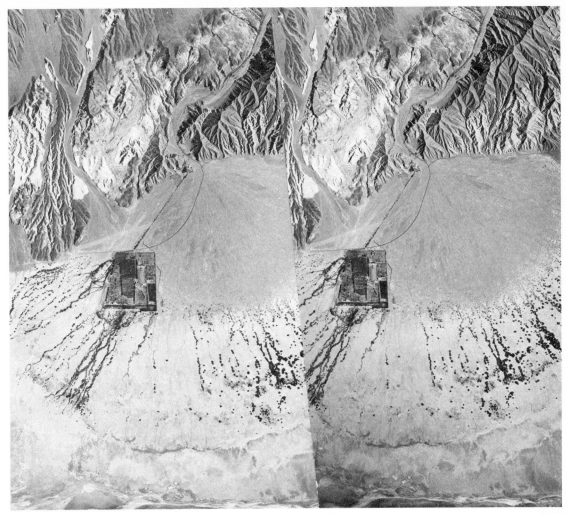

0   1   2 mi

1   2 km

FIGURE 14-18. Alluvial fan at Furnace Creek Ranch, Death Valley, California. The upper boundary of the streaks of vegetation marks the seep line. The seeps provide enough moisture to support vegetation along the channels. Scale 1:65,000. (*Photos by U.S. Geological Survey. Stereopair by Arthur D. Howard.*)

spect, it lacks the ability to penetrate clouds.

Radar is particularly useful in rain forests that are under clouds almost the entire year. Radar can penetrate all but the heaviest clouds and has provided maplike imagery of regions hitherto largely unmapped. Darien Province of southern Panama, a region of perpetual cloud cover, was mapped by radar in a few days to present the first clear and reasonably detailed view of this province.

The most grandiose radar project, involving preparation of maps and their environmental interpretation is Projeto Radam (Radar Mapping of the Amazon), initiated in 1970. The general intent of the program was to provide knowledge of the biologic and physical aspects of the Amazon basin prior to human utilization.

FIGURE 14-19. Nighttime thermal infrared imagery of the contact area of an alluvial fan and an old lake bed. The moist ground is warmer than the autumn nighttime temperatures of the dry soils and rock and shows up in lighter tones. The open water in the pond and canals appears almost white. Progressively darker tones indicate lesser moisture contents. (*Courtesy HRB-Singer Inc.*)

FIGURE 14-20. Ancient dune field, Newton County, Indiana. The dune field has been reduced to an almost level plain, with only the ground plan remaining. The well-drained dune sands contrast in tone with the darker less permeable soils around them. The arcuate forms (D) suggest original barchan or parabolic dunes. (*Stereopair prepared by Cornell University Center for Aerial Photographic Studies, Landform Series, vol. 6-ONR, 1951. Photos by U.S. Dept. Agriculture.*)

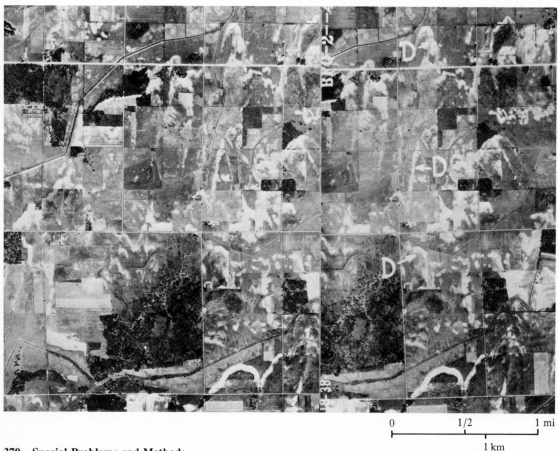

0   1 mi

1 km

FIGURE 14-21. **Drainage lines in flat terrain, Amazon basin, Brazil. Small tributaries (T) are clearly defined despite being completely hemmed in by high-crown trees. Scarplike sides of larger valleys (L) are the result of abrupt change from high-crown trees above flood level to low-crown trees on flood plains. (*Photo by Petroleo Brasileiro, Brazil [Petrobras].*)**

The project was scheduled to cover 4,800,000 sq km (1,850,000 sq mi).

The first phase of the project was to obtain radar imagery in sheets 1° by 1°30′ at a scale of 1:250,000 (Fig. 14-23). Each radar map thus embraces about 18,000 sq km (7000 sq mi).

The second phase was the organization of six scientific research sections: cartography, geology, vegetation, soils, geomorphology, and land use. The 1:250,000 radar maps serve as the basis for interpretation by each of the sections. Infrared photos at a scale of 1:130,000, where not obscured by clouds, were used to assist in the interpretation, as were multispectral photos at 1:70,000. Interpretation was spot checked in the field. If the work schedule is maintained, the entire Amazon basin will have been mapped and described in reconnaissance fashion and its potential land use proposed, all within 6 or 7 years. The results will consist of 21 reports with maps at a scale of 1:1,000,000. Most of the reports are in print.

The prospects for agriculture, the raising of livestock, lumbering, and plant-extraction activities, as tapping trees for rubber manufacture, are evaluated. Areas are recommended for

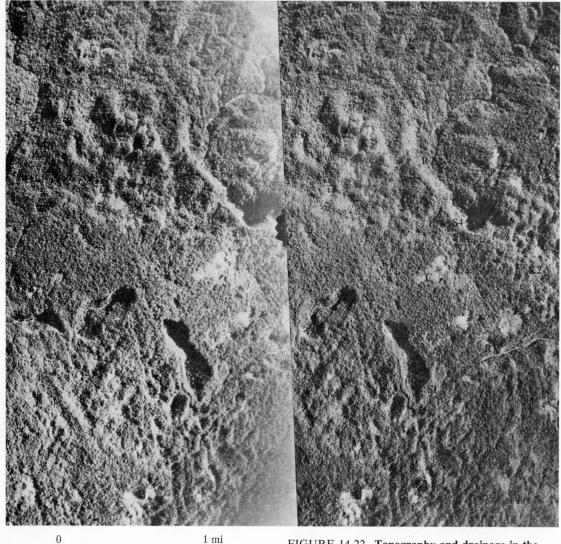

0   1 mi

1 km

FIGURE 14-22. **Topography and drainage in the rainforest of New Guinea, Indonesia. The limestone upland in the bottom half of the view shows many sinkholes. The river, R1, disappears underground to reappear on the floor of the sinks S1, S2, and S3. Advanced karst topography appears at K1 and K2, and a small outlying area at K3. The topography in the upper half of the view shows a normal array of divides and valleys. The streams, such as R2, flow toward the limestone upland where they disappear underground.** (*Courtesy Shell Development Company. Stereopair by Arthur D. Howard.*)

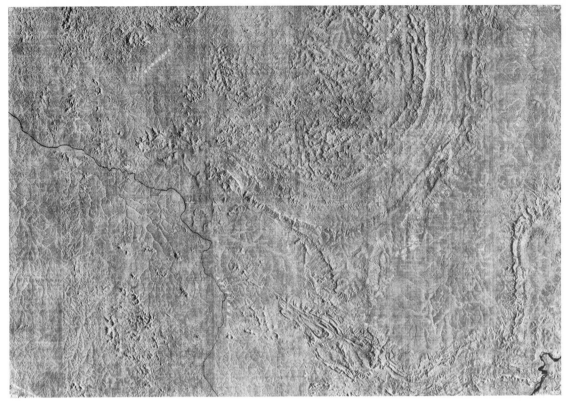

FIGURE 14-23. **Radar mosaic, Amazon basin, 600 km (375 mi) south of Manaus. The mosaic extends 1° in latitude and 1 1/2° in longitude and covers 18,000 sq km (7000 sq mi). The original size, prior to present reduction, was 44 by 66 cm (17 1/4 by 26 in.). These radar images have revealed the diversity of rock types and the complexities of geologic structure in little-known areas of the Amazon basin. (*Courtesy National Department of Mineral Production, Brazil.*)**

preservation as national reserves and parks, and all other areas are given ratings as to best usage, including the siting of industrial plants such as lumber mills, wood fabrication plants, and rubber treatment plants.

Considering the speed and low cost at which the task is being consummated, and admitting its reconnaissance nature, it is an invaluable preliminary step in rational development of the Amazon basin. Larger scale maps and studies will be undertaken as interest develops in particular sites.

Cold Environments

Two stereopairs illustrate the usefulness of three-dimensional views in environmental interpretation in cold or formerly cold regions.

Figure 14-24 shows an area in Kennebec County, Maine. Although there are a few clearly defined valleys V, stream erosion does not account for much of the surface aspect of the

area. For example, the many large water-filled depressions are obviously not the product of erosion by through-flowing streams. And careful examination in the vicinity of H will reveal an obscure hummocky topography with undrained depressions. The possibility of the topography at H being due to landsliding is remote. The final clue to the nature of the surface deposits is the winding ridge or esker E. Thus, the inference is that a former stream-eroded landscape is mantled by glacial deposits. The esker is a potential source of sand and gravel,

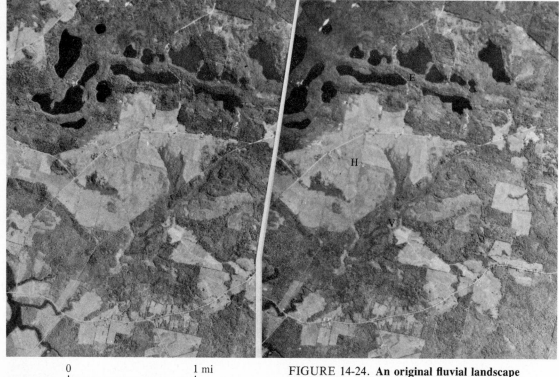

0   1 mi

1 km

FIGURE 14-24. An original fluvial landscape covered by glacial deposits. V—stream valley; H—hummocky ground; E—esker. (*Stereopair prepared from U.S. Geological Survey photographs by University of Illinois Committee on Aerial Photography. Annotation by Arthur D. Howard.*)

while the remainder of the area is presumably underlain by tight, impervious, and possibly bouldery till.

Figure 14-25 is a stereopair of a permafrost area in Alaska. The evidence for permafrost consists of the polygonal ground, the polygonal stream pattern, and the thaw basins B representing the melting of ice masses along the stream course.

Thermal infrared scanning has proved useful in locating snow-covered crevasses in glaciers, in mapping sea ice, in revealing frozen lakes and rivers under snow cover, and in delineating the shoreline of snow-covered, ice-bound shorelines. The upper view of Figure 14-26 is an ordinary panchromatic photo of a snow-covered shoreline; 8 in. of snow mantle rolling hills and sea ice. The relief is so slight that the shoreline could not be located either in overflights or in stereoscopic examination of

photos. The lower view is a thermal infrared image of the same area. The thermal reflection from the warmer, favorably oriented slopes enhances the relief effect, permitting precise mapping of the shoreline.

Radar is also useful in revealing the intricate surface patterns of sea ice and the presence of refrozen cracks and passageways of open water. The cracks and passageways show up dark in strong contrast to areas of older intervening ice.

Earthquake Environments

Two illustrations indicate that recent faulting, not expressed by topographic offsets, may be revealed by remote sensing even in relatively

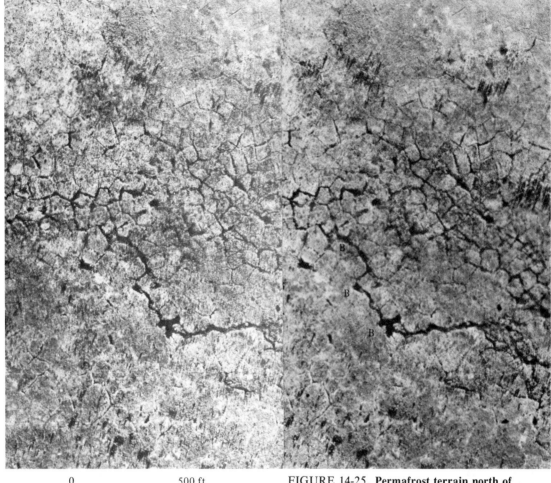

0 |———————————| 500 ft
                150 m

FIGURE 14-25. **Permafrost terrain north of Fairbanks, Alaska. Polygonal ground, polygonal stream pattern, and thaw basins (B) along stream are indicators of permafrost. Scale 1:5600. (*Photos by U.S. Air Force. Stereopair by Arthur D. Howard.*)**

flat terrain under concealing covers of soil and vegetation.

Figure 14-27 shows fault traces in the Amazon rain forest. The fracture traces are expressed by aligned segments of stream channels and by slight tonal difference in the forest. Such traces show up far better in vertical photos taken at low sun angles when shadows are enhanced. Radar imagery has a similar modeling effect.

Figure 14-28 is a thermal infrared image of a lineament in a cultivated area. Small temperature differences due to accumulation of mois-

ture along the fault account for the tonal difference.

Volcanic Environments

Recent volcanic features are obvious in an air photo; older ones less so. Figure 14-29 shows a fresh volcanic cinder cone at 1 with a lava flow at 2 extending from its base. The dark tone of both the cone and the flow indicate basaltic

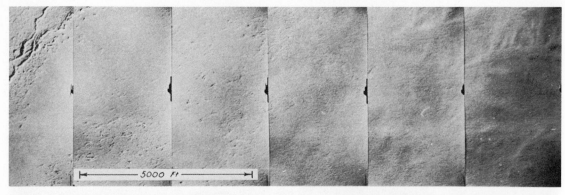

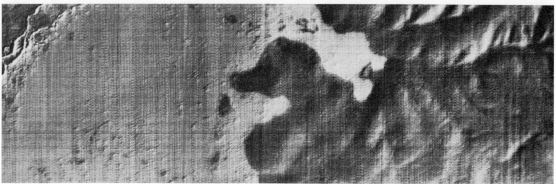

FIGURE 14-26. Panchromatic (upper) and thermal scanning infrared view (lower) of ice-bound coastline under 20 cm (8 in.) of snow. The ridges in upper left are pressure ridges of sea ice. Both views were taken in daylight at a sun angle of 21°. Canadian Arctic. (*Views taken on joint U.S.–Canadian Research Project "Bold Survey" and provided through U.S. Army Terrestrial Sciences Center, Hanover, New Hampshire.*)

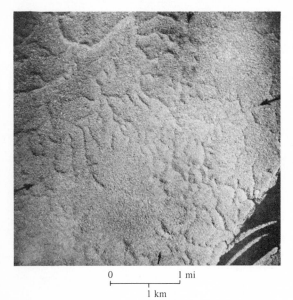

material. Note the larger, older, more-weathered, and less-obvious flow in the area extending west from 3. The lobes project south, indicating a source to the north of the photo. An old, breached volcanic cone at 4 and the possible remnant of a still older one appears between it and the fresh cone. In the area of 5, particu-

FIGURE 14-27. Fracture traces in the Amazon rain forest (shown by arrows). The parallel arcuate features in the lower right are levees standing above the inundated floodplain. (*Photo courtesy of Petrobras. Annotation by Arthur D. Howard.*)

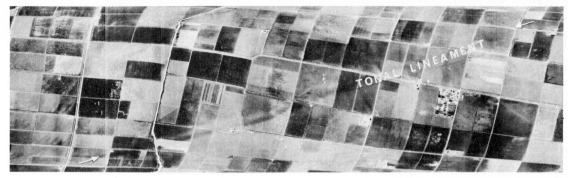

larly to the west and south, are flats and benches of sedimentary rock. The illustration demonstrates that centers of volcanic eruption may shift in time. There is no assurance that the next eruption in this area will come from the present cone.

Thermal infrared imagery is playing an increasing role in detecting warm areas in volcanic regions, and through recurrent satellite imag-

FIGURE 14-28. Thermal infrared imagery commonly reveals linear tonal anomalies along otherwise obscure fault traces. This lineament is in an area in southern California underlain by several thousand feet of sediment. The anomaly is presumably due to concentration of moisture along a fault. (*Courtesy HRB-Singer, Inc.*)

FIGURE 14-29. Volcanic landforms north of Flagstaff, Arizona. A fresh cinder cone and lava flow are at 1 and 2. The area to the left of 3 is an older lava from a source outside the photo. An older cone appears at 4 and sedimentary rocks at 5. (*Dept. Commerce, Bureau of Public Roads.*)

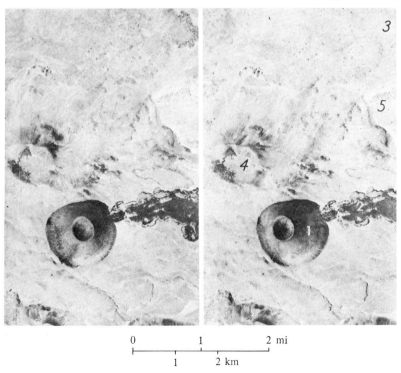

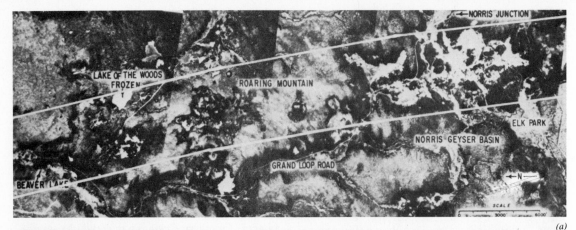

FIGURE 14-30. Conventional photography (*a*) and nighttime thermal infrared imagery (*b*). Yellowstone National Park. The white lines in (*a*) delineate a strip 2 to 13 km (1 to 7 mi) wide covered by the infrared imagery (*b*). The photos record white surface deposits in light tones whether presently warm or not. In the photos, unfrozen Beaver Lake appears dark, and frozen Lake of the Woods, light. The infrared imagery clearly distinguishes between the thermal areas (light) and other light-toned areas in the photographs and reveals some thermal areas that are dark in the photographs. (*J. H. McLerran and J. O. Morgan.*)

FIGURE 14-31. Panchromatic photography (*a*) and infrared imagery (*b*) of summit of Kilauea Volcano, Hawaii. Lighter tones in (*b*) indicate warmer surfaces. Craters A and D are inside the large caldera; crater C is outside. Slumped caldera wall at B. (*U.S. Geological Survey.*)

ery, changes in temperature and position that might be symptomatic of impending eruptions. Figure 14-30 illustrates the effectiveness of thermal infrared imagery in detecting warm spots. The thermal imagery outlines the warm areas clearly while the air photo records only the barren light-colored soils, which may be more extensive than the thermal areas.

In Figure 14-31, an air photo of the summit of Kilauea Volcano, there is no direct indication of the distribution of "hot spots." The thermal image, on the other hand, displays these vividly.

Springs have been discovered in nighttime thermal imagery. The springs show up light because they are slightly warmer than the surrounding ground. Warm springs, of course,

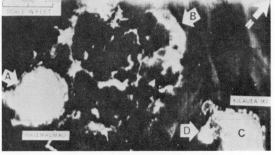

show even more clearly.

Infrared imagery is also used to detect smoldering fires in waste piles from coal mines and in the detection and mapping of smoldering fires in forest lands.

ADDITIONAL READINGS

American Society of Photogrammetry and Society of Photographic Scientists and Engineers: New Horizons in Color Aerial Photography, *Seminar Proceedings*, 1969.

Anderson, J. R., E. E. Hardy, and J. T. Roach: A Land-Use Classification System for Use with Remote-Sensor Data, U. S. Geological Survey Circular 671, 1972.

Anderson, R. R., and F. J. Wobber: Wetlands Mapping in New Jersey, *Photogrammetric Engineering*, April 1973.

Colwell, R. N., ed.: "Manual of Photographic Interpretation," American Society of Photogrammetry, Washington, D.C., 1960.

Estes, J. E., and L. W. Senger: "Remote Sensing: Techniques for Environmental Analysis," Hamilton Publishing Company, Santa Barbara, California, 1973.

Howard, Arthur D.: Drainage Analysis in Geologic Interpretation: A Summation, *Bulletin American Association Petroleum Geologists*, November 1967.

Morris, D. B., and P. H. A. Martin-Kaye: Remote Sensing of the Environment, *Endeavour*, September, 1973.

Pestrong, R.: Multiband Photos for a Tidal Marsh, *Photogrammetric Engineering*, May 1969.

Simon, I.: "Infrared Radiation," Van Nostrand Momentum Books, Princeton, N. J., 1966.

Simpson, R. B.: Radar, Geographic Tool, *Annals of the Association of American Geographers*, March, 1966.

Smith, J. J., Jr., ed.: "Manual of Color Aerial Photography," American Society of Photogrammetry, Washington, D.C., 1968.

Strandberg, C. H.: "Aerial Discovery Manual," John Wiley & Sons, Inc., New York, 1967.

University of Michigan, Institute of Science and Technology: Proceedings volumes, *Symposia on Remote Sensing of Environment*: First (1962, one volume, 1962), Second (1962, one volume, 1963), Third (1964, one volume, 1965), Fourth (1966, one volume, 1966), Fifth (1968, one volume, 1968), Sixth (1969, 2 volumes, 1969), Seventh (1971, 3 volumes, 1971), Eighth (1972, 2 volumes, 1972), and Ninth (1974, 3 volumes, 1974). (These proceedings volumes include articles covering applications of remote sensing to all facets of the environment.)

land-use
planning
and
implementation

Part of the San Mateo County Mid-Coastside Planning Area.
(*U.S. Geological Survey. Courtesy G. G. Mader.*)

Reactivation of old landslide, resulting in damage and litigation. Santa Cruz Mountains, California. (*G. G. Mader.*)

CHAPTER FIFTEEN

environmental law

INTRODUCTION

Environmental law, the newest area of American jurisprudence, is the law by which the interactions of people and the environment are governed. Inasmuch as law is a pragmatic response to the needs of an evolving society, it is constantly changing. Today, statutes, regulations, and court decisions are proliferating so rapidly that much of what is written here may need revision in a few years. Environmental law, however, necessarily operates within the broader legal system of the United States, and an understanding of the former requires familiarity with the latter.

FUNDAMENTAL LEGAL PRINCIPLES

The laws of the United States are multilayered. One layer is a durable body of principles (*constitutional law*) which changes relatively slowly. These principles govern the development and enforcement of all other federal or state laws. Although the Congress and the people may directly change this primary charter, the United States Supreme Court, as the final arbiter of its meaning, has the effective day-to-day power to modify it. Comparable power is vested in the supreme court of each state to interpret and modify the basic precepts of state law as embodied in state constitutions.

Another layer is *statutory law*, an accumulation of statutes enacted by Congress and the state legislatures. The elected representatives have the power to make or change any law, provided they do not violate the supreme laws of the land.

A third layer is the burgeoning body of *administrative law* propounded by federal and state administrative agencies. This regulatory law is, in theory, intended only to amplify and explain statutory law. In fact, it often strays into the legislative domain by pronouncing entirely new legal principles.

A fourth layer is *local or regional law*, enacted primarily by locally elected governing boards of cities, counties, or special districts. Each of these bodies has authority within the restraints of state law to enact ordinances or rules for its own electorate and territory.

The final layer is the *common law*, provided by federal and state courts whose primary duty is to interpret and apply federal and state law to the decision of factual disputes. These courts have no authority to legislate, but the effect of court decisions over time is to do exactly that. Each appellate court, in interpreting and applying existing law to the solution of a dispute, renders a decision which thereafter binds all the trial courts that are subject to its jurisdiction.

Under our system of government, only elected representatives have the power to make laws. Judges may only interpret laws and may not disregard clear and unambiguous meanings and thereby create new and different laws. Attempts by the judiciary to ignore express statutory language have often been quickly followed by new legislation to correct judicial abuses. Courts, however, may strike down as unconstitutional any statute which is in violation of the primary laws of the land.

Once laws are enacted, they are enforced in the courts, where the statutory language is dissected to see if it is applicable to the litigation at hand. Often the courts are guided or instructed in this task by formal pronouncements of administrative agencies which are required by statute to spell out the details in their regulations. Such is the role of the *Environmental Protection Agency*. The courts must follow these regulations unless they are inconsistent with statutes.

Notwithstanding the primacy of the legislative branch in initiating change, the courts have recently become innovators within the confines of their special discipline. Prior to the enactment of environmental laws, the courts used the ancient device of injunction to stop activities which defiled the land or offended sensitivities. Thus, to abate a public or private "nuisance," the courts, without aid of any particular statutory authority, may order offensive situations terminated. Persons who violate these injunctions may be jailed for contempt of court. In this fashion, and by the creation of other rights and remedies, the courts have for generations protected the community and the land from many harmful or offensive conditions.

LAW AND EARTH RESOURCES

Property Rights and Earth Resources

General Considerations. All the earth's "natural resources" are theoretically susceptible to use and to some form of ownership. Since the primary goal of environmental law is to control the use of resources, a clear understanding of property rights is essential.

The law has traditionally divided natural things into movable and immovable materials. Things that are affixed to the earth, such as buildings, trees, and crops, are regarded as immovable. In the language of the law, land is *real property*; and the natural components of the land (minerals), as well as things that are affixed to it or rooted in it, are also real property. All other things are *personal property*, or *chattels*. Thus, trees when cut, and minerals when mined, are transformed from real to personal property, with attendant changes in legal rights and interests.

The law has traditionally regarded fluids as unique materials entitled to special legal treatment. The early courts likened mobile fluids (particularly oil and water) to wild animals, which are not subject to usual property rights until reduced to possession by some individual. The air and ocean, however, are regarded as beyond the dominion of individuals and subject only to the laws of sovereign states or to agreements between states.

Present Laws. *Property Rights and Limitations (General).* Fundamental to all property rights is the concept that they cannot be lost at the whim of the state or a powerful neighbor. The Fifth Amendment of the United States Constitution embodies the concept thus:

> *Nor shall private property be taken for public use, without just compensation.*

A number of states adopted comparable provisions in their constitutions. Notwithstanding its omission from any state charter, the Fifth Amendment is a pervasive law of the land, enforceable by all federal and state courts.

Thus, if the federal government desires to convert a farmer's fields into an Army base, it must either convince the farmer to sell the land, or it must pay the farmer for taking it. If it elects to "take" the land, it must proceed to a determination of the market value of the land which must be paid to the owner. The same principles apply if the government usurps the air rights above the property for aircraft, or underground rights for tunnels, or strips of land for roads or utility lines.

In addition to this requirement of compensation for taking over land, the Fifth Amendment guarantees that no person shall be

> *deprived of life, liberty, or property, without due process of law.*

In addition, the Fourteenth Amendment guarantees that

> *no state shall make or enforce any law which shall abridge the privileges or immunities of citizens of the United States; nor shall any State deprive any person of life, liberty, or property, without due process of law; nor deny to any person within its jurisdiction the equal protection of the laws.*

On the other hand, it was never the intention of the law to free property from reasonable restraints and regulations. Land-use controls comparable to our present zoning laws and building codes were common in Elizabethan England. In the American colonies, urban dwellers were often compelled to construct their buildings in accordance with specified plans and materials to enhance the general welfare of the community.

The rationale for these seeming invasions of property rights was that the government had an overriding obligation to protect the interests of the community at large. It was therefore necessary that the government be empowered to exercise "police powers" in the preservation of the public order and promotion of the general welfare. In time of crisis, without liability for compensation, the federal, state, or local government may seize and use any property, burn or destroy buildings, or evict people from their homes. In time of war, the army may confiscate supplies; in a conflagration, firemen may destroy buildings to create fire lines; in times of flood, the city may occupy houses for use as hospitals; and so on.

These are relatively obvious, and generally acceptable, exercises of police powers. The more difficult cases are those in which the government interferes in the use of property, for example, by the establishment of zoning laws which restrict development to single-family dwellings on 2-acre parcels, health laws which outlaw the use of septic tanks, and safety laws which shut down old hotels that lack adequate fire doors. Nevertheless, these or comparable constraints upon the use of property have long been approved by every high court in the country.

The basic precepts about property in the United States today are therefore these: The government, any agency of government, and even delegated nongovernmental agencies such as public utilities, must justly compensate any owner for taking over or substantially interfering with private property. However, no such compensation need be paid if the taking or interference is reasonably necessary to protect or to enhance some larger public benefit. In the event of any dispute as to the purpose of the taking or the amount of compensation, the owner is entitled to due process in a court of law.

Property Rights Relative to Particular Resources. With regard to *land and minerals gen-*

erally, contemporary laws of the United States afford to any person, including "fictional persons" such as corporations, the right to acquire, use, and dispose of any tangible or intangible materials in any way they see fit, provided that such acquisition, use, or disposition does not harm some other public or private interest.

Thus, anyone may purchase a piece of land, provided it is for sale and not in some way co-opted by the government, and may use it in any way, provided the use does not unreasonably interfere with the rights of others. The purchaser may, furthermore, sell all or any part of it, and failing to so dispose of it, the purchaser may cause it to pass to others in accordance with the terms of his or her will.

All these considerations apply equally to any mineral component of the land. They also apply generally to things that are not part of the land, i.e., to virtually any kind of personal property.

Water rights apply to streams, rivers, and underground water. *Stream and river rights* are patterned after English tradition. In the humid Eastern United States, property owners along a stream or river enjoy *riparian rights.* These include the right to use the water for reasonable purposes upon the land to which the right applies. Hence, the water may not (in most riparian states) be diverted for use on nonriparian land; nor may its natural flow, its quantity, or its quality be materially diminished. These basic common law tenets prevail throughout the eastern and some midwestern states, but are in many instances superseded by state and local controls.

In the arid southwestern states, riparian rights made little sense because water rarely appeared in the channels. It was more important that the water be used when available, even if the user was not located along the water course. Thus, one who appropriated water for one's own use obtained the right to its continued use, in the same general amounts and for the same purposes. One who failed to continue such use lost one's *appropriative rights* by abandonment. On the other hand, as long as such use continued, the rights created thereby were superior to those of any subsequent user. Moreover, the water rights could be transferred separately from the land, but the transfer of land generally carried no special rights to waters in abutting channels. In addition, water could be appropriated by communities, as well as by individuals, and this practice was expanded to include all levels of government, including the federal government. Appropriations are today controlled by comprehensive state and federal laws and regulations which prescribe higher priorities for certain uses than others, and also control the use and disposition of the waters.

California and several other states with adequate overall rainfall developed a hybrid form of water rights. For example, in California the early settlers carried from the East traditional notions of riparian rights and applied them in pursuit of normal agrarian activities. Then came the miners, whose needs were very different. It mattered little to the miner in the Sierra foothills of the motherlode that his mining claim did not abut a stream channel. Without water he could not process ore, so he "appropriated" water from the nearest source. There thus evolved two very different and often contradictory principles of water law, which continue to trouble legal scholars today.

Underground-water rights received almost no attention in colonial days. The Eastern states were blessed with abundant river water, and there were few places where a well failed to yield ground water.

In many parts of the Western states the only available water was underground water. The courts in these states, therefore, extended principles comparable to appropriative water rights to ground water. Whoever first used water from beneath the ground, even though not on his property, obtained the right to continue such use. Other states have adopted the view that ground water is subject to ownership and use only by the owner of the property beneath which the water occurs. However, the property owner loses the right to any excess that flows beyond the property's boundaries. Many states impose regulatory rules recognizing that water is a limited resource upon which others may depend. Some states require permits for its use.

Oil, gas, and other fluids posed special problems because they could escape from the land. A number of state courts finally evolved two separate concepts to resolve the matter, both of which produce substantially the same results. One concept regards oil as a fleeting substance which the property owner is entitled to capture as it passes beneath the owner's property. The other states that oil, lie any other mineral, is part of the land itself, but, unlike other real property, rights to it may be lost if it is allowed to escape from the property. These basic principles have been extended to cover natural gas and other mineral fluids.

Air is now regarded as a precious substance in many parts of the United States, and laws relating to it have evolved accordingly. While it is still theoretically true that a property owner's rights extend upward indefinitely, the owner is presently entitled to claim only as much of the space above the land as he or she can reasonably use, and then only if such use does not interfere with air commerce. The Constitution grants to the Congress the power to "regulate commerce" among the several states, and this power extends to all navigable waterways and the navigable airspace. The airspace which is subject to federal control is virtually everything above a few hundred or a few thousand feet of the surface.

Even prior to the advent of environmental laws, no property owner was permitted to discharge fumes, smoke, or other particulates into the air so as to cause injury to other persons or property. Even unreasonable vibrations have been prohibited by modern court decisions. This concept also applies to the use of the airspace by others.

With respect to *oceans, rivers, and lakes*, Article I of the Constitution granted congress the power

> to regulate Commerce with foreign Nations, and among the several States, and with the Indian Tribes.

The Supreme Court very early held that inherent in this power was the power to regulate all navigable waterways which could be used in the conduct of interstate or foreign commerce. Thus, wherever commercial boats and vessels might travel, the federal power prevailed. The power encompassed small tributaries as well as major rivers, and it clearly extended to interstate lakes as well as ocean waters.

Controversies still rage over the rights of states to own or to regulate the use of lands beneath the oceans. A series of decisions by the High Court has gradually narrowed the scope of states' rights. The offshore rights of most states extend only 3 mi (5 km), while a few others extend up to "3 marine leagues" offshore, that is, about 9 statute miles (14 km).

Environmental Law: Restrictions and Liabilities

Historical Perspective. Environmental law is intended to retard, hopefully to correct, the adverse effects of human activities upon the natural environment. The primary intent is to curtail detrimental uses of property.

Such efforts are not new. *Courts of equity* were created in England in the Middle Ages to ensure equity, not blind adherence to archaic laws. These courts, sitting apart from courts of law, developed the *injunction*. An injunction is an order of the court that someone do, or cease doing, something. The failure to comply was punishable by fine or imprisonment.

This judicial tool was used with great effect to stop civil abuses which courts of law were largely powerless to atone for, except by monetary awards.

In addition, the medieval courts of law were developing responses to an age-old problem, *nuisances*. A nuisance was an interference with one's use or enjoyment of his land by the activities of another person outside the land. *Public nuisances*, which initially were chiefly obstructions of rights of way, ultimately came to include keeping animals, diverting stream waters, emitting noxious smoke, discharging fireworks, and the like. Since the activities were offensive to the public good, public nuisances were crimes, and the states early adopted penal laws to restrict them. Violations were subject to fines and imprisonment.

At the same time, activities which primarily impaired the use or enjoyment of private property were deemed to be *private nuisances*, for which the traditional relief was an award of damages to the injured owner. Thus, blocking access to private property, dumping sewage or silt upon it, fouling the air above it with offensive odors or dust, and threatening its structures with fire hazards were all private nuisances.

As the courts of law and equity merged in the United States, the twin remedies of injunction and damages became readily available to fully redress nuisance injuries in a single action. These remedies, together with fines and imprisonment for criminal infractions, have become the principal enforcement tools of the new environmental laws.

One might ask why these long-available legal tools have not been adequate to protect the environment. There are several answers to this question.

First, the problem was ill-perceived in the past. For example, the possible harm that might ensue from the discharge of wastes from a few early factories into the vast reaches of Lake Michigan was unrecognized. Second, many of the problems are of only local interest. Contaminants in a small stream in Massachusetts are of little concern to fishermen in California. Third, while the electorate in general, and special interest groups in particular, have had an effective voice in government, nature and the environment have had little representation and hence little legislative protection. Four, private litigants have not been legally equipped to represent the community at large. Private claimants could effectively represent only themselves and be recompensed for their own injuries. *Class actions* were virtually unknown. Moreover, the larger public interest could be represented only through governmental enforcement agencies, primarily prosecuting attorneys, and these agencies could enforce only the strict and limited language of the law. Finally, the courts, which in accordance with law and tradition, had bowed to legislative supremacy in addressing new ills, were reluctant to carve out new rights for new wrongs. In any case, judicial remedies were available only to the parties whose controversies were before the courts; the judges neither could, nor would, render advisory opinions in the abstract. Hence, it was difficult for the courts to effectively respond to pervasive threats to the environment.

Recent Developments. Attitudes and responses changed markedly in the 1960s. Expanding populations encroached dangerously on the natural sources of their own sustenance. Burgeoning factories polluted the rivers, the lands, and the atmosphere. A more enlightened citizenry was more receptive to greater controls and diminished freedoms in combatting these threats to the environment. And the advanced state of technology, as well as the seemingly limitless financial resources of the larger society, favored an assault on environmental degradation.

The result was a literal outpouring of federal and state statutes and regulations and a flood of public and private lawsuits to do battle against the common threat.

In rapid succession came the Federal Water Pollution Control Act Amendments of 1961, the Oil Pollution Act of 1961, the Clean Air Act of 1963, the Wilderness Act of 1964, the Water Quality Act of 1965, the Motor Vehicle Air Pollution Control Act of 1965, the Solid Waste Disposal Act of 1965, the Clean Air Act Amendments of 1966, the Department of Transportation Act of 1966, the Air Quality Act of 1967, the Wild and Scenic Rivers Act of 1968, the National Environmental Policy Act of 1969, the Water Quality Improvement Act of 1970, the Clean Air Amendments of 1970, the Environmental Improvement Act of 1970, the National Materials Policy Act of 1970, the Mining and Minerals Policy Act of 1970, the Noise Control Act of 1972, the Coastal Zone Management Act of 1972, the Marine Protection, Research and Sanctuaries Act of 1972, the Trans-Alaska Pipeline Authorization Act (1973), and the Energy Supply and Environmental Conservation Act of 1974. These statutes have in turn become sources of even more comprehensive administrative rules, regulations, and decisions. And, yet, *these are only the federal contribution to environmental law*. Each state, through comparable statutes and regulations, has made additional contributions.

California, for example, has enacted the California Environmental Quality Act of 1970, patterned after the National Environmental Policy Act of 1969. Prior to that, it created the San Francisco Bay Conservation and Development Commission (1965). It subsequently created the State Air Resources Board and various air pollution control districts throughout the state. It reorganized comparable boards and districts to deal with water quality. It most recently established a Coastal Zone Conservation Commission, primarily to protect the California coast through a coastal zone conservation plan. Yet these are only a few of California's laws to protect the environment. Other states have been equally active in environmental control.

It should be emphasized that the federal enactments are in general the supreme law of the land and may supervene any state laws to the contrary. This is not to say that state laws which come under the shadow of federal supremacy are automatically void. The Supreme Court has said that unless the federal enactments have "occupied the field" with comprehensive legislation on a given subject, or unless the Congress has indicated its desire to exclude state laws from the area, states may exercise jurisdiction in matters affecting their territories and citizens.

To examine a case in point, control of air pollution in California would properly seem the responsibility of the state, since the air pollution directly affects the health of its citizens. The United States Constitution assigns no authority to the federal government in such matters. Under what authority, then, has Congress enacted legislation to control air pollution in California, as well as in all the other states? The answer is that the language of the Constitution has been gradually more broadly interpreted by the Supreme Court to yield greater powers to the federal government. The authority to regulate interstate commerce has been expanded to authorize Congress to control virtually anything which materially "affects" interstate commerce. Thus, vehicular traffic on freeways in the Los Angeles air basin affects interstate travel, and vehicle pollution may therefore be controlled at the federal level. Similar constraints may naturally flow from this interpretation of delegated power over interstate commerce.

Nor is this the only source of such authority. Pollutants which are not generated by components of interstate commerce are regulated by powers found in Section 8 of Article I of the Constitution, wherein Congress is given the power to "provide for the common Defense and general Welfare of the United States." This provision has been interpreted as vesting in Congress federal police powers to protect the public health, morality, and well-being. Thus, polluted air spilling over from one state to another evokes the restraining authority of federal law, and hence it can give rise to federal controls on factory emissions.

On the other hand, Congress has not, in the language of the Supreme Court, "pre-empted" the field in air pollution. States are authorized, in fact virtually compelled, to adopt comprehensive pollution control measures. Moreover, states which desire to impose even stricter standards may generally do so if permitted by Congress.

Let us briefly consider a few of the comprehensive federal environmental laws.

Details of Federal Environmental Laws. *Clean Air Act.* Congress enacted the Clean Air Act of 1963, which, in the course of a decade, has gone through many modifications. As presently constituted, the Clean Air Act provides the following in a series of sections, or "Titles."

Title I—Air pollution prevention and control: This portion of the act encourages cooperation between federal, state, and local agencies in developing and implementing reasonable air quality standards. Civil and criminal enforcement measures are provided.

The administrator of the Federal Environmental Protection Agency is required to promote research into air pollution, its causes, and its prevention and control. He is directed to develop national standards for air quality. Each state is required to develop a plan for implementing the standards in each *air-quality control region* within the state. In the event a state fails to submit an acceptable plan within the time specified by the act, the Administrator is required to develop a plan for the state. In the event of a dispute between the federal and state governments, the controversy may be appealed to the federal courts.

The Administrator is also required to identify various *hazardous air pollutants.* Once such pollutants have been so identified, it is unlawful for any person to cause or permit them to be emitted from an existing source, or to construct a new source which is likely to emit them. Exemptions from these restrictions may be granted under limited circumstances by the administrator, and in a few instances by the President himself.

Enforcement of the various state plans and federal regulations is primarily the responsibility of the federal government. The Administrator is empowered to order any violator to comply with the state's plan. The state is simultaneously given the opportunity to compel compliance. Failing voluntary compliance or state enforcement, the administrator may commence a civil suit against the violator. The court is authorized to issue injunctions and impose fines or jail terms, or both. Persons who knowingly violate a state plan, refuse to obey the administrator's order, or violate the regulations with regard to hazardous pollutants or new stationary sources are subject to fines of up to $25,000 per day of violation, or 1 year in prison, or both.

In cases not involving specific violations, but where general air pollution endangers the health or welfare of the people within a state, the administrator must initiate a series of conferences among the affected parties aimed at resolving the problem. At the conclusion of the conferences, if the administrator believes that effective progress is not being made toward abatement of the problem, he must recommend remedial action. If there is noncompliance, the administrator may submit the case to a federal trial court for decision.

Title II—Emission standards for moving sources: The principal purpose of this portion of the Clean Air Act is to ensure rapid modification in the design and components of motor vehicles and aircraft, and in the composition of fuels, so as to materially reduce air pollution. These provisions are a response to two major problems: one, the national problem of air pollution; two, the problem of conflicting state standards for motor vehicles and aircraft. An important provision therefore prohibits any state from enforcing its own emission standards for motor vehicles or from adopting standards for aircraft emissions which are different from federal standards.

The administrator must establish emission standards for new motor vehicles and test procedures for assuring compliance by manufac-

turers. The Administrator may also develop standards for fuels and control or prohibit the manufacture or sale of any fuel which may endanger the public health or welfare or impair the performance of emission devices.

The act also creates a Low-Emission Certification Board to promote the development and use of nonpolluting motor vehicles.

Finally, the administrator is given the responsibility to develop and enforce aircraft emission standards.

Title III—General: Possibly the most important provision in Title III is that which authorizes suits by private citizens to enforce the act. It states that any person may commence a civil action on his or her own behalf (1) against any other person or entity (including the United States) who violates an "emission standard or limitation" under the act or who is in violation of any order of the administrator or of a state with respect to such a standard or limitation; or (2) against the administrator himself if he fails to perform any duty required by the act.

Another enforcement provision within Title III is the prohibition against government agencies entering into contracts with persons convicted of knowingly violating certain provisions of the act when the contract is to be performed at the facility which gave rise to the violation.

Title IV—Noise Pollution and Abatement Act of 1970: Title IV creates an Office of Noise Abatement and Control. The agency is mandated to study the effects of various kinds of noise on humans, wildlife, and property, and the administrator is to report the results to the President and the Congress. If the administrator determines that any federal agency is carrying on or sponsoring any activity which results in noise amounting to a public nuisance, or is otherwise objectionable, the agency is required to find possible ways to abate the noise.

The provisions of the Clean Air Act are of sweeping effect, reaching into almost every enterprise formerly thought to be of only private or local concern. This is the general pattern of the federal environmental laws. They strenuously regulate matters within the obvious purview of federal power; they then nudge states into action in matters still reserved to state discretion. In doubtful cases, the federal authority is usually asserted.

Water Pollution Control Act. The federal Water Pollution Control Act seeks to "restore and maintain the chemical, physical, and biological integrity of the Nation's waters." It sets goals such as the elimination of discharge of pollutants into navigable waters and the protection of fish and wildlife. The act establishes timetables for attainment of many of these goals. It requires that the administrator promote research into water pollution and that demonstration projects be instituted. It mandates the establishment and enforcement of national water-quality standards for interstate and intrastate waters and the correlative identification and regulation of pollutants and their sources. The act promotes federal-state cooperation in the achievement of most of the goals, but makes it clear that national standards must prevail. It imposes upon each state the responsibility to inventory all waters within the state, the water quality thereof, and the types and sources of pollution. Each state is required to adopt a plan to achieve the national standards, and, failing such adoption, the administrator may impose his own plan. As in the Clean Air Act, the Water Pollution Control Act deals with moving sources (i.e., ships) and stationary sources and requires the establishment of standards of performance in the control of pollutant discharges from new sources. The act is concerned with every substance which may have a deleterious effect upon any and all forms of life, both in and out of water, and expressly includes thermal pollution. The regulations apply to public agencies (federal, state, and local), as well as to private individuals and enterprises. To help defray the massive costs of compliance demanded by its provisions, the act authorizes substantial grants of federal aid to various levels of state and local governments to develop pilot or demonstration projects and to construct or upgrade water-treatment facilities.

As with the Clean Air Act, enforcement responsibility is primarily vested in the Administrator, but legal suits by private citizens are authorized. The means of enforcement are again administrative orders, court injunctions, and monetary penalties. In addition, the Water Pollution Control Act gains leverage from an old federal statute. The Rivers and Harbors Act of 1899 (otherwise known today as the "Refuse Act") had long controlled activities and conditions affecting navigable waterways. It delegated to the Army Corps of Engineers control of all construction, excavation, filling, or dumping in or about such waterways. A little-enforced provision made it illegal to discharge any refuse into any navigable waters, or any tributaries of such waters, except as allowed by permit from the Corps of Engineers. The permitting provisions of this old statute were given new life in the Water Pollution Control Act. Any facility which may result in any discharge into navigable waters must have a federal permit. No such permit will be granted unless the proper state or interstate agency certifies that the discharge will meet the standards specified in the act. In the case of discharges into ocean waters, development of guidelines and issuance of permits thereunder are the responsibility of the administrator.

National Environmental Policy Act. The Clean Air Act and the Water Pollution Control Act extend the reach of federal control into virtually every significant facility and activity in the nation. They are nevertheless limited in scope because they are tied to specific resources. The most pervasive of all federal environmental programs is that which is promulgated by the National Environmental Policy Act of 1969. The act gives promise (or threat, in the view of many) of fundamentally reordering state as well as national priorities and also of significantly recasting the most basic precepts of our national society, including the time-honored concepts of "free enterprise" and "private property." Many states have imitated this act.

The National Environmental Policy Act of 1969, NEPA for short, has as its purposes

to declare a national policy which will encourage productive and enjoyable harmony between man and his environment, to promote efforts which will prevent or eliminate damage to the environment and biosphere and stimulate the health and welfare of man; to enrich the understanding of the ecological systems and natural resources important to the Nation; and to establish a Council on Environmental Quality.

The act declares that it is federal policy to create and maintain conditions under which man and nature can exist in productive harmony and still fulfill the social, economic, and other requirements of present and future generations. Heretofore, *free enterprise* and *private property* have been predicated on the notion that persons and entities are free to do what they will as long as it does not adversely affect individual or collective rights of others. Thus, the ethical presumption has been in favor of any activity or use of property until contrary overriding interests were demonstrated. The consequence has been the largely uncontrolled use of the earth's manifold resources. NEPA recognizes that such unbridled use will inevitably result in degradation of the total environment and ultimately threaten human natural "life support systems." Thus, NEPA assumes ever greater controls over the use of resources in order to preserve and enhance the environment. On the other hand, the act recognizes that environmental protection for its own sake may defeat other legitimate human goals. Consequently, it mandates that the national policy also fulfill the social, economic, and other requirements of each generation and "achieve a balance between population and resource use which will permit high standards of living and a wide sharing of life's amenities."

The implementation of these goals and policies is the responsibility of the federal government and is to be accomplished through the activities, programs, and regulations of federal agencies. At first glance, it would seem that NEPA has influence upon only federal activities. The actual operation of the act is much more encompassing. The *permitting powers* of the government control a vast array of public and private activities, and each such activity is therefore within the scope of NEPA.

The key to the act is the seemingly innocuous provision which states that all agencies of the federal government shall

include in every recommendation or report on legislation or other major Federal actions signifi-

cantly affecting the quality of the human environment, a detailed statement of the responsible official on

(i) the environmental impact of the proposed action,

(ii) any adverse environmental effects which cannot be avoided should the proposal be implemented,

(iii) alternatives to the proposed action,

(iv) the relationship between short-term uses of man's environment and the maintenance and enhancement of long-term productivity, and

(v) any irreversible and irretrievable commitments of resources which would be involved in the proposed action should it be implemented.

It is this requirement of an *environmental impact statement* which has done most to redirect the focus of federal agencies in the direction of the environment. The stimuli for this redirection have come primarily from private citizens and environmental groups, which have brought the full weight of the federal legal system to bear on enforcement of NEPA goals and policies.

NEPA also created a Council on Environmental Quality. The council reviews the various programs and activities of the federal government for the purpose of determining the extent to which they contribute to the achievement of NEPA policies and recommends to the President national policies to improve environmental quality.

NEPA at first was viewed by many as an innocuous approach to environmental concerns, requiring only nominal compliance.

The Atomic Energy Commission was one of the first agencies to be hooked by NEPA's hidden barbs. Under a prior act of Congress, the AEC had been given the responsibility for licensing nuclear power plants. Following the enactment of NEPA, and after required investigations and public hearings, the AEC approved an application for a plant at Calvert Cliffs, Maryland. It decided that, although the licensing of nuclear power plants constitutes a "major Federal action significantly affecting the quality of the environment," it would not perform an environmental analysis for projects already in progress. Moreover, it decided to focus its environmental inquiry primarily on matters relating to radiation hazards and to rely for an assessment of other environmental effects upon evaluations by other agencies.

This administrative decision brought a prompt response by a local environmental group, and the case of *Calvert Cliffs' Coordinating Committee v. U.S. Atomic Energy Commission* ultimately came before the three-judge federal appellate court for the District of Columbia. The court said that the AEC must perform a full and independent environmental assessment of each nuclear power-plant project, balancing all pertinent environmental factors. It was to do this before authorizing any construction, and it could not rely upon determinations by other federal or state agencies that the project was environmentally acceptable.

The effect of that decision was to impose massive investigative responsibilities upon the AEC, requiring the acquisition of personnel and resources sufficient to do a comprehensive, systematic, interdisciplinary study of each project. As a result of this increased workload, notwithstanding the vast number of applications pending for nuclear power plants, very few permits have been issued since the date of that decision. Thus, in spite of the argument that other national interests, such as energy, may suffer from delay, no federal agency may disregard its responsibility to perform a complete environmental assessment of major federal actions within its purview.

Many case decisions have amplified that theme and enlarged the scope of NEPA. It is now clear that virtually every government program, or any activity requiring federal permits, must be preceded by a complete environmental study.

To illustrate, the Federal Bureau of Prisons and the General Services Administration decided to construct a federal courthouse annex in Manhattan, to consist of offices and a jail-correctional facility. In 1971, the GSA prepared a brief environmental statement which concluded that

> the impact of the proposed action will have no adverse effects on the environment, including ecological systems, population distribution, transportation, water or air pollition, nor will it

> be any threat to health or life systems or urban congestion.

Hence, the GSA decided that no impact statement was required. A neighborhood group disagreed and brought a class action to stop the project until a full environmental assessment had been performed. After several actions and counteractions, an appellate court decided that the agency had failed to adequately consider the project's potential for increasing crime in the area and had failed to give the public an opportunity to be heard. Further proceedings were stipulated to correct these deficiencies.

Considerable litigation has arisen over what constitutes a "major" federal action and what constitutes a "significant" effect on environmental quality. For example, an appellate court ruled that federal approval of a lease of Indian lands to a private corporation constituted a *major* federal action because the government might be held liable for damages incurred on the land and the federal action could have a *significant* impact on the environment through residential, recreational, and commercial development. A full environmental impact assessment was mandated.

In another example, the City of New York objected to rate increases tentatively approved by the Interstate Commerce Commission for the Penn Central Railroad. The appellate court dismissed the objections as premature but acknowledged that rate increases could divert traffic from trains to trucks, and hence lead to increased air pollution in the city. The ICC assured the court it would conduct a full environmental investigation before it gave final approval to the rate increases.

It should not be assumed from these decisions that plaintiffs almost always prevail. For example, local citizens tried to prevent the Navy and Marine Corps from staging a mock amphibious landing on the beaches within a state park. The court rejected the contention that NEPA applied, stating that the operation was not a major federal action and that the environmental damage was insignificant.

In another case, Gulf Oil Corporation challenged some of the regulations of the Emergency Petroleum Allocation Act of 1973, arguing that the agency had not prepared an environmental impact statement in connection with the regulations. Gulf asserted that the regulations would shift refining of crude oil from large, efficient refineries to smaller, less efficient refineries. The court denied relief on the grounds that the emergency nature of the 1973 act overrode the policies of NEPA.

Commercial advocates sometimes prevail, however. Several helium producers asked the court to enjoin the Secretary of the Interior from terminating a government contract for the purchase of helium without conducting an environmental impact study. The plaintiffs argued that if they failed to extract helium from natural gas prior to its use by consumers, it would be vented into the atmosphere and harm the environment. The court accepted this argument and enjoined termination of the contract pending compliance with NEPA.

By now, there is scarcely a resource or environmental value which is not in some way covered by federal law. The federal statutes cover almost every form of pollution presently recognized, including thermal, visual (aesthetic), and noise pollution, as well as mineral and chemical pollution.

State Environmental Laws. Notwithstanding the monumental scope and impact of the federal programs, the state environmental laws are generally the most effective in controlling the use and disposition of resources within their boundaries. The laws are a mixture of statutes, regulations, and judicial decisions.

A number of states have focused on specific environmental concerns, in some cases only those that are unique in their territories. All have responded to the federal mandates for clean air and water. A few, notably California, Montana, New Mexico, and Washington, have enacted "Little NEPAs," patterned after the federal model.

The California Environmental Quality Act (CEQA) sets environmental goals and policies. It then requires that all "discretionary projects proposed to be carried out or approved by public agencies" be first evaluated by the agency to ascertain if the project will have "a significant effect upon the environment." The determination must be made in accordance with guidelines prepared by the State Office of Plan-

ning and Research and must be embodied in an "environmental impact report" (EIR). "Public agencies" include state agencies and county and city governments; and the projects which must be so reviewed include enactments or amendments of zoning ordinances, the issuance of zoning variances, the issuance of conditional use permits, and the approval of tentative subdivision maps. The EIR guidelines require cognizance of at least the following considerations:

> (a) *A proposed project has the potential to degrade the quality of the environment, curtail the range of the environment, or to achieve short-term, to the disadvantage of long-term, environmental goals;*
> (b) *The possible effects of a project are individually limited but cumulatively considerable;*
> (c) *The environmental effects of the project will cause substantial adverse effects upon human beings, either directly or indirectly.*

The act then sets forth essentially the same considerations as NEPA for balancing competing interests, including assessment of alternatives.

The statute has sweeping effects in that there is scarcely an activity that is not subject to some zoning law or use permit. The cases generated by the act followed in rapid order, the landmark being the *Friends of Mammoth v. Board of Supervisors of Mono County.*

The *Mammoth* case involved a proposed construction project in the shadow of the Sierra Nevada Range. The project was to consist of a large condominium complex, specialty shops, and a restaurant, with some of the structures as high as 8 stories. Since the site was in a largely unpopulated wilderness area, environmentalists sought to prevent the development. After the Board of Supervisors approved the project, the Friends of Mammoth brought suit in state court, asserting that CEQA had not been followed. The Board of Supervisors argued that, under the then-existing language of the act, only "public works" projects, not private developments, were covered by CEQA. The lower courts agreed with the Board, but the California Supreme Court held that any issuances of a conditional use permit to a private party must be preceded by an EIR to determine if it will have a significant effect on the environment.

The fallout of that decision was profound. Local governments were bewildered by the enormity of their new responsibilities. New construction in the state almost ground to a halt while local agencies grappled with the mechanics of handling a flood of EIRs. The trauma gradually disappeared, and EIRs are now part of the everyday world of local government. The troubling language of CEQA was soon amended to give legislative force to the Supreme Court decision, but the amendment also provided relief to private projects that had been approved prior to the effective date of the amendment. This *grandfather clause* postponed much of the inevitable litigation that is now making its appearance.

The cases cover a wide range of projects. The city of Los Angeles was required to prepare an EIR prior to commencing a project for extracting underground water from Inyo County. Also in Los Angeles, ordinances which approved the drilling of oil wells were invalidated because environmental issues had not been considered and resolved. In another case, the proposed annexation of land to a city was ruled a project requiring an EIR. Courts have struck down zoning changes (e.g., agricultural to residential) because EIRs had not been prepared. Plans for subdivisions which require excess grading, environmental damage, or excessive traffic require prior EIRs.

Michigan, which enacted the Environmental Protection Act of 1970, authorizes any person to bring suit against any other private person (or entity) or public agency to protect the "air, water and other natural resources and the public trust therein from pollution, impairment or destruction." The plaintiff's case is bolstered by a presumption. Once the plaintiff initially proves that some harm envisioned by the statute has occurred or is likely to occur, the burden of proof shifts to the defendant, who can then prevail only if he or she can show that there is "no feasible and prudent alternative to defendant's conduct and that such conduct is consistent with the promotion of the public health, safety, and welfare."

The act has two features of special interest: one, the notion that the environment is a public trust; and two, that citizens as well as public agencies may seek aid of the courts to protect this common trust. Resulting cases include an attack on a subdivision project, a suit to stop open burning of trash, a challenge to a channelization project, a complaint to stop condemnation for a power line, a complaint to abate pollutants from a foundry, and a complaint to compel Indians to comply with fishing regulations.

Other states have accomplished similar results by statute. Still others have benefited from judicial contrivances, in the absence of statutes. The courts have been alert to the new spirit of environmental concern. Many have simply limbered up the old theory of nuisance. Some have fashioned new substantive rights out of familar concepts, such as the public trust theory embodied in the Michigan statute. Others have amplified enforcement of existing rights by giving new classes of persons *standing to sue*.

Standing means that no one is permitted to commence legal proceedings unless he or she has some reasonable connection with, or interest in, the subject matter of the suit. The rationale for this largely judicial policy is that the courts are intended to serve only those whose rights are involved. Thus, although one might be pained to see one's neighbor suffer, he or she cannot serve as surrogate to protect the neighbor's interests. Standing to sue may be expanded to fit the limits of recognized rights. Thus, if the courts determine that protection of the environment is a public interest, almost everyone has standing to prevent environmental damage.

This concept has a corollary. If the class of persons possessing standing is large, it is difficult, perhaps impossible, to include them all in legal proceedings. Hence, one or a few may speak for all, and represent the *class* in proper cases.

"Standing" and "class actions" have become increasingly familiar terms in environmental lawsuits. Most of the landmark cases bear the names of environmental associations which have sued in behalf of themselves and all other citizens similarly affected. They have an identifiable interest to be protected, and hence have standing. Without the active involvement of such groups, much of the vast body of environmental law would be largely untried and therefore ill-defined and incomplete. As noted earlier, courts may not render advisory decisions; they must decide real cases or controversies between actual litigants before them. Thus, without litigants there can be no case law; without standing, there are no litigants.

The twin concepts of standing and class actions gained new recognition and momentum during the decade of the 1960s. It soon became apparent that the stream of environmental cases could become a torrent, and possibly inundate the courts. Abruptly, the United States Supreme Court tightened the spigot, and the flow of cases began to slacken. In *Sierra Club v. Morton*, the United States Forest Service selected Walt Disney Enterprises, Inc., to construct and operate a year-round resort in Mineral King Valley in the Sierra Nevada Mountains. The Sierra Club sued to halt the project, claiming that they had "a special interest in the conservation and sound maintenance of the national parks, game refuges, and forests of the country." The Sierra Club relied for their asserted right on the provisions of the federal Administrative Procedures Act:

> A person suffering legal wrong because of agency action, or adversely affected or aggrieved by agency action within the meaning of a relevant statute, is entitled to judicial review thereof.

The Sierra Club was especially concerned about the construction of a road into the valley, alleging that it "would destroy or otherwise affect the scenery, natural and historic objects and wildlife of the park and would impair the enjoyment of the park for future generations."

The Supreme Court responded thusly:

> Aesthetic and environmental well-being, like economic well-being, are important ingredients of the quality of life in our society, and the fact that particular environmental interests are shared by the many rather than the few does not make them less deserving of legal protection. But the "injury in fact" test requires more than an injury to a cognizable interest. It requires that the party seeking review be himself among the injured.

The Court rejected the assumption that "the Club's longstanding concern with and expertise in" natural resources and the environment should give it standing as representative of the public. It concluded that the Sierra Club was without any other basis for standing, and the case was dismissed.

In a later, independent decision, the high court struck a further blow at such suits by requiring that in all federal class actions *each plaintiff* must show he or she was monetarily harmed by at least $10,000, the jurisdictional qualification which is required in most other federal suits.

The Right of Suit. In bygone times, the source of the law for all practical purposes was the king. This precept, known as *sovereign immunity*, was accepted as part of colonial law and incorporated into the fabric of our constitutional system. Today, governmental agencies may be sued to protect environmental interests, but such rights of action are generally by permission only. Courts have whittled away some of this hard shell of immunity, declaring, for example, that governmental activities that were really "proprietary," like those of any other property owner rather than governmental in nature, were subject to judicial scrutiny. But sovereign protection is effectively surrendered only by statute. Thus, the Tort Claims Act opened the federal government to suits for the negligent or intentional acts of its employees. A number of states have adopted similar measures.

In addition, the right of suit is often enhanced by so-called administrative procedures acts of the federal and a few state governments. These statutes give the right to judicial review of administrative decisions by governmental agencies, as was seen in *Sierra Club v. Morton*.

For environmental action cases the most effective aids to suit are often the inducements in the environmental statutes themselves. For example, in the Clean Air Act, private citizens are expressly authorized to sue any violators, including federal or state agencies, to compel compliance with the law. They may even sue the administrator of the EPA to compel him to enforce the law. Almost equally important, the successful (sometimes even unsuccessful) complainant may be compensated for attorney's fees and court costs in bringing the suit. This is indeed a powerful encouragement to volunteer groups, which are frequently without adequate funds to launch costly suits.

Another aspect of this authority to sue the government derives from rights inherent in the Constitution itself. These rights relate to another transgression called *inverse condemnation*.

Inverse Condemnation. The right of a citizen to own property is protected from governmental invasion by the Fifth Amendment to the Constitution. This is not to say that the government may not sieze property for the public good; rather it must pay the fair value of what it takes. The exception to this rule is the taking of property in the exercise of police powers. Condemnation, or *eminent domain*, then, is the process by which the government takes and pays for property. It involves the conscious and deliberate exercise of power to physically and legally acquire the property.

The problem is more difficult when the government neither intends nor cares to take the property but when a taking nonetheless results. This is the reverse of the normal situation, and is therefore known as *inverse condemnation*. For example, if military aircraft descend low over a chicken farm while landing at an adjacent airbase, this intrusion into the farmer's private domain, with attendant damage to his chickens, constitutes an invasion of his property rights which effectively results in a "taking." Suppose the government fails to institute proceedings to formalize the taking or to compensate the farmer; what can the farmer do? A right without a remedy is really no right at all. Hence, the courts have fashioned a remedy by permitting the farmer to sue the government to obtain damages for the amount of the taking. The farmer's case was actually presented to the court in 1946, wherein the Supreme Court, responding to the government's defense of lawful flight operations within the navigable airspace, said

> Though it would only be an easement of flight which is taken, that easement, if permanent and not merely temporary, normally would be the equivalent of a fee interest. It would be a definite exercise of complete dominion and control over the surface of the land. . . . The owner's right to possess and exploit the land—that is to say, his beneficial ownership of it—would be destroyed.

The Court therefore held that the plaintiff was entitled to be compensated for as much of his property as had been taken from him.

But this case is a relatively simple one. What of the farmer who wishes to give up farming and develop the land for other uses? Do the zoning laws which restrict land use to agriculture destroy the "right to possess and exploit the land"? The answer is almost universally no. What if the land were in the process of development for commercial activities in accordance with applicable zoning, and the local government "downgraded" the zoning to agricultural uses; would this constitute a taking? Or suppose, instead, they mandated that the property be preserved as "open space," with no productive use whatever? And suppose further that the municipality contrived to tax the property at the rate applied to other commercial properties? Would these acts constitute a taking? These present some of the vital issues in environmental law today.

Environmental law deals mainly with restrictions on the condition and use of property. Nowhere is this more evident than in the exercise of the planning function; and land use or resource management appears to be the order of the day.

One of the benchmark cases in this broad field was decided in 1922. Plaintiffs sought an injunction to stop underground mining operations by the defendant coal company in Scranton, Pennsylvania. Plaintiffs argued that a Pennsylvania statute, enacted in 1921, forbade the mining of coal in such a way as to cause subsidence of any residential structure and that the coal company's operations threatened to collapse their house. The coal company countered that plantiffs' title to the house derived from a deed given in 1878 by the coal company to plaintiffs' predecessors. Under the terms of the deed, the coal company parted only with the surface rights; it kept the underground rights for possible coal mining activities. The coal company therefore argued that the state statute interfered with the most valuable part of its interest in the property, and hence amounted to a taking.

The United States Supreme Court agreed with the coal company, holding that the mining operations could continue unless the company were compensated for its property. Justice Holmes said:

> Government hardly could go on if, to some extent, values incident to property could not be diminished without paying for every such change in the general law. As long recognized, some values are enjoyed under an implied limitation, and must yield to the police power. But obviously the implied limitation must have its limits or the contract and due process clauses are gone. One fact for consideration in determining such limits is the extent of diminution. When it reaches a certain magnitude, in most if not all cases there must be an exercise of eminent domain and compensation to sustain the act.

This decision, reflecting the philosophy of its time, served as the guide to lesser courts for more than a generation. The futile dissent of Justice Brandeis was, however, a bellweather for changing times. He said:

> Coal in place is land; and the right of the owner to use his land is not absolute. He may not so use it as to create a public nuisance; and uses, once harmless, may, owing to changed conditions, seriously threaten the public welfare. Whenever they do, the legislature has power to prohibit such uses without paying compensation; and the power to prohibit extends alike to the manner, the character, and the purposes of the use.

The Pennsylvania legislature, in order to provide some protection to its citizens in similar cases, adopted a modification of the original statute. The amendment, in reference to the same type of mining operations, added the qualifying language of "in such a negligent manner," as to cause surface damage. In 1970, the Federal Appellate Court was asked to prohibit mining operations under the new statute. The court ruled that the plaintiffs had failed to demonstrate that the mining operations were conducted negligently.

In 1962, Los Angeles enacted a zoning ordinance which prohibited rock, sand, and gravel operations in agricultural and residential areas. The owner of a gravel pit, which had been in operation since 1931, wanted to permit its tenant, the gravel operator, to continue the operation on the owner's adjacent property. The owner and tenant brought suit in the state court for a declaration of their rights under the zoning ordinance. The trial court found that the operation "can be conducted on the plaintiffs' said property with compatability to adjacent properties and with minimal detriment to the living amenities or health conditions of adjacent properties or in the general area and without probable depreciation in property values to the adjacent properties." However, the trial court also found that the area had a national reputation as a haven for sufferers from respiratory ailments, and the many sufferers who resided there contributed substantially to the economy of the area. The trial court decided that proper zoning for the area was a value judgment rightly left to the local government and therefore upheld the ordinance. On appeal, the Supreme Court, after reviewing many prior decisions dealing with zoning regulations throughout the country, concurred. It said that laws must be sufficiently flexible to respond to changing conditions, adding:

> As a corollary to this recognized principle of the capacity of the police power to meet the reasonable current requirements of time and place and period in history is the equally well settled rule that the determination of the necessity and form of such regulations, as is true with all exercises of the police power, is primarily a legislative and not a judicial function.

The Connecticut Supreme Court carried this logic further. A plaintiff land developer sought to enjoin the city from enforcing an ordinance under which it could compel developers to dedicate land in subdivisions for playground and parks. The court held that, since the activity of plaintiff, namely, construction of the subdivision, would increase the population and the consequent need for additional parks and playgrounds, the ordinance was a reasonable exercise of the police power and not a taking.

Other courts which have confronted comparable issues have gone both ways. But it seems clear that the trend is in the direction of upholding the legislative enactments. The question seems not so much *whether* the requirement of open space constitutes a taking, but rather *how much* is required before a taking occurs. A number of communities have attempted to increase open space or decrease population densities by increasing lot-size requirements. In one instance, a corporation applied for a subdivision permit. While it was pending, the city increased the minimum lot sizes. The New York Court of Appeals heard testimony that the soil and drainage conditions were such that septic tank spacing required 2-acre zoning to avoid water pollution. The court ruled that although the plaintiff corporation might be significantly injured (the restrictions increased the cost per lot by $3,560.00), the city had demonstrated a legitimate relationship between the ordinance and health requirements. The ordinance was therefore upheld.

On the other hand, a number of courts, on similar facts, have gone the other way. The New Jersey Supreme Court had no difficulty in striking down a zoning ordinance which was a thinly disguised taking of property for open space. The plaintiff, a sand and gravel operator, owned 60 acres of industrially zoned land in a 1500-acre swamp known as Troy Meadows. Two years after the plaintiff purchased this land, the zoning was changed to "indeterminate zone classification," pending determination of proper future use. In 1960 it was redesignated as the "Meadows Development Zone" and restricted essentially to agricultural and recreational activities. When plaintiff applied for a permit to continue its operations, it was denied because it was in violation of the ordinance. Unsuccessful in its efforts to have the zoning declared unconstitutional by the trial court, plaintiff appealed. The New Jersey high court ruled the ordinance invalid.

On the other hand, on somewhat comparable facts, the California Supreme Court upheld a state-mandated plan to protect San Francisco Bay from further land fills. It accepted at face value the legislature's findings that "the public has an interest in the Bay as the most valuable single natural resource of an entire region" and hence upheld the decision of the Bay Conservation and Development Commission in denying permits to plaintiffs to fill their tidelands.

These cases give a general indication of the direction of the law applicable to the legislative exercise of the planning function. It is clear that the trend is definitely in the direction of greater controls and restrictions upon the use or condition of property.

ADDITIONAL READINGS

Anderson, F. R.: "NEPA in the Courts: A Legal Analysis of the National Environmental Policy Act," Resources for the Future, Washington, D.C., 1973.

Burchell, R. W., and D. Listokin: "Future Land Use: Energy, Environmental, and Legal Constraints," The Center for Urban Policy Research, Rutgers, The State University of New Jersey, 1975.

Grad, F. P., G. W. Rathjens, and A. J. Rosenthal: "Environmental Control: Priorities, Policies, and the Law," Columbia University Press, New York, 1971.

Gray, O. S.: "Cases and Materials on Environmental Law," 2d ed., The Bureau of National Affairs, Washington, D.C., 1973.

Reitze, A. W., Jr.: "Environmental Law," North American International, Washington, D.C., 2d ed., 1972.

———: "Environmental Planning: Law of Land and Resources," North American International, Washington, D.C., 1974.

Rose, J. G., ed.: "Legal Foundations of Environmental Planning: Cases and Materials on Environmental Law," The Center for Urban Policy Research, Rutgers, The State University of New Jersey, 1974.

Yannacone, V. J., Jr., B. S. Cohen, and S. G. Davison: "Environmental Rights and Remedies," The Lawyers Co-Operative Publishing Company, Rochester, N. Y., Bancroft-Whitney Co., San Francisco, 1972.

CHAPTER SIXTEEN

land-use planning and geology

INTRODUCTION

In this chapter, we will first consider the planning process and the determinants of land use and then examine how treatment of geologic hazards and resources can be effectively included in land-use planning.

LAND-USE PLANNING PROCESS

Planning for future land use is carried out in the private and public sectors, but with differences in emphasis. In the private sector the emphasis is usually on planning for one type of land use, such as a development of homes or an industrial park. In the public sector, the concern is more with the interrelationship of all land uses. The public sector does, however, become involved in the detailed review of private development projects.

Geology plays an important role in private and public planning. Because government is the major influence in land-use planning, we will focus on public planning. Many of the principles are equally applicable to planning by private interests.

The land-use-planning process can be divided into four steps as shown in Figure 16-1. A discussion of each follows.

399

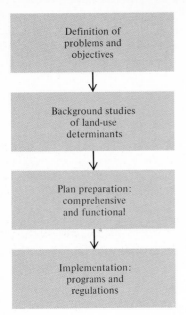

FIGURE 16-1. **Flow sheet of land-use-planning process.**

Definition of Problems and Objectives

A broad definition of problems is the first step in planning. Common problems are inadequate highways, lack of housing, insufficient recreation areas, inadequate water supply, flooding, landsliding, and air pollution. A responsible governmental agency or group must first formulate long-range objectives. The objectives may simply specify present inadequacies and where improvements are needed. For example, a city might decide that all housing should be safe from flooding. That statement can provide the starting point for a full study of the causes of flooding and the subsequent preparation of plans and programs to avoid or prevent flooding. The governmental agency must decide whether to proceed further in dealing with the problem.

Background Studies

Once it is agreed that a problem exists and action is needed, the next step is to study the basic groups of factors affecting land use in the area. Studies are made of these *land-use determinants: economic, social, political, and physical.*

The planning agency usually acquires such information over a period of years. For example, basic maps are prepared and maintained, covering all pertinent aspects of the geologic environment. The maps may show flood-prone areas, earthquake-susceptible areas, landslide areas, water resources, sand and gravel resources, and mineral deposits. In addition, specialized studies may be made of specific problems.

Plan Preparation

Plan preparation involves more implicit definitions of the objectives broadly stated in step one, the preparation of a proposed plan and/or alternate plans, and critical review of the plan or plans. Alterations are usually made before the plan is finally adopted. The land-use plans that a governmental agency prepares are either comprehensive or functional.

Comprehensive plans deal with all aspects of land use and circulation, not with selected issues. They are concerned with total contemplated growth and changes resulting from economic, social, political, and physical forces.

Comprehensive plans generally project ahead 20 to 30 years, although they also provide a basis for making shorter range plans of 1 to 5 years. They exclude detail which cannot be worked out at the time the plan is prepared and which is not appropriate to a general policy statement.

Comprehensive plans need not cover the entire area for which a governmental agency has responsibility. They may cover smaller areas based on natural or cultural boundaries. Examples are watersheds, central business districts, and manufacturing areas. These restricted plans are "area" plans.

Functional plans deal with single aspects of the environment such as a park system or drainage system for the entire area under control of the governing agency. Functional plans are usually based on comprehensive plans, but provide for additional data and policy relative to the particular subject. Functional plans often deal with natural processes and resources.

Implementation

The final and most critical step is implementation. Successful implementation involves (1) imaginative leadership, (2) enactment of carefully conceived regulations, (3) taxation, (4) public purchase and construction, and (5) the review of proposed private and public projects. While each of these is important, the most widely used are carefully conceived regulations and project review, each of which is discussed below.

Regulation. The power to enact and enforce planning regulations is based on the *police power* discussed in Chapter 15. The term "police power" as used in environmental discussions refers to the power of government to enact and enforce regulations to protect the public health, safety, and general welfare. The police power is reserved to the states, and the states delegate the power to local governments. Traditional means of local regulation under the police power include zoning and subdivision regulations and housing and building codes. These regulations are often the most effective means by which local governments control growth and development.

The nature and extent of controls under the police power may change for two basic reasons. First, courts interpret the law in terms of contemporary society, and thus some regulations acceptable today would not have been legally acceptable in earlier years. Second, experts are continually proposing new regulations to deal with new problems, to accommodate to advances in research, or to respond to new public concerns. A key element of concern in environmental earth sciences is the enactment of regulations that recognize the importance of basic geologic factors.

Project Review. Project review refers to appraisal and either approval or denial of public and private projects by governmental agencies. Project applications may involve zoning, subdivision, or building regulations.

The review of projects under the National Environmental Policy Act and parallel state acts has increased the importance of the review process. Development is no longer regarded as primarily a private matter. It now requires public participation in evaluating effects of a proposed development on all aspects of the environment.

LAND-USE DETERMINANTS

The planner views urban growth and change as ongoing processes resulting in a continuously changing environment. The environment at any one time reflects the impact of many land-use determinants (economic, social, political, and physical). It is necessary for the planner to learn how and why the environment is changing and to attempt to manipulate the determinants to achieve the goals of the community. Unfortunately, the ways in which land-use determinants affect the environment and interact with one another are extremely complex.

A brief illustration of the interplay of the land-use determinants in a local situation was presented in Chapter 1. A review of the determinants will provide insight into the relative role played by geology, one of the physical determinants. Figure 16-2 shows the interrelationships of land-use determinants and land-use patterns. Note that as the environment is altered, it may lead to modifications of the original determinants. It should also be remembered that the determinants themselves change with changing values and technology.

Economic

The economist sees land as a commodity in the marketplace. In the perfect economic model, all land is brought into optimum use through the pricing mechanism. It is assumed that buyers and sellers of land know all economic factors that bear on their actions and make sound decisions.

Many economic theories attempt to explain why present land-use patterns are as we find them. These theories emphasize that locational decisions are based on optimal economic return. For instance, a factory, a retail store, and a refinery will each seek a different site based on the activity's needs. These theories are a great help in explaining why land uses occur as they do. They fall short, however, in that deci-

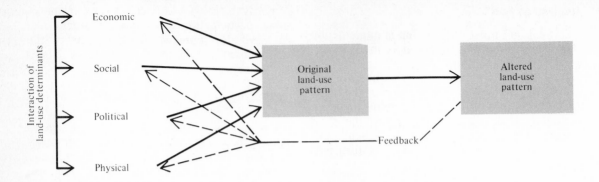

FIGURE 16-2. **Interrelationships of land-use determinants and land-use patterns.**

sion makers do not always have perfect information and therefore do not always make the best economic decisions. Also, social and political considerations are often inadequately considered in decision making.

Social

The social scientist is concerned with social values and behavior. He or she is interested in how different segments of society organize themselves, formally and informally, and how they interact with other segments of society, as well as the actions of individuals in terms of individual values and behavior patterns. The social scientist has developed many useful theories which attempt to explain changes in population characteristics, location, density, and values. This information is vital to any planner who is to prepare land-use plans that will attempt to satisfy the needs of all residents.

Political

Political determinants express the public interest. They include not only prevailing law, but the roles and actions of elected officials and politically oriented action groups as well. Political decisions will determine whether and how the physical environment is to be altered. One of the most important aspects of the political determinants is the police power. This is because local regulations under the police power are the most effective means by which land use is controlled.

Physical

Physical determinants are vital in determining land use. The physical environment is the total of all natural and cultural physical features. Our principal concern is with the natural environment, especially the geologic environment, and how it conditions and responds to human activities.

From a land-use-planning point of view, there are two basic ways of categorizing *geologic environments*—as *hazards* and as *resources*. The challenge to humanity is to respect the hazards and conserve the resources. People must understand the natural processes and live within the constraints they impose.

GEOLOGIC ENVIRONMENTS AND THE DECISION MAKER

In planning for land use, the natural environment can be most effectively dealt with if the planner understands the aggregate of land forms that make up the landscape, as well as the natural processes which bring about change. In addition, the planner should understand the particular relevance of the features and processes to land-use planning. Finally, the planner must know when to view a feature as a resource and when to view it as a hazard.

Resources such as soils and mineral deposits need to be developed properly if their maximum value to society is to be realized. Soils, however, may also be hazards. Some soils shrink and swell while others slide or creep downhill, causing various degrees of damage. On the other hand, landslide terrain is often a resource to be converted to park land or left in open space. Thus, from a planning point of view, it is useful to categorize terrain and geologic deposits as either hazards or resources, or both.

There has been a tendency in this country for planners and other decision makers to ignore geologic hazards. In many instances, information concerning existing hazards has not been published or—if published—has not been made known to the decision makers. Success in decision making requires that all pertinent data be made available and in a form that can be readily understood.

Geologic environments provide two types of geologic resources from the planning point of view: those that must be extracted and those that are utilized in place. The extractive resources have been considered in Chapter 11. Examples of resources used in place are valley floors, coastal environments favorable for urbanization, and mountains, deserts, and beaches which are attractive for recreational uses. Other resources used in place are in part extractive, such as watersheds, ground-water aquifers, and soils from which water and crop nutrients are extracted.

In summary, plans for the physical environment must satisfy human needs to the greatest extent possible. In this role, the planner must deal with the economic social, political, and physical determinants of land use. The planner must also recognize and operate within existing governmental constraints so that the plans have a good chance of being approved. The geologic environment is a very important part of the physical environment with which the planner must cope. Most parts of the geologic environment have either resource or hazard potentials. Some have a dual potential. The following sections illustrate how some geologic hazards and geologic resources have been dealt with in land-use planning.

PLANNING FOR GEOLOGIC HAZARDS

To live in harmony with the changing earth, people must understand the nature and rate of its changes and adapt their method of habitation accordingly. It is appropriate at this point to consider how people at present react to hazards.

Responses to Hazards

There are five basic ways in which people respond to geologic hazards:

1. *Avoidance.* The most obvious response to a potential hazard is to avoid it. Thus, one would not lightly put buildings in an area subject to flooding or landsliding, or astride an active fault.

2. *Stabilization.* Some hazards can be stabilized by proper engineering measures as described in Part Two. Many of these measures are expensive and may not be economically justifiable.

3. *Provision for safety in structures.* In some instances it is possible to provide enough structural safety in a building or other structure to ensure its security. For instance, buildings can be built on piles to thwart periodic flooding or, by special construction methods, to withstand violent earthquakes. As with the stabilization methods, the cost must be economically justifiable.

4. *Limitation of land use and occupancy.* The type of land use, such as agricultural or residential, can be regulated in keeping with the potential hazard. In addition, the occupancy, such as the number of persons per acre, can be adjusted to the degree of hazard.

5. *Establishment of warning systems.* Some hazards can be forecast, thereby allowing time for emergency action. Floods, hurricanes, seismic sea waves, and some volcanic eruptions are of this type. With increasing research, earthquakes may be added to the list. Warning systems have proven especially effective where hazards are confined to certain areas from which people can be evacuated, such as floodplains and shores.

Where there is plenty of land for human activities, the easiest course of action is to avoid the hazard. However, many large communities are already situated in hazardous environments. Thus, the other four approaches are commonly used, alone or in combination.

The degree of safety achieved in responding to geologic hazards is often a function of how much is invested in precautionary measures. The question is often asked, "How much is it worth to provide protection against a certain hazard?" To answer this question, there has to be a decision as to the degree of risk those involved are willing to take. For example, the risk to life and property on floodplains can be reduced to zero if all construction and human activity are prohibited. The costs in human productivity and in "living space," however, may be too high. To many, the risk of flooding once in a 100 years is an acceptable level of risk. A similar question is often posed in earthquake country: "How much additional money should be spent in construction of a building to withstand the highest magnitude but extremely rare earthquake?"

Acceptable level of risk has been considered systematically with respect to only a few types of hazards. Even if it is impossible to make precise computations, a general appraisal of the risk inherent in a decision should be considered by the decision-making body.

In the discussions of specific hazards that follow, the concept of acceptable risk should be kept in mind. We will limit our consideration to three hazards: flooding, fault hazards, and landsliding. These hazards provide opportunities to demonstrate many of the concerns of responsible planning. Finally, an example will be provided of a planning response to a complex geologic setting.

Flooding

Humans have historically been attracted to waterfront sites in locating cities. As these cities have grown, they have expanded over flood-prone areas. The federal government has recognized the widespread nature of the problem, and in 1968 it adopted a National Flood Insurance Program. This program offers improved insurance coverage for areas meeting certain minimum flood-control standards.

It is common practice to provide physical protection for all floods up to the magnitude of the 100-year flood. The selection of the 100-year flood implies a certain level of acceptable risk.

There are four major methods of reducing the potential physical and financial impact of flooding: engineering works, land-use policies and regulations, warning systems, and insurance. The main concern of land-use planning is with the first two methods. The engineering approaches have been considered in Chapter 3. Land-use policies and regulations require further comment.

Under land-use policies and regulations, conscious decisions must be made to ensure that occupied land will not be subject to unacceptable flooding and that human activities will not lead to flooding elsewhere. A proper approach to flood control requires comprehensive as well as local planning, reflected in zoning, subdivision, and building regulations.

To provide effectively against flood hazards in the general plan, it is important to have maps documenting flood areas and frequency of flooding. With such information, the flood hazard can be considered in relation to the other land-use determinants—economic, social, and political. In some instances, it may be necessary to allow development on a floodplain, but with necessary engineering precautions. A bridge across a river is a good example of a necessary improvement. Other uses, such as residential developments, however, may not justify the expense of providing adequate flood control. Hence, in preparing general plans, parts of floodplains may be better left in open space or used for golf courses, athletic fields, or parks.

In the preparation of general plans, consideration should be given to upstream uses that will help minimize the frequency and extent of downstream flooding. Uses which minimize paving and disturbance of the natural terrain will help the watershed absorb moisture and reduce runoff.

Regulation of land-use and construction practices in areas of flooding is a normal use of police power. Zoning regulations may prohibit buildings in areas subject to flooding, regulate the types of land use permitted, and require floodproofing of approved building. Figure 16-3

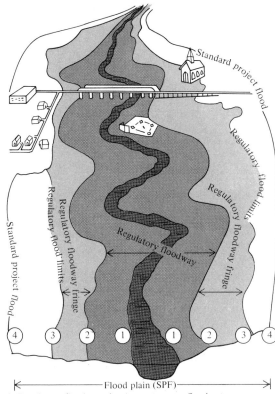

1. Regulatory floodway—kept open to carry floodwater—no building or fill.
2. Regulatory floodway fringe—Use permitted if protected by fill, flood proofed or otherwise protected.
3. Regulatory flood limit—Based on technical study—outer limit of the floodway fringe.
4. Standard project flood (SPF) limit—Area subject to possible flooding by very large floods.

FIGURE 16-3. A classification of riverine flood-hazard areas. (*From U.S. Water Resources Council, "Regulation of Flood Hazard Areas to Reduce Flood Losses," vol. 1, parts I–IV. Courtesy Wisconsin Dept. Natural Resources, 1971.*)

indicates one method of classifying the areas subject to flooding along a river and indicates in general the types of regulations that might be applied. The boundaries of such flood areas are added to official zoning maps and thus effectively control future development.

Fault Hazards

Faults produce a variety of effects (including earthquakes), as described in Chapter 8. Each requires different planning responses. Perhaps the easiest hazard to deal with is ground rupturing. However, even this presents difficulties because of a variety of complicating factors including:

1. *Uncertain recurrence interval.* Because of the long intervals between major earthquakes along any one fault, there is a paucity of useful planning data. The lack of data makes it difficult to justify specific plans and regulations governing land use along and near a fault and to get political backing for these. It is hoped that earthquake prediction will improve to the extent that responsible predictions can be made.

2. *Width of fault zone.* In many instances the fault zone may be very wide, up to 1 mi (1.6 km) or more. Without a basis for predicting where in this zone the next rupture is likely to occur, it is difficult to justify restrictive plans and regulations.

3. *Existing development.* Development has already taken place in many fault valleys and on many fault traces. It would be futile to attempt to rid such areas of all human habitation. The difficulty of convincing the public of the need is compounded by the lack of precise data on the location and imminence of the hazard.

In spite of all this, fault hazards must be taken into account and the best decisions made based on whatever data are available. In California, homes, hospitals, and schools are located astride active faults. This dangerous laxity at the local level caused the state of California to enact a Geologic Hazard Zones Act in 1972. This act requires the state geologist to map certain active faults and make the maps available to local governments. The act and ensuing regulations prohibit construction of buildings for human habitation across such faults. They also require local governments to review all developments within the "special studies zones" along the faults, normally 0.40 km (0.25 mi) or less in width, and certify that no undue hazards from faulting are present. An excerpt from one of the state maps is included as Figure 16-4.

The town of Portola Valley in the San Francisco Bay Area went further than the California state law and adopted special building setback lines along the most recent San An-

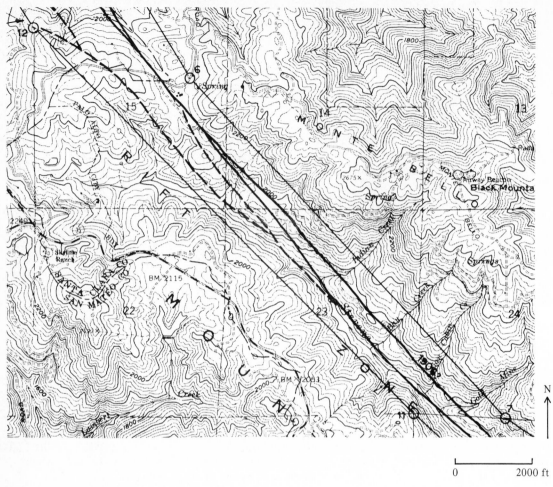

0   2000 ft

N

3   4   SPECIAL-STUDIES ZONE BOUNDARIES
⊙————⊙   These lines define the special studies zone. The numbers
               assist in defining segments of the zone.

1906   C   POTENTIALLY ACTIVE FAULTS
————┴— — —   Faults considered to have been active during Quaternary time: solid line where accurately located,
                 long dash where approximately located, short dash where inferred, dotted where concealed; query (?)
— — — — —··?·   indicates additional uncertainty. Evidence of historic offset indicated by year of earthquake-associated
                 event or C for displacement caused by fault creep or possible fault creep.

FIGURE 16-4. **Portion of California earthquake
special studies zone along the San Andreas Fault in
Santa Clara and San Mateo Counties.** (*Slightly
modified from California Division of Mines and
Geology, "Special Studies Zones, Mindego Hill
Quadrangle," 1974.*)

dreas Fault traces. These lines are included as a part of the zoning ordinance. The regulations provide that no building for human occupancy be constructed across the fault trace or within 15 m (50 ft) of it. In the next 22 m (75 ft) on either side of the trace, only 1-story, wood frame, single-family residences are permitted. Beyond these limits, special studies of the possibility of fault offset may be required when deemed necessary. Where the fault is mapped as "inferred," increased setbacks are required until the locations of the faults are mapped more precisely. A portion of the fault setback map is included in Figure 16-5 in simplified form.

Landslides

Planning in landslide terrain is even more difficult than in flood-prone areas or earthquake terrain. The difficulties stem from two basic factors:

1. Landslide terrain is often made up of a complex of landslides which require detailed mapping to determine specific boundaries for use in planning and regulation.

2. Landslides often have different degrees of potential for movement, from slight to extreme. Detailed investigations are necessary to classify the slides, and even then, considerable

FIGURE 16-5. **Simplified map of fault setbacks required by zoning regulations in Portola Valley, California. No construction for human occupancy is allowed within the 15-m (50-ft) setbacks. Uses are restricted in all setback zones. (*G. G. Mader, et al., "Land Use Restrictions along the San Andreas Fault in Portola Valley, California," 1972. Proceedings International Conference on Microzonation for Safer Construction Research and Application, vol. II, Seattle, Washington, Oct.–Nov., 1972.*)**

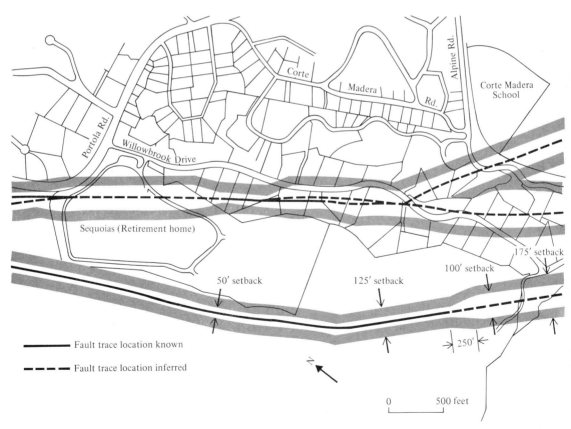

doubt as to the potential for movement will exist.

Geologic stability maps as described in Chapter 2 have been developed to assist planners and others in recognizing and dealing with landslide terrain at an early stage in the planning process. With such a map, it is possible to prepare general plans that assign land uses compatible with inherent instabilities. Such assignments must consider acceptable risk and financial outlays to stabilize the landslides or protect the installations.

The town of Portola Valley, in response to a stability map prepared by a geologist, adopted a policy for land use in landslide areas. A com-

mittee, composed of earth scientists, an engineer, an attorney, and a city planner, advised the town council on the matter and addressed the matter of acceptable risk as follows:

Assuming that the town cannot afford a slope failure involving substantial damage to residential structures, roadways, etc., the committee should approach the town's hillside problems from a conservative viewpoint as far as acceptable risk is concerned.

FIGURE 16-6. **Table adopted by town council, Portola Valley, California, as policy guide to city actions with respect to developments in areas with different land-stability characteristics.** (*G. G. Mader, "Earthquakes, Landslides, and Public Planning," Cry California, Journal of California Tomorrow, 1974.*)

CRITERIA FOR PERMISSIBLE LAND USE IN PORTOLA VALLEY

	LAND STABILITY SYMBOL	ROADS		HOUSES (parcel acreage)			UTILITIES	WATER TANKS
		Public	Private	¼-Ac	1-Ac	3-Ac		
MOST STABLE	Sbr	Y	Y	Y	Y	Y	Y	Y
	Sun	Y	Y	Y	Y	Y	Y	Y
	Sex	[Y]	Y	[Y]	Y	Y	Y	[Y]
	Sls	[Y]	[Y]	[N]	[Y]	[Y]	[Y]	[N]
	Ps	[Y]	[Y]	[N]	[Y]	[Y]	[Y]	[N]
	Pmw	[N]	[N]	[N]	[N]	[N]	[N]	[N]
	Ms	[N]	[N]	N	N	N	N	N
	Pd	N	[N]	N	N	N	N	N
	Psc	N	N	N	N	N	N	N
	Md	N	N	N	N	N	N	N
LEAST STABLE	Pf	[Y]	[Y]	(covered by zoning ordinance)			[N]	[N]

LEGEND:

	Y	Yes (construction permitted)
	[Y]	Normally permitted, given favorable geologic data and/or engineering solutions
	N	No (construction *not* permitted)
	[N]	Normally *not* permitted, unless geologic data and/or engineering solutions favorable
	S	Stable
	P	Potential movement
	M	Moving
LAND STABILITY SYMBOLS: (as used on geologic hazards map)	br	bedrock within three feet of surface
	d	deep landsliding
	ex	expansive shale interbedded with sandstone
	f	permanent ground displacement within 100 feet of active fault zone
	ls	ancient landslide debris
	mw	mass wasting on steep slopes, rockfalls and slumping
	s	shallow landsliding or slumping
	sc	movement along scarps of bedrock landslides
	un	unconsolidated material on gentle slope

The background statement in the resolution adopting the policy guide in Figure 16-6 describes some of the basic provisions proposed:

The Town Council of Portola Valley realizes the extreme importance of geologic data in many decisions which face the town. It also realizes that geologically hazardous conditions exist in extensive portions of the town. While results of highly detailed geologic studies might justify detailed restrictions on the use of some lands, such studies are not now available for most of the town. The geologic maps which have been prepared by the town, however, are based on the study of aerial photographs, field investigations and other available geologic studies, and portray geologic conditions with considerable accuracy. Given this level of data, the Town Council finds it appropriate to adopt these maps as policy, to have them serve as guidelines for administering the affairs of the town, and to modify them from time to time as better information becomes available. It is the Town Council's intention that these maps and related land-use policies shall be employed as guides in all decisions to which they are relevant and shall be adhered to unless modifications or deviations are permitted as provided for herein.

A Complex Geologic Problem

Portola Valley presents an example of a complex geologic setting dealt with effectively in the planning process. Geologic data were used to advantage, and a planned community district was approved for a 177-ha (438-acre) residential development. As explained in Chapter 17, this special zoning for large developments allows greater flexibility in design.

In the application for community zoning, the developer was required to submit a map showing the local geology (Figure 16-7) and a map showing "relative" slope stability (Figure 16-8).

The slope-stability map indicated three main categories of land—stable, potentially moving, and moving. The classification was concerned with only the current situation and not with changes that might result from development.

The stable land, mainly in the eastern half of the property, had only minor geologic problems which could be dealt with by appropriate foundation designs if foundations did not rest on bedrock. The potentially moving and moving land was earmarked for additional geologic and soils study before final development proposals were to be submitted. The largest area of potentially moving and moving land was identified as a composite landslide. Scarps which were believed to be landslide related were also mapped. Finally, strips of potential fault rupture 30 to 60 m (200 to 200 ft) wide were identified along the apparent trace of the San Andreas Fault.

The report recommended that where moving ground was shallow, the shallow soil was to be removed or stabilized before construction. Where moving ground was deeper, further detailed studies were recommended with a view toward employment of special construction methods. The report also recommended that scarps be more carefully investigated and no house built astride them.

With geologic data in hand, and a zoning ordinance which permitted clustering of houses in limited areas with large surrounding open spaces, the developer prepared a design for detached single-family houses on the more stable ridge areas (Figure 16-8). Proposed lots were away from the landslide scarps, away from areas of deep landsliding, and outside the 30- to-60-m-(100- to-200-ft) wide zone along the faults.

Individual subdivision maps, to be filed later by the developer for the eight sections of the subdivision, were to include detailed geologic and soils reports involving subsurface study. At that time, the detailed design of the lots and more detailed siting of houses were to be considered.

The plan, as finally approved, included clusters of lots, each about 0.30 ha (0.75 acre) in size, making a total of 199 lots on 177 ha (438 acres) of land. Thus, approximately 45 percent of the development was in lots and streets and 55 percent in open space. The 199 lots were only 4 lots fewer than the maximum number permissable under the zoning ordinance. The flexibilities of the zoning regulations, which permitted the developer to cluster homesites in a planned community development on the more stable lands, meant no serious reduction in the number of building sites, even though the area has many geologic problems.

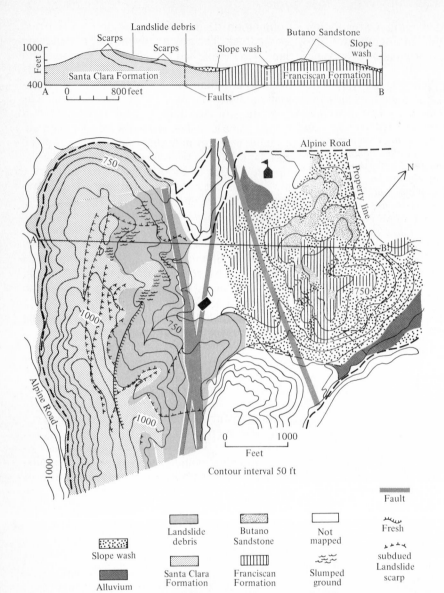

FIGURE 16-7. **Reconnaissance geologic map showing faults, bedrock, and thickness of soil. Cross section shows subsurface conditions inferred from surface data.** (*G. G. Mader and D. F. Crowder, An Experiment in Using Geology for City Planning—The Experience of the Small Community of Portola Valley, California, in "Environmental Planning and Geology," Proceedings of the Symposium on Engineering Geology in the Urban Environment, U.S. Geological Survey and Housing and Urban Development, 1969.*)

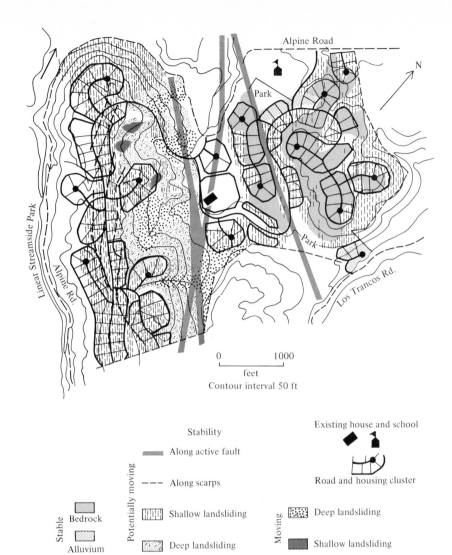

Stability

Along active fault

Along scarps

Shallow landsliding

Deep landsliding

Potentially moving

Stable

Bedrock

Alluvium

Deep landsliding

Shallow landsliding

Moving

Existing house and school

Road and housing cluster

FIGURE 16-8. **Stability map with proposed residential development superimposed. Houses are clustered on ridges to conform to town general plan and avoid major unstable areas. (***G. G. Mader and D. F. Crowder, An Experiment in Using Geology for City Planning—The Experience of the Small Community of Portola Valley, California, in "Environmental Planning and Geology," Proceedings of the Symposium on Engineering Geology in the Urban Environment, U.S. Geological Survey and Housing and Urban Development, 1969.***)**

A preliminary subdivision layout for the same property had been prepared in 1956 (Figure 16-9). That plan showed the entire parcel covered by roads and lots, with no attention paid to geologic problems. If the plan had been followed, at least 15 homes and some main roads would have been built on actively moving ground and active fault traces.

PLANNING FOR OPTIMUM USE OF GEOLOGIC RESOURCES

From a land-use-planning point of view, the basic resource issues require that policy be established at federal and state levels. Lesser levels of government could then guide growth with some assurance that their contributions would add to the overall plan. Because efforts toward such federal and state approaches are only in their beginning stages, we shall consider examples from several states of resource issues at more local levels.

FIGURE 16-9. **Residential development proposed in 1956, which, if built, would have resulted in a number of homes on faults and landslides.** (*G. G. Mader and D. F. Crowder, An Experiment in Using Geology for City Planning—The Experience of the Small Community of Portola Valley, California, in "Environmental Planning and Geology," Proceedings of the Symposium on Engineering Geology in the Urban Environment, U.S. Geological Survey and Housing and Urban Development, 1969.*)

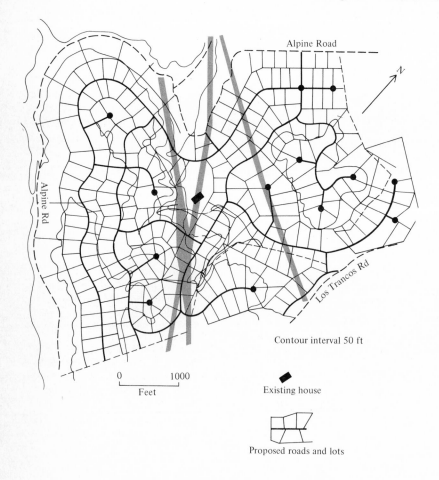

Alpine Road

Alpine Rd

Los Trancos Rd

Contour interval 50 ft

0   1000
Feet

◆ Existing house

Proposed roads and lots

Conservation Plans in California

California state planning law require cities and counties to include consideration of resources in their general plans. These requirements are found in sections of the planning law dealing with the conservation element and the open space element. Excerpts from the state law follow:

The conservation element in a city or county general plan in California must provide for

the conservation, development, and utilization of natural resources including water and its hydraulic force, forests, soils, rivers and other waters, harbors, fisheries, wildlife, minerals, and other natural resources. That portion of the conservation element including waters shall be developed in coordination with any countywide water agency and with all district and city agencies which have developed, served, controlled or conserved water for any purpose for the county or city for which the plan is prepared. The conservation element may also cover:

1. The reclamation of land and waters.

2. Flood control.

3. Prevention and control of the pollution of streams and other waters.

4. Regulation of the use of land in stream channels and other areas required for the accomplishment of the conservation plan.

5. Prevention, control, and correction of the erosion of soils, beaches, and shores.

6. Protection of watersheds.

7. The location, quantity and quality of the rock, sand and gravel resources.

The open space element in a city or county general plan in California must address the following types of open space:

1. Open space for the preservation of natural resources including, but not limited to, areas required for the preservation of plant and animal life, including habitat for fish and wildlife species; areas required for ecologic and other scientific study purposes; rivers, streams, bays and estuaries; and coastal beaches, lakeshores, banks of rivers and streams, and watershed lands.

2. Open space used for the managed production of resources, including but not limited to, forest lands, rangeland, agricultural lands and areas of economic importance for the production

of food or fiber; areas required for recharge of ground water basins; bays, estuaries, marshes, rivers, and streams which are important for the management of commercial fisheries; and areas containing major mineral deposits, including those in short supply.

3. Open space for outdoor recreation, including but not limited to, areas of outstanding scenic, historic and cultural value; areas particularly suited for park and recreation purposes, including access to lakeshores, beaches, and rivers and streams; and areas which serve as links between major recreation and open-space reservations, including utility easements, banks of rivers and streams, trails, and scenic highway corridors.

4. Open space for public health and safety, including, but not limited to, areas which require special management or regulation because of hazardous or special conditions such as earthquake fault zones, unstable soil areas, flood plains, watersheds, areas presenting high fire risks, areas required for the protection of water quality and water reservoirs and areas required for the protection and enhancement of air quality.

These state requirements have caused local governments to identify and establish policy with respect to resources. While the decision of whether to preserve or exploit a resource is largely left up to local government, the law has at least required that the main issues be addressed. A major benefit, therefore, has been to bring resource issues into the public decision-making arena.

San Francisco Bay Conservation and Development Commission

The history of San Francisco Bay has largely been one of artificial filling and expanded use of its shores for industry, housing, and open dumps and landfills. The rate of destruction was not recognized by local governments. Therefore, the state passed legislation in 1965 establishing the San Francisco Bay Conservation and Development Commission, giving it authority to prepare a regional plan for the Bay and to review all developments to require that they conform to the plan. The adopted plan has two primary objectives:

1. To protect the Bay as a natural resource for the benefit of present and future generations

2. To develop the Bay and its shoreline to their highest potential with a minimum of filling

The plan prohibits all fill unless absolutely necessary to provide waterfront locations for vital uses and where already filled land is not available. Such vital waterfront uses include port expansions, industries that require waterfront locations, and recreational uses such as parks, beaches, and trails. In addition, the margins of the Bay are restricted to uses to which a waterfront location are essential. Thus, uses unrelated to the Bay such as dumps, housing tracts, and most industries are no longer permitted on the edge of the Bay.

McHenry County, Illinois

The McHenry County Regional Planning Commission has been making active use of data on soils and geologic resources for years. With the aid of the Soil Conservation Service, State Geologic Survey, and the local soil and water conservation district, the Regional Planning Commission compiled a set of basic soil and geologic data maps and prepared various interpretive maps from these basic maps. Some of the interpretive maps deal with soil limitations for building foundations, agricultural uses, and septic waste-disposal systems, and others are concerned with water hazards.

In order to easily convey the technical findings on the interpretive maps, the county developed a universal "traffic light" color system. Green is used to indicate "no" or "slight" limitations for a potential use. Yellow indicates "moderate" limitation, or problems which can be economically overcome. Red shows "severe" limitations, or problems that cannot be economically overcome. The system, of course, allows the ratings to reflect the harm a potential use might cause a natural system, such as an aquifer, as well as the hazard to the proposed land use.

Southeastern Wisconsin Regional Planning Commission

The Southeastern Wisconsin Regional Planning Commission encompasses Milwaukee and the area to the west and north. A study of the Milwaukee River Watershed of 1790 sq km (690 sq mi) was undertaken as background material for a comprehensive plan for the watershed. It uncovered many potential problems, including loss of wetlands which had served as natural flood-control basins, loss of wildlife habitats, diminishing areas of soils suitable for septic tanks, problems of flooding, eutrophication of lakes, a need for more recreational land, and a need for additional residential land.

One of the significant findings of the study was the identification of primary "environmental corridors." These are generally elongate areas which encompass the best remaining elements of the natural resource base and include "lakes and streams and associated shorelands, wildlife habitat areas, areas containing rough topography and significant geological formations, and the best remaining potential park and related open-space sites." These corridors make up 23 percent of the watershed but contain more than half of the major lake shorelines, major stream channels, wetlands, woodlands, and wildlife habitats. By identifying these corridors, the planners made a strong case for a major preservation effort.

EVALUATING THE NATURAL ENVIRONMENT

While studies of individual facets of the natural environment are useful, it is often necessary to consider several aspects simultaneously in making land-use plans. Thus, if an area is deemed suitable for a particular land use based on one factor such as freedom from flooding, it may be unsuitable based on another factor such as landsliding. Also, it may be necessary to relate these natural environmental factors to cultural factors, such as transportation routes and utilities, and to social and economic factors. These requirements lead to development of methods that allow the handling of a variety of data in a systematic manner with high degrees of flexibility. Several approaches to this problem will be discussed shortly.

Another basic aspect of environmental studies for planning purposes reflects a difference in orientation. The difference stems from whether the study is being made to determine the ability of land to support development, or to determine the extent to which the ecological systems can be tampered with and still survive. Thus, one approach tends to emphasize the problems of development; the other, problems of ecologic balance. We will consider both.

Studies to Determine Capability for Development

A land capability study that included computer analysis is the Foothills Environmental Design Study prepared for the city of Palo Alto, California. The first part of this study consisted of an inventory of factors relating to: planning, the real estate market, ecology, visual qualities, recreation, soils, and geology. After these factors had been mapped in some detail, it became necessary to combine the factors so they could be considered in toto for any part of the area. In order to accomplish this, the 3000-ha (7500-acre) planning area was broken into 8-ha- (20-acre-) square grid cells. The data from each of 23 basic data maps were then encoded as follows for each grid cell.

It was decided that each factor would have no more than five categories, such as five steepness-of-slope categories, and that a rating of 1 would indicate low capability and a rating of 5 would indicate a high capability. The ratings were all with respect to a single possible land use, residential development. Because some factors were more important with respect to residential development than others, it was decided that each factor would be assigned a weight of from 1 to 10. The weighted ratings for a single factor for a single cell could then be obtained by multiplication and could range from 1 to 50.

With the weighted ratings assigned and added, the maximum value in a cell for all the factors turned out to be 480 and the minimum value 94. The list of factors and weighted ratings is shown in Table 16-1, and the corresponding map showing capability ratings is presented in Figure 16-10.

The map of capability ratings shows how the various 8-ha (20-acre) cells compare in terms of capability for residential development based on 23 stated factors and with certain

assumptions as to rating and weighting factors used. The map, therefore, is a synthesis of factors and provides a guide to the planner. Other land use determinants, such as economic, social, and political, may justify construction in some areas that may not be otherwise suited to development. For instance, a geologically hazardous area may have to be crossed by a road in order to provide for fire protection. In such an event, however, the planner is alerted to the need for special precautions in installing the road and he can minimize intrusion in such areas.

A similar study was prepared for the city of Monterey, California, for a 2100-ha (5200-acre) area that was under consideration for annexation. The city wanted to develop information that would be useful in dealing with the following problems: annexation, freeway and major roadway connections, sanitary sewer and flood control, future land use, and zoning with emphasis on areas of preservation either by prohibition of construction or limited density. The study considered a variety of factors including slope, geology, visual aspects, drainage, vegetation, utilities, and roads. Steep slopes, visual dominance, and fault lines were indicated as constraints on future development and were superimposed one on the other on the map shown in Figure 16-11. The map indicates the areas of steep slopes (generally over 30 percent), visual dominance, and the fault lines. The areas of visual dominance are the ridges and slopes most widely seen from the surrounding areas. This map indicated that development should be restricted to the relatively limited remainder of the property. Of course, the other factors included in the study impose further limitations.

Studies to Determine Optimum Land Use

Ian McHarg, a noted landscape architect, supports the view that natural and human factors combine to provide "natural or intrinsic uses" for property. He starts his studies with an inventory of the major natural features of an area. These characteristically include climate, geology, landscape, hydrology, soils, vegetation, and wildlife. The studies may provide information on factors such as productive soils;

Table 16-1. Factors and weighted ratings of the Palo Alto Foothills Environmental Design Study

Planning and Market Factors	Weighted Rating	Visual and Recreation Factors	Weighted Rating
Size of ownership		Views	
Less than 50 acres	4	None	3
51–150 acres	8	Middle or short range	9
151–250 acres	12	Distant	12
251–350 acres	16	Panoramic	15
Over 350 acres	20	Visibility of study area from Palo Alto east of El Camino Real	
Average slope		Visible	3
Over 50 percent	10	Not visible	15
31–51 percent	20	Visibility of Foothills Park from study area	
16–30 percent	40	Not visible	3
0–15 percent	50	Visible	15
Proximity to present development		Proximity to Foothills Park and Upper Stevens Creek Park	
Over 1 mile	1	Over 1 mile	2
¾–1 mile	2	Within 1 mile of Upper Stevens Creek Park	4
½–¾ mile	3	Within 1 mile of Foothills Park	10
¼–½ mile	4		
Within ¼ mile	5		

Geologic and Soils Factors	Weighted Rating

Planning and Market Factors (cont.)	Weighted Rating
Time distance from freeway interchange	
Over 20 minutes	8
16–20 minutes	16
11–15 minutes	24
6–10 minutes	32
Within 5 minutes	40
Improved access dependent on other political jurisdictions	
Dependent	4
Not dependent	20
Present utilities services	
Gas only	2
Sewer or water	4
Sewer or water and gas	6
Sewer and water	8
Sewer, water, and gas	10

Geologic and Soils Factors	Weighted Rating
San Andreas Fault zone	
Within zone	7
Not within zone	35
Other fault zones	
Within zone	3
Not within zone	15
Landslides	
Within slide area	6
Not within slide area	30
Natural-slope stability	
Poor	5
Fair	15
Good	25
Cut-slope stability	
Poor	4
Fair	12
Good	20
Excavation difficulty	
Heavy	2
Normal	10
Soil suitability as fill	
Poor	2
Fair	6
Good	10
Soil-erosion potential	
Severe	6
Moderate	30
Soil-expansion potential	
High	3
Moderate	9
Low	15

Ecological Factors	Weighted Rating
Vegetative cover	
Chaparral, coastal scrub, and urban cultural landscape	5
Rural cultural landscape	10
Fir forest	15
Transition woodland and grassland	20
Oak parkland	25
Proximity to surface-water features	
Over 300 feet	3
Within 300 feet	15
Air pollution	
Heavy	2
Light	10
Fire hazard	
High	8
Medium	24
Low	40

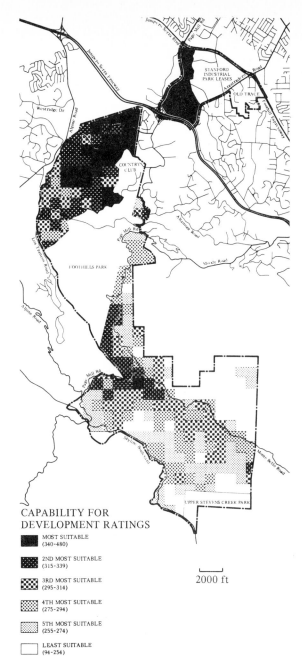

CAPABILITY FOR
DEVELOPMENT RATINGS

MOST SUITABLE
(340–480)

2ND MOST SUITABLE
(315–339)

3RD MOST SUITABLE
(295–314)

4TH MOST SUITABLE
(275–294)

5TH MOST SUITABLE
(255–274)

LEAST SUITABLE
(94–254)

2000 ft

FIGURE 16-10. **Land-capability ratings for residential development of foothills in Palo Alto, California. Each grid cell is 8 ha (20 acres).** (*Livingston and Blaney, "Open Space vs. Development, Final Report to the City of Palo Alto, Foothills Environmental Design Study," 1971.*)

important coal, limestone, or other deposits; the relative abundance of water in rivers and aquifers; the extent of forests, oyster banks, or areas of wilderness; relative accessibility; historic sites; and areas of great natural beauty. Because regions differ from one another geologically, there is regional variation in the resources considered. However, there is also a relative consistency within each region due to climate, topography, hydrology, and soils. Thus, it is not difficult to interpret this information in terms of the dominant, intrinsic resource or resources.

In his study of the Potomac River basin, McHarg grouped compatible land uses. He pointed out, for instance, that

> a single area of forest may be managed either for timber or pulp; it may be simultaneously managed for water, flood, drought, erosion control, wildlife and recreation; it may also absorb villages and hamlets, recreational communities and second homes.

Having thus derived a range of land uses which reflect the intrinsic character of the resource, he prepared maps which show his evaluation of the inherent suitability of lands for a range of possible land uses. With respect to the Potomac River basin, he concludes

> that mining, coal and water-based industry offer the maximum opportunity in the Allegheny Plateau, with forestry and recreation as subordinate uses. In the Ridge and Valley Province, the recreational potential is dominant, with forestry, agriculture, and urbanization subordinate. In the Great Valley, agriculture is the overwhelming resource, with recreation and urbanization as lesser land uses. The Blue Ridge exhibits only a recreational potential, but of the highest quality. The Piedmont is primarily suitable for urbanization with attendant agriculture and nondifferentiated recreation. The Coastal Plain exhibits the highest potential for water-based and related recreation and forestry, and a lesser prospect for urbanization and agriculture.

McHarg concludes that this is a method by which the characteristics of a region may be determined: "The place must be understood to be used and managed well. This is the ecological planning method."

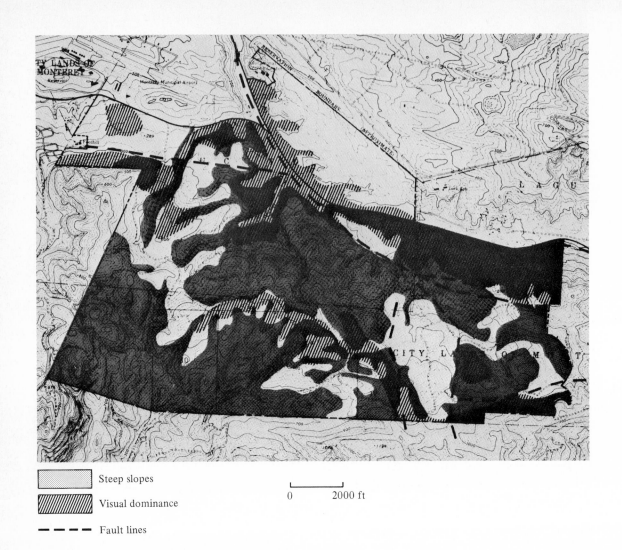

Steep slopes

Visual dominance

0   2000 ft

Fault lines

FIGURE 16-11. **Map showing major constraints on future development for a 2100-ha (5200-acre) area west of Monterey, California. (*Dept. City Planning, Monterey, Calif., 1971, "Work Ranch-West, Planning Study," 1971.*)**

Many of McHarg's study maps consist of transparent plastic overlays on which conditions are shown in shades of gray. Thus, when the overlays are superimposed, the darker tones indicate the areas less suitable for a land-use option and the lighter tones indicate the areas more suitable. An example of this technique is shown in Figure 16-12. This map indicates the suitability of Staten Island, New York, for urbanization based upon slope, forested areas, poor surface drainage, poor soil drainage, areas susceptible to erosion, and areas subject to flooding.

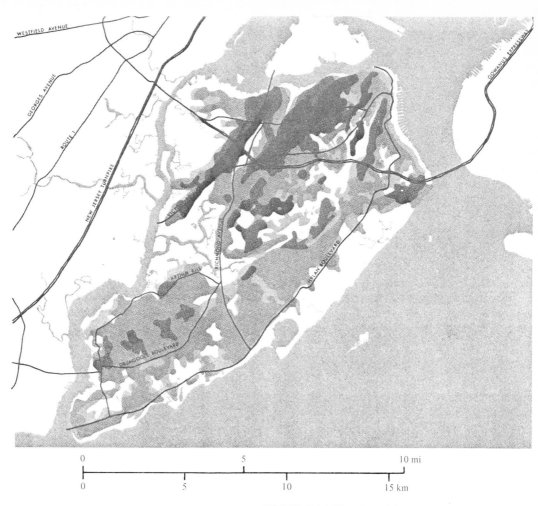

FIGURE 16-12. **Map of Staten Island, New York, showing, in lighter tones, areas suitable for urbanization.** (*From "Design with Nature," copyright 1969 by Ian McHarg. Reprinted by permission of Doubleday & Co., Inc.*)

ADDITIONAL READINGS

Chapin, F. S.: "Urban Land Use Planning," University of Illinois Press, Urbana, 1965.

International City Manager's Association: "Principles and Practice of Urban Planning," 4th ed., Washington, D.C., 1968.

McHarg, I. L.: "Design with Nature," Natural History Press, New York, 1969.

Nichols, D., and Campbell, C., eds.: "Environmental Planning and Geology," *Proceedings of the Symposium on Engineering Geology in the Urban Environment*, Assoc. Eng. Geol. Annual Meeting, 1972, U.S. Geological Survey and U.S. Dept of Housing and Urban Development, 1972.

University of Texas, Bureau of Economic Geology: Environmental Geologic Atlas of the Texas Coastal Zone, 7 volumes, 1972.

William Spangle and Associates: Application of Earth Science Information in Urban Land-Use Planning, State-of-the-Art Review and Analysis, *NTIS*, Springfield, Va., 22161, Report No. USGS-GD-74-038, PB 238 081, 1974.

CHAPTER SEVENTEEN

private development: planning and implementation

INTRODUCTION

Chapter 16 described some of the basic principles of land-use planning, with special emphasis on geological hazards and resources. The present chapter discusses planning and implementation of private development within the framework of local community land-use planning and emphasizes the role of geology. Chapter 18 will describe a student group project in environmental planning.

The procedure followed, from conception of a project to completion of construction and sale or lease, is a complex and lengthy one. We will consider a hypothetical residential development, but one that embodies many of the problems encountered in real life. We will follow this development from beginning to end, pointing out along the way how geologic information is used, what specialists are involved, and how the developer interacts with local government. The case is idealistic in that we assume that our developer employs the proper staff and consultants at all times and acts on their advice.

Although our example emphasizes concerns of a physical nature, only selected physical factors are considered, and many vital matters such as financing and sales programs are hardly touched upon. A hypothetical example is justified even though geologic problems, types of development, and local govermental regulations vary from place to place, because the implementation process is basically the same throughout the country.

The role of the developer has changed in recent years. The process of residential land subdivision formerly consisted only of acquiring a parcel of land, filing a plan for its layout into blocks and lots, and selling the lots to buyers. In modern practice, the creation of residential lots is part of a larger package that may include single-family residences, townhouses, apartments, commercial property, and land improvements with paved streets, installed utilities, and recreational areas. The subdivision, not the lot, is now the basic unit of land development. The objective is increasingly the creation of a total community designed for good living.

It is the developer who initiates the land-development process and controls its progress until the homeowners take title. The developer usually represents a group of investors who provide the capital and undertakes securing the land, planning the development, building it, and selling the finished product. All this must be done within local physical, social, economic, and governmental restrictions.

THE PLANNING-REGULATION-DEVELOPMENT PROCESS

Compliance with governmental regulations requires that our developer proceed through an orderly progression of steps. Figure 17-1 illustrates the planning-regulation-development process at the local governmental level.

We see at the top of the diagram that "long-term decisions" affecting "large areas" and "general concepts" are made first, and these lead to "short-term decisions" for "small areas" and "specific details" within the larger plan. These considerations apply to each of the three types of subject matter in the figure: "development," "plans and regulations," and "geology."

The developer is first concerned with the compatibility of his proposal with "general community goals" as expressed in the local

FIGURE 17-1. **The planning-regulation-development process. Geologic data are necessary at every step.** (*G. G. Mader and D. F. Crowder, 1971, An Experiment in Using Geology for City Planning—the Experience of the Small Community of Portola Valley, California, in D. R. Nichols and C. C. Campbell, eds., "Environmental Planning and Geology."*)

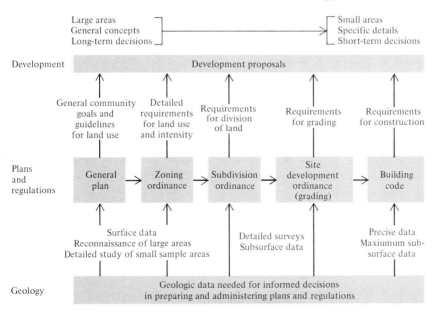

"general plan." The developer is also concerned with the broad geologic setting for the development. Later in the process, the developer will need highly detailed geologic information as he designs buildings. Thus, the developer makes more refined decisions as he gets closer to the construction stage.

The series "plans and regulations" illustrates how a community can develop broad concepts in the general plan and then adopt regulations to help implement the plan. The "general plan" sets forth the broad goals of the community for the next 20 to 30 years. The "zoning ordinance" sets forth detailed limitations on the use of land and buildings. The "subdivision ordinance" governs the manner in which land is divided and sold and also sets forth the requirements for improvements to be made by the developer, such as streets, water lines, and other utilities. The "site development ordinance" (grading) governs the manner in which land is modified by grading. Finally, the "building code" stipulates requirements for the construction of houses, offices, and virtually all other construction. Proposals for development, however, are made only after market analysis has shown a need and indicates economic feasibility.

MARKET ANALYSIS

J. A. North of North Development Company, our fictitous builder, has decided to undertake a new project. He has been building residential developments south of San Francisco, but because of the shortage of available land, he has decided to look to another area. Having heard that considerable land is available for development north of the city, he decides to investigate. His initial contacts reveal considerable construction activity. Discussions with banking officials, realtors, builders, and city officials lead him to consider a project in the area.

Before progressing further, Mr. North needs certain information that can best be supplied by a market analysis. In short, he must learn the characteristics of the housing demand. He wants to know who his prospective home buyers are, what their preferences are, what their levels of income are, how many children they have, and other characteristics of the market. Once he has determined the market demand, he will choose a site and plan a development to fit the market.

An economist specializing in market analysis is retained to study the housing market in the area. The economist must determine what kind of a housing development would fit the local housing need. This requires a study of population characteristics such as age, occupation, income, family size and car ownership. The inventory of present housing will include dwelling type, age, condition, size, vacancy, and sales and rental values. It will take into account existing substandard housing and any local plans and programs for demolition or rehabilitation that may displace people and create a need for housing. It will include local preferences, prejudices, and customs, and what new building materials, techniques of construction, and merchandising and management are acceptable in the local market.

A number of construction indicators to be derived from the market analysis are important. For example, at what prices can new housing be absorbed? What are the features in new homes that sell (garage, number of bathrooms, etc.)? Other important market factors are tax rates and assessments, levels of public services, quality of schools, community facilities and services, and availability of open space. Finally, the market analysis considers the direction in which urban growth is occurring and city policy regarding new growth.

The results of the market study indicate a solid demand for detached single-family residences and townhouses by persons of middle income. With this evidence of housing need, Mr. North now feels confident enough to start looking for a site for his housing development.

SITE SELECTION

Site selection involves locating land available for sale and suitable for Mr. North's contemplated residential development. He now looks for available properties in the 200-acre size range, the size of development he is used to undertaking. He finds three properties that appear to fit his needs and appear to be desirable for development. He knows from past experience, however, that he must now obtain professional help to make certain he selects the most suitable piece of property.

It is hoped that Mr. North is acquainted with the range of specialists shown in Table 17-1 and appreciates that their expertise may be needed. For site selection, he is particularly concerned with city plans and regulations that apply to the several available sites, the requirements for improvements such as streets, water, and other utilities, and potential geologic problems.

Mr. North can engage the needed specialists independently or can go to a large firm or firms that employ a wide range of specialists. For this project, he decides to engage three firms which can provide the range of specialists he believes to be needed. A consulting geological firm is engaged which has a seismologist, engineering geologists, and soils engineers. An engineering firm is selected which has civil

Table 17-1. Selected list of specialists involved in land development

Specialist	Definition	Types of Input to Developer
Seismologist	A specialist in earthquakes	Advises as to earthquake potential
Geologist	A specialist in earth materials and processes	Determines the geologic hazards, and advises the civil engineer
Engineering geologist	A geologist concerned with the effects of geology on engineering works	Advises on the potential impact of geologic materials, structures, and processes on engineering projects, and vice versa. Advises on engineering solutions to geologic problems
Civil engineer	An engineer who designs and constructs public and private works	Designs and supervises construction of roads, utility lines, parking areas, etc., in subdivision
Soils engineer	A civil engineer who specializes in soil problems	Advises on the engineering properties of soils, the bearing capacity for buildings, and potential shaking and failure in earthquakes
Hydraulic engineer	An engineer who applies engineering principles and methods to the control, conservation, and utilization of water	Advises on methods of handling subsurface and surface-water problems
Structural engineer	An engineer who analyzes and designs engineering structures of adequate strength to withstand expected stresses	Advises on designs for buildings and other structures to assure sufficient strength under normal and unusual conditions, such as earthquakes
Architect	A person trained to design buildings. Emphasis is on meeting needs of building users and aesthetics. Understands basic engineering but relies on engineers	Normally has responsibility for design of buildings and receives advice from structural engineer
Landscape architect	A person trained to adapt the works of man harmoniously to the natural environment	Designs the open space around buildings, advises on building siting and site plans, and is involved in basic subdivision layout
Planner	A person trained to understand the urban environment and prepare plans and programs for future growth and change which fit within the urban framework and are consistent with the public interest	Advises on integration of development with community and related plans and regulations; may become involved in basic layout and design of development; May serve as development coordinator

engineers, hydraulic engineers, and structural engineers. A design firm is employed which has architects, landscape architects, and planners.

Mr. North asks that the planner, engineer, and geologist make quick appraisals of the sites. The planner assesses each of the sites to determine if the contemplated residential development is consistent with the general plan of the city and with the city's zoning regulations. If they are not consistent, he must determine if modifications of the city plans and regulations are justified and might be favorably received by the community. He is concerned with the quality of the several sites in terms of proximity to city services such as schools, parks, and libraries as well as the availability of necessary commercial facilities to meet the everyday shopping needs of the future residents. In summary, he wants to make certain that the proposed development will fit well into the urban complex and have lasting value.

The civil engineer investigates the availability of the necessary services including water, sewage disposal, storm drains, streets, electricity, gas, and telephone. If major extensions of these services are necessary in order to serve the development, the cost to the developer could be prohibitive. He will also be concerned with the standards included in the local subdivisions ordinance. These standards will indicate the types of improvements necessary, which in turn affect the cost of the improvements.

The geologist is asked to look into the geology of the sites to determine if they have any special geologic problems and if any preventive or remedial actions will be necessary. The geologist must therefore examine available geologic maps, air photos, and other data and make reconnaissance field investigations. He will be particularly concerned with problems such as landslides, flooding, and earthquake-related hazards.

The first of Mr. North's three sites (site A) is adjacent to existing subdivisions, close to town, and on rolling farmland. Site B is relatively flat, at the edge of a river, and also close to town. Site C is a few miles from town in attractive oak-studded hills. Each site is analyzed on the basis of the factors shown in Table 17-2.

Table 17-2. Factors to be considered in site selection. (Adapted from J. R. Mckeever. See Additional Readings)

 1. Accessibility and transportation
 2. Location and approaches
 3. Size of development
 4. Land cost
 5. Physical characteristics
    a. Topography and shape of the site
    b. Drainage
    c. Soils
    d. Geology
    e. Tree Growth
 6. Utility services
    a. Water supply
    b. Sewerage
    c. Storm drainage
    d. Electricity and gas
    e. Site-improvement policies
 7. Site environment
    a. Land use
    b. Streets with heavy traffic
    c. Dampness, smoke, and views
    d. Flooding
    e. Fire hazard
    f. Airports and their effects on residential developments
 8. City services and community facilities
    a. Fire and police protection
    b. Schools
    c. Schools, recreation, and open space— planning and development policies
    d. Waste disposal and street service, such as sweeping and snow removal
    e. Community facilities
 9. Municipal regulations
    a. General plan
    b. Zoning
    c. Subdivision regulations
    d. Building codes
10. Consultation with local officials
11. Methods of purchasing land in large acreage
    a. General
    b. Options and purchase contracts
    c. Release clauses

Site B is found by the geologist and hydraulic engineer to have significant flooding potential. They agree that flood prevention would be prohibitively expensive. Furthermore, the geologist points out that the site is on loose alluvium within 1 mi of an active fault and that potential damage from earthquake shaking would have to be considered in the structural designs. The planner reports that the land is restricted in the general plan of the community to agricultural and recreational uses. He believes these are appropriate use designations and sees no purpose in seeking to have their role in the community plan changed. Therefore, site B is rejected.

Site C is in a very attractive physical setting. The geologist reports, however, that it includes a number of landslides. The engineer reports that the nearest utility lines are 1 mi from the property and that the cost of extending them into the property would be very high. The planner reports that while the general plan recognizes eventual residential use for the property, current zoning calls for agricultural use and no change in designation can be expected for at least 10 years.

Site A is given a good report by the geologist. He has found some expansive soils on the site, but careful foundation work and drainage can solve the problems at reasonable cost. The engineer reports that all necessary utilities are available at the boundaries to the property and that there should be no problems in extending the utilities into the property. The planner indicates that a development of single-family homes and townhouses would be consistent with the local general plan. He points out that the lands are only marginally suitable for agriculture and that the property would probably be rezoned from the current agricultural designation for the intended development. Again, provisions must be made to protect the development from earthquake damage. Much of the site, however, is on firm bedrock and should not experience intense shaking. Thus, site A turns out to be the best choice for the North development.

PLANNING THE DEVELOPMENT

Once the site has been selected, the next step for Mr. North is to prepare a plan for the development that will serve as a basis for obtaining a planned community (PC) zoning designation from the city. This type of zoning, variously named, is usually applicable only to large developments and allows greater flexibility in design than is appropriate for small developments. In the development plan, Mr. North must develop the concept for the entire project in sufficient detail to be certain the project is feasible. He must also demonstrate to the community that the project will fit harmoniously into the overall city plan. He does not need to prepare plans of such accuracy or detail as are needed for subdivision approval, because problems of subdivision arise later.

Mr. North again goes to his team for assistance. His civil engineer will at this point be responsible for preparing an accurate contour map of the topography showing existing natural and cultural features, including streams, trees, fence lines, buildings, roads, and utility lines. The civil engineer's survey indicates that maximum relief is 30 m (100 ft). He discovers, however, that one of the fence lines is inaccurately placed on the original map, and the development site includes a fine stand of oak trees which had been presumed to be off the site. The topographic survey also indicates a depression that appears subject to ponding. These new bits of information will be taken into consideration in the design stage.

The engineering geologist and soils engineer working together make a brief study of the site, largely using surface data and aerial photographs. A few test pits are dug to check the depth of expansive soils. The expansive soils are found to be 0.3 to 0.6 m (1 to 2 ft) deep over much of the property. House foundations can be designed to extend below these depths, but substantial expansive material will need to be removed along planned roads. Some of the swales between hills have very deep expansive soils, and these sites may have to be avoided or bridged in construction if possible (see Figure 2-32). The study also reveals the existence of a landslide area which must be avoided. The results of the study confirm the usefulness of the greater part of the site, but indicate areas to be avoided.

Mr. North now asks the planner, landscape architect, and architect to work together in developing the best project design possible. The planner is particularly concerned with the harmonious relationship of the development to the surrounding community. Thus, he recommends careful integration of the new street system with existing street patterns, the preservation of the wooded knoll as open space both for the development project and the surrounding area, and the location of the attached townhouses with their density of population nearest shopping facilities and the neighborhood school for maximum convenience. He recommends that 75 percent of the developable land be apportioned to single-family detached residences and 25 percent to townhouses, a proportion that would be consistent with the community general plan and the market analysis.

The landscape architect, after analyzing the form of the terrain, existing trees, soil and geologic maps, and views of and from the property, prepares a proposed layout scheme in consultation with the other team members. (In some instances, this preliminary layout might be prepared by a planner or a civil engineer.) The landscape architect is particularly concerned that there be minimum disturbance of the natural features of the site and thus tries to visualize the finished development as it would appear when all plantings have matured.

The architect considers the results of the market analysis which indicate preferences for dwelling types and, with the developer, prepares descriptions of suitable types of housing. He then makes preliminary designs of prototype houses and checks to see that they conform to the findings of the other members of the team.

Through team cooperation, the areas of deep expansive soil and landsliding have been avoided, a grading design to eliminate the ponding problem has been developed, and the houses have been assured of good views owing to the careful layout of the streets with respect to the terrain.

During the preparation of the plan, the civil engineer constantly works with the team to make certain that the design is consistent with good engineering practice. For example, the engineer makes certain that the inclination of sewer lines is correct, that an economical road and water system can be developed, and that intersections, grades, and curves are safe for vehicles.

After a month of work, the team has prepared a conceptual plan which shows a development of 750 single-family detached houses and 250 townhouses. A system of open spaces has been proposed, and the development is carefully integrated with surrounding residential areas.

The developer is now ready to submit his plan to the community to secure the required planned community (PC) zoning. Along with the zoning request, he must submit a draft environmental impact report (EIR) as required under the California Environmental Quality Act. This act, modeled after the National Environmental Protection Act, was discussed in Chapter 15.

The draft EIR is often prepared by firms specializing in environmental matters or by teams of specialists. In this case, Mr. North has already utilized a broad range of specialists who have much of the information necessary for the preparation of the draft EIR. The draft EIR must identify the environmental impacts of the proposed development and steps that will be taken to ameliorate the deleterious effects. The report deals with such diverse aspects of the site and development as the archeology, geology, climatology, fauna and flora, aesthetics, economic and social impacts, pollution (including noise pollution), hydrology, historic features, and possible alternate uses for the land.

While the development team has been in close contact with city officials, especially the planning staff and engineering department, it is now time for the application for PC zoning and the draft EIR to be formally submitted to the city. As a first step in the city review, the application for PC zoning is set for public hearing before the planning commission, and the draft EIR is circulated for review.

The city planning staff has the responsibility for coordinating the staff review of both the PC zoning application and the draft EIR. Thus, each of these is submitted to other city staff including the city engineer, health officer, fire chief, and director of parks. Also, nearby property owners, utility companies, and other affected groups are informed of the proposed development and the draft EIR and are given the opportunity to review the documents and make comments at a public hearing before the planning commission.

After comments on the draft EIR have been received by the city planning staff, a final EIR is prepared by the city planning staff for presentation to the planning commission. The planning commission consists of residents of the community appointed by the elected city council. The commission members are not professional planners, but come from all walks of life. The city planning staff, which advises the commission, is made up of professional planners.

The EIR is an information document which has as its purpose the introduction of all relevant information regarding a development with respect to the environment. It is not binding on the city but is used as background information in making decisions. The planning commission must accept the final EIR before action can be taken on the PC zoning.

At the public hearing on the PC zoning and the EIR, the city planning staff and developer make presentations to the commission. The staff indicates that the proposed development is in conformity with the city general plan and has responded to the EIR. Therefore, the staff recommends approval. The citizens living in the vicinity of the proposed development approve of the development but are concerned with the impact of the construction work on the area in terms of noise, dust, and soil erosion during the rainy season. In response to their concerns, the city staff indicates that when the details of subdivision are proposed (the next step in the development sequence), limitations will be put on hours of operation and appropriate constraints imposed for dust and erosion control.

Because all actions to amend the zoning ordinance are actually legislative actions, the proposed PC zoning must also be subject to public hearing before the city council. Thus, a duly advertised public hearing is held. No new testimony is presented to the city council which therefore concurs with the planning commission proposal that the North Development PC zoning be approved.

At this point, Mr. North knows that the basic provisions of his development are acceptable to the city. He can now move forward to more detailed considerations. His next step is to have a subdivision map prepared.

SUBDIVIDING THE LAND

This step involves the preparation of detailed plans for subdivision. The plans provide for the precise layout of streets and lots. The subdivider must also engineer the road improvements and all utilities to be installed, including water lines and storage tanks, electrical lines, sewage lines, gas lines, and telephone lines. The subdivider must pay for all these improvements before lots may be sold. Because of the large investment in improvements, Mr. North decides to proceed with his development one part or unit at a time. In this way, the total initial or "front-end" cost will be kept down. As he sells homes in one unit, he will develop succeeding units.

For this phase of the project, the civil engineer is the key professional assisting the developer. If more detailed studies of the environment are needed, they may be undertaken at this time. It is decided that further geological exploration and soils analysis are required. Subsurface exploration is recommended to provide better information on the depth of expansive soils so that the amount of soil to be excavated for road beds can be determined and places for disposal selected. Also, it is recommended that the boundaries of the landslide be carefully mapped so that roads, houses, and utilities will not be constructed within its confines.

The civil engineer, with input from the geologist and soils engineer, then proceeds with the subdivision design. The engineer has the job of translating a part of the plan approved for the PC zoning to a subdivision map. In doing this, the civil engineer will most likely discover that some adjustments will be necessary and will of necessity consult with the planner, landscape architect, and architect.

When the subdivision map is completed, it is subject to approval by the planning commission, which is advised by the city planning staff just as it was advised relative to the PC zoning. Also, the final subdivision map, accompanied by bonds to ensure installation of improvements as required of the developer, is reviewed and approved by the city council.

After approval of the final subdivision map by the council, the map is recorded with the county recorder. Once this has been done, the developer can legally sell lots as shown on the map. The buyer is protected by the bond that the improvements will be provided as shown on the subdivision map. Because Mr. North wants to build the housing units and sell them with the lots, he will delay his sales until the houses are completed.

SITE DEVELOPMENT (GRADING)

The next step in the planning-regulation-development process (Figure 17-1) is the grading of the site. Grading for the subdivision as a whole is done as part of the general improvement of the subdivision and under the provisions for standards and inspections set forth in the city site development ordinance. The grading of each lot for a particular housing unit is a second step in the grading process and is done after the basic subdivision grading has been completed. Grading for individual housing units is also controlled by provisions of the site development ordinance.

Grading plans are normally prepared by civil engineers, with input from engineering geologists and soils engineers. Architects and landscape architects often advise on grading designs to assure that the grading will enhance the proposed development and will be aesthetically pleasing. Aesthetically pleasing grading involves blending the cut and fill surfaces with the natural terrain in such a way that they appear as natural as possible.

CONSTRUCTION OF HOUSING

With the first unit of the development now subdivided, Mr. North is anxious to build some houses and sell them. Not until the first house is sold will there be any financial return on the development.

The architect now becomes the key specialist. He must prepare detailed building plans indicating the exact construction details of the houses. In fact, he will have been working on house designs during the subdivision stage. He must design houses that will fit the housing market, be economical to build, and meet the detailed requirements of the building code. Also, if it is anticipated that federally insured home loans are to be used by purchasers, his designs must meet federal requirements.

Mr. North's development is only 1 mi from an earthquake fault. At this stage, he therefore asks the soils engineer and structural engineer to report on potential earthquake intensities and advise the architect on special structural requirements that should be included in the building. Also, the soils engineer is asked to consult on proper foundation designs in areas of expansive soils.

The landscape architect now helps in the siting of buildings on the land and develops planting designs; hence, the landscape architect should be familiar with the various types of plants that can grow in the area so he can make sensible recommendations. It is becoming increasingly desirable to utilize native (indigenous) plants, since they are adapted to the climate and soil and require little maintenance.

When the housing plans are completed, they are submitted to the building department of the city for review. The building department finds them to be in conformity with the building code and the PC zoning. Therefore, building permits are issued to the North Development Company, allowing construction of the homes in the first subdivision unit.

The city building department's job is not finished at this point. Each step in the house construction must be inspected and approved by the department. This inspection and approval assures that the houses are built as planned.

ADDITIONAL READINGS

Mckeever, J. R., ed.: "The Community Builders Handbook," Anniversary Edition, Urban Land Institute, Washington, D.C., 1968.

Nichols, D. R., and C. C. Campbell, eds.: "Environmental Planning and Geology," published cooperatively by U.S. Geological Survey, U.S. Dept. of Interior, and Office of Research and Technology of U.S. Dept. of Housing and Urban Development, 1971.

Subdivision on coastal terrace,
Half Moon Bay area, California.

CHAPTER EIGHTEEN

student group project on environmental land-use planning

INTRODUCTION

Purpose

This chapter presents comprehensive land-use plans conceived by nontechnical students after assimilating the background provided by this text. The plans blend environmental constraints and the techniques and judgments of land-use planning. The area chosen was the community and environs of Half Moon Bay, California, 32 km (20 mi) south of San Francisco, and known as the San Mateo County Mid-Coastside.

The project is described for two reasons. First, such projects have provided stimulating educational experiences for our students, and similar projects can be undertaken at other institutions. Second, whereas this text is primarily concerned with the geologic aspects of land-use planning, the project involved many nongeologic factors.

Land-Use Sketch Plan and Objectives

A *land-use sketch plan and statement of objectives* is prepared explaining in general terms the goals and objectives of the proposed development plan. The plan allocates land for different purposes consistent with the goals and objectives.

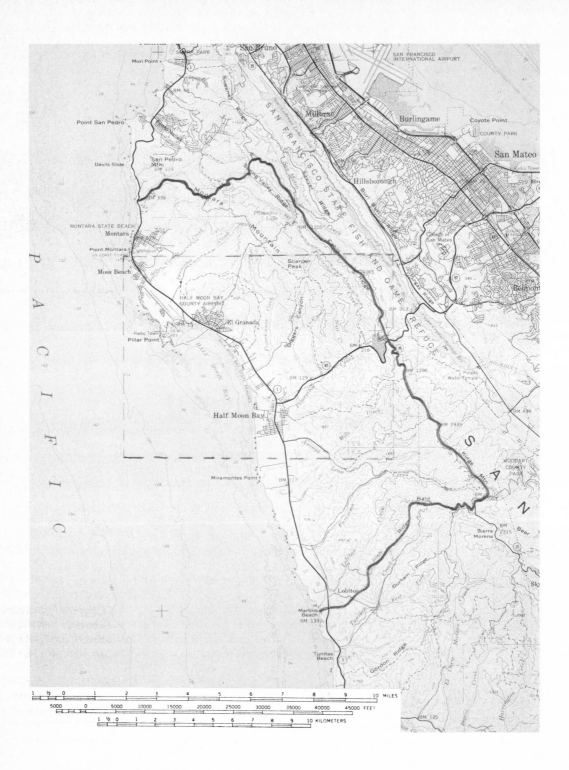

If the community decision makers find the sketch plan and objectives inconsistent with their perception of the community future, it is returned for revision. If they find them acceptable, the planners prepare a more detailed plan.

Half Moon Bay Area

Figure 18-1 shows the planning area outlined on a portion of the 1:250,000 U.S. Geological Survey topographic map of the San Francisco Bay region. Figure 18-2 is an aerial photograph of most of the planning area and its surroundings. The reconnaissance maps and plans presented in this chapter are restricted to the area outlined by dashed lines in Figure 18-1. They have been generalized from student maps and work sheets.

The study area extends from the crest of the Santa Cruz Mountains, locally exceeding 600 m (2000 ft), to the sea (Figure 18-1). The figure also shows the area of intense development northeast of the planning area on the San Francisco Bay side of the Santa Cruz Mountains.

The coastal area in the vicinity of Half Moon Bay is spectacularly beautiful. Marine terraces extend inland from the sea cliffs to the foot of the Santa Cruz Mountains. The terraces and lower slopes are devoted to agriculture or are covered by grass and chapparal. The mountains are covered by coniferous forests. Summer weather is cool, windy, and foggy, and winter weather is also cool and windy, but usually with clear blue skies. The population in 1970 was between 10,000 and 12,000, with most of the labor force engaged in agriculture. Because of the sparse population and beautiful scenery, the area is a popular "Sunday" recreational area for residents of San Francisco and other communities.

FIGURE 18-1. *(Opposite.)* **The San Mateo County Mid-Coastside area. The margins of the planning area are shown by the irregular solid line. The total length of the planning area along the coast is 23 km (14 mi). The rectangular area enclosed by dashed lines is that part of the total area illustrated by the maps and plans of Figures 18-3 to 18-16.**
(Topographic map by U.S. Geological Survey.)

Great population pressures exist in San Francisco to the north and in the Bay area to the east. Yet, the Santa Cruz Mountains have shielded the area from development. Parts of Route 1 extending north to San Francisco are cut into steep cliffs and are almost impossible to maintain. The road is repeatedly cut by landslides at Devils Slide just north of the project area (see Figure 18-1). Route 92 to San Francisco Bay is overloaded with weekend recreational traffic.

Expensive real estate developments have bypassed Half Moon Bay in favor of Santa Cruz, Carmel, and Monterey farther south. Half Moon Bay is a low-to-moderate-income farming community. Its few tract homes are inexpensive and economically designed. The area has more fog, wind, and inclement weather than Santa Cruz, Carmel, and Monterey, which may explain why expensive resort development has favored the latter communities.

The time seemed ripe for accelerated development of the Half Moon Bay area when the study was undertaken in the spring of 1971. Population pressures and land costs were making it necessary for increasing numbers of people to seek homes outside of San Francisco and the San Francisco Bay communities. Large developers were assembling sizeable holdings. One development had already opened with homes in the $60,000 class, considerably more expensive than earlier developments. Relocation of Route 1 inland from Devils Slide and expansion to four lanes were being planned despite opposition from conservationists. Expansion of Route 92 to four lanes was also being considered. Thus, accelerated development seemed imminent, and the area presented a ripe subject for the exercise described herein.

Organization of the Study

The study involved three sequential phases, each of which will be described.

The first phase, a *basic data reconnaissance*, involved library, field, and laboratory study to assemble basic data, to present these data on maps at a scale of 1:24,000, to prepare data reports, and to present the findings orally to the entire group. The class was organized into 16 groups of 1 to 3 students for this phase.

FIGURE 18-2. **Aerial photograph of the San Mateo County coastal area. The restricted area in the vicinity of Half Moon Bay used for the reconnaissance maps is outlined. (*U.S. Geological Survey*.)**

In the second phase, the basic data were used to prepare *land-capability ratings* for residential and agricultural uses. These resulted in colored maps showing land capability for the two development options. The class was divided into groups of four or five students for this phase.

The third phase involved the preparation of a *sketch plan and objectives*. The groups organized for the land-capability ratings were continued for this effort. Each group prepared a sketch plan consisting of maps and text and presented the results orally for criticism by students, faculty, and invited planners.

BASIC DATA RECONNAISSANCE

Introduction

The topics listed in Table 18-1 were selected for investigation and data reconnaissance. Topics I to X relate to the natural environment; topics XI to XVI, to human considerations. These will

Table 18-1. Environmental factors investigated by student teams in data reconnaissance. Liberties were taken in the groupings to minimize the number of agencies that each team would have to contact

Natural Environment

I. Topography
  1. Slope declivity
  2. Slope direction
  3. Elevation and relief
II. Climate
  1. Rainfall
  2. Temperature
  3. Wind
  4. Humidity
  5. Fog
  6. Air quality
III. Plant communities
IV. Geology and geologic hazards
  1. Rocks and structures
  2. Aquifers
  3. Mineral resources
  4. Landslides
  5. Seismic hazards
V. Soils
  1. Soil depth
  2. Water-holding capacity
  3. Permeability
  4. Shrink-swell behavior
  5. Agricultural capabilities
  6. Septic-tank suitability
VI. Drainage
  1. Flooding
  2. Stream erosion
  3. Sedimentation
VII. Ocean and shoreline
  1. Waves and currents
  2. Longshore drifting
  3. Upwelling
  4. Areas of erosion
  5. Areas of deposition
  6. Water quality, pollution, and salinity
VIII. Wildlife
  1. Land animals
  2. Marine animals
IX. Hydrology
  1. Existing water supplies
  2. Existing sewage disposal and its effects on water supplies
  3. Streamflow
  4. Ground-water potential
  5. Sewage disposal potential and hydrologic effects
  6. Water-table elevation
  7. Subsidence due to ground-water withdrawal

  8. Water quality
X. Aesthetics and visual aspects
  1. Coastal views
  2. Hillside environment
  3. Existing development

Cultural Features and Factors

XI. Land Use
  1. Residential
  2. Commercial
  3. Agricultural
  4. Recreational
  5. Other uses
XII. Land ownership
  1. Major owners and their holdings
  2. Access to waterfront
  3. Plans of major developers
XIII. Transportation and utilities
  1. State highways
  2. City and county roads
  3. Harbor
  4. Airport
  5. Solid-wastes disposal
  6. Gas and electricity
  7. Water and sewage lines

Economic and social factors

XIV. Population and economics
  1. Population size, characteristics, and projections
  2. Housing
  3. Economic base
  4. Potential economic base
  5. Employment
  6. Tax base and city and county finances
  7. Williamson Act (tax relief for agriculture)
XV. Recreation and tourism
  1. Parks and trails
  2. Harbors and marinas
  3. Beaches

Public Interest Factors

XVI. A. Government and special interest groups
  1. Government agencies
    a. State
    b. County
    c. City
  2. Local agency formation commission
  3. Special interests (Sierra Club, Coastal Alliance, etc.)
  B. General plans and zoning
  1. San Mateo County
  2. Half Moon Bay
  3. Harbor district

Table 18-2. Reconnaissance guide. The reconnaissance study groups are numbered as in Table 18-1

Student Group	San Mateo County — Planning Department	County Engineer	Parks and Recreation	Local Agency Formation Commission	U.S. Geological Survey — Housing and Urban Development Study	Engineering Geology	Water Resources	State of California — Division of Highways	Beaches and Parks	Fish and Game	Water Resources	Agricultural Extension Service	City of Half Moon Bay — City Manager	City Engineer	Other Federal Agencies — Soil Conservation Service	Corps of Engineers	Weather Bureau	Districts — Harbor	Coastside Water	Miscellaneous — Assoc. of Bay Area Governments	Conservation Coordinators	Planning Consultants	University — Plant Ecologist	Animal Ecologist	Water Supply Engineer
I					×												×								
II																									
III	×				×	×						×									×		×		
IV																									
V												×			×										
VI						×										×									
VII	×												×			×		×		×					
VIII		×			×		×			×	×	×				×			×					×	×
IX																									
X	×																								
XI	×																								
XII	×																								
XIII	×							×						×				×	×						
XIV													×												
XV			×	×					×				×					×							
XVI	×																			×		×			

Additional information sources: Attorney general, airport, chamber of commerce, school districts, sewage disposal districts, and utilities.

be discussed with the reconnaissance results. The students were to use readily available data. Although there was little time for original field mapping, field orientation and spot checking were encouraged.

Table 18-2 is a guide to data sources. The rows refer to the 16 report groups in Table 18-1; the columns indicate local agencies from which data might be obtained.

Reconnaissance Results

Selected results of the reconnaissance studies are presented below. The materials have been abstracted from the student reports with little modification or comment. Only those parts of maps that apply to the rectangular area shown in Figure 18-1 are herein reproduced.

Topography. Figure 18-3 shows a portion of the slope declivity map generalized from a U.S. Geological Survey map. The percentage slope classes are those commonly used by land-use planners, but degree equivalents are added. The map shows the relatively flat slopes on the coastal terrace and on the floor of Pilarcitos Canyon and the steeper slopes elsewhere.

Table 18-3 from the student report describes the characteristics of the different slope categories in Figure 18-3 and the suitabilities for different types of development. Construction costs increase with slope.

In addition to the slope map, two other

FIGURE 18-3. **Reconnaissance map showing slope categories. Small areas of one category within another are neglected to avoid confusion.**

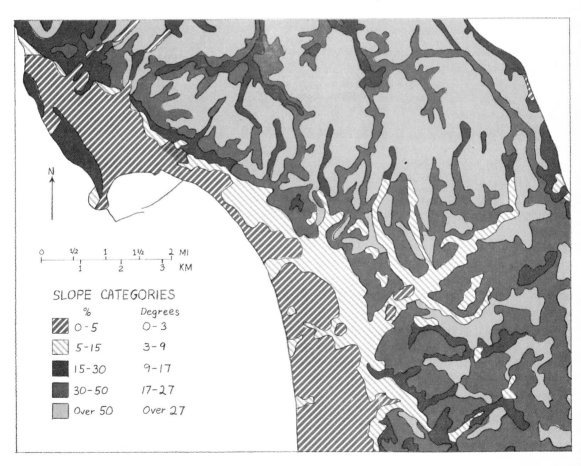

Table 18-3. Characteristics of the slope categories in Figure 18-3 and their suitability for land development

Slope Category	Characteristics and Suitabilities
0 to 5 percent (0 to 3°)	Almost level. Suitable for urban and agricultural development. Part susceptible to flooding and part with poor drainage
5 to 15 percent (3 to 9°)	Moderately sloping. Too steep for airports or most heavy industry. Irrigation restricted but suitable for dry farming. Good drainage. Good setting for residential development
15 to 30 percent (9 to 17°)	Too steep for most cultivation. Erosion problems. Slopes up to 20 percent suitable for crops such as artichokes and brussel sprouts. Also suitable for limited light industry, detached houses, high-rise apartments, institutional complexes, and intensive recreational facilities
30 to 50 percent (17 to 27°)	High-rise apartment clusters and large-lot residences appropriate. Low density required. Suitable for low-intensity recreation and summer resorts. Forest and grazing lands
Over 50 percent (over 27°)	Generally too steep for real real estate development. Best restricted to wildlife, forestry, and limited grazing

maps based on topography proved useful in the planning. One portrayed elevation zones, the other, the directions in which slopes face. The latter is important with respect to microclimate, hydrology, vegetation, and ground stability. While such data may be derived from topographic maps, the separate maps enable the planner to appraise more readily the topographic settings of the various land parcels.

Climate. The climate of the area is important for its bearing on recreational appeal, residential development, water supply, vegetation, and

agriculture. We have noted that the area has been bypassed by expensive recreational and residential developments in favor of areas with more suitable climates. On the other hand, the climate accounts for redwood groves in the mountains and special agricultural crops, such as flowers, artichokes, and brussel sprouts, in the lower areas.

Some basic climatic data are given in Table 18-4. On the whole the climate is invigorating, cool, and free of pollution.

Plant Communities. Figure 18-4 shows agricultural development on the coastal terraces, the coastal scrub, chaparral, and grassland along the lower slopes of the Santa Cruz Mountains, and the conifer and broadleaf forests along the higher slopes.

The student report identifies seven plant communities:

1. *Marine vegetation.* Abundant and ecologically varied intertidal floras which have been depleted by human collecting.

2. *Coastal strand vegetation.* An extremely fragile plant community of the coastal dunes and beaches requiring special protection.

3. *Coastal scrub.* Dense low brush on coastal bluffs, marine terraces, and coastal hills.

4. *Chaparral.* A dryland scrubby vegetation covering the slopes of the Santa Cruz Mountains above the coastal scrub and below the forests.

5. *Grasslands.* Generally the result of land clearing and grazing. Much of the grassland would revert to coastal scrub and chaparral if left ungrazed.

6. *Conifer forest.* Redwood and redwood–Douglas fir forests on the higher slopes where rainfall is greater. The summer fog helps maintain the moist environment. Redwood logging can be especially disruptive because it leads to greatly accelerated soil erosion. Furthermore, aeration and sunlight that result from logging, modify the moist cool environment these trees require.

7. *Mixed broadleaf evergreen forests.* Oak-madrone, Oak-buckeye, and Douglas fir–tanbark-oak assemblages in drier areas, commonly on the fringes of the redwood forests.

Table 18-4. Climatic data for the Half Moon Bay area

Average January temperature	9.8°C (49.7°F) (coldest month)
Average September temperature	15.1°C (59.2°F) (warmest month)
Average annual precipitation	62.2 cm (24.5 in.) (mainly from October to April)
Average growing season	363 days
Average number of partly cloudy days	120
Average number of clear days	150
Average number of cloudy days	100
Average wind velocity	8 to 16 km (5 to 10 mi) per hour
Average relative humidity	70 percent
Lowest recorded temperature	−2.8°C (27°F)
Highest recorded temperature	34°C (93°F)
Summer fog is common	

FIGURE 18-4. **Reconnaissance map showing vegetation types.**

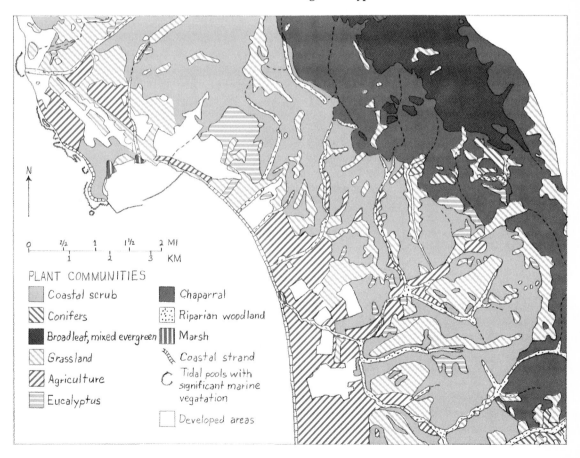

PLANT COMMUNITIES

- Coastal scrub
- Conifers
- Broadleaf, mixed evergreen
- Grassland
- Agriculture
- Eucalyptus
- Chaparral
- Riparian woodland
- Marsh
- Coastal strand
- Tidal pools with significant marine vegetation
- Developed areas

Geology and Geologic Hazards. The student group assembled data on the geology of the area and evaluated the geologic hazards. Figure 18-5 shows landslide susceptibility and landslide deposits, and Figure 18-6 shows seismic hazards based on the susceptibility of the ground to shaking and the location of known faults.

The student landslide map shows the following categories:

I Areas least susceptible to landsliding
II Low susceptibility to landsliding
III Moderate susceptibility to landsliding
IV Moderately high susceptibility to landsliding
V Highest susceptibility to landsliding

The map shows that virtually the entire southern part of the area inland from the coastal terraces is landslide-prone.

Figure 18-6 shows known faults and delineates areas of different degrees of stability under ground-shaking. Downfaulting on the east side of the Seal Cove Fault accounts for the long ridge extending south from Moss Beach to Pillar Point. The fault, which extends southeast beyond the planning area, has also affected the ground-water situation, as discussed later. The ground-shaking potentialities were judged largely on the basis of the geologic materials.

The students recommended an open-space zone at least 30 m (100 ft) wide along each fault

FIGURE 18-5. **Reconnaissance map showing landslide susceptibility and landslide deposits.**

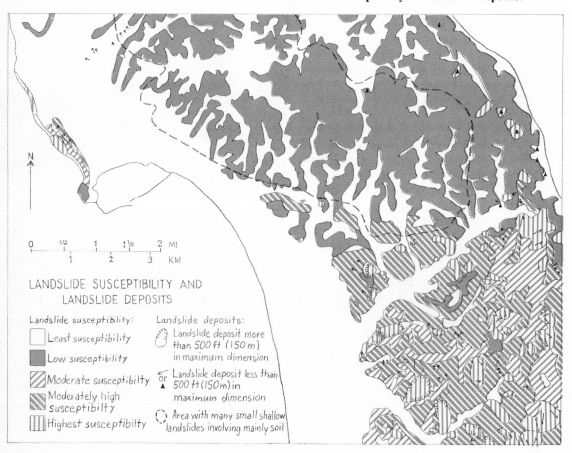

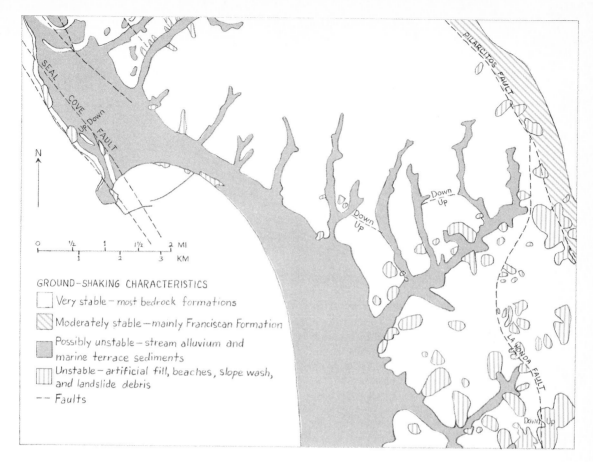

GROUND-SHAKING CHARACTERISTICS

☐ Very stable – most bedrock formations

▨ Moderately stable – mainly Franciscan Formation

▧ Possibly unstable – stream alluvium and marine terrace sediments

▥ Unstable – artificial fill, beaches, slope wash, and landslide debris

–– Faults

FIGURE 18-6. **Reconnaissance map showing seismic hazards and ground-shaking characteristics.**

trace based on the likelihood that future displacements will occur at the sites of past displacements. They also advised that those living in the area have earthquake insurance because of the active faults.

Soils. The most useful soils information came from the "San Mateo Area Soil Survey, California," published by the U.S. Soil Conservation Service in 1961. Similar reports have been prepared for many areas in the United States.

The students first prepared a map showing soil types and associations. Additional maps and reports dealt with:

1. Water-holding capacity
2. Permeability
3. Shrink-swell soil characteristics

4. Agricultural soil capabilities
5. Septic-tank-filter field limitations
6. Resistance to erosion

Drainage. The group studying drainage regarded flood probabilities of primary importance. Figure 18-7 delineates areas having a 1 in 100 chance on the average of being flooded during any given year.

Ocean and Shoreline. Figure 18-7 also includes the findings of "the ocean and shoreline" group. Primary concern was the effect of erosion on coastal development. The figure shows coastal erosion rates in the vicinity of Half Moon Bay.

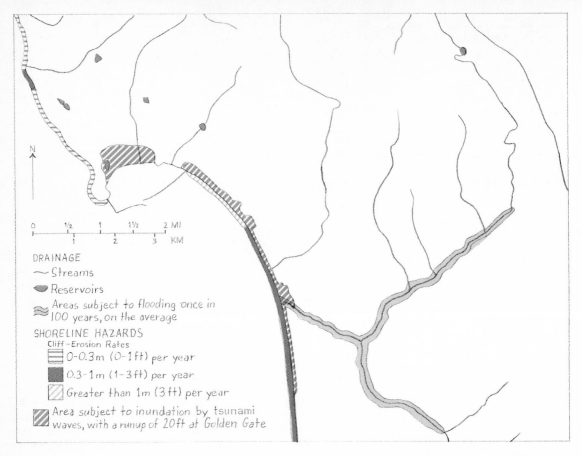

Within the figure legend:

DRAINAGE
~ Streams
◥ Reservoirs
≈ Areas subject to flooding once in 100 years, on the average

SHORELINE HAZARDS
Cliff-Erosion Rates
▤ 0-0.3 m (0-1 ft) per year
▨ 0.3-1 m (1-3 ft) per year
◪ Greater than 1 m (3 ft) per year
◪ Area subject to inundation by tsunami waves, with a runup of 20 ft at Golden Gate

FIGURE 18-7. **Reconnaissance map showing flood-prone areas, cliff erosion rates, and areas susceptible to inundation by a given tsunami.**

Cliff erosion ranges from 0.15 m (0.5 ft) to 1.4 m (4.6 ft) per year. The figure also shows areas that would be inundated by a tsunami with a runup of 6 m (20 ft) at Golden Gate 32 to 40 km (20 to 25 mi) to the north.

The student report considers a real estate development on the marine terrace at Moss Beach. Assuming an economic design life of 50 years, 24 m (80 ft) of the property would be lost to erosion if the cliffs were left unprotected. The developer has two options: to leave the sea cliff unprotected against erosion and subdivide only the inland portion of the site, or to provide protection that will minimize sea cliff retreat and subdivide the entire parcel.

Wildlife. The report on wildlife describes the animal life in the study area. The fauna of the chaparral habitat alone is impressive. Among the mammals are deer, cougars, bobcats, coyotes, foxes, raccoons, opossums, weasels, and skunk and a variety of smaller forms, including rabbit, gopher, shrew, mice, and bats. Reptiles include a variety of snakes and lizards, and amphibians are represented by salamanders and toads. The list of birds is large, including hawks, vultures, owls, quail, sparrows, finches, hummingbirds, jays, wrens, towhees, and waxwings.

The abundance of some of these species is surprising. For example, there are 30 to 50 California quail per 40 ha (100 acres) and 60 to 100 deer per 2.6 sq km (1 sq mi).

The report emphasizes the fragility of the faunal assemblages in the face of human development. For example, only Pilarcitos and Purisima Creeks still have seasonal runs of silver salmon and steelhead, whereas all the creeks had such runs in the past. The report urges that representative samples of all habitats be preserved to maintain the local species.

Hydrology. Annual mean precipitation varies from 62.2 cm (24.5 in.) at the coast to 114 cm (45 in.) at the crest of the Santa Cruz Mountains. The precipitation and other data were used to prepare a climatic water budget in order to estimate the gross water yield. Figure 18-8 is a graphical presentation of the water budget. It shows that the moisture surplus is generated in January, February, and March; soil moisture is depleted in April, May, and June; soil moisture is at wilting in July, August, September, and October; and soil moisture storage is restored in November, December, and January.

The Half Moon Bay weather station is at an elevation of 18 m (60 ft) above mean sea level. Because the planning area also includes mountains, the students determined the climatic water budget for Ben Lomond, a mountain station farther south in the Santa Cruz Mountains. By weighting the data for the lowland station and the mountain station according to the watershed elevations and studying streamflow records, it was concluded that the gross water yield is 12,300 ha•m (100,000 acre•ft) per year for the planning area. This then is the maximum water supply that can be developed under ideal conditions.

Figure 18-9 is a hydrologic map of the area. With the exception of a few small lakes and reservoirs, the coastal area contains no surface freshwater storage: virtually all water used for agriculture is pumped from the ground. Half Moon Bay urban water is imported from the San Francisco Municipal Water District's Pilarcitos Reservoir in a valley in the Santa Cruz Mountains (see Figure 18-1) and transmitted via a pipeline which follows the natural drainage down Pilarcitos Canyon.

A considerable part of the broad, low coastal terrace in the planning area is underlain by the Purisima Formation, with a thin covering of terrace sediments. Much of this formation is fine-grained, with a permeability too low for the development of productive wells. The overlying terrace materials, although sandy and permeable, are too thin to yield significant amounts of water.

We have noted the Seal Cove Fault in Figure 18-6. The ground on the northeast side of the fault is depressed relative to that on the southwest. Therefore, the top of the Purisima Formation is at a lower elevation to the northeast, and the overlying coarse permeable terrace deposits are thick enough to constitute the Half Moon Bay Terrace Ground Water Basin. Figure 18-9 shows the area in which wells yield 11 cu m per hour (50 gpm) to 114 cu m per hour (500 gpm). Only a few wells, however, yield more than 18 cu m per hour (80 gpm), and the total aquifer yield is only about 107 ha•m (870 acre•ft) per year, restricted to agricultural uses.

Permits and application approvals have been issued over the years to appropriate more than 27,000 ha•m (219,000 acre•ft) of water per year. Thus, legal rights to water exceed by far the gross water yield.

FIGURE 18-8. **Climatic water balance.**

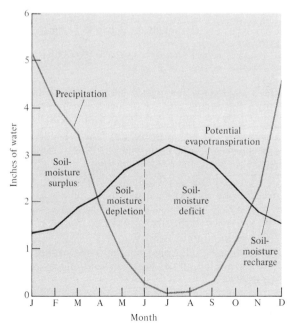

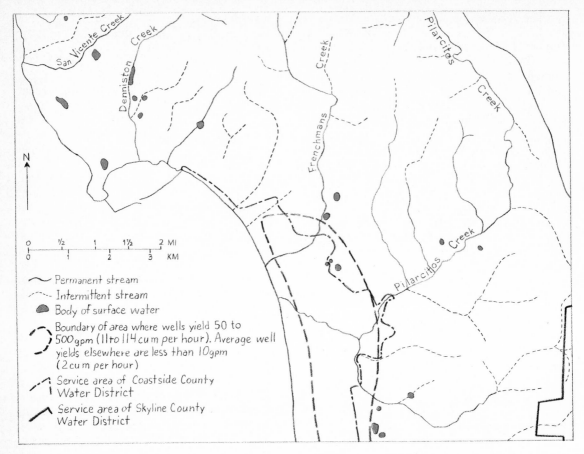

FIGURE 18-9. **Reconnaissance map showing hydrology.**

The Coastside Water District purchases 3000 cu m (2.4 acre•ft) per day from the San Francisco Municipal Water District Lake Pilarcitos Reservoir. It pumps an additional 750 cu m (0.6 acre•ft) per day from wells along Pilarcitos Creek during the wet season. Supply problems will develop unless additional water sources are developed to accommodate population increases.

Ground water in the Half Moon Bay Terrace Ground Water Basin is developed too intensively to risk contamination by septic tank effluents, and sediments elsewhere are too impermeable to accept effluents. Thus, the use of septic tanks is not feasible.

Aesthetics and Visual Aspects. Figure 18-10 deals with aesthetics and visual aspects. It

shows some of the major factors which could affect the visual impact of a prospective development. They help determine how visible construction will be, from where it will be visible, and how difficult it will be to blend the construction into the natural setting.

The visual elements and auxiliary data mapped by the group are:

1. Areas with substantial tree cover, useful for concealing structures.

2. Forest-canopied roads, to be preserved.

3. Important roadside viewing areas, to be preserved.

4. Areas at or below the levels of the main roads and from which they can be seen.

5. Areas above the levels of the main road and from which they can be seen.

6. Major visual corridors that provide outstanding scenic views, and should be preserved.

The students also indicated "points of focus." A point of focus is a scenic landform such as Pillar Point, a commercial center such as downtown Half Moon Bay, or a recreational center such as the marina-restaurant development at Half Moon Bay Harbor.

The students felt that each community should maintain its own identity rather than fade out into peripheral subdivisions that merge with the next community.

The student report was critical of some of the newer developments which consisted of uncoordinated pockets of intense development. In some high-density developments, architectural variation meant little more than varying the trim paint on garage doors. In places, mobile home parks obliterate the view of the ocean from the main highway and from other subdivisions.

The charm of some of the older developments illustrates the aesthetic importance of architectural variation.

The report suggests a number of aesthetic guidelines:

1. No real estate development should be permitted on the beaches or adjacent bluffs.

FIGURE 18-10. **Reconnaissance map showing visual elements.**

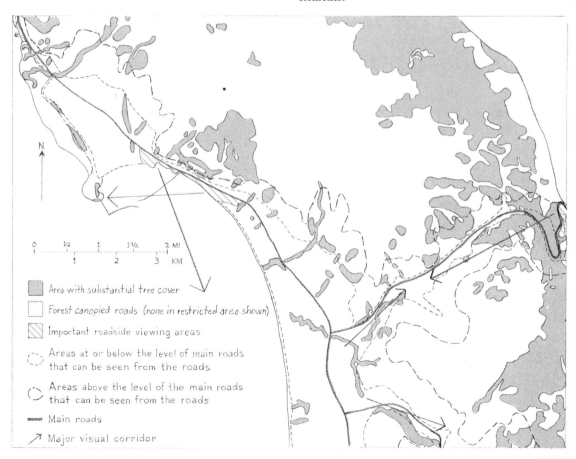

N

| 0 | ½ | 1 | 1½ | 2 MI |
| 1 | | 2 | | 3 KM |

▨ Area with substantial tree cover

☐ Forest canopied roads (none in restricted area shown)

▨ Important roadside viewing areas

⌒ Areas at or below the level of main roads that can be seen from the roads

⌒ Areas above the level of the main roads that can be seen from the roads

━ Main roads

↗ Major visual corridor

2. No development should be allowed to substantially interfere with the view of the ocean from the major highways as presently located.

3. Developments on the conspicuous ridges and westward facing ridges should be low density in areas of natural vegetative cover.

4. Extensive use should be made of artifically emplaced vegetative screens.

5. Architectural variety at reasonable cost should be stressed.

6. Development should include natural open space.

7. Materials and designs should blend homes with the surrounding terrain.

8. High-rise development should be discouraged to minimize visual obstruction.

9. Distinctive natural features should be preserved. For example, the marsh at the base of Pillar Point and the marine tide pools contain flora and fauna not found elsewhere in the area and should be preserved.

Land Use. Figure 18-11 shows land use in the vicinity of Half Moon Bay in 1972. Residential developments have encroached considerably upon agricultural land since 1956. However, much agricultural acreage has since reverted to open space for a number of reasons. First, much land peripheral to present housing has been removed from agriculture for new housing

FIGURE 18-11. **Reconnaissance map showing land use in 1972.**

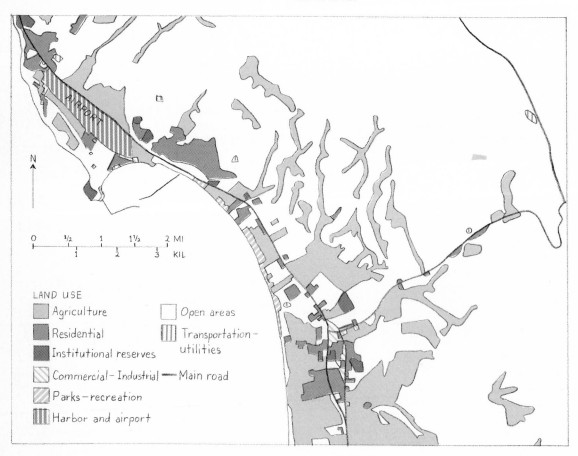

but is as yet undeveloped. The sale of agricultural land for subdivision is also related to decreasing incomes for agricultural products and increasing expenses, including rising taxes in response to increasing land values. Rising taxes have contributed not only to the removal of large areas from potential alternate uses but also from the community tax base.

The land-use group was pessimistic about prospects for controlled growth. Public planning groups are underfinanced as compared with large land development organizations. Available environmental data of all types are inadequate. Market and tax factors have more impact than planning. Property owners, particularly farmers and speculators, influence land use more than do county and city agencies. In short, "public planners are playing catch-up with ongoing patterns of land use."

The study group recommended the use of greenbelt planning rather than community growth by annexation of peripheral acreage. To carry out this concept, the city and county would have to determine appropriate use. Open space would have to be provided for by purchase and zoning. Floodplains, areas of landslides, and strips along faults would be included in the open spaces. Agricultural and recreational parcels would be included where feasible. Unoccupied and low-density land in the foothills might be added to the open space. The purposes of the greenbelt plan are to prevent urban sprawl and to preserve open space. Community boundaries would be fixed by greenbelts, and each community would be planned as to ultimate size and development as a coherent unit. No unincorporated areas would be permitted such as those which have had to be periodically annexed to Half Moon Bay.

Land Ownership. The pattern of land ownership in the project area varies from numerous undeveloped subdivision lots 7.6 by 30.5 m (25 by 100 ft) to large holdings of thousands of acres. Much of the coastal terrace has been subdivided, but the hillsides and a few key areas near the shore are still in large holdings. The largest landowner, Deane and Deane, a Los Angeles development organization, owned 3300 ha (8200 acres), plus options to buy other properties.

Deane and Deane made an in-depth study of the geologic, environmental, and aesthetic factors in the area. They visualized a development potential of 15,000 to 18,000 units for their holdings. They planned a first development with homes in the $45,000 to $60,000 price range, a country club with golf courses, and a "fishing village" which would consist of a marina and fresh water canals lined with eighteenth century shops with apartments above. The students concluded that the Deane and Deane developments were well planned. Deane and Deane's main problem was to upgrade the Half Moon Bay area if they were to sell their higher priced homes. The community of Half Moon Bay would have to decide whether it wished to change to a higher income residential community.

Transportation and Utilities. At the time of the study, 25.5 percent of the population commuted, using either Highway 1 to San Francisco or Highway 92 to the Bay communities. These highways are inadequate for carrying increased commuter traffic. If the area is to become a residential community for San Francisco and the Bay area, these two-lane highways would have to be expanded. The most effective way of discouraging excessive development is to leave the highway capacities unchanged.

If Route 1 is expanded to freeway status and follows its present route down the coastal terrace, it will continue to cross large portions of developable land, pass close to residences, and act as an impediment to recreational access to the shore. Therefore, the student report recommended a route following the base of the foothills, even though this would be more expensive.

The San Mateo County Harbor District operates the harbor in Half Moon Bay. A market analysis has shown increased need for berths for private and commercial boats, and expansion plans are being implemented. However, a key concern is the competition from the "fishing village" marina proposed by private developers.

The most urgent utility problem was sewage disposal. Table 18-5 shows the design figures for the three coastal sanitary districts, city of Half Moon Bay, Granada, and Montara. The

Table 18-5. Design figures for the San Mateo Mid-Coastside sanitary district. None of the facilities was designed to handle wet-weather flows, which consist of domestic sewage and leakage into the system from storm runoff. Instead, wet-weather flows bypass the sewage disposal plants and discharge into the ocean untreated

	City of Half Moon Bay Sanitary District	Granada Sanitary District	Montara Sanitary District
Design capacity (gallons per day)	300,000	300,000	500,000
Average dry-weather flow (gallons per day)	240,000	200,000	190,000
Average wet-weather flow (gallons per day)	1,000,000	1,000,000	1,200,000
Area served (acres)	650	600	810

average dry-weather flow was 80 percent of the capacity of the Half Moon Bay sewage disposal facility. Substantial volumes of sewage bypassed the facility during wet weather and during peak dry weather days. Consequently, there was a temporary moratorium on building permits until the city could provide additional sewage treatment capacity.

The El Granada and Half Moon Bay facilities provide primary treatment that includes screens, settling ponds, chlorination, and in the case of El Granada, chemical precipitation. Only 25 to 35 percent of the BOD (biochemical oxygen demand) is satisfied, and only 60 percent of the suspended solids are removed. The Montara Sanitary District uses only screens and chlorination.

Despite its poor quality, the effluent from all facilities is disposed of in the ocean. The distances from the shore of the outfall sewers are 9 m (30 ft) at Montara at a mean depth of 2 m (7 ft), 90 m (300 ft) for El Granda at a mean depth of 1.5 m (5 ft), and 400 m (1300 ft) for Half Moon Bay at a mean depth of 9 m (30 ft).

An official study of sewage problems along the coast of San Mateo County proposed a two-step remedy. The first involved interconnection of the El Granada and Half Moon Bay districts and the enlargement and inclusion of secondary treatment at the Half Moon Bay facility. This would satisfy 85 to 95 percent of the BOD and remove most of the suspended solids. The first step could be completed in 1 year at a cost of $2 million.

The second step involved tertiary treatment of all sewage at a capital cost of $6.25 million. The officials of Half Moon Bay believe that tertiary-treated water can be sold for agricultural and golf course irrigation.

Population and Economics. The San Mateo County Planning Department in 1971 (see Additional Readings) listed eight factors affecting population growth in the county:

1. The supply of land suitable for building was diminishing.
2. Available land was expensive.
3. Fewer people were moving into the San Francisco Bay Area because of economic conditions.
4. Taxes were high.
5. The market for homes was depressed.
6. People were discouraged by the congestion and pollution.
7. There was decreasing migration to California.
8. The national birth rate was diminishing.

The 1970 Mid-Coastside population was estimated at 10,000 to 12,000. Because of the above factors, its rate of growth was not expected to be as large as might be expected for an area peripheral to San Francisco and San Francisco Bay.

Several population projections were available. One was based on the assumed holding capacity of the land. It considered the potential developable land, development status, and an assumed density of development. It indicated that water supply, sanitary sewerage facilities, and highways will affect the growth rate for the next 20 years. The following possibilities were contemplated:

1. Population could increase to 15,000 in the next 20 years, with minimal or no improvement in existing water and sewerage systems.

2. The population could grow to 21,000 with development of a supplemental water supply and with interim enlargement of the capacities of the existing sanitary districts. This is the maximum population until additional facilities bring water into the region and a new sewage disposal plant is built.

3. The lack of freeways connecting Half Moon Bay to centers of employment in San Francisco and the Bayshore side of San Mateo County will also limit population growth. Based upon the capacities of existing Routes 1 and 92, the maximum number of cars that can commute from Half Moon Bay during the rush hour period is 4500. It is unlikely that the population can exceed 23,000 to 29,000 until freeways are provided.

4. With provisions for additional water, the construction of a new sewage treatment plant, and the improvement of Routes 1 and 92, the population in the region could expand to 80,000 to 100,000.

The consensus by various government agencies and their consultants is for a population of 35,000 by 1990. The student report concluded that a population of 35,000 to 40,000 would be an appropriate estimate for the year 2000.

The obvious factor that separates the economy of the area from that of its external markets, suppliers, and employers is distance and poor transportation. All freight transport is by truck over inadequate Routes 1 and 92. This in part accounts for the small amount of industry and commerce. The economy is oriented toward specialty crop agriculture with some commercial recreation and fishing.

Productive activities such as vegetable growing, cattle ranching, forestry, and commercial recreation cannot support the cost of the land because of taxes, interest, and land speculation. The present productive use of the land is an attempt by developers and other investors to defray part of their costs (taxes and interest) with rentals from land-use activities. Either the cost of holding the land will have to decline or the land investment sector of the economy will experience financial distress.

The students could foresee no prospects for an adequate economic base for employment in the study area. The economic future seems dependent upon the presence of a substantial commuting population.

Recreation and Tourism. The students found the project area to be particularly suited for recreational uses. It has many beaches, a wide variety of popular game fishes, interesting communities of marine and terrestrial life, sheltered ocean access and harborage, a mixture of flat open space and rolling hills, undeveloped valleys, canyons suited for reservoir construction, as well as concentrations of redwoods, Douglas fir, oak, and other interesting and aeshetically pleasing plant communities. It is only minutes away from the rapidly expanding Metropolitan Bay Area, creating an increasing demand for recreational and urban development.

Efforts to provide recreational facilities have not kept pace with the growing demand. The state and county governments have given the highest priority to the acquisition of recreational land in the face of rapidly increasing land values and have postponed major developments until all proposed acquisitions have been made.

Local recreational facilities for the residents, such as community and neighborhood parks, playgrounds, and cultural centers, are inadequate. Proper planning will require reserving space at strategic locations for these facilities, as well as for regional facilities.

Government, Special Interest Groups, General Plans, and Zoning. The creation of a viable growth plan will require cooperation among federal, state, regional, county, and municipal agencies, as well as developers and special interest groups. The highest level of cooperation, however, is not always achieved because of differences in opinion regarding the merits of alternate plans, finances, and competitive attitudes. Therefore, one aim of the student group was to determine the relative degrees of authority of the various agencies, as well as their growth concepts and plans. Table 18-6 is a list of the governmental groups concerned with planning for the San Mateo Mid-Coastside together with their structures and functions.

Figure 18-12 shows governmental jurisdictions in the Half Moon Bay area. These jurisdictions include the city of Half Moon Bay and the Half Moon Bay Harbor district. The map shows the Half Moon Bay sphere of influence as delineated by the San Mateo Local Agency Formation Commission (LAFCO). If urbanization is intensified, the area within the sphere of influence would ultimately be annexed to the city of Half Moon Bay. The map also shows the "1000-yd (915-m) permit area" which came within the jurisdiction of the Coastal Commission, as discussed in the Epilogue of this chapter. The 1000-yd permit area lies within the Half Moon Bay sphere of influence.

LAND-CAPABILITY RATINGS

Introduction

In considering the capabilities of different land parcels for given land uses, favorable and unfavorable factors must be considered. It is advantageous to have a system for evaluating all these characteristics in common units. A land capability map can then be prepared for each land-use option, showing the relative capabilities of various land parcels.

Consider, for example, two sites. Site A is in flat terrain, near rail transportation with adequate public water supply, in an area of sparse vegetation, but with swelling soils. Site B is in hilly terrain, far from rail transportation, in an area of attractive woodland, stable soils, but inadequate water supply. What are the capabilities of the two sites for each of the following three land-use options: residential, heavy industry, and specialty-crop agriculture?

Table 18-6. Governmental groups concerned with planning of the San Mateo Mid-Coastside, their structures and functions

I. Association of Bay Area Governments (ABAG)
   The Association of Bay Area Governments (ABAG) brings together county and city governments to deal with regional problems on a cooperative, coordinated basis. ABAG has little more than advisory and informative power over development.
II. San Mateo County
   A. Controls all unincorporated lands
   B. Planning commission
      Regulates land use by master plan, zoning, and subdivision ordinances
   C. Regional planning committee
      1. Coordinates long range plans and integrates governmental policies
      2. Performs advisory functions without enforcement powers
III. San Mateo Local Agency Formation Commission (LAFCO)
   A. Its purpose is to prevent urban sprawl and encourage orderly function of local governments
   B. Approves formation and dissolution of special districts and incorporation of new communities
   C. Policies in regard to Half Moon Bay
      1. Established the Half Moon Bay sphere of influence as shown in Figure 18-12. Unincorporated areas within the Half Moon Bay sphere of influence may be annexed only to Half Moon Bay
      2. Supports restriction of future development to areas already urbanized
      3. Supports consolidation of El Granada and Montara sanitary districts
IV. City of Half Moon Bay
   Under state law, the city of Half Moon Bay through its city council, city manager, and planning commission has the power to deal with issues concerning development within the city limits.
V. Harbor District
   A. An autonomous entity funded from San Mateo County taxes
   B. Controls development and operation of harbor areas
VI. Coastal Commission
   This was not in existence at the time of the student study. It is discussed in the Epilogue to this chapter.

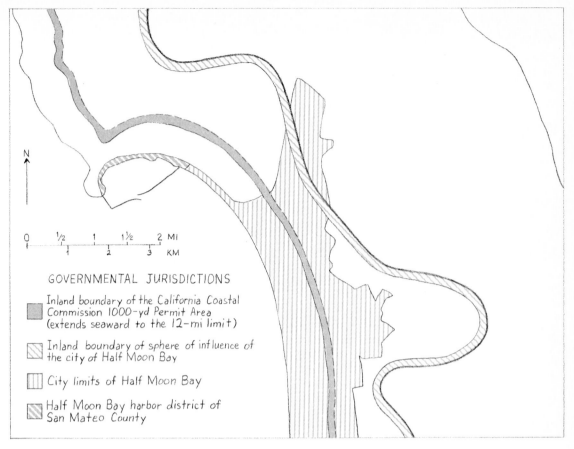

GOVERNMENTAL JURISDICTIONS

- Inland boundary of the California Coastal Commission 1000-yd Permit Area (extends seaward to the 12-mi limit)
- Inland boundary of sphere of influence of the city of Half Moon Bay
- City limits of Half Moon Bay
- Half Moon Bay harbor district of San Mateo County

FIGURE 18-12. **Reconnaissance map showing governmental jurisdictions.**

Residential. The flat terrain and public water supply would facilitate economical residential development at site A. The swelling soils would require expensive foundation work. The presence of rail transportation would be of no consequence if bus transportation and good roads were available. The sparse vegetation might deter home buyers for aesthetic reasons.

The hilly terrain and woodland at site B would be very attractive as home sites. The savings resulting from the stable soil might balance the costs of digging water wells or extending water mains and the increased development costs on hilly terrain.

Site B would probably be more attractive than site A for residential development. A land-capability rating system, such as will be described shortly, would almost certainly also assign a higher value to site B for residential development, but on a more objective basis.

Heavy Industry. Site A would be preferable for heavy industry because of the flat terrain, rail transportation, and public water supply. Plant cover and aesthetics would be of little import to heavy industry. Foundations of large structures would extend below the swelling soils. Thus, site A should have a higher land-capability rating for this land-use option.

Specialty-Crop Agriculture. Site A provides flat and sparsely vegetated terrain, easily cleared and farmed. Rail facilities would be available for shipping crops, and water supply would be available for irrigation (if not too expensive). Soil swelling would be of little consequence. Site B, with its steep slopes and woodland, would be hard to clear and work. Irrigation and additional shipping costs could constitute problems. Thus, site A would have a higher land-capability rating for this land-use option.

Preparation of Land-Capability Ratings. The procedure for rating land capabilities was introduced in Chapter 16. The students applied this procedure to the Half Moon Bay project. The procedure involved:

1. *Preparing and encoding* environmental data
2. Assigning *capability values*
3. *Weighting these values*
4. Computing *land-capability ratings*

Preparing and Encoding Environmental Data

The first step was to prepare a transparent grid on a scale of 1 in. = 2000 ft (2.5 cm = 610 m) to be compatible with the 1:24,000 base maps used in the project. Grid squares (cells) were 1/2 by 1/2 in. (1.3 by 1.3 cm), thus representing an area 1000 by 1000 ft (305 by 305 m). Later, this grid was superimposed on each of the data maps, and the environmental characteristics of each grid cell were coded using index numbers plotted directly on the overlay.

The next step involved establishing land-use options. For this academic exercise, the study was restricted to determining the potential of the area for residential and agricultural uses. Additional land-use options would have been considered in the preparation of a comprehensive plan.

It was then necessary to decide on the environmental factors that would influence capability for the specified land-use options. For residential and agricultural land uses, the students selected nine environmental factors: slope, vegetation, soils, ground stability, wildlife, aesthetic qualities, size of land parcels, climate, and access.

Each of the nine environmental factors was subdivided into no more than five subclasses

and given *index numbers* from 1 to 5. Although the index numbers are for identification only, the subclasses should be arranged in a logical sequence, as from gentle to steep or small to large. For example, the index numbers in Table 18-7 indicate the different subclasses of slope and vegetation. In this case, the index numbers were arranged in order of increasing slope and increasing density of trees.

Assigning Capability Values

Capability values were assigned to each index number for each land-use option. A larger capability value indicates that the land is better suited for the particular land-use option because of the given environmental factor.

Our capability values ranged from 5 to 0 and were defined as follows:

5—Very high
4—High
3—Moderate
2—Low
1—Very low
0—Not suitable

Where a 0 capability value renders land useless for the particular land use option, the total land capability value for that cell should be 0 regardless of how suitable the other environmental factors might be. For example, land being considered for the residential land-use option would receive a capability rating of 0 for ground stability if it were underlain by an active fault. That land should not be developed for residences no matter how favorable the other environmental factors might be.

Table 18-7 also shows the capability values for slope and vegetation and the two land-use options. It was decided that a slope exceeding 50 percent is not appropriate for either residential or agricultural development. The capability value of 0 means that cells with slopes exceeding 50 percent will not be allocated to either of these use options regardless of how favorable other factors might be. Slopes of 0 to 5 percent were deemed to have very high capabilities for either residential or agricultural use and were given capability values of 5 as shown in the table.

Table 18-7. Preparation of weighted capability values for two environmental factors, slope and vegetation, and two land-use options, residential and agricultural, for the Half Moon Bay area

Environmental Factor	Subclass	Index Number	Residential Land-Use Option				Agricultural Land-Use Option			
			Weight	× Capability Value	=	Weighted Capability Value	Weight	× Capability Value	=	Weighted Capability Value
Slope	0–5%	1	4	5		20	5	5		25
	5–15%	2		4		16		3		15
	15–30%	3		2		8		2		10
	30–50%	4		1		4		1		5
	50%+	5		0		0		0		0
Vegetation	Grass	1	3	1		3	1	5		5
	Chaparral	2		2		6		3		3
	Grass-oak	3		5		15		2		2
	Oak	4		4		12		1		1
	Conifer	5		3		9		1		1

Grass-oak vegetation was judged to be the most attractive for residential development and received a value of 5 for aesthetic reasons. Grass and chaparral were deemed least attractive and possible fire hazards, and received lower ratings. The vegetation capability values for agricultural use were based largely on costs of clearing the land.

Weighting

The relative importance of the environmental factors depends on the contemplated land use. For example, slope is important in planning for heavy industry but is less important in planning open space. Thus, for each contemplated land use, a weighting, or measure of relative importance, must be assigned each environmental factor. In the present study, the weighting is scaled from 5 to 0 as follows:

5—Very high importance
4—Highly important
3—Moderately important
2—Low importance
1—Very low importance
0—No importance

To illustrate the procedure, consider two grid cells being appraised for agricultural use, the first with grass cover and the second with conifer cover. In Table 18-7 the respective capability values are 5 and 1, based largely on costs of clearing the land. However, land clearing is a one-time operation that is relatively inexpensive when amortized over the life of an agricultural development. Therefore, the presence of vegetation is relatively unimportant in considering agricultural land use in this area, and the students assigned a relatively low weight of 1. Even though the overall rating is low, there are variations depending on the kinds of vegetation, as indicated by the capability values. When the individual capability values are multiplied by the weighting value, we get *weighted capability values*. These range from 1 to 5 in the agricultural land-use option.

In contrast to the vegetation factor, slope is very important in agriculture. A slope exceeding 50 percent cannot be farmed, whereas a slope of less than 5 percent can be farmed easily using large machinery. Because slope is so important, it was assigned a weight of 5. When the weight and capability values are multiplied, the weighted capability values for the slope factor in agricultural land use range from 0 to 25, as shown in the table.

Land-Capability Ratings

A weighted capability value was obtained for each grid cell for each land-use option and for each environmental factor. These were added for all the environmental factors to obtain a *land-capability rating* for each land-use option for each cell. For example, Figure 18-13 shows a land-capability map for part of the Half Moon Bay area for residential development based on the nine environmental factors listed above. When the students added the nine numbers representing the weighting capabilities for each cell, they obtained a residential land-capability rating which was then classified according to Table 18-8.

For the cell in the upper left-hand corner of Figure 18-13, the nine weighted capabilities gave a land-capability rating between 81 and 100 when added. This cell is therefore classified as having moderate capability for residential development. The cell in the upper right-hand corner had 0 land-capability rating and has no capability for residential development because slopes are greater than 50 percent. The cell in the lower right-hand corner also had 0 land-capability rating and had no capability for residential development because of active faulting or landsliding. It is noteworthy that the parts of the area with the highest land capability ratings in Figure 18-13 are those where development is already largely concentrated.

The land-capability ratings should be used only as aids to preliminary planning. Careful consideration of economic, social, political, and physical factors not included in the capability ratings are essential to the preparation of final plans. Finally, computers are admirably suited to the storage and analysis of environmental data and the preparation of land-capability ratings.

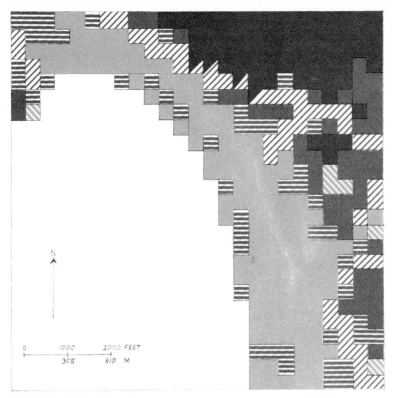

LAND-CAPABILITY RATINGS FOR RESIDENTIAL DEVELOPMENT

Rating	Capability for residential development
Greater than 100	High
80-100	Moderate
50-79	Low
1-49	Very low
0	No capability because slopes exceed 50 percent
0	No capability because of active faults and/or landslides

FIGURE 18-13. Land-capability ratings for residential development based on the following nine environmental factors: slope, vegetation, soils, ground stability, wildlife, aesthetic qualities, size of land parcels, climate, and access. The smallest squares represent the grid cells for which weighted capability values were computed.

LAND-USE SKETCH PLAN AND OBJECTIVES

Introduction and Definitions

The first step in translating acquired information into action is its incorporation into a *preliminary land-use sketch plan and objectives*. This was the final assignment in the student project.

Sketch Plan. The *sketch plan* is a preliminary allocation of land for different uses. The plan should reflect appropriate goals and objectives.

Table 18-8. Classification of land capability for residential development in the Half Moon Bay area, according to the land capability ratings

Class	Land-Capability Rating	Land Capability for Residential Development
1	Greater than 100	High
2	81–100	Moderate
3	50–80	Low
4	1–50	Very low
5	0	No capability because slopes are greater than 50 percent
5	0	No capability because of active faults and/or landslides

It should attempt to satisfy all legitimate interest groups.

The student sketch plan for the San Mateo Mid-Coastside area was prepared as an overlay to the 1 in. = 2000 ft (1 cm = 240 m) base map. The students were instructed to include at least the following categories of information:

1. The distribution of residential areas by density classification, acreage, and population.

2. A pattern that reveals some organizational concept. For example, residential areas may consist of contiguous or separated neighborhoods, each with its own school, shopping center, and park.

3. Commercial centers by type (neighborhood, community, or regional) and approximate areal extent.

4. Major institutional uses (government center, special institutions, etc.).

5. Recreation areas (region-serving and community-serving).

6. Agricultural areas.

7. Open space.

8. Industrial areas (if any).

9. Airport.

10. Circulation (freeways, thoroughfares, arterials, and major collector roads).

Each plan was accompanied by a report describing goals, objectives, policies, and standards. In addition, each report discussed the plan concept, its philosophy and justification, implementation methods, and its feasibility.

Goals, Objectives, Policies, Standards. *Goals* are very general statements about the purpose of the plan in future development of the area during the planning period. For the Mid-Coastside area, the planning period was taken as 30 years.

Objectives are more detailed statements with respect to major aspects of the plan such as recreation, agriculture, and industry.

Policy statements explain how the objectives will be achieved. Some policies will be indicated in the sketch plan diagram, others in the report.

Standards are quantitative criteria applicable to residential densities, acreage requirements for commercial areas, open space requirements, and the like.

Examples of Student-Prepared Land-Use Sketch Plans and Objectives

Plan Based on Multiple-Nuclei Development. Figure 18-14 shows a land-use sketch plan for the Mid-Coastside area that attempts to preserve the open-space–agricultural aspect by dispersing the design population in multiple nuclei.

The first goal expressed by this student group was to develop the Half Moon Bay planning area as a "rural-residential community" of 40,000 people. The second goal was to preserve the recreational and open-space aspects of the community within a framework of multiple-nuclei developments set off by intervening greenbelt-agricultural open spaces.

The objectives of the student group that aimed at preserving the agricultural-open space aspect of the area were:

1. Reserve the maximum amount of suitable area for recreational uses.
2. Eliminate building along the shoreline wherever possible.
3. Encourage the development of recreational uses and open space in ways that preserve the natural character of the area.
4. Provide functional, aesthetically pleasing, urban residential development.
5. Provide recreational access to the area.
6. Preserve, where possible, the present agricultural areas.

The policies emphasized housing for the elderly, young married couples, small families, and childless couples, in view of current trends toward small families. Some examples of proposed policies are:

1. Large developments should be encouraged to allow uniform quality of architecture and design.
2. Condemnation proceedings should be utilized to obtain large tracts of land for rational development.
3. Geologic study of hazards should be contracted for by each developer.

FIGURE 18-14. **Land-use sketch plan based on preservation of the open-space–agricultural aspect by dispersing population in multiple nuclei.**

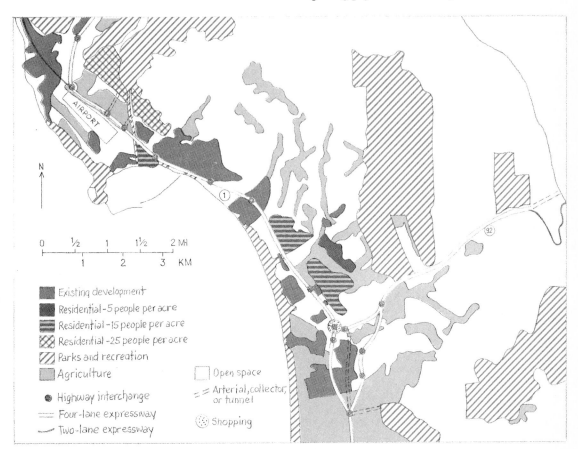

4. Cluster-type residential areas should be encouraged wherever possible to allow for maximum open space.

5. Pedestrian pathways and bicycle paths should be required in large developments.

6. Nucleus development should continue in order to prevent urban sprawl.

7. The location of clusters of development should be encouraged within the confines of natural boundaries as in canyons.

8. Residential development should be restricted to flat terrain.

9. Heavy development should remain clustered around the intersection of Highways 1 and 92.

10. Route 92 over skyline summit should remain a two-lane roadway.

11. All development (other than grazing in a few specified areas) should be prohibited in areas designated by the Regional Planning Commission for purchase by San Mateo County in the 1980 Plan for San Mateo County.

12. Creek beds and floodplains should be incorporated into the greenbelts as recreation areas.

Residential standards provide for density zoning and for ordinances to prevent construction in geologically hazardous locations. It is recommended that large developers be required to set aside 15 percent of their land for open space. Maximum distances of residences from schools, playgrounds, public parks, and shopping centers are specified.

Access standards refer to the freeways, expressways, feeder roads, and greenbelt requirements. This student group eliminated the present Highway 92 interchange along the mountain crest (Figure 18-14) in order to discourage development along the skyline.

Recreation standards specify 1 acre (0.4 ha) of playground, 1 acre (0.4 ha) of local park, and 1 acre (0.4 ha) of playfield or recreation center per 800 people. Minimum sizes specified are 5 acres (2 ha) for a playground, 5 acres (2 ha) for a local park, and 25 acres (10 ha) for a playfield or recreation center.

Plan Based on Widely Dispersed, Low-Density Development. A second student group expressed this goal: "to maintain the rural character of

the Half Moon Bay study area as a low-density commuter area while at the same time preserving the recreational and environmental qualities."

The group originally intended to limit growth to three nodal clusters and to use regulatory zoning to impose open space or agricultural buffer zones between them. However, they felt that existing land-use patterns and ownership boundaries prevented adherence to a nodal concept. Instead, Figure 18-15 shows widely dispersed, low-density development along the coastal terrace, with the hillsides free of development.

A main objective of this group was to keep the population below 30,000 by regulatory zoning and limited access. Another objective was to develop a more self-sufficient community with light industry, agriculture, and commercial recreation.

Plans Concentrating Development at a Small Number of Nodes. Most of the study groups concentrated the design population at a small number of nodes separated by greenbelts. This maintained the natural and rural character of the landscape. At the same time, the separate communities could be planned to provide the amenities of urban living.

Figure 18-16 is one of the plans based upon restricting development to a small number of nodes. The goal of this group was to retain the small-town and rural atmosphere of the Half Moon Bay area while accommodating a significant increase in resident population and tourist trade. A development objective was to provide safe and pleasant areas for well-planned residential, commercial, and industrial developments of diverse types, but within well defined areas so as to minimize sprawl and promote a sense of community to the residents. The group recommended that residential development be encouraged in areas (see Figure 18-16) which by virtue of slope, stability, aesthetics, and other considerations are best suited to this use.

Policies relating to geologic hazards require that no building be permitted within 30 m (100 ft) of the trace of an active fault, or within the zone of influence of an active landslide. Furthermore, land of poor geologic stability will require study by an appropriate agency having

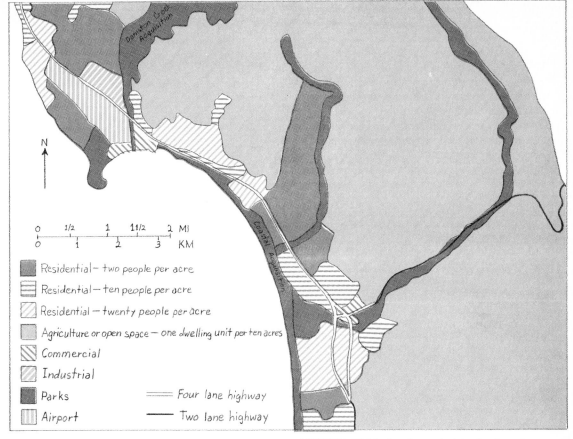

Legend:
- Residential — two people per acre
- Residential — ten people per acre
- Residential — twenty people per acre
- Agriculture or open space — one dwelling unit per ten acres
- Commercial
- Industrial
- Parks
- Airport
- Four lane highway
- Two lane highway

FIGURE 18-15. **Land-use sketch plan based on widely dispersed low-density development.**

geologic expertise before construction is permitted.

A wide spectrum of objectives, policies, and standards were prepared by all groups to cover many aspects of development, industry, agriculture, recreation, access, transportation, and the natural environment.

EPILOGUE

As of 1975, the Half Moon Bay area looked much as it did at the time of the student study in 1971. However, there have been a number of changes in the private and governmental areas.

Deane and Deane, who provided the development inpetus of 1971, has withdrawn from the area, returning the land to Half Moon Bay Properties, which is pursuing a less active development program. Limited sewage capacity, economic considerations, and the Coastal Commission review process have contributed to reduce development for the time being. Also, the uncertain future of an adequate water supply still looms.

The recently created Coastal Commission has used its permit powers to review all developments within 1000 yd (915 m) of the shoreline while it prepares a comprehensive coastal plan. This plan is aimed at preserving the coast of California as a natural resource and allowing only those uses that protect the natural aspects. The aims of the commission were enacted into law in 1976 by the California legislature.

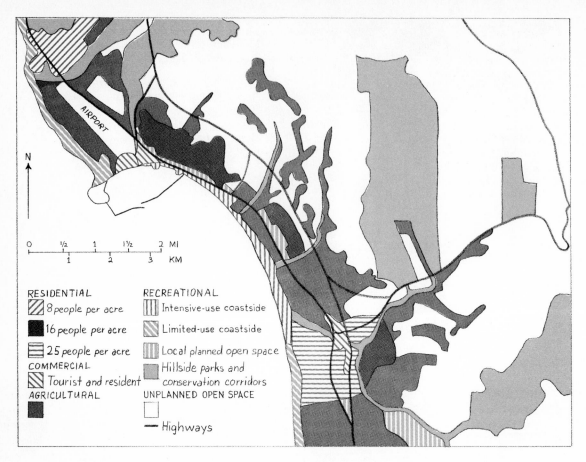

RESIDENTIAL
- 8 people per acre
- 16 people per acre
- 25 people per acre

COMMERCIAL
- Tourist and resident

AGRICULTURAL

RECREATIONAL
- Intensive-use coastside
- Limited-use coastside
- Local planned open space
- Hillside parks and conservation corridors

UNPLANNED OPEN SPACE

— Highways

FIGURE 18-16. **Land-use sketch plan based on restricting development to a small number of development nodes.**

The Harbor district has received pressure from several groups to reduce the scope of harbor expansion. It is now working on scaled-down plans.

San Mateo County has made a resource management zone district of much of the mountainous terrain in the Half Moon Bay area. The zoning is aimed at preserving this terrain as open space.

The debate as to the appropriate population size is still going on. Some groups wish to keep the two access roads at two lanes each, thereby limiting the commuting population. Other groups wish to increase the road capacity to four lanes to accommodate future growth.

The future of the Half Moon Bay area has not yet been decided. The various groups with a stake in the future still disagree on major issues.

ADDITIONAL READINGS

Association of Bay Area Governments: Ocean Coastline Study, Supplemental Report IS-5, Berkeley, Calif., 1970.

——: Preliminary Planning Policies: *ABAG Ocean Coastline Planning Program*, Berkeley, Calif., 1972.

California Division of Mines and Geology: Geologic Map of California, San Francisco Area, Bull. 149, San Francisco, Calif., 1961.

Chapin, F. S., Jr.: "Urban Land Use Planning," 2d ed., University of Illinois Press, Urbana, 1965.

Local Agency Formation Commission, A Water Pollution Control Study for the Half Moon Bay

Basin: Jenks and Adamson, Palo Alto; Frahm, Edler, and Assoc., Redwood City; Wilson, Jones, Morton, and Lynch, San Mateo.

Marshall Kaplan, Gans, and Kahn, and Frahm, Edler and Assoc.: General Plan Program, Mid-Coastline Area, Half Moon Bay, California, San Francisco, 1971.

Patri, T., D. C. Streatfield, and T. J. Ingmire: Early Warning System; The Santa Cruz Mountains Regional Pilot Study, *Dept. of Landscape Architecture*, University of California, Berkeley, 1970.

San Mateo County Planning Department: Population Growth in San Mateo County: Trends and Prospects, Redwood City, Calif., 1971.

U.S. Geological Survey and U.S. Dept. of Housing and Urban Development: Program Design for San Francisco Bay Region Environmental and Resources Planning Study, Menlo Park, Calif., 1971, 123 pp.

U.S. Soil Conservation Service: San Mateo Area Soil Survey, California, *U.S. Dept. Agriculture*, 1961.

Williams, J. C., and H. C. Monroe: "The Natural History of the San Francisco Bay Area," McCutchan Publishing Corp., Berkeley, Calif., 1970.

conversion tables

ABBREVIATIONS USED IN TEXT, ALPHABETICALLY ARRANGED

bbl	barrel	l	liter
bpd	barrels per day	lb	pound
cfs	cubic feet per second	m	meter
cm	centimeter	mi	mile
ft	foot	ml	milliliter
g	gram	mm	millimeter
gal	gallon	oz	ounce
gpd	gallons per day	psi	pounds per square inch
gpm	gallons per minute	qt	quart
ha	hectare	sec	second
in.	inch	μ(or μm)	micron (micrometer)
kg	kilogram	yd	yard
km	kilometer	yr	year

LENGTH

Metric	U.S.		U.S.	Metric
1 mm = 0.001 m	= 0.039 in.		1 in. =	= 25.4 mm
1 cm = 0.02 m	= 0.394 in.			= 2.54 cm
1 m = 0.001 km	= 39.37 in.		1 ft = 12 in.	= 0.305 m
	= 3.281 ft		1 yd = 3 ft	= 0.914 m
1 km = 1000 m	= 0.621 mi		1 mi = 5280 ft	= 1610 m
	= 3280 ft			= 1.61 km

AREA

Metric	U.S.
1 sq cm	= 0.155 sq in.
1 sq m	= 10.764 sq ft
	= 1.196 sq yd
1 are	= 0.025 acre
1 ha (100 ares)	= 2.471 acres
1 sq km	= 247 acres
	= 0.386 sq mi

U.S.	Metric
1 sq in.	= 6.452 sq cm
1 sq ft	= 0.093 sq m
1 sq yd	= 0.836 sq m
1 acre	= 40.5 ares
	= 0.405 ha
1 sq mi	= 640 acres
	= 259 ha
	= 2.59 sq km (ha)

VOLUME

Metric	U.S.
1 cu cm	= 0.061 cu in.
1 cu m	= 35.315 cu ft
	= 1.308 cu yd
1 cu km	= 0.239 cu mi
1 ha · m	= 8.108 acre · ft

U.S.	Metric
1 acre · ft	= 0.1233 ha m
1 cu in.	= 16.387 cu cm
1 cu ft	= 0.028 cu m
1 cu yd	= 0.765 cu m
1 cu mi	= 4.17 cu km

CAPACITY

Metric	U.S.
1 cu m = 1000 liters	= 264 gal
1 liter = 100 cu cm	= 61.025 cu in.
	= 0.035 cu ft
	= 0.264 gal

U.S.	Metric
1 cu in.	= 0.016 liter
1 cu ft	= 28,316 liters
1 gal	= 3.785 liters
1 bbl (42 gal)	= 159 liters

WEIGHT

Metric	U.S.
1 g	= 0.035 oz
1 kg	= 2.205 lb
1 ton (metric)	= 2205 lb
	= 1.102 short tons

U.S.	Metric
1 oz	= 28.35 g
1 lb = 16 oz	= 453.59 g
1 ton (short) = 2000 lb	= 907.18 kg
	= 0.907 ton (metric)
1 bbl oil (average)	= 140.0 kg

PRESSURE

Metric	U.S.
1 kg per sq cm	= 14.223 psi
1 kg per sq m	= 0.205 lb per sq ft

U.S.	Metric
1 lb psi	= 0.070 kg per sq cm
1 lb per sq ft	= 4.882 kg per sq m

WATER EQUIVALENTS: VOLUME AND WEIGHT

U.S.		Metric
1 gal = 0.134 cu ft		= 8.34 lb
1 cu ft = 7.48 gal		= 62.5 lb
1 acre · ft = 325,851 gal	= 43,560 cu ft	
		= 1235 cu m

RATE OF FLOW
U.S. Equivalents

1 cu ft per sec (cfs)	= 449 gal per minute (gpm)
	= 724 acre·ft per year
1 acre·ft per day	= 325,900 gal per day (gpd)
1000 gal per minute (gpm)	= 192,000 cu ft per day
	= 4.42 acre·ft per day

U.S.	Metric
1 cu ft per sec (cfs)	= 2450 cu m per day
1 cu ft per min	= 0.472 liter per second
1 acre·ft per year	= 3.38 cu m per day
	= 0.1233 ha · m per year
1 billion gpd	= 3,790,000 cu m per day

WORK AND POWER

1 foot-pound (ft · lb)	= work done in moving a 1-pound weight through 1 foot of distance in the direction of the force.
1 joule	= 0.738 ft · lb
1 watt	= 1 joule per second
1 horsepower	= 550 ft · lb per sec
	= 746 watts
1 kilowatt	= 1000 watts
1 megawatt	= 1,000,000 watts
	= 1000 kilowatts

TEMPERATURE AND HEAT

1°centigrade (°C) = $\frac{9}{5}$°Fahrenheit (°F)

1°Fahrenheit = $\frac{5}{9}$°centigrade

To convert centigrade readings to Fahrenheit, use the formula:

$$°F = \tfrac{9}{5}°C + 32$$

To convert Fahrenheit readings to centigrade, use the formula:

$$°C = \tfrac{5}{9}(°F - 32)$$

1 calorie	= heat required to raise the temperature of 1 gram of water 1 degree centigrade
1 Btu (British thermal unit)	= heat required to raise the temperature of 1 pound of water one degree Fahrenheit
	= 252 calories
1 barrel of crude oil	= 5,000,000 Btu
1 cu ft of natural gas	= 1000 Btu
1 short ton of coal	= 24 to 28 million Btu
1 kwh of electricity	= 3400 Btu

index

Page references in **boldface** refer to illustrations; page references in *italic* refer to definitions or more comprehensive explanations of important terms or concepts.